Electronic Instrumentation

JOHN A. ALLOCCA

ALLEN STUART

Reston Publishing Company, Inc.
A Prentice-Hall Company
Reston, Virginia

To my children, Jennifer and Jerry;
my mother and father, Dorothy and
Frank; my brother, Frank; and my friend,
Jon-Ellyn.

John A. Allocca

To Ida and Max, Janice, Bernice, and
Manuel.

Allen Stuart

Library of Congress Cataloging in Publication Data

Allocca, John A.
    Electronic instrumentation.

    Includes bibliographies.
        1. Electronic instruments.   2. Medical electronics.
I. Stuart, Allen.   II. Title.
TK7878.4.A45   1983   621.381   82-25020
ISBN 0-8359-1633-2

Editorial/production supervision and interior design
by Barbara J. Gardetto

© 1983 by
Reston Publishing Company, Inc.
A Prentice-Hall Company
Reston, Virginia 22090

10   9   8   7   6   5   4   3   2   1

Printed in the United States of America

Contents

2

TRANSDUCERS ASSOCIATED WITH ELECTRONIC INSTRUMENTATION 37

3

DATA-ACQUISITION SYSTEMS 88

4

ANALOG-TO-DIGITAL CONVERTERS 114

5
DIGITAL INSTRUMENTATION
SYSTEMS   156

6
DIGITAL-TO-ANALOG
CONVERTERS   198

7
ELECTRONIC INSTRUMENTATION
WAVEFORM GENERATION   229

15
MICROPROCESSORS AND MICROCOMPUTERS 518

16
SPECIALIZED BIOELECTRONIC INSTRUMENTATION 589

Preface

The first nine chapters of this first edition of *Electronic Instrumentation* have been written as a text for a two- or three-year electrical engineering college curriculum. The next six chapters are intended for students taking a course in computer science or engineering. The textbook can also be adapted for technical institutes or as an elective in a community college. Chapter 16 introduces the student to *bio*electronic instrumentation. The goal of the writers is to introduce their audience to the latest in instrumentation in industry and medical applications. The text also serves as a handbook for practicing engineers and technologists.

Chapter 1 covers the instruments for analog and digital circuit elements. Chapter 2 stresses instrumentation transducers. Chapter 3 introduces the reader to data-acquisition systems. Chapters 4 through 6 cover analog signal conversion to a digitized waveform and rerouting of that signal to an analog or digital readout. Chapter 7 stresses electronic instrumentation waveform generation and spectrum analyzers and Chapter 8 emphasizes the latest in oscilloscopes. Chapter 9 covers analog and digital readout devices and Chapters 10 through 15, computer-aided systems, computer-based systems including sensors, electronic counters, data-processing systems, and microprocessors and microcomputers. Chapter 16 introduces the student to specialized health-care technology. Each chapter includes review questions and references.

The authors wish to thank the Keithley Instrument Co., Datel-Intersil Inc., Global Specialities Corp., Sencore, International Business Machines Corp., Hewlett-Packard Co., and the many other manufacturers who have made significant contributions to this text. Without their help, this book would not have been possible.

The authors finally acknowledge Mr. David Dusthimer, Electronic Technology Editor, and Mrs. Linda MacInnes of Reston Publishing Co. for their support of this project.

John A. Allocca
Allen Stuart

1

Electronic Instruments Used to Measure Analog and Digital Circuit Elements

1.1
INTRODUCTION

In any electronic circuit, we can monitor voltage, current, or impedance. We can use a probe and attach it to points of an electronic circuit. The voltage and current signal can be read on an analog or digital multimeter. Today, we use the digital multimeter because considerable progress has been made in the field of digital technology. Digital multimeters provide a yardstick with which we can quantify the actual voltage, current, and resistance or impedance relationships of any circuit.

Current through any circuit is measured by an analog or digital indicating instrument called an *ammeter* connected in series with the circuit to be measured. The ammeter is connected such that the current to be measured passes through the instrument. Therefore, the ammeter must be capable of carrying this current without damage or excessive loading effects. If the circuit is drawing current in the milliampere range (10^{-3} amperes), it is useless to make the measurements with a 0 to 1 ampere meter.

The direct-current ammeter has two terminals marked (+) and (−). If the meter is connected such that current flows into the (+) terminal, the meter will read properly up scale. If the direction of current flow is the opposite, the meter will read down scale. This can be corrected by simply reversing the two wires connected to the ammeter.

The voltage across any element or two points is measured by an indicating instrument called a *voltmeter* connected directly across (in parallel) the element or two points. The voltmeter, since it is connected in parallel (Figure 1.1), must have an input resistance large enough such that it does not affect the voltage value being measured. The analog voltmeter must be such that the instrument will not be damaged and the pointer should deflect as full scale as possible for highest accuracy. Such a problem does not exist with a digital multimeter. Digital multimeter measurements should be made as close as possible to full scale to minimize errors.

A good rule of thumb to use in making voltage measurements and to ensure a loading error of less than 5% is to utilize a voltmeter with an input resistance at least 20 times greater than the resistance of the element whose voltage we are measuring. The input resistance of the analog voltmeter or volt-ohm-milliampere meter (VOM) is generally the ohms per volt multiplied by the maximum scale voltage or maximum scale voltage divided by the voltmeter current sensitivity. Digital multimeters have an input resistance typically of 10 megohms ($M\Omega$) on dc volts.

In making any electrical or electronic measurement, the analog instrument used will have some specified accuracy. The accuracy is usually specified as a percent accuracy of full-scale reading. A meter that has a specified accuracy of $\pm 2\%$ will have a maximum error of 0.02×100 volts $= \pm 2$ V on the 100-V scale and a maximum error of $0.02 \times 10 = \pm 0.2$ V on the 100-V scale; the voltage error may be 12 V $\pm$ 2 V, so the actual percent accuracy is $2 \times (100/12) = 16.7\%$. If the voltmeter had a 20-V scale that was used, the maximum error would be $\pm 0.02 \times 20 = \pm 0.4$ V. The percent error for a 12-V reading would be $\pm \left(\frac{0.4}{12}\right) \times 100 = 3.33\%$ instead of a maximum error of 16.7% on the 100-V scale.

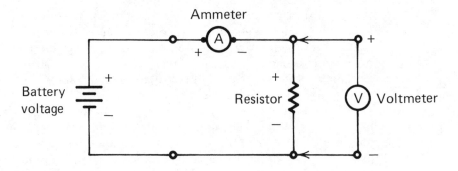

FIGURE 1.1
Direct-current ammeter and voltmeter connection in a series circuit.

1.2
THE DIGITAL MULTIMETER [a]

The capability to measure and digitally display volts, ohms, and milliamperes [1,2] has been the basis of an entire industry for better than 15 years. While this is a relatively short period, the digital multimeter (DMM) industry has advanced at an incredible rate.

From the introduction of the first digital voltmeter in 1963 to the most complex systems meter of today, the digital multimeter business has always been on the leading edge of technology. Each new breakthrough in technology has resulted in products that perform better and cost less.

Following the introduction of the DMM, the industry went through a phase common to most newly born industries. Many companies went off in different directions, each trying to develop successful products. Methods such as mechanical decade counters driven by servomotors were developed. While such events now seem amusing, it is typical of a fledgling industry until a proved direction is established.

The DMM industry established the direction it was to follow in the late 1960s. At this time, the John Fluke Manufacturing Company introduced its first DMM, the 8100, whose price was high and performance not startling. The significance of this event was that a company known for quality test equipment entered the DMM business (see Figure 1.2).

During the late 1960s and early 1970s, electronic technology improved rapidly: the first operational amplifier was introduced; field-effect transistors became affordable; light-emitting diode (LED) readouts and digital-logic integrated circuits (ICs) became available; and large-scale integration (LSI) became practical.

Technology again intervened when LSI companies produced two-chip sets that would handle analog-to-digital (A/D) conversion. Firms such as Intersil and Siliconix began to offer reasonably priced precision A/D converters that functioned with a minimum of external components. These LSI A/D converters offered advantages in parts count and cost over the discrete A/D, which was used in the 8000A. Companies such as Keithley Instruments began to offer a complete line of DMMs based on these LSI A/D converters, which were low in cost and offered excellent performance.

Fluke, beginning to feel the effects of competition in the DMM marketplace, again made a move that would shake up the DMM business. They approached Intersil, a large manufacturer of FETs and monolithic A/D converters; together they developed a monolithic A/D converter that would become the basis for virtually every handheld DMM made today.

---

[a] The material in Section 1.2 is from George B. Tuma, "Digital Multimeters Historical Overview," *Measurements and Control*, Dec. 1980; and "Keithley Electronic Measurement Instrumentation Catalog Guide," courtesy of Keithley Instrument Co., Cleveland, Ohio.

Until this time A/D converters were based on NMOS technology. This meant high power consumption, which precluded battery operation. They also required regulated power supplies.

The Intersil A/D converter (known as the ICL 7106) (see Figure 1.3) is a $3\frac{1}{2}$-digit single-chip A/D converter. Integrated in CMOS for low power consumption, it features direct liquid crystal drive and on-board power-supply regulators. It was designed to require a minimum of external components, and provides more than enough accuracy for $3\frac{1}{2}$-digits.

Simultaneously, a complete line of low-power op-amps and precision references became available from Intersil. These further facilitated the design of handheld DMMs. Intersil granted one-year exclusive rights to the

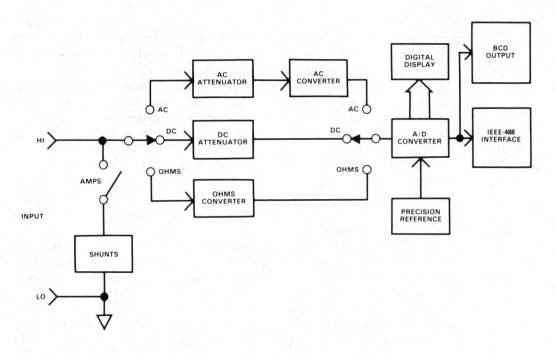

FIGURE 1.2
The specifics may vary, but all DMMs are designed around this basic architecture. The analog-to-digital (A/D) converter is the heart of every DMM design. Through one of a number of conversion techniques, a voltage input is accurately displayed on a digital readout. Circuitry is then developed to convert ohms, ac volts, or amperes into proper A/D input voltage. (Courtesy of Keithley Instrument Co.)

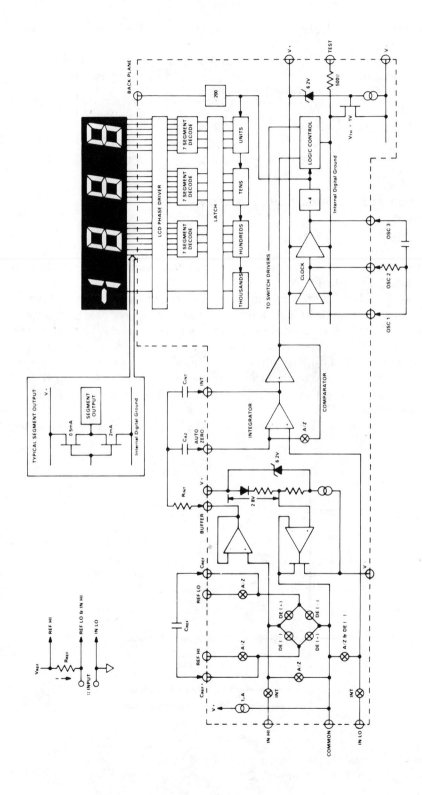

FIGURE 1.3

The Intersil ICL 7106 is a single-chip A/D converter that runs directly from a 9-V battery. Driving a liquid crystal display, the power consumption is about 1 mA. The input configuration can be used to measure ohms with no additional components. (Courtesy of Datel-Intersil, Inc.)

READING UNCERTAINTY

VOLTAGE LEVEL	% OF FULL SCALE	3½ DIGITS ±(0.1% + 1d)	4½ DIGITS ±(0.1% + 1d)
1V	5%	1.1 %	0.2 %
5V	25%	0.3 %	0.12 %
10V	50%	0.2 %	0.11 %
19V	95%	0.15%	0.105%

FIGURE 1.4
Reading uncertainty. (Courtesy of Keithley
Instrument Co.)

7106 A/D converter, and Fluke immediately introduced the 8020, the
world's first handheld DMM.

When the 7106 became available to the rest of the industry, it was a
situation almost unique to the electronics industry; every DMM manufacturer had available the same design for a battery-powered DMM. At the
heart of every DMM is the A/D converter, which converts an analog input
signal to some kind of digital output. The digital output can be a 7-segment
display, IEEE-488 standard interface, or the binary coded decimal (BCD)
output. The A/D converter is largely responsible for the performance characteristics of any DMM.

Low-cost handheld DMMs use a single-chip, $3\frac{1}{2}$-digit A/D converter.

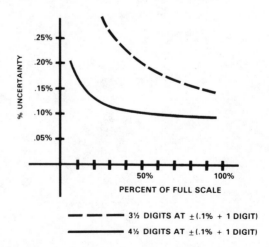

FIGURE 1.5
Percent of uncertainty versus percent of full scale.
(Courtesy of Keithley Instrument Co.)

High-accuracy $5\frac{1}{2}$-digit meters use a microprocessor-controlled discrete A/D converter. There are $4\frac{1}{2}$-digit meters, which are a tradeoff between $5\frac{1}{2}$-digit performance and $3\frac{1}{2}$-digit price. The $4\frac{1}{2}$-digit A/D converters may be discrete or LSI circuits (see Figure 1.4).

DMM accuracy (Fig. 1.4) is usually rated as a percent of reading, plus or minus one digit. This one-digit uncertainty can create some interesting effects, particularly at signal levels much lower than full scale. Figure 1.5 compares two generic DMMs. One is $3\frac{1}{2}$ digits; the other is $4\frac{1}{2}$ digits, and the accuracy will be assumed to be 0.1% ± 1 digit for both meters.

Sensitivity is also an important parameter in selecting a DMM. Sensitivity of 1 microvolt (μV) is generally available in $5\frac{1}{2}$-digit DMMs. The $4\frac{1}{2}$-digit DMMs are offered with 1-, 10-, or 100-μV sensitivities. The $3\frac{1}{2}$-digit DMMs are generally service-oriented instruments, so 100-μV sensitivity is sufficient.

DMMs invariably require some type of 7-segment readout. Light-emitting diodes (LEDs) are popular for line-powered instruments because they are easy to drive and provide good visibility in dimly lit areas. However, high power consumption combined with poor visibility in direct sunlight makes LEDs unsuitable for portable service instruments.

Liquid-crystal displays (LCDs) are much better suited to portable instrumentation. LCD operation involves aligning molecules with a polarizer to prevent light from passing through. They require very little power, owing to the fact that they generate no light of their own. Also, by their very design, they work best in bright light. The disadvantages of LCDs are that complex waveforms are required to drive them, and they have slower response, particularly at low temperatures.

Proper filtering is critical for quiet A/D performance. For this reason, it is important to reject as much external interference as possible. There are two specifications that describe noise rejection: the normal mode rejection ratio (NMRR) and the common mode rejection ratio (CMRR).

Normal-mode noise is noise that is mixed with the incoming signal. Most normal mode noise is line frequency noise, although noise of other frequencies is not uncommon. The amount of rejection while making dc measurements is specified as normal mode rejection in decibels (dB). This is the ratio of peak-to-peak input noise to peak-to-peak output reading. The integrating A/D techniques of all Keithley DMMs have inherently high NMRR at line frequency, since they average over an integral number of line cycles. A low-pass input filter is added to reject normal mode noise over a continuous ac frequency band.

Common-mode noise is so named because it appears in common to both the high and the low input terminals. High CMRR is obtained by maintaining high isolation between LO and earth ground.

Once the A/D converter has been established, signal conditioning is a matter of converting the input signal (dc or ac voltage or resistance) to the proper dc level.

Ohms can be measured by sourcing a precision current through the unknown resistor and then measuring the voltage drop across the resistor. Lower-cost DMMs source the current through the measuring test leads, terminating at the HI–LO inputs of the DMM. This two-wire ohms system works well for $3\frac{1}{2}$- and most $4\frac{1}{2}$-digit DMMs. However, the *IR* drop in the test leads alone can cause accuracy problems at $5\frac{1}{2}$ digits. This led to the development of the four-wire ohms circuit. Four-wire ohms allows one pair of test leads to be the current source conductors and the other pair to be the sense lines that measure the voltage drop across the unknown resistor. The only current flowing in the sense leads for this configuration is the input current of the A/D converter, which is in the picoampere range.

Keithley Instruments has been the only company to offer a professional-grade handheld DMM (see Figure 1.6).

Further significant cost reductions require that LSI manufacturers integrate more of the DMM architecture. Autoranging would eliminate the need for external range switches. An integrated ac converter (rms or otherwise) would further reduce the components needed. This DMM-on-a-chip idea will probably be realized in the 1980s.

1.3
MICROPROCESSORS AND THE DIGITAL MULTIMETER[b]

The influence of microprocessors on the electronics industry is ubiquitous. In the DMM industry, microprocessors have seen most application in higher-cost instruments, that is, instruments with extended resolution and accuracy, which find use in situations such as data logging or high-speed data acquisition. These types of applications are well suited to computer control, and a growing number of microprocessor-based, computer-interfaceable instruments are available.

These "systems" instruments are generally $6\frac{1}{2}$ digits, with the capability of taking at least 30 readings/second at $4\frac{1}{2}$ digits. Accuracy is on the order of 60 parts per million (ppm). Through an IEEE 488 standard interface, these instruments are programmable in both range and function. Thus, with nothing more than a systems DMM and a desktop-type computer, a relatively low cost, fully programmable, automated test system now can be assembled.

Microprocessors are widely used for *bus interfacing*. They serve another more important role in the design of $6\frac{1}{2}$-digit instruments with 1-μV resolution and 50-ppm accuracy: they compensate for many analog-circuit deficiencies that would otherwise make the design of a low-level instrument difficult if not impossible (see Figure 1.7).

For example, $3\frac{1}{2}$- and $4\frac{1}{2}$-digit A/D chip sets use an analog auto-zero

---

[b] The material in Sections 1.3 to 1.10 is used courtesy of Keithley Instrument Co., Cleveland, Ohio.

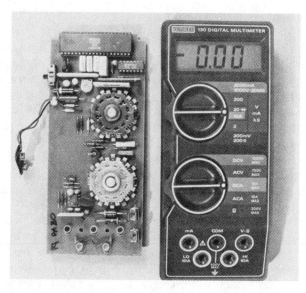

FIGURE 1.6
Keithley Model 130 has a small number of components. (Courtesy of Keithley Instrument Co.)

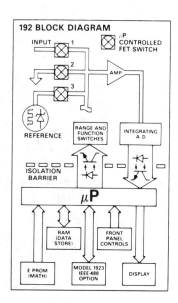

FIGURE 1.7
Keithley 192 systems DMM, $6\frac{1}{2}$ digit, features instant autoranging and range-function programmability and it could be the basis for a fully automated test system. (Courtesy of Keithley Instrument Co.)

scheme to compensate for their op-amps, which tend to drift with time and temperature. These analog auto-zero loops have never been easy to design beyond $4\frac{1}{2}$ digits owing to noise limitations and gain errors of the systems themselves.

Using a microprocessor, it is easy to internally short the input to the A/D converter, take a zero reading, take a reading of the input signal, and then subtract the zero reading from the input to obtain the correct reading. It is as though, at every conversion, the test leads are shorted, and the zero reading is noted and then subtracted from the measured input. This approach can be used even to 10-nV resolution, as in the Keithley Model 181 nanovoltmeter.

Microprocessors can be used also for digital filtering. At 1-μV levels, the inherent noise of amplifiers, resistors, and other components necessitates large amounts of filtering. Traditional analog filters are expensive, drift with temperature, and tend to have long settling times. Using a microprocessor, however, any type of filter can be implemented. Fast settling time, repeatable performance, and guaranteed stability characterize digital filters.

Microprocessors can even be used to accomplish an *autocalibration* cycle. Using a stable, precision reference as the input, the microprocessor can be used to introduce the proper scaling constants to ensure that the A/D converter is calibrated properly. See Chapter 15 for further details on microprocessors.

The day may come when instruments no longer have internal adjustments but, rather *calibration memories*. During calibration, a memory can be programmed to compensate for resistor inaccuracies, amplifier offsets, and the like. A calibration cycle such as this would be fully automated, always accurate, and require no human interaction.

Future systems instruments promise to improve upon their already excellent performance. Each new instrument provides more accuracy for lower price, faster data-acquisition times, and wider ac bandwidth.

Systems DMMs also are becoming capable of doing more complicated programming functions within themselves. These capabilities are limited, however, compared with the capability available when even a low-end computer is connected across the IEEE 488 microprocessor interface.

1.4
THE BENCH DMM

Bench DMMs have not received much development attention recently and promise to receive less as time goes on. They compete with both handhelds and systems DMMs. Virtually every DMM manufacturer already has a complete line of bench DMMs; quality bench meters are available at $3\frac{1}{2}$, $4\frac{1}{2}$, and $5\frac{1}{2}$ digits. Sensitivity for these meters runs between 1 and 100 μV. Some have averaging ac converters; others are true root mean square (TRMS) devices.

Few applications can be described for a bench DMM that cannot be filled by an existing DMM. Furthermore, calculated mean time between failure for the modern bench DMM is now about 10 years; replacement rates will be low.

Of the bench DMMs that have been introduced in recent years, some attempt to fill small gaps in the market, and others try to generate interest by adding "bells and whistles" such as "conductance" and "dB." Any new bench DMM will have to be a superior product in terms of price and performance to generate much interest in today's bench DMM market.

The intense competition among DMM manufacturers has resulted in a marketplace highly beneficial to the DMM user. As the price of handheld meters continues to fall, an industrial grade handheld DMM will be within the reach of every technician or repairman. As electronics becomes a bigger part of the average life-style, a consumer market for handhelds may even arise. Innovations in the area of high-resolution, high-precision handhelds will soon make benchtop precision available in a portable package.

As for bench instruments themselves, products to fill virtually every need are already available. Principal improvements in this area will come in terms of existing capabilities available at a lower price. DMMs may also combine with other test instruments, such as frequency counters or function generators, to form universal service appliances.

1.5
THE SYSTEM DMM

System DMMs (Figure 1.8) have not been around long enough to allow the definition of a generic systems product to emerge. Higher-end systems products are still improving in terms of resolution, accuracy, and speed. As microprocessors become more powerful, perhaps an IEEE interface will no longer be necessary; the DMM will be able to provide self-contained sourcing and measuring capability. Of the three basic DMM categories (handheld, bench, and systems), systems DMMs are undergoing the most improvements in terms of absolute performance increases.

Electronic measurement of voltage and current with a digital multimeter depends mainly on the following characteristics:

1. Range
2. Accuracy
3. Input impedance
4. Frequency response
5. Maximum input
6. Power supply line variations
7. Stability, resolution, and calibration
8. Other features

A $6\frac{1}{2}$-digit unit [3] can be used as a programmable digital multimeter

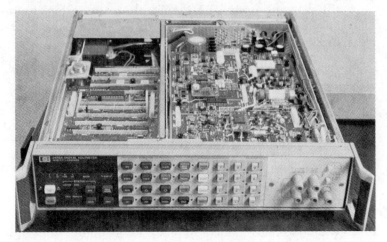

FIGURE 1.8
Hewlett-Packard systems meter is "Cadillac" of systems DMMs. Use of multiple microprocessors in products of this complexity requires the efforts of many development groups. (Courtesy of Hewlett-Packard Co., Palo Alto, Calif.)

system via use of JFET, special converters, and a single processor chip to provide generic functions via the IEEE 488 microprocessor interface.

1.6
ELECTROMETERS

The electrometer is a refined dc multimeter [4]. The input characteristics permit it to perform voltage, current, resistance, and charge measurements using operational amplifiers with an input resistance above 10^{14} Ω and an offset current of 5 x 10^{-14} amperes (A) or lower.

When an ammeter is conected to the source, as in Figure 1.9, a current (I_s) flows through the source resistance equivalent ammeter circuit. The indicated current (I_M) is equal to the current that would flow in the circuit if the ammeter were not inserted (I_{AB}), minus errors due to the voltage burden V_M, shunt currents I_{SH}, currents generated in the interconnection I_E, and the meter uncertainty, U_M. If V_M, I_E, and I_{SH} are small, I_M is simply I_{AB} plus or minus the specified meter uncertainty, U_M.

From Figure 1.9, we can derive the following equations:

$$I_i = I_S - I_{SH} - I_E \qquad (1.1)$$

$$I_S = \frac{V - V_M}{R_S} \qquad (1.2)$$

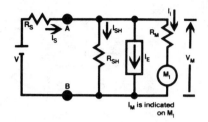

FIGURE 1.9
Current measurement equivalent circuit and error
calculations. (Courtesy of Keithley Instrument Co.)

Thus

$$I_M = \frac{V}{R_S} - \frac{V_M}{R_S} - I_{SH} - I_E \tag{1.3}$$

where

$$I_M = \text{indicated current}$$

$$\frac{V}{R_S} = I_{AB} \text{ to be measured}$$

$$I_{SH} = \text{shunt current error}$$

$$I_E = \text{generated current error}$$

$$U_M = \text{specified ammeter uncertainty}$$

Digital multimeters generally utilize a "shunt technique" for current measurement, shown in Fig. 1.10. In this configuration, the maximum voltage burden V_M is typically 200 mV; in some cases it may be 2 V.

From Figure 1.10, we can determine the following equation:

$$V_M = V_R = I_i R_M \tag{1.4}$$

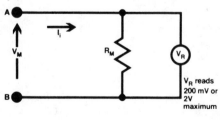

2-1: $V_M = V_R = I_i R_M$

FIGURE 1.10
Shunt ammeter circuit. (Courtesy of Keithley Instrument Co.)

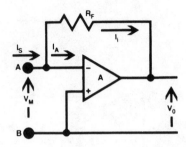

FIGURE 1.11
Feedback ammeter circuit. (Courtesy of Keithley
Instrument Co.)

This is also the configuration used in the NORMAL mode of electrometers.

At sensitive current levels, feedback ammeters are usually used. This configuration, used in picoammeters and the FAST mode of electrometers, is shown in Figure 1.11. In this case, as Eq. (1.3) indicates, the voltage burden V_M is equal to the output voltage (usually 200 mV or 2 V) divided by the amplifier gain A, which is typically 5×10^4 to 10^6. Thus, V_M is usually in the microvolt region. Picoammeters and electrometers designed for measuring low currents typically have a maximum specified voltage burden of 1 mV or less. This low voltage burden is beneficial in two ways. See Fig. 1.11, from which we can deduce the following equations:

$$I_s = I_i + I_A \tag{1.5}$$

Thus if $I_A \ll I_s$,

$$I_s = I_i$$

Then

$$V_0 = -I_i R_F \tag{1.6}$$

and

$$V_M = \frac{-V_0}{A} \tag{1.7}$$

Equivalently,

$$R_M = \frac{V_M}{I_i} = \frac{I_i R_F}{A I_i} = \frac{R_F}{A} \tag{1.8}$$

First, errors caused by inserting the ammeter in the circuit are greatly reduced, as shown by the following calculations and Figure 1.12.

From Eq. (1.2), $I_s = (V - V_M)/R_s$ and $I_{AB} = V/R_s$, and we may calculate the error:

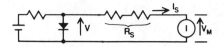

FIGURE 1.12
Voltage burden error circuit. (Courtesy of Keithley
Instrument Co.)

$$\frac{I_s - I_{AB}}{I_{AB}} = \frac{I_s}{I_{AB}} - 1 = \frac{V - V_M}{V} - 1 \tag{1.9}$$

$$\text{Error} = \frac{-V_M}{V} \times 100 \text{ as a percent}$$

Thus if $V = 700$ mV, (the forward voltage drop of a silicon diode), the burden for a typical DMM is $V_M = 200$ mV. The error is as follows:

$$\text{Error} = \frac{-200 \text{ mV}}{700 \text{ mV}} = \times 100\% = -28.6\%$$

For a typical picoammeter, $V_m = 1$ mV, the error is as follows:

$$\text{Error} = \frac{-1 \text{ mV}}{700 \text{ mV}} \times 100 = -0.14\%$$

Second, the requirement for a high shunt resistance is significantly reduced, as shown by the following calculations and Figure 1.13.

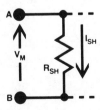

FIGURE 1.13
With the circuit shown above we can calculate
the minimum shunt resistance ($R_{SH\ min}$) required
to maintain I_{SH} below a given ($I_{SH\ max}$). (Courtesy
of Keithley Instrument Co.)

Using Figure 1.13, we can obtain the following equations:

$$I_{SH} = \frac{V_M}{R_{SH}}$$

(1.10)

Thus

$$R_{SH\ min} = \frac{V_M}{I_{SH\ max}}$$

(1.11)

For example, if $I_{SH\ max}$ is to be 10^{-11} A, for a typical DMM,

and $V_M = 200$ mV

then, $R_{SH\ min} = \dfrac{0.2\ \text{V}}{10^{-11}\ \text{A}} = 2 \times 10^{10}\ \Omega$

For a typical picoammeter, $V_M = 1$ mV, and therefore

$$R_{SH\ min} = \frac{0.001\ \text{V}}{10^{-11}\ \text{A}} = 10^8\ \Omega$$

Generated currents may be created in the source, the interconnection, or the ammeter, as illustrated in Figure 1.14. I_{SE} consists of all unwanted currents in the circuit under test. I_{SE} represents currents generated in the interconnection due to triboelectric, piezoelectric, or electrochemical phenomena.

Triboelectrically generated currents result from the creation of charges at the interface between a conductor and an insulator, due to frictional forces at the interface, as in the case of a cable that is moved. The mechanism involved is one of rubbing off electrons, creating a charge unbalance, and thus a current flow. Low-noise cables are available that contain a conductive coating (usually graphite) at the metal–insulator boundary, which reduces this effect significantly.

Using Figure 1.14, we can obtain the following equations, with $I_{SE} = $ dc, ac, or noise currents generated in the source, $I_{CE} = $ currents generated in the

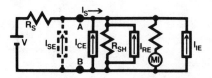

FIGURE 1.14
Generated currents. (Courtesy of Keithley Instrument Co.)

cabling or interconnection, and I_{RE} = noise current generated by R_{SH}. I_{IE}, in Figure 1.14, is the specified input current of the picoammeter. For any resistor,

$$I_{\text{noise rms}} = \sqrt{\frac{4KTF}{R}} \qquad (1.12)$$

where

$$k = \text{Boltzmann's constant}$$
$$T = \text{temperature, °K}$$
$$F = \text{noise bandwidth}$$

For F = 1 hertz (Hz) and T = 300°K, the noise generated by R_{SH} is

$$I_{RE} = \frac{6.5 \times 10^{-10} \text{ Ap-p}}{\sqrt{R_{SH}}}$$

with Ap–p being the peak-to-peak amperage. For example, if R_{SH} = 10^{10} Ω,

$$I_{RE} = 6.5 \times 10^{-15} \text{ Ap-p} \qquad (1.13)$$

Piezoelectric currents are generated when mechanical stress is applied to certain insulating materials. The effect occurs in ceramics and other crystalline material. Teflon® and some other plastics used for insulated terminals and interconnecting hardware exhibit a *space charge effect*, wherein an applied force creates a change in capacitance and thus a charge redistribution. The behavior is the same as for piezoelectric materials; force creates a current.

Ionic chemicals can create weak "batteries" between two conductors, owing to electrochemical action. For example, epoxy printed circuit boards that are not thoroughly cleaned of etching solution, flux, and so on, can generate voltages of a few volts or currents of a few nanoamperes between two (unconnected) conductors. All interconnected circuitry should be cleaned with methanol or Freon to avoid currents due to this effect.

In addition to these phenomena, Ohm's law indicates that leakage resistances between low-current conductors and nearby voltage sources can create significant error currents. For example, if a printed circuit element has a leakage path with a resistance of 10^9 Ω to a nearby 15-V supply terminal, a current of 15 nA will be generated, as shown in Figure 1.15(a).

To keep this current below 1 pA, the leakage resistance would have to be above 1.5×10^{12} Ω. This high resistance is difficult to maintain in many situations. To eliminate such stringent insulation resistance requirements, guarding techniques may be used, as shown in Figure 1.15(b). The theory of guarding is simple: surround the sensitive input with a conductor (the guard) connected to a low-impedance point that is at (virtually) the same potential. In feedback ammeters, the guard point is input LO. In electrometer voltage or NORMAL mode measurements, the guard is the X1 output.

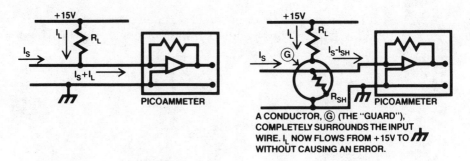

(a) Unguarded circuit   (b) Guarded circuit

FIGURE 1.15
Leakage current. (Courtesy of Keithley Instrument
Co.)

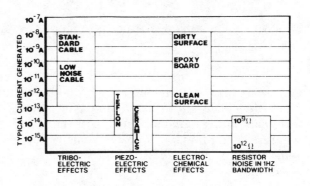

FIGURE 1.16
Approximate magnitudes of various generated
currents. (Courtesy of Keithley Instrument Co.)

In addition to these dc currents, any resistance generates a random
noise current described by Eq. (1.12). This is a theoretical measurement
limit for any given temperature and bandwidth.

I_{IE}, the input current of the picoammeter or electrometer, is given in the
specifications of the instrument. Whereas DMMs may have a few nanoamperes of input current, picoammeters and electrometers have input currents of 10^{-12} to 10^{-17} A.

Rough magnitudes of the preceding current-generating effects are indicated in Figure 1.16. Specific values for a given interconnection circuit or
source are quite dependent on surface condition, moisture levels, and stress
and vibration levels. Specific magnitudes are best determined empirically,
by measuring the current with a picoammeter or electrometer.

1.7
HIGH-RESISTANCE ELECTROMETER
MEASUREMENTS

A low-current measurement may be made to determine the value of a high resistance, as shown in Figure 1.17(a). In this *constant-voltage method* of measuring high resistances, the applied voltage V is set to a desired level by a low-noise source and the resistance calculated by Ohm's law, as shown in Eq. (1.14). The measurement concepts and considerations are the same as for low-current measurements, discussed previously. Figure 1.17(b) shows the constant-current method.

In Figure 1.17(a) we can deduce the following:

$$R_X = \frac{V}{I_M} \tag{1.14}$$

In Figure 1.17(b) we can deduce the following:

$$R_X = \frac{V_M}{I} \tag{1.15}$$

In the *constant-current method*, the current through R_X is fixed by a current source, I, and the voltage, V_M, is measured with a high input resistance voltmeter (i.e., an electrometer). This configuration is also useful for plotting semiconductor *IV* characteristics and measuring breakdown voltages at low currents, using the electrometer as a current source.

In the voltage or resistance-measuring mode, the electrometer amplifier consists of a noninverting amplifier whose effective input resistance is so high that the most significant limit is the insulation resistance of mechanical hardware connected to the input, such as the input connector. Instruments such as the Keithley Model 642 have over 10^{16} Ω of input resistance.

The equivalent measuring circuit of an electrometer ohmmeter is

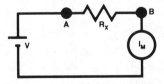

(a) Voltage source

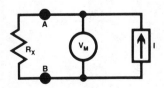

(b) Current source

FIGURE 1.17
Measurement of high resistances. (Courtesy of Keithley Instrument Co.)

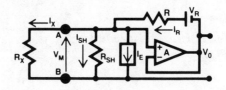

FIGURE 1.18
Electrometer resistance measuring circuit.
(Courtesy of Keithley Instrument Co.)

shown in Figure 1.18. As in the case of low-current measurements, R_{SH} is the net effective shunt resistance across input terminals A and B, and I_E is the sum of all generated currents, due to the effects discussed earlier. In practice, the limiting factor is usually the degree to which the shunt resistance R_{SH} can be made much greater than the effective source resistance.

From Figure 1.18 we can obtain the following:

$$V_0 = V_M = I_X R_X \tag{1.16}$$

$$I_X = I_R - I_E - I_{SH} \tag{1.17}$$

$$I_{SH} = I_X \frac{R_x}{R_{SH}} \tag{1.18}$$

Thus for $I_{SH} \ll I_X$

$$R_{SH} \gg R_X$$

R_{SH} is determined by the resistance of the insulation in the cabling and interconnection hardware. Figure 1.19 provides a rough indication of resistances obtained using various insulating materials.

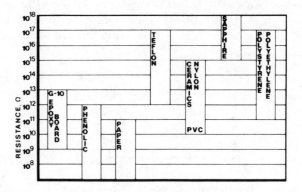

FIGURE 1.19
Approximate resistance of various insulating materials (between 2 points). (Courtesy of Keithley Instrument Co.)

1.8
VOLTAGE MEASUREMENTS FROM HIGH-RESISTANCE SOURCES

When making voltage measurements with an electrometer, the amplifier output $(X1)$ is made available and may be used as a guard for some of the insulators connected to the input, thus reducing shunt current I_{SH}, since the voltage across insulators located between the input and the guard is the output voltage divided by amplifier gain, rather than the full output voltage (which is equal to the input voltage). Note also that the techniques discussed earlier for maintaining high shunt resistance, guarding the input and minimizing generated currents, are applicable to measurements of high-resistance voltage sources, such as pH meters, biological sources, and the like.

As the preceding discussion indicated, electrometers and picoammeters are designed to make the measurement of low currents and voltages from high-resistance sources not only possible but quite practical as well. The feedback ammeter and noninverting unity-gain amplifier configurations provide convenient means of reducing errors and long time constants due to shunt resistance and capacitance in the interconnection, as well as low voltage burden errors. The remaining sources of error are primarily currents generated by a variety of physical phenomena. While these are difficult to quantitatively analyze, they can be directly measured, using the electrometer or picoammeter in the circuit.

The Keithley 642 is a direct reading MOSFET-based electrometer for sensitive measurement of current, voltage or charge. Its current measurement range is 10aA to 200nA with typically less than 1mV voltage burden. Voltage reading capabilities are 10μV to 10V with 10,000TΩ input resistance. Charge can be measured from 800aC to 100pC. Maximum sensitivity is obtained when using the Charge mode to integrate current applied to the input. All functions and ranges are easily selected on the mainframe front panel using color-keyed pushbuttons.

High resolution. The digital format permits observation of small changes in large signals with resolution to 1 part in 20,000 at the digital display. The FEEDBACK output has low noise and non-linearity of only about 5ppm. This wide dynamic range (from noise level to maximum output) provides constant gain over the full span, and eliminates the need for range changing.

State-of-the-art design. The 642 uses a specially packaged dual monolithic MOSFET with compensated temperature coefficient in a guarded package. Variations in ambient temperature do not cause significant errors because compensation circuitry is individually adjusted for each FET to give a voltage coefficient of only 30μV/°C.

In the design of the remote head, the active input volume has been minimized. Less than 15 ionization current pulses per hour are observed.

FIGURE 1.20
Keithley Model 642 electrometer. (Courtesy of Keithley Instrument Co.)

1.9
OTHER ELECTROMETER USES

Electrometers are also used with the following:

1. Photomultipliers (discussed in Chapter 2)
2. Resistivity measurements
3. Transistor voltage breakdown tests
4. Ionization chambers

With digital displays and push buttons, the latest electrometers [6] make some analog measurements that cannot be done by average digital multimeters and skirt many sneaky errors within DMM range. The Keithley Model 642 detects current as low as 100 aA and as high as 40 μA with a voltage burden of from 20 μV (see Figure 1.20). The input resistance is 10,000 teraohms (TΩ) and features $4\frac{1}{2}$ digits, high sensitivity, and analog outputs.

1.10
NANOVOLTMETERS

Measurements of dc signals with resolutions of one to a few hundred nanovolts are easily made with nanovoltmeters and null detectors. Certain phenomena that would be insignificant at ordinary voltage levels must be considered at this level since they may limit the usable resolution of the measurement. An understanding of the nature of these phenomena and of the principles used to minimize them is extremely useful in making meaningful measurements.

The Keithley Model 181 is a microprocessor-based digital nanovoltmeter with 10-nV sensitivity, built-in IEEE microprocessor interface and $5\frac{1}{2}$- to $6\frac{1}{2}$-digit resolution. The front panel, with status annunciators and push-button zero control, is designed for ease of use. Its fast (0.5 s) response time, excellent stability (10 nV/°C), low noise (30 nV p−p), ranges up to 1 kV, and low cost provide new standards for nanovoltmeters of the 1980s.

The Keithley Model 148 is the most sensitive voltmeter available. It has 1-nV sensitivity and utilizes FETs instead of subminiature tubes in the low-noise amplifier.

The nanovoltmeter is useful in measurements of (1) thermoelectricity, (2) grounding loops, and (3) magnetism.

After the number of dissimilar metal junctions has been reduced as much as possible, the circuit performance can be further improved by reducing the temperature gradients within the circuit. This can be done by placing the remaining junctions near one another and by providing good thermal contact with a common heat sink. Most good electrical insulators are good thermal insulators as well; that is, they have very low thermal conductivity. Certain materials are available that combine good electrical

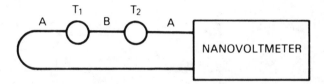

FIGURE 1.21
Thermal electromotive force. (Courtesy of Keithley Instrument Co.)

insulation with high thermal conductivity, such as hard anodized aluminum, beryllium oxide, specially filled epoxy resins, sapphire, and diamond.
By using these materials together with a massive metallic heat sink, the
temperature gradients across the remaining junctions will be minimized, as
will the thermal emfs generated (see Figure 1.21).

The thermal emf developed by dissimilar metals A and B in a series
circuit is:

$$E_{AB} = Q_{AB}(T_1 - T_2) \tag{1.19}$$

where

E_{AB} = thermal emf

Q_{AB} = thermoelectric power of material A with respect to B, $\mu V/°C$

T_1 = temperature of the B to A junction, °C or °K

T_2 = temperature of the A to B junction, °C or °K

In addition, if the equipment is allowed to warm up and reach thermal
equilibrium, and if the ambient temperature can be held constant, any
remaining thermal emfs will also be constant and can be compensated for
by use of the zero suppression of nanovoltmeters.

The motion of a conductor in a magnetic field, even one as weak as the
earth's, can cause significant spurious signals in nanovolt measurements.
The principal means for reducing this source of error is to reduce the area
enclosed by the circuit. Minimizing the motion of the various parts of the
circuit will also help. It may be necessary to provide some form of magnetic
shielding. The most useful materials for this are special alloys with a high
permeability at low flux densities, such as Mumetal.

Troublesome magnetic fields may be generated within the circuit by
conductors carrying large currents. By using twisted pairs of lines to carry
large currents, the magnetic field generated may be largely canceled out
(see Figure 1.22).

The voltage developed due to a field passing through a circuit enclosing
a prescribed area is

$$E_B = \int_0^A \frac{\partial \vec{B}}{\partial t} \, dA \tag{1.20}$$

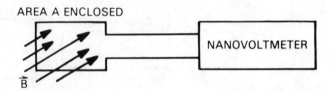

AREA A ENCLOSED

FIGURE 1.22
Magnetic fields. (Courtesy of Keithley Instrument Co.)

where

A = integral to be evaluated over the area enclosed by the circuit

$\vec{B}$ = magnetic field intensity, webers/m² or 10^{-4} gauss

Frequently, a complete measuring system will have several points connected to earth ground. The power source, the experiment itself and the indicating instrument may all be grounded. If a small difference in potential exists between these points, a large ground current may circulate, causing unexpected voltage drops to occur. In making low-level dc voltage measurements, it is desirable to have a single ground at one point. If the power source, the indicating instrument, and other parts of the circuit are well isolated from earth ground, the most appropriate single point may be chosen with ease (see Figure 1.23).

Input voltage to the nanovoltmeter is

$$E_{IN} = E_S + I_I R_I \qquad (1.21)$$

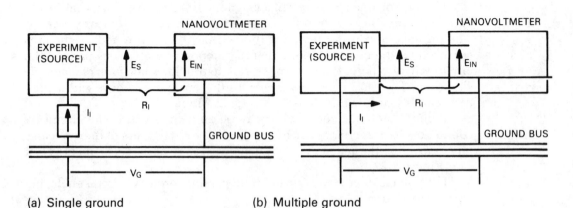

(a) Single ground

(b) Multiple ground

FIGURE 1.23
Grounding configurations for single or multiple grounds. (Courtesy of Keithley Instrument Co.)

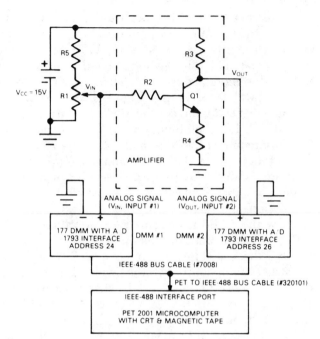

FIGURE 1.24
Amplifier circuit and acquisition–analysis system.
(Courtesy of Keithley Instrument Co.)

where

E_S = source voltage or desired voltage

I_I = current passing through input LO connection due to common-mode currents in source (magnitude is typically in nanoamperes)

R_I = resistance of input LO connection (typically around 100 mΩ or 0.1 Ω)

$I_I R_I$ is insignificant compared to E_S. The common-mode impedance is typically in megohms. Thus V_G generates I_I of only nanoampere range.

1.11
EXPERIMENTAL APPLICATION OF THE DIGITAL MULTIMETER

[c]There are hundreds of applications of digital multimeters in electronic instrumentation. We will concentrate on a few considerations. Let us first consider Figure 1.24.

---

[c]The following two paragraphs are from Physica 1 Microcomputer Controlled Data Acquisition and Analysis Systems. Reference: S. D. Senturia and B. D. Wedlock, *Electronic Circuits and Applications,* John Wiley & Sons, Inc., New York, 1975. Courtesy of Keithley Instrument Co., Cleveland, Ohio.

Construct the bipolar transistor, common-emitter amplifier circuit as shown in the upper part of Figure 1.24. Then install the first 1793 IEEE 488 bus interface into the first Keithley 177 DMM following the instructions of the 1793 manual. The switches on this Keithley 1793 should be set to addressable mode, address 24. The first 177/1793 combination is referred to as DMM 1, as shown in Figure 1.24. Install the second 1793 into the second 177 DMM and set the 1793 to addressable mode, address 26. This second 177/1792 combination is referred to as DMM 2, as shown in Figure 1.24. Interconnect DMMs 1 and 2 with the 7008 IEEE 488 bus cable. Then connect them to the IEEE 488 interface output port of the PET 2001 micro-computer with the 320101 cable, as shown in Figure 1.24. Set the DMM ranges to 20 V. Analyze the gain for the amplifier circuit and acquisition–analysis system.

[d]Use the DMM to measure the true values of the resistors for the circuit of Figure 1.25(a) and (b). Remember to apply $+V_{cc}$ and $-V_{cc}$ to the op-amp, which is discussed in detail in Chapter 3. Record these in the DATA table. Construct the op-amp circuit shown in Figure 1.25. For the input, use the signal generator to produce a 1-kHz sine wave of 1-V peak-to-peak amplitude (1 V p−p) with 0-V dc offset. Use the oscilloscope to monitor V_{IN} and V_{OUT} simultaneously. Sketch the input and output waveforms in the DATA table. Set the DMM to ac volts and measure the amplitude of V_{IN} and V_{OUT}. Record these in the DATA table.

NOTE: The DMM does not measure peak-to-peak voltage. It measures root-mean-square (rms) voltage. For a sine wave, $V_{rms} = 0.707\ V_{p-p}$. With an oscilloscope, one can measure true peak-to-peak voltage. See Chapter 8 for a detailed discussion of oscilloscopes.

Next, leaving R_1 unchanged, use Eq. (1.22) in Figure 1.25b to calculate the value of R_2 needed to yield a gain of -10. Select a resistor for this new value of R_2, measure the true value, and record this in the DATA table. Replace R_2 in the circuit with the new value and input the same signal as before. Measure and record the amplitude of V_{IN} and V_{OUT} using the DMM set to ac volts. Monitor V_{IN} and V_{OUT} with the oscilloscope. Adjust the dc offset of the signal generator in the positive direction until a large amount of clipping occurs. Sketch the output waveform in the DATA table. Repeat for negative dc offset.

Measure the values of the resistors of Figure 1.26. Construct the circuit and input a 1 V_{p-p} 1 kHz sine wave with 0 V dc offset from the signal generator. Monitor V_{IN} and V_{OUT} simultaneously on the oscilloscope. Sketch the input and output wave forms in the DATA table. Measure and record the amplitudes of V_{IN} and V_{OUT} using the DMM and record these in the DATA table.

---

[d]The following three paragraphs are from Electrical Engineering Experiment 3, Linear Op Amp Circuits. Reference: S. D. Senturia and B. D. Wedlock, *Electronic Circuits and Applications*, John Wiley & Sons, Inc., New York, 1975. Courtesy of Keithley Instrument Co., Cleveland, Ohio.

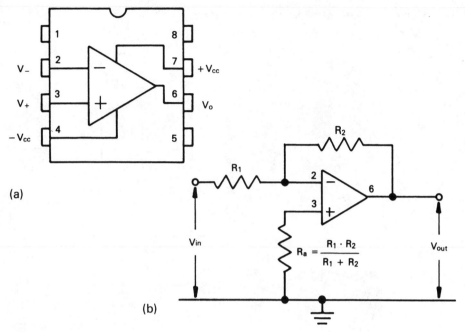

(a)

(b)

FIGURE 1.25
(a) Configuration of an op-amp. LM741N op-amp.
(b) Inverting op amp. Mathematically, the closed
loop gain, G_{CL}, is

$$G_{CL} = \frac{V_{OUT}}{V_{in}} = \frac{-R_2}{R_1} \tag{1.22}$$

(Courtesy of Keithley Instrument Co.)

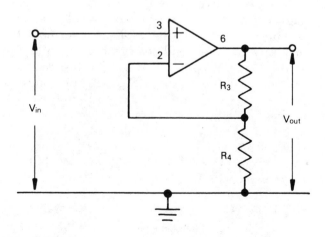

FIGURE 1.26
Noninverting op-amp.

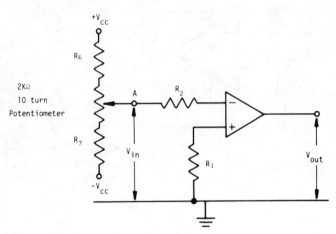

FIGURE 1.27
Comparator.

[e]Consider the comparator used in A/D converters circuitry and discussed in detail in Chapter 4. When an operational amplifier (op-amp) is used with no feedback, the gain (called the open-loop gain) is on the order of 10^4 to 10^5. The output voltage (V_{OUT}) is governed by the relationship

$$V_{OUT} = A(V_+ - V_-) \tag{1.23}$$

where A is the open-loop gain, V_+ is the noninverting input voltage, and V_- is the inverting input voltage. V_{OUT} will saturate at $+V_{CC}$, the positive supply voltage, which is typically $+15$ V, if $A(V_+ - V_-)$ is greater than $+V_{CC}$. Similarly, V_{OUT} will saturate at $-V_{CC}$, -15 V, if $A(V_+ - V_-)$ is more negative than $-V_{CC}$. With no feedback, V_{OUT} will saturate for very small voltage differences between the two inputs. This is the basis of the comparator, shown in Figure 1.27. It provides a way to distinguish which of two voltages is more positive. The two voltages are input to the op-amp, one to the inverting input (V_-) and one to the noninverting input (V_+). If V_+ is greater than V_-, then V_{OUT} is $+15$ V. If V_+ is less than V_-, then V_{OUT} is -15 V.

Construct the circuit of Fig. 1.27. Remember to apply the supply voltages $+V_{CC}$ ($+15$ V) and $-V_{CC}$ (-15 V) to the op-amp. Use the oscilloscope to monitor V_{OUT}. Connect the DMM across point A to ground to monitor V_{IN}. The DMM should be set to dc volts. Set V_{IN} to $+2$ V. Then decrease V_{IN} slowly while watching V_{OUT} on the oscilloscope. Record the value of V_{IN} that causes V_{OUT} to switch from $-V_{CC}$ to $+V_{CC}$ in the DATA section. Repeat the experiment with V_{IN} starting at -2 V and increasing the voltage.

[e]The following three paragraphs are from Electrical Engineering Experiment 1-34 #4, Non Linear Op Amp Circuits. Reference: S. D. Senturia and B. D. Wedlock, *Electronic Circuits and Applications*, John Wiley & Sons, Inc., 1975. Courtesy of Keithley Instrument Co., Cleveland, Ohio.

Replace the potentiometer input with the signal generator at point A. Set the signal generator to produce a 4-V $_{\text{p-p}}$ sine wave at 1 kHz with 0 V dc offset. Monitor V_{IN} and V_{OUT} on the oscilloscope. Sketch these in the DATA section. Increase the signal generator dc offset to +1 V. Sketch the corresponding input and output waveforms. Note the relative positions of V_{IN} and V_{OUT}. Raise the dc offset above +2 V and note what happens.

1.12
OTHER IMPORTANT ELECTRONIC INSTRUMENTS[f]

There are many other electronic instruments used to measure circuit elements. Some still use analog readout devices. Many electronic instruments have been digitized with a digital readout device. The aim of this section is to introduce the reader to the most important devices with their uses, photos, and diagrams where appropriate.

The component measurements of the absolute magnitude of the impedance, phase angle, capacitance, C, inductance, L, resistance, R, the dissipation factor, D, and figure of merit, Q, can be made with the traditional manual null measurement technique to the completely automatic microprocessor controlled systems type.

Impedance measurements with a range of 0.001 Ω to greater than 10^6 Ω can be made with the proper instrument. In the bridge technique, circuit conditions required to achieve a balance or null detector are detected and processed to an analog readout device. The voltage–current method uses Ohm's law in that a constant voltage or current is applied to the unknown. The converse current or voltage is an indication of the unknown circuit element value. The Q method uses series resonant circuitry to obtain Q and, indirectly, inductance, capacitance, and resistance. Today's technology yields automatic digital readout systems [10] with accuracy exceeding the less sophisticated manual bridges.

Figure 1.28, the GenRad 1657, is an automatic, microprocessor-based bridge designed to measure R, L, C, D, and Q at better than three measurements per second. Basic accuracy for R, L, and C is 0.2%.

A five full-digit LED readout is displayed for R, L, and C. Four full digits are dedicated to D and Q. This means better resolution and, therefore, better measurement confidence and reliability.

This digibridge's built-in test fixture uses guarded Kelvin measurement techniques for both axial and radial lead components. This eliminates the contributions of any shunt admittances or series impedances whether you use the built-in test fixture or the optional extender cables. The result is a

---

[f]Section 1.12 appears here in part by courtesy of GenRad.

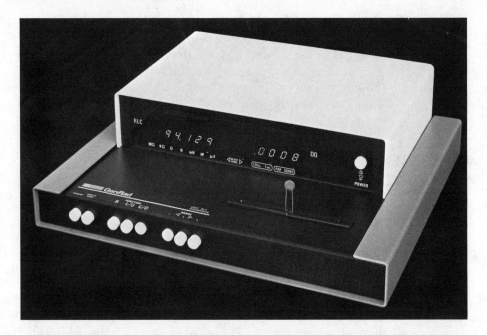

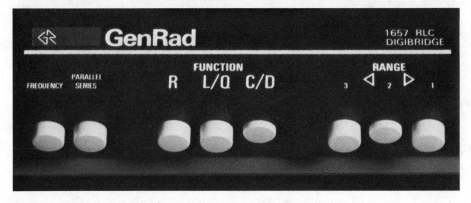

FIGURE 1.28
GenRad digibridge Model 1657. (Courtesy of
GenRad.)

higher level of test accuracy. The operator has a choice of two sinusoidal
test frequencies: 1 kHz and either 120 or 100 Hz.

As another flexibility feature, series or parallel measurement modes are
operator-selected across the full measurement range of every test parameter.

The 1657 offers measurements in three ranges, each range covering two
full decades of measurement capabilities. This feature is made possible by
the 1675's automatic decimal point positioning. This feature, plus the use

of a full digit for the most significant digit, allows D measurement from 10^{-4} to 10.

Microprocessor-directed ranging takes the guesswork out of setting the correct range, too. Lighted arrows on the front panel indicate which range button should be pressed. As a result, the optimum range is automatically identified. This means better resolution and better accuracy. There is no chance for operator mistakes, since no operator decisions are required.

Specifications for the 1657 follow:

1. **Measurement parameters:** Measures R series or parallel; L and Q series or parallel; C and D series or parallel. All measurement modes are push-button selectable.

2. **Measurement speed:** Greater than three measurements per second, unqualified.

3. **Test frequencies:** 1 kHz and 120 Hz. Also 100 Hz in place of 120 Hz. Push-button selectable.

4. **Measurement ranges:** Push-button selectable.

   a. Three ranges for R, L and C (multiples of 100)

      - R = 0.001 Ω to 99.999 MΩ
      - L = 0.0001 mH to 9999.9 H
      - C = 0.0001 nF to 99999. μF

   b. One range for D and Q

      - D = 0.0001 to 9.999
      - Q = 0.01 to 999.9

5. **Display:** R, L, and C, five full digits (99999), LED display with automatic decimal point positioning. D and Q, four full digits (9999), LED display with automatic decimal point positioning.

FIGURE 1.29
4800A HP vector impedance meter. (Courtesy of Hewlett-Packard Co., Palo Alto, Calif.)

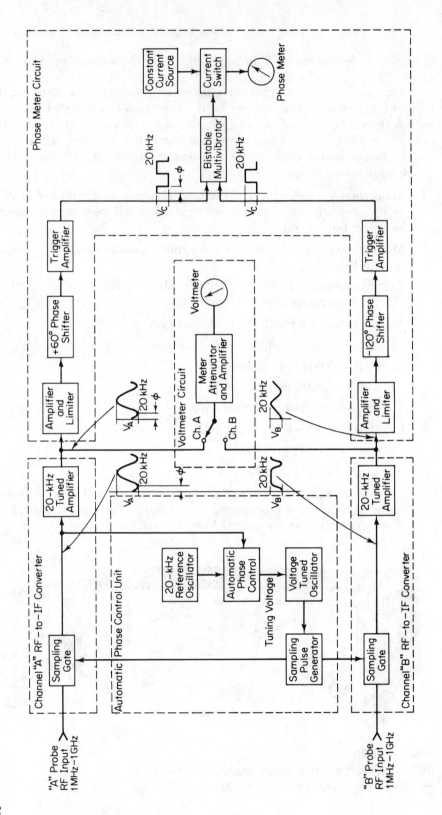

FIGURE 1.30
Block diagram of the vector voltmeter. (Courtesy of Hewlett-Packard Co., Palo Alto, Calif.)

32

6. **Applied voltage:** 0.3 V rms maximum.

7. **Supplied:** Power cord.

8. **Power:** 90 to 125 or 180 to 250 V, 48 to 62 Hz. Voltage selected by rear-panel switch. 25 W maximum.

9. **Mechanical:** Bench model.
   - Dimensions ($W \times H \times D$): 14.78 × 4.4 × 13.5 in. (356.35 × 101.97 × 330.20 mm).
   - Weight: 12.5 lb (5.7 kg) net, 22 lb (10 kg) shipping.

10. Accuracy: R, L, and $C = \pm 0.2\%$ of reading covering the following ranges of value:

Mode	Frequency	Min	Max
R	120 Hz or 1 kHz	2.0 Ω	1.9999 MΩ
L	1 kHz	200 μH	199.99 H
L	120 Hz	2.0 mH	1999.9 H
C	1 kHz	200 pF	199.99 μF
C	120 Hz	2.0 nF	1999.9 μF

The GenRad 1657 is used in engineering laboratories and small incoming inspection departments. Other microprocessor units such as the GenRad 1658, 1687-B, and 1688 serve production testing, semiconductor measurements, material research, component design, and evaluation.

Digital impedance instruments vary from manufacturer to manufacturer and are available for frequency ranges of 1 to 1000 MHz.

The vector impedance meter shown in Figure 1.29 makes simultaneous measurements of impedance and phase angle over a range from 5 Hz to 500 kHz. The unknown component is simply connected across the input terminals of the instrument, the desired frequency is selected by turning the front panel controls, and the two front panel meters indicate the magnitude of the impedance and the phase angle.

The vector impedance meter yields also Q and inductor values by using the following:

$$Q = \frac{fo}{\Delta f} = \frac{WL}{R_P} \tag{1.24}$$

where

$$\Delta F = \text{bandwidth of the tuned series circuit}$$
$$fo = \text{resonant frequency}$$
$$WL = \text{inductive reactance}$$
$$R_P = \text{total resistance in the series circuit}$$

A vector voltmeter (Figure 1.30) measures the signal amplitudes of two points in a circuit and simultaneously measures the phase difference between the voltage waveforms at these points. The vector voltmeter is used for the following:

1. Amplifier gain and phase shift measurements
2. Complex insertion loss measurements
3. Filter transfer function measurements
4. Two-port network measurements
5. Very high frequency applications

Other electronic instruments with a digital readout presently available include the following:

1. Digital line frequency meter with a 0.01-Hz resolution, measurement from 40 to 500 Hz, five large $\frac{3}{8}$ in. light-emitting diodes (LEDs), portable. Can be used to check and adjust power line frequencies at 115 V ac or 230 V ac. Application includes checking line frequencies in laboratories, factories, generating stations, hospitals, and other standby power sources. This unit (model 7245FST) is manufactured by Transcat of Rochester, New York.

2. Clamp-on digital meters (measuring voltage, ampere, and resistance), digital clamp-on wattmeters, and portable kilowattmeter / kilowatt-hour meters are also available from Transcat of Rochester, New York. The specifications of the clamp-on wattmeter include:

FIGURE 1.31
CB42 digital RF wattmeter lets you read extremely small changes in the transmitter output while troubleshooting. (Courtesy of Sencore, Inc.)

a. Ranges: High, 100 W to 199.9 kW
   Low, LOW to 19,990 W

b. Accuracy: ±2% of reading ±1 digit

c. Phase angle error: ±1 degree

d. Display: 4-digit LCD

e. Input voltage: 90 to 280 V ac; additionally 90 to 560 V ac for dual-range versions

3. Most CB technicians use a radio-frequency wattmeter to tune up the transmitter after it is repaired. The transmitter output is fed to a 50-Ω dummy antenna load to provide a reference load and to prevent the possibility of damaging the transmitter during this test. The RF wattmeter in the CB42 (Figure 1.31) provides this dummy load and a direct-reading digital readout of the RF power. The CB42 wattmeter offers an advantage over analog-type meters for troubleshooting because it has 10 times better resolution. The CB42 will show output power to within 0.01 W while an analog meter is usually limited to 0.1 W.

1.13
REVIEW QUESTIONS

1. Discuss the loading effects of voltmeters.

2. Draw a block diagram of a digital multimeter and discuss the function of each part.

3. Discuss the accuracy of a digital multimeter.

4. Discuss the function and uses of an electrometer.

5. Draw an operational amplifier. Discuss how a digital multimeter is used to measure the input and output voltages and currents.

6. Discuss methods of measuring resistances on the order of 0.1 Ω.

7. Discuss the uses of measurements made with a nanovoltmeter.

8. Discuss the uses of measurements made with a picoammeter.

9. List three experimental applications of the digital multimeter.

10. Discuss the techniques used to measure: inductance, capacitance, Q, watts, ac voltage, and ac current.

1.14
REFERENCES

1. George B. Tuma, "Digital Multimeters," *Measurement and Control*, Dec. 1980.

2. Keithley Measurement Instrumentation Catalog and Buyers Guide, Keithley Instrument Co., Cleveland, Ohio, 1981.

3. Thomas J. DeSantis, "Digital Multimeter Satisfies Bench and Systems Needs," *Electronics*, Nov. 6, 1980.

4. *Electrometer Measurements*, revised second edition, Keithley Instrument Co., Cleveland, Ohio, March 1977.

5. Dry Circuit Testing Simplified, Product Notes, Keithley Instrument Co., Cleveland, Ohio, 1974.

6. Ken Reindel, "New Electrometers Reach Beyond Usual DMM Range," *Electronic Design*, 127–131, Oct. 29, 1981.

7. Charles L. Garfinkel and Robert J. Erdman, "Non-Measurements in Electricity," *Industrial Research*, July 1974.

8. Phil D'Angelo, "Structured Analysis Simplifies Modern Software Design," *Electronic Design*, 159–168, Sept. 3, 1981.

9. Phil D'Angelo, "Good Software Depends on Proper Testing and Management," *Electronic Design*, 187–190, Oct. 29, 1981.

10. Hewlett-Packard Measurement/Computation, Electronic Instruments and Systems, Hewlett-Packard Co., Palo Alto, Calif., 1982 (published yearly).

2

Transducers Associated with Electronic Instrumentation

2.1
INTRODUCTION

Transducers (sensors) measure pressure, force, velocity and acceleration, flow, sound, temperature chemical parameters such as partial pressures, pH, electrical impedance, and many more. Recording electrodes are also an integral part of what we call sensors today. It is the intent of this section to introduce the electronic engineer to the principles entailed in the operation of sensors.

The transducer (sensor) has to be physically compatible with its own function. In the operation of a transducer there are eight other important parameters:

1. The operating principles used.

2. External voltage and/or current applied to the transducer to make it work.

3. The electrical output of the transducer.

4. Repeatability of the transducer to reproduce output readings under all environmental conditions.

5. Stability of the transducer to be operative during its operating and storage life.

6. Reliability; if the transducer is dropped, its operation will continue.

7. Leads from the transducer should be sturdy and not be easily pulled off.

8. The rating of the transducer should be sufficient so that it will not break down.

The factors influencing the type of transducer usage and quality of measurements include the following:

1. Nonlinearity effects

2. Hysteresis effects

3. Temperature effects

4. Load alignment effects

5. Calibration

6. Component limitations

7. Physical size

2.1.1 Transducer Characteristics[a]

2.1.1.1 Pressure. A large fraction of all electronic measurements is concerned with pressure. In medical applications, absolute pressures less than atmospheric are expressed in millimeters of mercury (mm Hg) or torrs or inches of water or pounds per square inch absolute or dynes per square centimeter.

Pressures above atmospheric are commonly expressed as pounds per square inch absolute (psia) when referred to a perfect vacuum, gage pressure (psig) when referred to ambient atmospheric, and differential pressure when referred to some arbitrary reference pressure. See Table 2.1 for pressure transducer definitions.

Table 2.2 gives a pressure conversion chart for millimeters of mercury, inches of mercury, pounds per square inch, pounds per square foot, inches of water, feet of water, and centimeters of water.

The majority of pressure measuring instruments with electrical output, however, are transducers operating on the principle that the deflection or deformation resulting from the balance of pressure and elastic forces may be used as a measure of pressure. Commonly used electrical sensing elements include metallic and semiconductor strain gages, potentiometers, piezoelectric elements, variable capacitance and variable inductance de-

---

[a] The material in Section 2.1.1 is from *ISA Transducer Compendium*, 2d ed. © Instrument Society of America, 1969.

TABLE 2.1
Pressure transducer terms

Note: An italicized word appearing in a definition indicates that it has a definition elsewhere in the list.

ABSOLUTE PRESSURE: Pressure measured relative to a perfect *vacuum*. It is usually expressed in *pounds per square inch* absolute (*psia*).

ABSOLUTE PRESSURE TRANSDUCER: A device containing a *vacuum* reference such that it measures *absolute pressure* either of the local ambient or of a pressure source piped to the transducer.

ACCURACY: The ratio of the transducer output voltage *error band* to the transducer output voltage *span*. It is usually expressed as percent of span (% of span).

ALTIMETRIC PRESSURE TRANSDUCERS: (See *barometric pressure transducer*.)

BAR: A unit pressure measurement equal to one million dynes per square centimeter. One (*psi*) is equal to 68.947 *millibars (mB)*.

BAROMETRIC PRESSURE TRANSDUCER: An *absolute pressure transducer* measuring the local ambient pressure. When airborne, such devices may be used as *altimetric pressure transducers*.

BEST STRAIGHT LINE: The best straight line is chosen so that the actual response curve contains three points of equal, maximum deviation. This deviation is the magnitude of the *linearity error band*. The slope of this line represents the *sensitivity* of a pressure transducer.

BURST PRESSURE: The maximum pressure that can safely be applied to an *absolute pressure transducer* that assures no leakage of the pressurized fluid to the surrounding ambient.

COMMON MODE PRESSURE—MAXIMUM: The maximum pressure that the lower pressure source, relative to which a *differential pressure transducer* measures *differential pressure,* can apply without changing the transducer's performance beyond the specified *accuracy* and *interchangeability*. This is sometimes called *maximum line pressure*.

DIFFERENTIAL PRESSURE: The pressure difference measured between two pressure sources. It is usually expressed in *pounds per square inch* differential (psid). When one of the sources is a perfect vacuum, the pressure difference is called absolute pressure. When one of the sources is the surrounding ambient of the measurement device, the pressure difference is called *gage pressure*.

DIFFERENTIAL PRESSURE TRANSDUCER: A device which measures the *differential pressure* between two pressure sources piped to its inputs.

ERROR BAND: The maximum error in output signal measured relative to the expected or previously measured value at the worst case point within the specified calibrated pressure *range*. That error is then assumed to be capable of existing at any point within the specified calibrated range either above or below (plus or minus error) the expected value.

FEET OF WATER (ft H_2O): A unit of pressure measurement equal to 0.434 *pounds per square inch*. One *psi* equals 2.3067 ft H_2O.

GAGE PRESSURE: *Differential pressure* wherein one of the two pressure sources is the surrounding ambient of the measurement device. It is usually expressed in *pounds per square inch* gage (psig).

GAGE PRESSURE TRANSDUCER: A *differential pressure transducer* wherein one pressure source must be the surrounding ambient.

HYSTERESIS: The *error band* associated with approaching a pressure point within the calibrated range first with increasing pressure and then with decreasing pressure.

INCHES OF MERCURY (in Hg): A unit of pressure measurement equal to 0.491 *pounds per square inch*. One *psi* equals 2.036 in Hg. The standard weather report barometer readings are given in these units in the United States.

INCHES OF WATER (in H_2O): A unit of pressure measurement equal to 0.03613 *pounds per square inch*. One psi equals 27.68 in H_2O.

INTERCHANGEABILITY: The *error band* associated with the difference between the actual output of any given transducer at room temperature with the specified nominal output. Within that *error band* one transducer may be substituted for any other transducer of that type.

LINEARITY: The *error band* expressing the ability of a transducer to approach a perfect straight line output voltage response to the pressure or temperature input over the specified *range*.

LINE PRESSURE: (See *commom mode pressure*.)

(Continued on following page)

TABLE 2.1 (Continued)

MICRON (μ): (See torr.)

MILLIMETER OF MERCURY (mmHg): (See *torr.*)

MILLITORR (mt): (See *torr.*)

OVERALL ACCURACY: The *error band* associated with comparing any manufactured transducer with the nominally perfect transducer over the calibrated temperature and pressure *ranges* in such a manner as to combine worst case errors due to *temperature coefficients, linearity, hysteresis* and *repeatability*. Temperature errors usually account for most of the total *error band*.

POUNDS PER SQUARE INCH (psi): The primary unit of pressure measurement in the English System.

PROOF PRESSURE: The maximum pressure that can be applied across the sensing diaphragm of a pressure transducer without changing the transducer's performance beyond the specified accuracy and *interchangeability*. This is sometimes called maximum operating pressure.

RANGE: The end point pressure values (pressure range) or temperature values (temperature range) 'that correspond to the calibrated voltage output limits.

REPEATABILITY: The *error band* expressing the ability of a transducer to reproduce output signal values, at a specific temperature and pressure, after exposure to any pressure and temperature within the specified ranges.

SEAL PRESSURE: The maximum pressure that can be applied to a *gage pressure transducer* that assures no leakage of pressurized fluid to the surrounding ambient.

SENSITIVITY: The ratio of a change in transducer output signal to a change in measured pressure. It is also the best straight line. It is usually expressed in volts per *psi* (V/*psi*), or mV per $^{\circ}$K (mV/$^{\circ}$K).

SPAN: The arithmetic difference between the transducer output signal at the specified maximum limit of the temperature or pressure *range* and the transducer output signal at the specified minimum limit of the temperature or pressure range. It is usually expressed in volts (V).

STABILITY: The *error band* expressing the ability of a transducer to reproduce with time output signal values at a specific, constant pressure and temperature value within the specified ranges.

STANDARD ATMOSPHERE (Satm): The accepted standard atmospheric pressure at sea level (zero altitude) is 14.696 *psi* or 406.79 *in* H_2O or 33.9 *ft.* H_2O or 29.921 *in Hg* or 1.0132 *bar* or 760 *torr* absolute.

TEMPERATURE COEFFICIENT (TC): The maximum deviation of transducer voltage output signal from its value at any point within the specified pressure range at room temperature, due to varying the temperature to values other than room temperature within the specified temperature range, divided by the specified temperature range. The TC is usually expressed in percent of *span* per degree centigrade (% of span/$^{\circ}$C). In most transducers, the temperature error is substantially different for a temperature increment in one portion of the specified temperature range than in another. It is, therefore, often useful to segment the temperature range, specifying more than one TC. For some transducers, it is also useful to separate the TC of offset from the TC of sensitivity. This practice is not useful for NSC pressure transducers because TC of sensitivity is comparatively negligible, such that it may be safely assumed that the entire TC is TC of offset. The error due to TC has the largest influence on the overall accuracy of a pressure transducer compared to all other errors.

TORR: A unit of pressure measurement equal to one *millimeter of mercury (mmHg)*. One *(psi)* equals 51.715 torr. The *millitorr (mt)* (equal to a μmHg) sometimes called a *micron (μ)*, is the universal standard of vacuum measurement.

VACUUM: A perfect vacuum is the absence of gaseous fluid. From a practical standpoint 0.1 torr (mmHg) defines a vacuum for National pressure transducers.

VACUUM RANGE: The *vacuum* range of *absolute pressure* measurement includes those *absolute pressures* between a perfect *vacuum* and one *Satm*. The *vacuum range* for *gage pressure* measurement includes those values between a perfect *vacuum* and the surrounding ambient pressure.

VACUUM TRANSDUCER: A transducer scaled for pressure measurement in the *vacuum range*. Although *absolute pressure transducers* are most often used for this purpose, this is not universally the case. Therefore, care must be taken to specify whether an *absolute vacuum transducer* or a *gage vacuum transducer* is required. Very often an inverted output is desired for a *vacuum transducer;* that is, one where the output signal is of minimum value at one *Satm* for *absolute pressure transducers* or *ambient pressure* for *gage pressure transducer* (corresponding to no *vacuum*) and of maximum value at perfect *vacuum* (corresponding to maximum *vacuum*).

---

vices, and differential transformers; these are discussed in Sections 2.2 to 2.6. A few transducers employ more esoteric principles: magnetostriction, ionization, photoelectricity, vibrating wire, electrokinetic potential, and so forth.

The electrical output of almost all the transducers listed is in the form of analog signals. The increasing use of digital computers for processing data produced by transducers has made conversion of the transducer output to digital signals necessary. Moreover, practical problems of information transmittal in the presence of extraneous noise make it highly advantageous to perform conversion to digital form as close to the transducer as possible. Transducers are being developed that incorporate digital encoders integrally in the body of the transducer, and developments of this kind may be expected to increase greatly as the use of integrated circuits becomes more sophisticated.

A variety of techniques has been applied to effect analog-to-digital conversion. Two types may be recognized: those that act on the mechanical output resulting from the conversion of pressure to displacement, and those that act on an electrical variable. Conversion principles used in the former include coding discs and strips read by photoelectric array. The latter are essentially digital voltmeters, digital impedance meters, and frequency counters. It should be recognized that the addition of integral electronics to a transducer may impose more stringent limits on the operating environment than would apply to the sensing element alone.

2.1.1.2 Force and Torque.

The transducer must either sense force or torque and convert that phenomenon to an electrical or mechanical indication of the quantity being measured. As an example, a force transducer in physiology determines the isotonic contraction of muscles. Force is either a push or pull exerted by a body, whereas torque is a force that tends to produce a rotation. An unbalanced force acting on a body will cause either a distortion or a displacement of that body, depending upon the rigidity of the body. The distortion or displacement occurs in accordance with Hooke's and Newton's laws governing the behavior of elastic or nonelastic bodies. The conversion of this distortion (deformation) or motion to an electrical signal provides the means to determine the value of the force. The instrument or device that performs this conversion is the *force* or *torque transducer*. The SI unit of force is the newton, and the SI unit of torque is the newton-meter.

A great many transducers are commercially available for monitoring force or torque. They accomplish the conversion of these phenomena to an electrical signal by making use of the basic laws to effect a change in a parameter of an electrical circuit. Typical of the kind of parameters that are changed owing to these phenomena are resistance, inductance, and capacitance. In addition, changes in some of the electromechanical properties of the materials themselves (permeability, dielectric strength, magnetostriction, electrostriction, and the piezoelectric effect) are also employed to

TABLE 2.2
Pressure conversion chart

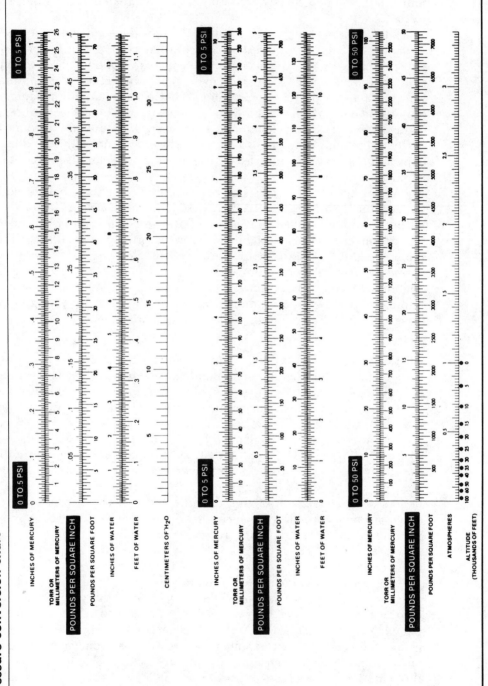

Courtesy of Validyne Corp.

produce signals related to the physical change or motion of the sensing device.

Most force transducers presently in use are designed for the measurement of load or weight. The transducer, or load cell, is inserted between the body to be weighed and a reference platform. The deformation of the load column of the transducer results in a strain at its surface. Either foil or wire strain gages are bonded to the surface of the load-bearing column; these will be discussed in the strain gage presented in Sections 2.2 through 2.4.

2.1.1.3 Motion. Motion transducers convert linear and angular displacement and their time derivatives (velocity, acceleration, and jerk) to electrical signals. Each motion parameter may be directly sensed and converted in one or more steps to the desired electrical signal, or it may be derived by differentiation or integration from one of the related motion parameters.

Motion may be measured relative to a reference point on a material object or with respect to inertial space. In the latter case, the measurement is sometimes referred to as *absolute.* Since absolute displacement and absolute velocity are undefined, measurements of these parameters independent of reference objects can be made only by integration of acceleration. The integration is often accomplished electrically or mechanically. Absolute measurement of acceleration is based on Newton's second law for constant mass. That is, force is mass times the change in velocity per change in time.

Velocity. A measurement of velocity is most readily made by means of a transducer that follows the motion and whose output system has a response proportional to the first time derivative of the displacement, that is, to the velocity. The most commonly used electrical devices involve the relative motion of a coil with respect to a magnetic field. It is required that the magnetic flux linking the coil change as a result of the motion. This may be accomplished either by having a small coil move in a nonhomogeneous magnetic field of large extent or by having a long coil move in a very concentrated magnetic field. In either case, nonlinearity effects will occur unless the displacement range is limited, as by provision of mechanical stops.

A mass isolated from accelerating forces can be used as a reference against which to measure changes in velocity, since the velocity of such a mass will remain constant. Such an isolated mass may be approximated by a spring–mass system whose undamped natural frequency is well below the frequency corresponding to the rate of change of velocity of interest.

Acceleration. Acceleration is measured by means of a spring–mass system, with the spring fixed to the case of the instrument, which is in turn attached to the structure to be tested. It is readily shown that, for frequencies which are low with respect to the undamped natural frequency of the system, the displacement of the mass with respect to the frame of the instrument is proportional to the acceleration imparted to the frame. A measurement of acceleration can thus be converted to a measurement of relative displacement. The most commonly used transducers include potentiometers, bonded and unbonded wire strain gages, devices based on changes in self-inductance, linear variable differential transformers, vibrating wires, and capacitive, piezoresistive, and piezoelectric devices. For special applications such as ballistocardiography, electrical differentiation of the output of a relative-velocity meter may also be advantageous. The requirements for accelerometers suitable for vibration and shock measurements generally include a high resonant frequency, extremely rugged construction free of subsidiary resonances, very low response to accelerations at right angles to the sensitive axis, a mass that is negligible in its reaction upon the structure to be tested, small temperature effects, small sensitivity to sound and to ambient pressure changes, and low electrical noise generation by motion of the connecting cable. These characteristics may be achieved in a number of designs of piezoelectric accelerometers, but always at the expense of the ability to follow very low frequency acceleration changes.

Acceleration transducers for tremor measurements in physiological studies have been used.

Jerk. The characteristic of motion that is of interest under conditions of rapidly changing forces is sometimes the time rate of change of acceleration, or jerk. As indicated by the definition, jerk is expressed in grams per second, meters per second, or feet per second. The major applications for jerk measurements have been in connection with physiological measurements, where it appears that discomfort and injuries from transient motions correlate well with the measured jerk; and for ballistocardiographic measurements in which considerable additional detail is brought out by time differentiation of the more usual velocity or acceleration traces.

An output proportional to jerk is readily obtained by electrical differentiation of the output of an accelerometer. This process is particularly simple for piezoelectric accelerometers, because these devices represent capacitive sources that require only a resistive load to perform the required differentiation.

2.2
STRAIN GAGE TRANSDUCER AS A
DISPLACEMENT DEVICE

Remembering from basics that the resistance of a wire is directly proportional to its length and resistivity and inversely proportional to its cross-sectional area, if we stretch a wire, its resistance will change. This simple theory is used as the basis for the class of transducers called strain gages.

Specifications

Pressure range	−50 to +300 mmHg
Maximum over-pressure	5000 mmHg
Sensitivity*	50μ V/V/cm +1%
Volume displacement	0.04 mm³/100 mmHg pressure, approximately
Bridge resistance	350 ohms, nominal
Non-linearity and hysteresis*	±1.5 mmHg at 0 to 300 mmHg ±0.1 mmHg at 0 to 10 mmHg
Zero balance*	±15 mmHg
Thermal coefficient of sensitivity (typical)*	0.015%/°F
Thermal coefficient of zero (typical)*	0.12 mmHg/°F
Operating temperature	−65° to +175°F
Electrical leakage AC	2 μa maximum at 115V rms 60 Hz
DC	100 megohms minimum at 50V DC

*For 7.5V excitation

Insulation	Withstands 10,000V DC
Rated excitation voltage	7.5V DC or AC through carrier frequency
Maximum excitation voltage	10V DC or AC through carrier frequency
Weight	1.6 oz (45.4 grams)
Size	
Length	2.21 inches (56 mm)
Maximum diameter of base	.710 inch (18 mm)
Cable length	15'

Domes

TA1011D	Disposable Diaphragm Dome with Linden fittings
TA1010D	Disposable Diaphragm Dome with Luer Loks
TA1011	Reusable Dome with Linden fittings
TA1010	Reusable Dome with Luer Loks

Disposable Pressure Monitoring Kit

TAK1048D	Assembled kit includes TA1010D Disposable Diaphragm Dome, Sorenson Research Intraflo*, 1 each 12'' and 36'' pressure tubing, 3 each 3-way stopcocks, 4 venting caps, male, 2 venting caps, female.

Principles of Isolation, Model P23 ID

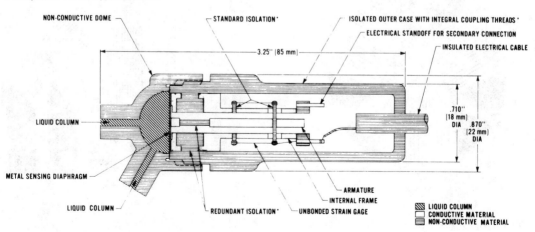

FIGURE 2.1

Internal structure of Gould/Statham P23 ID transducer. (Courtesy of Gould, Inc., Measurement Systems Division, Oxnard, Calif.)

If the conductor is stretched so that it is halfway between the length at rest and the length where its tensile strength is exceeded, and it parts (call this point R at length L), a simple strain gage has been created.

The types of resistance strain gages currently used as transducer elements include the unbonded metallic-filament strain gage, the bonded metallic-foil gage, and the bonded piezoresistive or semiconductor gage.

The unbonded strain gage elements are made of one or more filaments of resistance wire stretched between supporting insulators. The supports

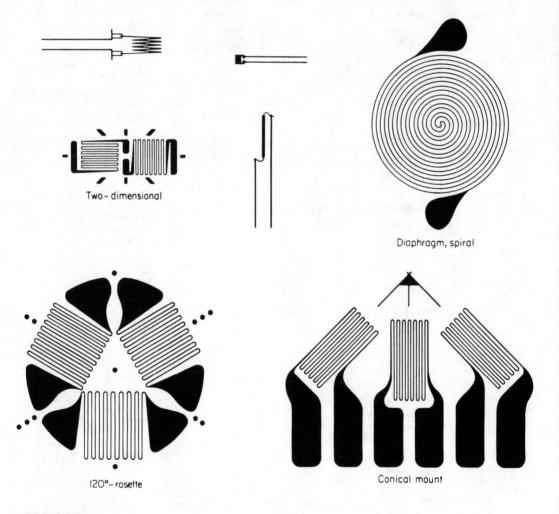

Two-dimensional

Diaphragm, spiral

120°-rosette

Conical mount

FIGURE 2.2
Typical foil strain-gage configurations. (From Harry E. Thomas, *Handbook of Biomedical Instrumentation and Measurements,* Reston Publishing Co., Reston, Va., 1974)

are either attached directly to an elastic member used as a sensing element or are fastened independently, with a rigid insulator coupling the elastic member to the taut filaments. The displacement (strain) of the sensing element causes a change in the filament length, with a resulting change in resistance. See the internal structure of the Gould/Statham P23 ID transducer of Figure 2.1 for use of an unbounded strain gage.

Foil gage configurations are shown in Figure 2.2. Foil gages can measure tensile, compressive, or torsional stresses, and are used in transducers such as binding biomechanics of teeth.

The piezoresistive strain gage exhibits more than 100 times the unit resistance change of a foil gage for any given strain. This means that if semiconductor gages are connected as the arms of a Wheatstone bridge, a very large output voltage can be produced, eliminating the need for subsequent amplification. However, such large resistance changes produce large unbalances in a Wheatstone bridge with constant-voltage excitation, resulting in very nonlinear outputs. This problem can be solved by exciting the bridge from a constant-current supply. A constant-current supply contains more complex circuitry than a constant-voltage regulator and may not always be as readily available.

Actually, the resistance change of a semiconductor strain gage as a function of strain is not completely linear over its total strain range. This results in nonlinearity in some transducers, even with constant-current excitation. Foil gage transducers require amplification because of low bridge output. The linearity of the output signal is not a problem.

2.3
APPLICATIONS OF BLOOD-PRESSURE TRANSDUCERS

[b] In a biological system where an artificial heart or other heart-assist devices are used, the central aortic pressure waveform, as well as other pressure waveforms of the aortic tree, should be detectable. Intracranial pressure of the brain can also be detected, as well as other physiological pressures. This aids the physician in his work. To determine the pressure, a *pressure transducer* is used. A number of electromechanical systems are possible:

1. A catheter with a pressure transducer terminated at the receiving or back end. A catheter is defined here as a flexible or rigid hollow plastic tube.

2. A catheter with a pressure transducer terminated at the sending or front end.

---

[b] This paragraph is based, in part, on Philip Kantrowitz, "Transducer Development for the Artificial Heart or Heart Assist Devices," *JAES*, vol. 17, no. 5, pp. 539–549, 1969.

3. A pressure transducer floating in the blood stream.

4. A pressure transducer fixed to a semirigid surface in an artificial heart device.

In all cases, the pressure transducer should be small, rigid, reliable, nontoxic, hypoallergenic, and should respond from dc to approximately 1 to 2 kHz.

The piezoresistive strain gage element is presently the most popular device when used as a transducer for monitoring pressures of the artificial heart or heart assist devices, primarily because it is a low-impedance device on the order of 300 to 2000 Ω and because it can be used with a carrier amplifier to minimize electric and mechanical noise. The output of the strain-gage pressure transducer can be fed to one leg of an ac bridge at the input of the carrier amplifier. The ac bridge is driven by a high-frequency oscillator (i.e., 25 kHz). The carrier signal is then amplified by a high-gain ac integrated-circuit amplifier and demodulated. The output of the phase-sensitive demodulator is connected to a filter, which is fed to a readout device such as an oscilloscope and/or recorder.

Piezoresistive devices can best be understood by considering their theoretical aspects. *Piezoresistive effect* is the name given to the change in electrical resistivity that occurs with the application of stress, or distributed force per unit area, to an appropriate material. Many materials exhibit the effect to some degree.

In semiconductors, the piezoresistive effect is unusually large. The optimum piezoresistive effect depends on physical form, type of material, its resistivity, or doping, and crystallographic orientation, in the room temperature range. Of about 24 varieties of *P*-type silicon characterized to date, the optimum form for a medical pressure transducer scheme appears to be a slender filament with resistivity of 1.0 Ω-cm and a 111 orientation in *k*-space.

For a thin semiconductor element, the stress-gage factor is related to the strain-gage factor by the scalar expression

$$G = Y\Pi\hat{e} \qquad (2.1)$$

where G is the strain-gage factor or fractional change in resistance per fractional change in longitudinal strain, Y is Young's modulus, and $\Pi\hat{e}$ is the coefficient of the longitudinal stress-gage factor (i.e., fractional change in resistance divided by the stress). One can define the Young's modulus as $Y = T_0/S$, where S is the longitudinal stress. Then $H\hat{e}\ T_0$ can be written as

$$\Pi\hat{e}T_0 = \Pi\hat{e}\ YS = \frac{\Delta\rho}{\rho} \qquad (2.2)$$

where ρ denotes the resistivity of the material.

The conventional expression for resistance of a thin wire is $R = \rho L/A$, where L is length and A is cross-sectional area. Let $\Delta L/L$ denote the resistance change due to a change in length, $\Delta A/A$ denote the resistance change due to a change in cross-sectional area, and $\Delta\rho/\rho$ represent the change in specific resistance. If R/R denotes the unit change of resistance, then

$$\frac{\Delta R}{R} = \frac{\Delta L}{L} - \frac{\Delta A}{A} + \frac{\Delta\rho}{\rho} \tag{2.3}$$
$$= S + 2\sigma S + \Pi\hat{e}(YS)$$
$$= (1 + 2\sigma + Y\Pi\hat{e})S$$

where σ is Poisson's ratio. This expression states that $\Delta R/RS$ is a constant and is defined as the gage factor of a semiconductor element. The gage factors of the semiconductor element are dependent upon electrical resistivity, crystal orientation, and ambient temperature. For the case of semiconductors containing fewer than 10^{17} carriers/cm^3, the gage factor is markedly temperature and strain dependent.

2.4
STRAIN GAGE CALIBRATION TECHNIQUE

In this calibration technique, the face of the transducer diaphragm is exposed to a hydrostatic head, which is measured in mm Hg and also by a Statham P23Db series type of transducer standard. The initial calibrations were performed at room temperature (about 70°F) to check for linearity of

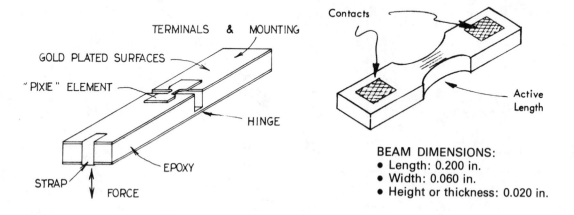

BEAM DIMENSIONS:
- Length: 0.200 in.
- Width: 0.060 in.
- Height or thickness: 0.020 in.

FIGURE 2.3
Endevco Pixie strain gage device with dimension.
(Courtesy of Endevco Corp., San Juan Capistrano, Calif.)

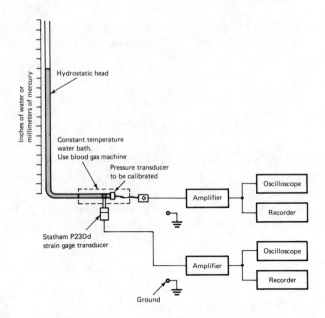

FIGURE 2.4
Schematic of calibration techniques.

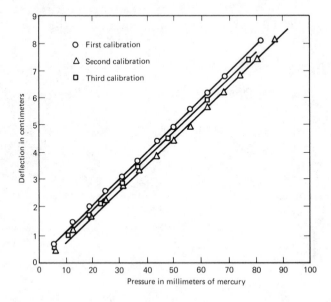

FIGURE 2.5
Pressure versus deflection for a strain gage transducer.

the device, and then a second calibration was performed at 37°C or body temperature, bathing the hydrostatic head in front of the transducer and the transducer in the IL blood gas machine's constant temperature bath. The calibration curves using the transducers in Figure 2.5 follow. It will be noted that the output from strain gages can be quite linear at least to 150 mm Hg. The strain gages can be also subjected to a negative pressure head and, although not drawn, the output again can remain linear.

The gain of the amplifiers can be adjusted so the device yields full-scale deflection for about 150 mm Hg pressure, and, again, the amplifier gain can vary from 10 to 50 depending upon the mounting of the strain gage shown in Figure 2.3. With a similar conventional strain gage mounting, the gain would be on the order of 1000 or more. The Statham P23Db strain gage transducers operate at a gain of approximately 250 for full-scale deflection for 100 mm Hg.

A schematic of the calibration procedure is shown in Figure 2.4 and pressure versus deflection curves are shown in Figure 2.5.

2.4.1 Temperature Considerations for Strain Gage Transducers

From the Endevco specification sheet for pixie chips, it can be seen that the resistance change due to temperature alone is $\frac{1}{2}\%/°F$. For a nominal 500-Ω chip, this means that there will be a resistance change of about 2.5 $\Omega/°F$. The temperature can vary over the range from 33° to 39°C or about 11°F change. This temperature drift will cause a resistance change of (2.5 $\Omega/°F$) × 11°F, or about 26 Ω. In the circuits of Figures 2.6 and 2.7, the basic 500-Ω bridge setup is used. The output drift due to this temperature increase (or decrease) will be approximately 0.11 V. Figure 2.6 shows the Wheatstone bridge with one active element. Figure 2.7 shows the Wheatstone bridge with two active elements.

The simple Wheatstone bridge circuits in Figures 2.6 and 2.7 point out the fact that temperature compensation cannot be accomplished by one pixie element alone. By placing a dummy element in an arm adjacent to the active element, temperature compensation can be accomplished; however, note not to place the chips in the opposite arms of the bridge. The output of the bridge can be fed to integrated circuitry.

Two active chips possessing identical temperature characteristics can be placed in adjacent arms if one is stressed in tension and the other in compression. This is referred to as *back-to-back* mounting.

The transducer temperature characteristics have to be specified. That is, how much drift is due to temperature? For a full-scale reading of 150 mm Hg the temperature-compensated transducer was found to drift 75 mm Hg from room temperature to constant water bath temperature (37°C). That is, there will occur with this specific transducer 75 mm Hg/150°C, or 5 mm Hg/°C drift, or approximately 3% of full-scale per degree Celsius.

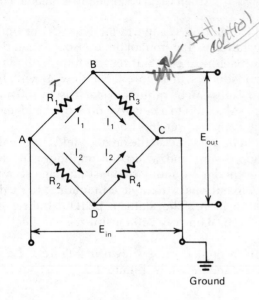

Input voltage, $E_{in} = 10$ V

All resistance elements:

$$R_1 = R_2 = R_3 = R_4 = 500 \; \Omega$$

To obtain output voltage, E_{out}, use Kirchhoff's law.

$$E_{out} = V_{BC} + V_{CD} = V_{BC} - V_{CD}$$
$$E_{out} = V_{in} - I_1 R_1 - V_{in} + I_2 R_2$$
$$E_{out} = I_2 R_2 - I_1 R_1$$

Since $I_2 = I_1$ and $R_2 = R_1$,

$$E_{out} = 0 \text{ V}$$

(a)

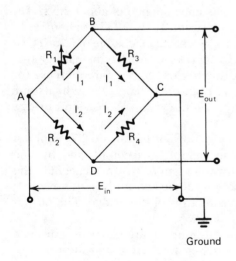

Input voltage, $E_{in} = 10$ V

$R_1 = 600 \; \Omega$ and

$R_2 = R_3 = R_4 = 500 \; \Omega$

To obtain output voltage, E_{out}, use Kirchhoff's law.

Start with

$$E_{out} = I_2 R_2 - I_1 R_1$$

$$E_{out} = \frac{10 \text{ V} \times 500 \; \Omega}{500 \; \Omega + 500 \; \Omega} - \frac{10 \text{ V} \times 600 \; \Omega}{500 \; \Omega + 600 \; \Omega}$$

$$E_{out} = (10 \text{ mA})(500 \; \Omega) - (9.1 \text{ mA})(600 \; \Omega)$$

$$E_{out} = -0.46 \text{ V}$$

Note: $I_2 = \dfrac{10 \text{ V}}{1000 \; \Omega}$ A and $I_1 = \dfrac{10 \text{ V}}{1100 \; \Omega}$ A

(b)

FIGURE 2.6
Wheatstone bridge with one active element:
(a) one active element, unstrained; (b) one active
element, strained, or increase in temperature.

2.4.2 Sensitivity of the Strain Gage Transducer

Sensitivity is the measure of electrical output of a chip for so many volts input when stressed by a mechanical load or pressure. The sensitivity of the device will naturally change depending upon the transducer mounting.

Shielding against electromagnetic noise in the environment was impractical and compelled the use of a strain gage and carrier amplifier combination.

The strain-gage element (Figure 2.3) is called an Endevco *pixie beam*. It is a slender filament of highly doped *p*-type silicon, with a resistivity of 1 Ω-cm and a gage factor of 150 to 180, mounted on a copper-clad substrate. The active fibers are approximately 9 mils by $\frac{1}{2}$ mil. The unique properties of thin fibers come into play to give this beam great strength, over 250,000 psi. The sensitive fibers are mounted on a lightweight copper-clad substrate and are coated with a gold layer to prevent corrosion.

The gage resistance for the element used is 400 to 750 Ω. The current through the gage is 9 mA, and maximum operating current is 15 mA. Maximum sensitivity (resistance change per gram of force) varies from approximately 1.4 to 2.4%. The electrical resistance changes only approximately $\frac{1}{2}\%/°F$, but temperature tends to stabilize by self-heating of the chip. According to the manufacturer, the element is not damaged by temperatures from $-65°F$ to approximately 200°F. For a short time (say 30 minutes) elevated temperatures such as 300°F can be tolerated.

The output of the strain gage shown in Figure 2.3 is approximately 13 mV/100 mm Hg at 5-V dc excitation. This corresponds to an acoustic sound pressure level of -140 decibels (dB) re 1 v/microbar. Note that -60 dB re 1 V/μbar equals 6.85 V/psi, since 1 μbar = 1 dyne/cm² and 1 dyne/cm² = 14.6×10^{-6} psi.

The element is linear ±1% to at least 5 g of applied pressure. Theoretically, monocrystalline silicon is a perfectly elastic material with no plastic region at temperatures below 1000°F. Thus any hysteresis in a strain-gage installation is generally a function of the cement, or hysteresis in the stressed structure.

Data indicate that semiconductor strain gages exhibit better fatigue life than conventional metallic gages. Tests on standard semiconductor strain gages in excess of 10,000,000 cycles have been performed with no failure, at strains on the order of ±500 microstrains.

The change in resistance is easily determined. Since the gage factor G equals $\Delta R/RS$, $R = \rho L/S$; this value may be substituted for R in the equation for G. Then

$$\Delta R = G\rho LS/A \qquad (2.4)$$

Temperature effects are not included. With $G = 170$, $\rho = 0.04$ Ω-in., $L = 0.005$ in., $S =$ longitudinal strain $= 0.024/170 = 140$ μin./in./g, and $A = 4.5 \times 10^{-6}$ in.², $\Delta R = 10$ Ω.

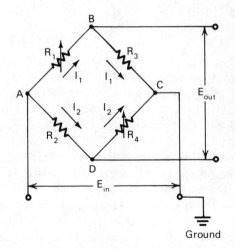

Input voltage, $E_{in} = 10$ V

$R_1 = R_2 = R_3 = R_4 = 500\ \Omega$

To obtain output voltage, E_{out}, use Kirchhoff's law.

$$E_{out} = V_{BC} + V_{CD} = V_{BC} - V_{DC}$$
$$E_{out} = V_{in} - I_1 R_1 - V_{in} + I_2 R_2$$
$$E_{out} = I_2 R_2 - I_1 R_1$$

Since $I_2 = I_1$ and $R_2 = R_1$

$$E_{out} = 0\ V$$

Ground

(a)

Input voltage, $E_{in} = 10$ V

$R_1 = R_4 = 600\ \Omega$

$R_2 = R_3 = 500\ \Omega$

$$E_{out} = I_2 R_2 - I_1 R_1$$

$$E_{out} = \frac{10\ V \times 500\ \Omega}{500\ \Omega + 600\ \Omega} - \frac{10\ V \times 600\ \Omega}{500\ \Omega + 600\ \Omega}$$

$$E_{out} = -\frac{10 \times 100}{1100}$$

$$E_{out} = -0.91\ V$$

Note: Double the drift due to temperature.

Ground

(b)

FIGURE 2.7
Wheatstone bridge with two active elements:
(a) two active arms, unstrained, opposite arms of bridge; (b) two active arms, opposite arms of bridge, strained or increase in temperature; (c) two active arms, adjacent elements, strained or increase in temperature; (d) two active arms, adjacent arms, strained or increase in temperature.

54

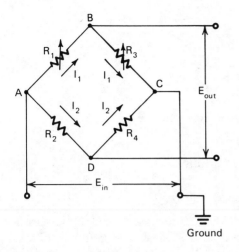

Input voltage, E_{in} = 10 V
R_1 = R_2 = 600 Ω
R_3 = R_4 = 500 Ω

E_{out} = $I_2 R_2 - I_1 R_1$

$E_{out} = \dfrac{10\ V \times 500\ \Omega}{500\ \Omega + 500\ \Omega} - \dfrac{10\ V \times 600\ \Omega}{600\ \Omega + 600\ \Omega}$

E_{out} = 0 V

(c)

Input voltage, E_{in} = 10 V
R_1 = R_3 = 600 Ω
R_2 = R_4 = 500 Ω

E_{out} = $I_2 R_2 - I_1 R_1$

$E_{out} = \dfrac{10\ V \times 500\ \Omega}{600\ \Omega + 500\ \Omega} - \dfrac{10\ V \times 500\ \Omega}{600\ \Omega + 500\ \Omega}$

E_{out} = 0 V

(d)

FIGURE 2.7 (Continued)

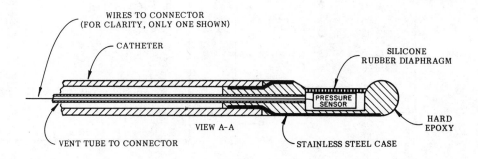

FIGURE 2.8
Circuit under constant-voltage conditions.

The pixie beams shown in Figure 2.3 are essentially modulating devices rather than reciprocal generators. External power from a dc current source can be used with the element. Thus a large modulation of the external power is produced from very small deflections of the semiconductor beam.

Consider, in Figure 2.8, a constant-voltage source, with E_b defined as the battery voltage and R_s the series resistor in series with $R + \Delta R$, the strain gage element. E_o, the output across the strain gage is as follows:

$$E_o = E_b \frac{R + \Delta R}{R + \Delta R + R_s} \qquad (2.5)$$

FIGURE 2.9
Millar MIKRO-TIP® catheter pressure transducer.
(Courtesy of Millar Instruments, Inc., Houston,
Texas)

With $E_b = 9$ V, $R + \Delta R = 500\ \Omega$, and $R_s = 500\ \Omega$, $E_0 = 9 \times 500 / 1000 = 4.5$ V. The current I flowing through the strain gage will then be given by $I = 4.5/500 = 9$ mA.

It should be noted here that as a safety precaution the current going through the pixie beam should be approximately $33\frac{1}{3}\%$ lower than that actually specified, since the temperature of the silicon chip may affect parts in contact with the pixie beam.

The average power P_0 developed by the strain gage will be as follows:

$$P_0 = I^2 R = 81 \times 10^{-6} \times 500 = 40.5\ \text{mW}$$

The pixie beam is usually mounted in compression for maximum strength, but it may also be mounted in tension.

Millar Instruments, Inc., of Houston, Texas, has developed the MIKRO-TIP® catheter pressure transducers (Figure 2.9). These pressure transducers were developed originally for use within the cardiovascular system, where single sensor measurements provided high-fidelity recordings suitable for accurate computation of (dP/dt) and $(dP/dt/P)$. These high fidelity requirements are essential for many computer analyses where the fluid–catheter systems introduced too many errors in terms of amplitude phase and overall frequency response. Multisensor catheter transducers permit high-fidelity measurements from several locations with the introduction of but a single catheter.

Kulite Semiconductor Products of Ridgefield, New Jersey, has developed the LQ-125 and LQ-30-125 series for artificial prosthetic and other industrial pressure devices (see Figure 2.10). Using the specifications in Figure 2.11, one can design the use of this as a pressure device for other applications, especially where steady-state accuracy is important. Manufacturing data and specifications will vary, and newer technology will provide improvements in electronic instrumentation technology.

2.5
CAPACITANCE AS A DISPLACEMENT TRANSDUCER

[c] Displacement can also be measured by using the relationship of electrical capacitance due to the separation and area of the metal plates of a capacitor. Mathematically, the expression for capacitance is

$$C = 0.225\frac{KA}{t} \qquad (2.6)$$

---

[c] The following four paragraphs are excerpted from Harry E. Thomas, *Handbook of Biomedical Instrumentation and Measurements*, Reston Publishing Co., Reston, Va., 1974.

LQ-125 and LQ-30-125 Specifications

SPECIFICATIONS	LQ-125-5 LQ-30-125-5	LQ-125-10 LQ-30-125-10	LQ-125-25 LQ-30-125-25	LQ-125-50 LQ-30-125-50	LQ-125-100 LQ-30-125-100	LQ-125-200 LQ-30-125-200	LQ-125-500 LQ-30-125-500
Rated Pressure	5 psi	10 psi	25 psi	50 psi	100 psi	200 psi	500 psi
Max. Pressure	20 psi	20 psi	100 psi	100 psi	200 psi	400 psi	1000 psi
Nominal Output at Rated Pressure	50 mV	100 mV	65 mV	75 mV	100 mV	100 mV	100 mV
Bridge Excitation (DC or AC)	20 V	20 V	10V	10V	10V	10V	10V
Input Impedance Output Impedance	4000 ohms (min.) 3000 ohms (max.)		1500 ohms (min.) 1200 ohms (max.)		400 ohms (min.) 500 ohms (max.)		
Zero Balance	±5% FS	±3% FS					
Combined Non-Linearity & Hysteresis	±0.5% FS (max.)						
Repeatability	0.25%						
Compensated Temp. Range	80°F to 180°F (25°C to 80°C) Any 100°F (55°C range within operating range on special order						
Operating Temp. Range	0°F to 250°F (−20°C to 120°C) Temperatures to 350°F (175°C) available on special order						
Change of Sensitivity with Temp.	±2.5%/100°F (55°C)						
Change of No-Load Output with Temp.	±3% FS/100°F (55°C)	±2% FS/100°F (55°C)	±1% FS/100°F (55°C)				
Natural Frequency (approx.)	70 KHz	70 KHz	100 KHz	130 KHz	160 KHz	200 KHz	350 KHz
Acceleration Sens. Perpendicular Transverse	0.002% FS/g 0.0004% FS/g	0.001% FS/g 0.0002% FS/g	0.0005% FS/g 0.0001% FS/g	0.0004% FS/g 0.00008% FS/g	0.0002% FS/g 0.00004% FS/g	0.0002% FS/g 0.00004% FS/g	0.00008% FS/g 0.000016% FS/g
Resolution	Infinite						

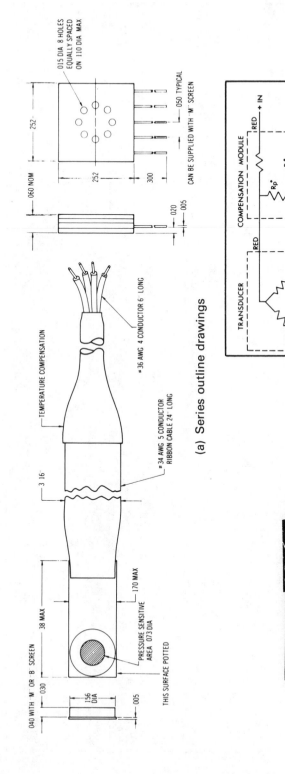

LQ-30-125

015 DIA 8 HOLES EQUALLY SPACED ON 110 DIA MAX

252

060 NOM

252

300

050 TYPICAL

020

005

CAN BE SUPPLIED WITH "M" SCREEN

LQ-125

TEMPERATURE COMPENSATION

#36 AWG 4 CONDUCTOR 6" LONG

3 16"

#34 AWG 5 CONDUCTOR RIBBON CABLE 24" LONG

170 MAX

38 MAX

040 WITH "M" OR "B" SCREEN

030

PRESSURE SENSITIVE AREA 073 DIA

156 DIA

005

THIS SURFACE POTTED

(a) Series outline drawings

(c) Wiring drawing

COMPENSATION MODULE

RED + IN

GREEN + OUT

BLACK IN

WHITE OUT

R_p^*

R_s^*

TRANSDUCER

RED

BLUE

GREEN

BLACK

WHITE

*THESE RESISTORS ARE SOMETIMES CONNECTED IN THE LOWER LEG BETWEEN THE BLACK, GREEN AND BLUE WIRES.

LQ-Series

(b) Pictorial

FIGURE 2.10

The LQ-125 and LQ-30-125 series can be used for medical artificial prosthetic devices. (Courtesy of Kulite Semiconductor Products, Inc., Ridgefield, N.J.)

VQ-250, VQ-500, SVQ-500 Specifications

SPECIFICATIONS	VQH-250-5 VQH-500-5 SVQH-500-5	VQH-250-10 VQH-500-10 SVQH-500-10	VQL-250-25 VQL-500-25 SVQL-500-25	VQS-250-50 VQS-500-50 SVQS-500-50	VQS-250-100 VQS-500-100 SVQS-500-100	VQS-250-200 VQS-500-200 SVQS-500-200
Rated Pressure	5 psi	10 psi	25 psi	50 psi	100 psi	200 psi
Max. Pressure	20 psi	20 psi	50 psi	100 psi	200 psi	400 psi
Output-Nom.	50 mV	100 mV	75mV	85mV	100 mV	100 mV
Excitation	20 V	20V	10 V	10 V	10 V	10 V
Impedance—Nom.	5000 ohms	5000 ohms	1500 ohms	500 ohms	500 ohms	500 ohms
Zero Balance	±5% FS	±3% FS	±3% FS	±3% FS	±3% FS	±3% FS
Combined Non-Linearity & Hysteresis	±0.5% FS	±0.5% FS	±0.5% FS	±0.5% FS	±0.5% FS	±0.5% FS
Repeatability	0.25%	0.25%	0.25%	0.25%	0.25%	0.25%
Compensated Temp. Range	80°F to 180°F (25°C to 80°C) Any 100°F range within the operating range on request					
Operating Temp. Range	0°F to 250°F (−20°C to 120°C) Temperatures to 350°F (175°C) available on special order.					
Change of Sensitivity with Temp. (Note 1)	±0.02%/°F	±0.02%/°F	±0.02%/°F	±0.02%/°F	±0.02%/°F	±0.02%/°F
Change of No-Load Output with Temp.	0.02% FS/°F	0.01% FS/°F	0.01% FS/°F	0.01% FS/°F	0.01% FS/°F	0.01% FS/°F
Natural Frequency (approx.)	70 KHz	70 KHz	100 KHz	130 KHz	160 KHz	200 KHz
Acceleration Sens. Perpendicular Transverse	0.004 % FS/g 0.0008% FS/g	0.002 % FS/g 0.0004% FS/g	0.0008% FS/g 0.00016% FS/g	0.0006 % FS/g 0.00012% FS/g	0.0004 % FS/g 0.00008% FS/g	0.0003 % FS/g 0.00006% FS/g
Resolution	Infinite	Infinite	Infinite	Infinite	Infinite	Infinite

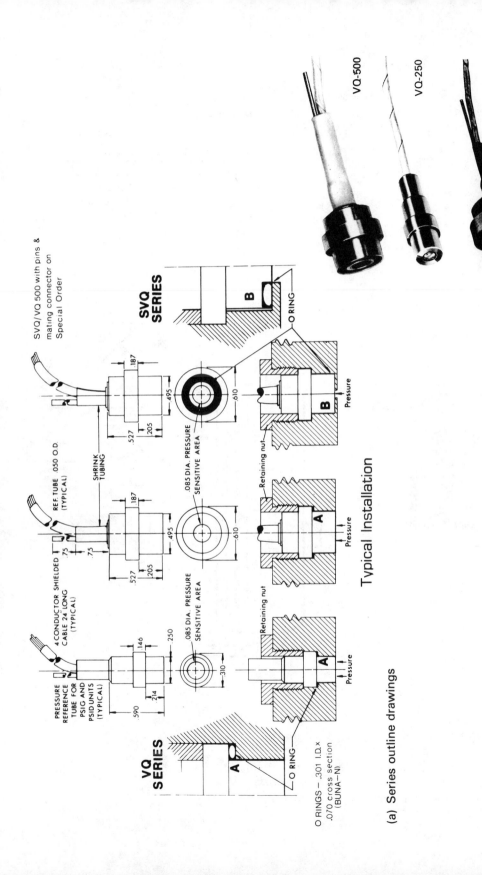

SVQ/VQ 500 with pins & mating connector on Special Order

Typical Installation

(a) Series outline drawings

FIGURE 2.11
The VQ-250, VQ-500, and SVQ-500 series.
(Courtesy of Kulite Semiconductor Products, Inc., Ridgefield, N.J.)

(b) Pictorial

VQ-500

VQ-250

SVQ-500

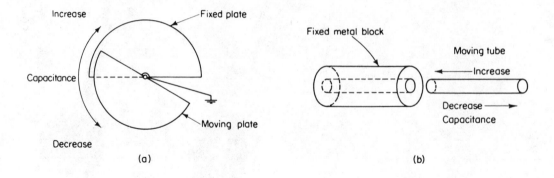

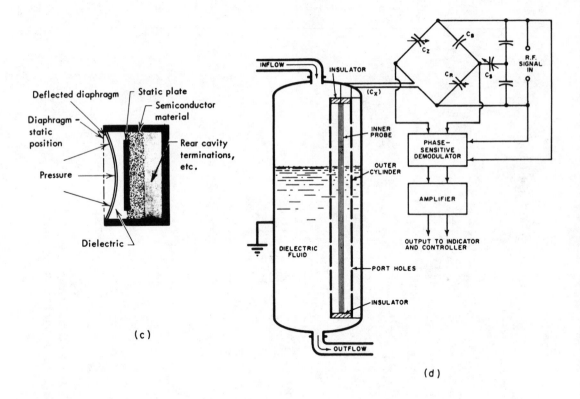

FIGURE 2.12

Capacitive transducers: (a) plate area change;
(b) plate separation change; (c) commercial unit;
(d) immersed electrodes. (From Harry E. Thomas,
*Handbook of Bioelectronic Instrumentation and
Measurements,* Reston Publishing Co., Reston,
Va., 1974)

where

C = capacitance in picofarads

K = dielectric constant

A = area of dielectric

t = dielectric thickness

From Eq. (2.6), it can be seen that the capacitance can be varied by changing either the area of the plates or the thickness of the dielectric, or both.

The basic capacitor transducer consists of two conducting plates separated by a dielectric. Ignoring fringing effects, transduction is effected by (1) a change of distance between plates, (2) a change of plate area, and (3) chemical or physical changes in the dielectric. Change of plate area is illustated in Figure 2.12(a), where the angular positioning of the semicircular plates delivers capacitive changes, which are applied to resonant circuits for signal processing. Change of distance between plates is used in small displacement measurements. Figure 2.12(b) shows a piston-actuated diaphragm that changes microinch motions through capacitance changes. Nonlinear capacitance changes can be compensated for by appropriate mechanical linkages and circuitry adjustments in the signal processing. Figure 2.12(c) illustrates the construction of a commercial cell using the change in capacitance principle actuated by pressure.

In level transducers for conducting liquids such as mercury, electrolytes, or water, the change in distance effect is also used, since the upper surface of the liquid acts as the movable plate with respect to the outside container.

Transducers using the change of dielectric principle are commonly found in nonconducting liquid level measuring devices. When an immersed probe electrode is used or when the capacitance between electrodes consists of two concentric cylinders, the change in capacitance as the liquid rises and falls is essentially dependent upon the liquid dielectric

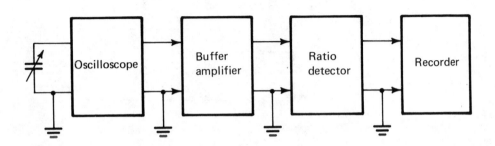

FIGURE 2.13
Capacitive micrometer displacement measuring system.

constant versus that of air [see Figure 2.12(d)]. An analogous arrangement is used in blood-pressure manometers.

Another use of the capacitor as a displacement device is the capacitive micrometer shown in Figure 2.13. In the capacitive micrometer design, it is necessary to connect the variable capacitance to a coil such that the resulting network controls the system resonance. An FM oscillator and a buffer amplifier limit the amplitude of the oscillator output as well as isolate the oscillator circuitry from the input of the ratio detector. The ratio detector converts the FM signal to an output potential that is proportional to the frequency but unaffected by the amplitude of the output. The output of the ratio detector is fed to a recording device.

2.6
LINEAR VARIABLE TRANSFORMER TRANSDUCER AS A DISPLACEMENT DEVICE

Another displacement transducer is a linear variable transformer, commonly known as a *differential transformer*. It is again relatively simple in construction but complex in operation. Looking at it schematically, as it appears in Figure 2.14, as two equal secondary windings wound equidistant from a primary winding, the coupling between the two secondaries and the primary is now dependent upon the position of the core in relation to the primary and each secondary. The secondary windings are connected to the output in such a manner that their outputs oppose each other. When

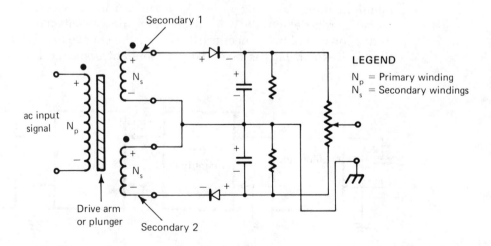

FIGURE 2.14
Linear differential transformer transducer with secondary voltage output equal in value.

the core is placed so that it is exactly centered over the primary winding, the magnetic coupling to each secondary is equal, the outputs of the secondary are equal, and, because they oppose each other, they cancel. The circuit arrangement (Figure 2.14) is quite similar to a ratio detector, but in this case the inductance is variable and not the frequency.

When the core is displaced so that it cuts more of secondary 1 than secondary 2, the output of the first secondary is larger than the output of the second secondary, and the total output becomes positive. If the coil slug is moved so that the second secondary is coupled more closely to the primary, it results in a net negative output voltage.

The linear variable differential transducer can be excited by an ac carrier amplifier with a higher output than the conventional strain gage.

2.7
PIEZOELECTRIC TRANSDUCER AS A DISPLACEMENT DEVICE

Although the piezoelectric effect has been known for more than 75 years, it is only during the last 30 years that practical piezoelectric transducers have become common. Piezoelectricity means simply "pressure" electricity. If particular types of crystals are squeezed along specified directions, an electric charge will be developed by the crystal.

Crystals for bending, shear, or compression modes can be designed for a particular medical application. Examples include piezoelectric microphones for detecting heart sounds and blood pressure.

Significant advances in the performance characteristics of motion, force, and pressure transducers have been made in the past decade by use of new piezo materials. New transducers are available that measure high amplitudes and also provide high outputs at small amplitudes. This is achieved by using improved sensing elements, including newer piezoelectric ceramics, piezoresistive materials, and special designs by comparison. Other transducers have limited dynamic ranges and are more susceptible to shock, vibration, temperature, and humidity.

The charge developed in piezoelectric transducers is proportional to the piezoelectric constant of the material, and to the applied stress. The constant depends upon the mode of operation employed. Although quartz crystals are used in some units, the man-made ceramics are now popular, since they exhibit higher piezoelectric constants, provide higher outputs, and are less susceptible to environmental effects, such as case strains and transverse forces or motions. Lead-zirconate-titanate ceramics and other proprietary ceramics are used extensively as pressure transducers.

A microphone amplifier (Figure 2.15) has been developed in conjunction with the U.S. Post Office for use with an electret (ceramic) transducer to replace the carbon transmitter.

Dual polarity operation is accommodated by an on-chip bridge. Full

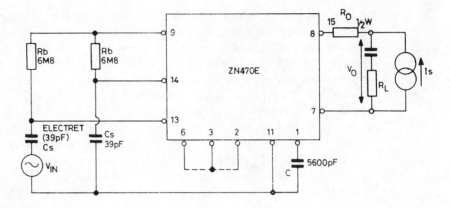

FIGURE 2.15
Typical electret microphone amplifier. (Courtesy
of Ferranti Electric, Inc., Commack, N.Y.)

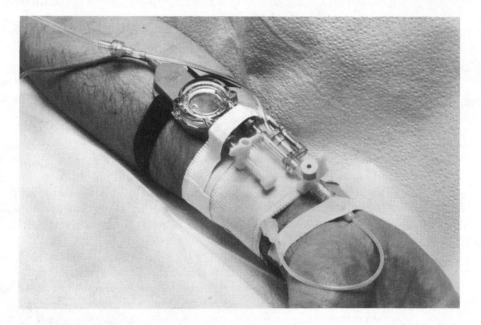

FIGURE 2.16
The Hewlett-Packard 1290A is a new, rugged,
lightweight physiological pressure transducer that
uses quartz crystal sensors to measure patients'
blood, gastrointestinal, intrauterine, intracranial,
and other physiological pressures. (Courtesy of
Hewlett-Packard Co., Palo Alto, Calif.)

lightning-surge protection is given by on-chip components, thus eliminating the need for an external surge-suppression diode. The high input impedance makes it suitable for use with high- or low-impedance microphones that provide a high output voltage.

A new rugged and lightweight physiological pressure transducer (Figure 2.16) that uses quartz crystal sensors to measure patients' blood, gastrointestinal, intrauterine, intracranial, and other physiological pressures is made by Hewlett-Packard. Several distinct advantages are offered the hospital staff by the quartz crystal technology used in the HP 1290A transducer. It is more rugged and durable than conventional transducers, which use thin-metal diaphragm sensors. It can withstand hard day-to-day usage and handling, and can be gas sterilized and even scrubbed with brush and detergent, all without affecting the transducer's accuracy.

Ultrasonic transducers use piezoelectric probes. Figure 2.17 gives information on ultrasonic probe materials and characteristics.

2.8
POTENTIOMETRIC TRANSDUCER AS A DISPLACEMENT DEVICE

Linear displacement can be achieved by using a pot (slide bar). The motion of the slider results in resistance change that can be made linear depending on the way the resistance wire is wound. A potentiometric transducer is diagrammed in Figure 2.18.

The mechanical resistive elements rely upon the resistance variation produced by a mechanical input to its movable slider. Typical construction of the resistance requires precision elements. Figure 2.19 shows how a spirometer bellows is mechanically linked to a potentiometer slider in order to directly transduce respiratory movement.

2.9
RESISTANCE THERMOMETER AS A TEMPERATURE DEVICE

The resistance thermometer uses the fact that resistivity, and therefore resistance, changes with temperature in a known manner. The resistance of a material can be calculated approximately by

$$R = R_0 (1 + \alpha_1 T + \alpha_2 T^2) \tag{2.7}$$

where

R_0 = resistance at 0°C

α_1 and α_2 = constant characteristics of the material from which resistance is made

Commercial ultrasonic transducers are mostly of three general types:
- Natural quartz crystals
- Lithium sulphate monohydrate crystals
- Polarized crystalline ceramics (such as Barium titanate or PZT)

Some of their general properties are listed below:

Property	Quartz	Lithium sulphate	Polarized ceramics
Chemical stability	Very good	Good up to 165°C	Good up to 300°C
Insolubility	Very insoluble	Very poor. Easily dissolves in water	Practically unaffected
Hard wear	Very good	Very fragile	Low mechanical strength
Stability with time	Very good	Very good	Fair
Operational efficiency	Poor	Fair (but a good receiver)	Very good
Freedom from mode conversion	Very poor	Little effect	Some effect
Use at low voltage	Poor, especially at low frequencies	Good	Very good

Electrical and acoustical data for piezoelectric transducer materials

	Crystal material					
	Quartz	Barium titanate	Lithium sulphate	Lead metaniobate	Lead zirconate-Titanate	Lithium niobate
Velocity of sound, V_x (m/s)	5.7×10^3	5.0×10^3	5.45×10^3	2.75×10^3	3.0×10^3	4.8×10^3
Acoustic impedance, ρV_x (kg m^{-2}/s^{-1})	1.52×10^7	3×10^7	1.12×10^7	1.6×10^7	2.25×10^7	2.26×10^7
Dielectric constant, e	4.58	1.35×10^3	10	2.25×10^2	1.5×10^3	
Electromechanical coupling coefficient (thickness model), k	0.1	0.4	0.35	0.42	0.675	0.68
Piezoelectric pressure constant (thickness model), g (V m/N)	5.8×10^{-2}	1.275×10^{-2}	17.5×10^{-2}	3.7×10^{-2}	2.44×10^{-2}	
Curie temperature, °C	576	115—150	75	570	350	1210

The following results are for a particular size of crystal, ie 20 mm diam, 6 MHz, air backed radiating into water, applied voltage 50 V (rms), striking a perfect reflector 1 mm^2.

Electrical input impedance (Ω)	990	2.95	483	9.72	1.59	
Transformer turns ratio, crystal to 75-Ω load	3.64:1	0.198:1	2.54:1	0.36:1	0.146:1	
Crystal capacitance, C_x (pF)	26.8	9000	61.2	2725	16670	
Power radiated, p (W)	0.65	6910	13.4	1227	27400	
Resultant voltage from received echo (V)	0.087	1.72	1.135	1.16	3.93	
Loop gain with respect to quartz	1	20	13	13.5	45	
Resultant voltage for 1 W of received ultrasound	1.8	0.4	5.5	0.6	0.4	

FIGURE 2.17

Ultrasonic probe materials and characteristics. (Reprinted from *Quality Technology Handbook,* published by IPC Science and Technology Press Ltd., Guildford, Surrey, England, with permission of the United Kingdom Atomic Energy Authority and Harwell Public Relations Group)

Disadvantages	Advantages

Disadvantages

a. Usually large size
b. The resolution is finite in most cases
c. High mechanical friction
d. Limited life
e. Sensitive to vibration
f. Develops high noise levels with wear
g. Requires large force–summing member
h. Low frequency response
i. Large displacement required

Advantages

a. High output
b. Inexpensive
c. Easily serviced
d. Easy to excite and install
e. May be excited with AC or DC
f. Wide range of functions
g. No amplification or impedance matching is necessary

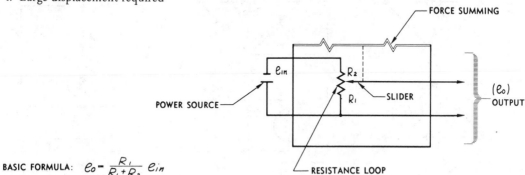

BASIC FORMULA: $e_o = \dfrac{R_1}{R_1 + R_2}\, e_{in}$

FIGURE 2.18
A potentiometer transducer. (From "Introduction to Transducers for Instrumentation." Courtesy of Gould, Inc., Medical Products Division, Oxnard, Calif.)

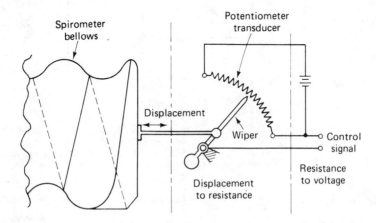

FIGURE 2.19
Potentiometer transducer in spirometer showing basic movement. (From Harry E. Thomas, *Handbook of Biomedical Instrumentation and Measurements,* Reston Publishing Co., Reston, Va., 1974)

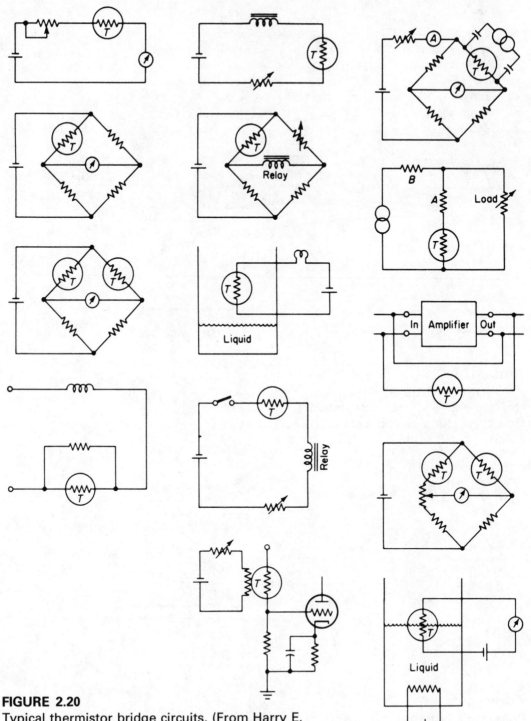

FIGURE 2.20
Typical thermistor bridge circuits. (From Harry E. Thomas, *Handbook of Biomedical Instrumentation and Measurements,* Reston Publishing Co., Reston, Va., 1974)

As shown in Eq. (2.7), the resistance of a resistance thermometer increases as the temperature increases, and vice versa. The resistance for various values of R can be found by known values of R_0, α_1, and α_2.

The resistance of the resistance thermometer is generally determined by using some type of Wheatstone bridge. For platinum, α_1 is 3.96×10^{-3} and α_2 is 5.83×10^{-6}. The temperature range for platinum is $-200°C$ to $850°C$, copper has a temperature range of $-200°C$ to $260°C$, and nickel has a temperature range of $-80°$ to $320°C$.

2.10
THERMISTOR AS A TEMPERATURE DEVICE

Thermistors are semiconductors whose resistance changes with temperature, and therefore they can be used as industrial or medical transducers. Although their existence was known for about 150 years, they were not extensively used until about 1940. They are commonly made of sintered oxides of manganese, nickel, copper, or cobalt, and are available in disc, wafer, rod, bead, washer, and flake form, with power-handling capabilities from a few microwatts up to 25 W. Standard units have a high temperature coefficient of resistance and are produced with resistances ranging from a few ohms to 100 MΩ.

Thermistors can be connected in series–parallel arrangements for applications requiring increased power-handling capability. High-resistance units find application in measurements that employ long lead wires or cables. Thermistors are chemically stable and can be also used in nuclear environments. Their wide range of characteristics also permits them to be used in general industrial applications, consumer appliances and household applications, medicine, laboratory and scientific applications, and communication applications. All thermistors should meet the performance requirements of MIL-T-23648 specifications. Figure 2.20 shows typical thermistor bridge circuits.

Standard precision interchangeable thermistors are epoxy encapsulated and color coded. Stiff wire is placed in the tube so that with slight finger pressure the probe can be bent to any configuration to place the thermistor exactly where it is desired.

It should be emphasized here that the Wheatstone bridge circuits for the thermistor are analogous to that of the strain-gage elements.

Perhaps the most widely used device for industrial and military applications is the thermistor bolometer. The thermistor contains a semiconductor substance with a large negative temperature coefficient. The thermistor itself is composed of a thin flake of nickel, cobalt, and magnesium oxide mounted on a sapphire backing. The thermistor undergoes a change in resistance of about 4.0% for each degree Celsius. Its response time is fast, although not as rapid as that of photoconductive devices.

2.11
THERMOCOUPLE AS A TEMPERATURE DEVICE

The thermocouple (TC) is a temperature transducer that develops an emf (Seebeck effect) that is a function of the temperature difference between its hot and cold junctions. Thermocouples made of base or noble metals are commonly used to measure temperatures from near absolute zero to about +3200°F, while special units are available for temperatures to +5600°F.

Most modern thermocouple systems employ electrical cold junction compensation to simulate a reference temperature. Electrical cold junction compensator cards are available for common thermocouple materials. These cards are compact, reliable, and consistently accurate, and require no operator attention or maintenance. They also are well adapted to zero suppression, an electrical technique for greatly improving resolution and accuracy in measuring small changes in temperature. Figure 2.21 shows a self-powered thermocouple reference junction.

[d] Figure 2.22 shows the Keithley 871 thermocouple (TC) based digital thermometer. It offers 0.25% accuracy, temperature measurements to 1370°C, dual inputs, selectable scales (°C/°F), and 0.1° resolution up to 200°C.

Unlike thermistors, resistance temperature detectors (RTDs), and semiconductor devices, TCs have all the following advantages:

1. Wide temperature range.

2. Wide choice of sensors, from off-the-shelf probes to custom wires.

3. Sensor interchangeability, as TCs are characterized by standard National Bureau of Standards tables.

4. Widest range of applications, from microbiology to jet turbines.

The Keithley 871's type K (NiCr–NiAl) TCs offer good chemical inertness, physical durability, and wide temperature span. Where fast thermal response is required, the Keithley 871 can achieve reading response times limited only by its digitization time of 400 ms.

Keithley offers a wide range of handheld probes designed to give users good value and quality. Cold junction electronic circuitry automatically compensates for ambient temperature changes. A special integrating analog-to-digital converter (A/D) corrects for TC nonlinearity during each digitization cycle. The resulting conformity to the standard TC voltage table is within 1°C over the entire temperature span of the instrument.

In a thermocouple, when the junction is blackened and placed so that

---

[d] The material in the following four paragraphs is from the *Electronic Measurement Instrumentation Catalog and Buyers Guide* and is used courtesy of Keithley Instruments Co., Cleveland, Ohio.

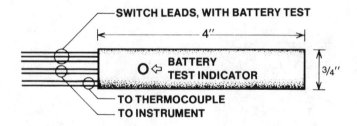

SWITCH LEADS, WITH BATTERY TEST

4″

O ⇦ BATTERY TEST INDICATOR

3/4″

TO THERMOCOUPLE
TO INSTRUMENT

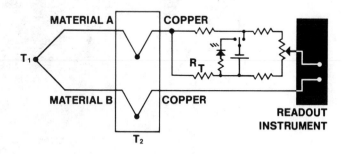

MATERIAL A

COPPER

T_1

R_T

MATERIAL B

COPPER

READOUT
INSTRUMENT

T_2

FIGURE 2.21

Unit contains a self-compensating electrical bridge network. This system incorporates a temperature-sensitive resistance element (R_T), which is in one leg of a bridge network and thermally integrated with the cold junction (T_2). The bridge is energized from a mercury battery. The output voltage is proportional to the unbalance created between the preset equivalent reference temperature at (T_2) and the temperature of the hot junction (T_1). In this system, the reference temperature is 0°C or 32°F. As the ambient temperature surrounding the cold junction (T_2) varies, a thermally generated voltage appears and produces an error in the output. However, an automatic equal and opposite voltage is introduced in series with the thermal error. This cancels the error and maintains the equivalent reference junction temperature over a wide ambient temperature range with a high degree of accuracy. By integrating copper leads with the cold junction, the thermocouple material itself is not connected to the output terminal of the measurement device, thereby eliminating secondary errors. (Courtesy of OMEGA ENGINEERING, INC., Stamford, Conn.)

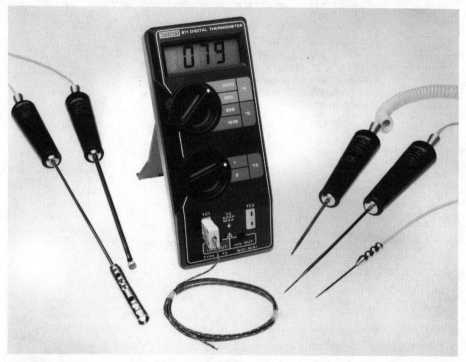

FIGURE 2.22
Keithley 871 digital thermometer. (Courtesy of
Keithley Instrument Co., Cleveland, Ohio.)

infrared radiation can impinge upon it, the device functions as an infrared
radiation detector. A thermopile is merely a series of thermocouples suitably connected so as to provide a greater output voltage or current. These
devices, however, have a relatively poor response time.

2.12
PHOTOEMISSIVE TUBE AS A
LIGHT DEVICE

Light transducers are devices that convert electromagnetic radiation to an
electric signal. The three major light transducers include photoemissive
cells (Section 2.12), Photovoltaic cells (Section 2.13), and photoconductive
cells (Section 2.14).

The photoemissive transducer (or phototube) is a device that produces
current variations resulting from light striking its cathode. The two main
categories are gas filled and hard vacuum. Ionization in the former adds
sensitivity, but the latter generally has an electron multiplier within the

bulb to give added range of sensitivity. Photoemissive devices are larger than most other phototransducer types, relatively complicated, less reliable, and require a power supply.

Figure 2.23(a) shows typical single and twin multiplier phototubes. Figure 2.23(b) shows the mechanics of a photoelectric conversion in a pressure transducer.

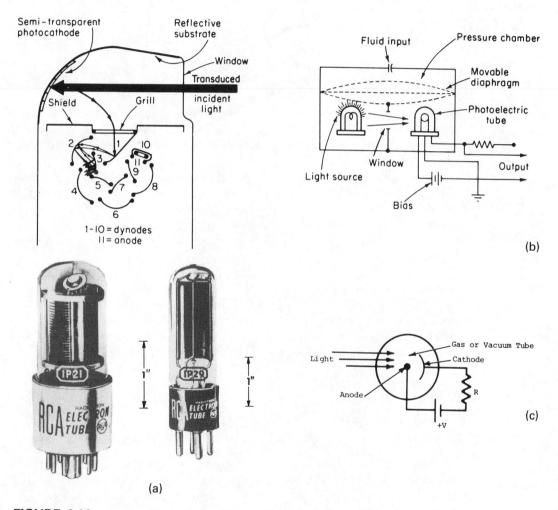

FIGURE 2.23
Photomultiplier tube in a pressure transducer:
(a) typical units; (b) mechanics of photoelectric
multiplication. (c) The photoemissive tube circuit.
(From Harold E. Thomas, *Handbook of Biomedical Instrumentation and Measurement,* Reston Publishing Co., Reston, Va., 1974)

The vacuum photodiode is relatively insensitive but has an extremely high frequency response, on the order of 1 MHz, and is therefore useful for measuring high-intensity, rapidly varying light energy. Its spectral sensitivity is approximately 10^{-3} to 10^{-6} W of light energy of 10^{-3} to 10^{-6} lumens. Below this intensity the energy cannot be measured.

2.13
PHOTOVOLTAIC CELLS AS LIGHT DEVICES [e]

Photovoltaic cells are semiconductor devices that convert radiant energy directly into a voltage. These devices are called "solar" cells since they are frequently used to convert radiation from the sun into electrical power. Silicon, gallium arsenide phosphor, cadmium sulfide, and selenium are materials frequently used in photovoltaic cells or photocells. The potential produced by the cell is small (about 0.6 V for silicon in sunlight) and is not a linear function of illumination for an open circuit.

Photovoltaic transducers are cells that possess a self-generation junction. When light is absorbed near a P-N junction, new mobile holes (positive charges) are released and a potential difference is developed. The mechanism and silicon cells are shown in Figure 2.24. Selenium cell currents as a function of load resistance and illumination are illustrated in Figure 2.24(b).

Silicon and gallium arsenide phosphor use the photovoltaic effect, the generation of a voltage across a P-N junction when the junction is exposed to light. Photocells have been used in such devices as computer card readers and smoke detectors and are now finding use in exposure meters, analysis equipment, and radiation measurement of response speed and linearity. Some major features of photocells include the following:

1. High-speed response
2. Excellent linearity
3. Small dark current (low noise)
4. Wide spectral response and mechanical ruggedness

Figure 2.25(a) shows a cross section of a photocell with the dark condition band model given in part (b). Since this device is in thermal equilibrium, the P layer and N layer Fermi levels are equal, and a voltage gradient develops in the depletion layer by virtue of the contact potential (potential barrier).

When radiation is allowed to strike the photocell, the internal electrons become stimulated. If the stimulating radiation is of high enough level, i.e., larger than the band gap, Eg, electrons will be pulled into the conduction

---

[e] The material in Section 2.13 is used courtesy of Hamamatsu Corp., Middlesex, N.J.

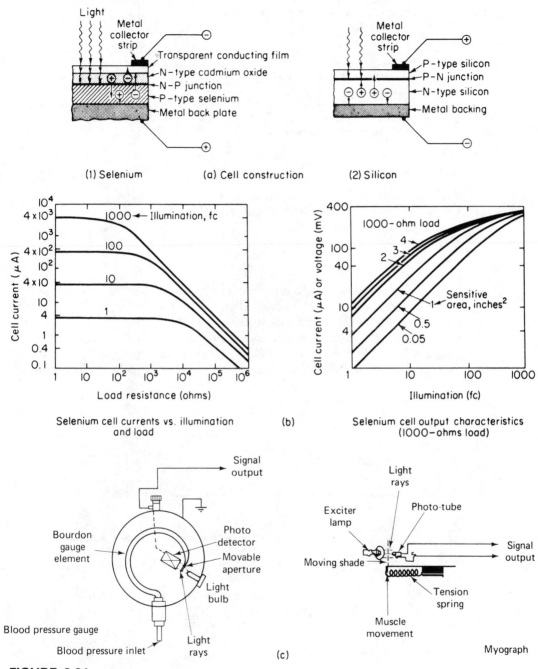

(1) Selenium   (a) Cell construction   (2) Silicon

Selenium cell currents vs. illumination and load   (b)   Selenium cell output characteristics (1000-ohms load)

(c)   Myograph

FIGURE 2.24
Photovoltaic transducers: (a) mechanism and
construction details; (b) currents and output voltage versus illumination; (c) photoelectric transducer action. (From Harry E. Thomas, *Handbook
of Biomedical Instrumentation and Measurement,*
Reston Publishing Co., Reston, Va., 1974)

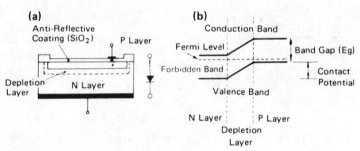

FIGURE 2.25
Photocell in a dark condition. (Courtesy of Hamamatsu Corp., Middlesex, N.J.)

band, leaving behind positive holes. These electron–hole pairs are generated throughout the N and P material and the depletion layer. In the depletion layer, electrons and holes drift toward the N and P material layers, respectively. In addition, electrons in the P layer diffusion length and holes in the N layer diffusion length diffuse toward the depletion layer, causing a charge to accumulate in the N and P layers. [see Figure 2.26(a) and (b)].

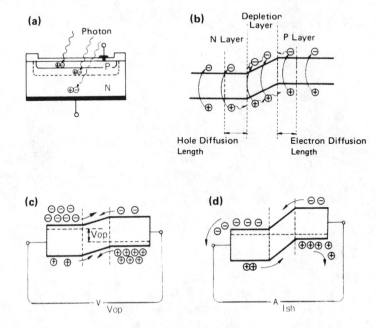

FIGURE 2.26
Photocell with incident radiation present.
(Courtesy of Hamamatsu Corp., Middlesex, N.J.)

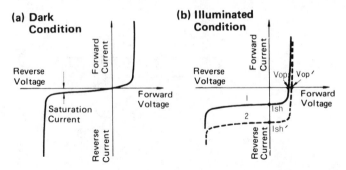

FIGURE 2.27
VI characteristics. (Courtesy of Hamamatsu Corp.,
Middlesex, N.J.)

This is known as the *photovoltaic effect*. Sensitivity can be improved by making the P layer as thin and transparent as possible and making the depletion layer as wide as possible.

As a natural reaction to the lower P-N potential barrier, electrons and holes drift toward the P and N materials, respectively, reducing the accumulated charge. For this reason continued application of radiation will not result in unlimited buildup of charge; the potential barrier is reduced until apparent charge transfer ceases. This potential barrier change, termed

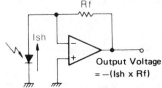

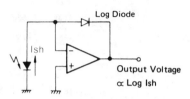

FIGURE 2.28
Operational amplifier (op-amp) connection examples. (Courtesy of Hamamatsu Corp., Middlesex, N.J.)

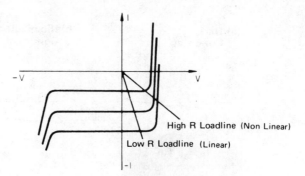

FIGURE 2.29
Load lines. (Courtesy of Hamamatsu Corp., Middlesex, N.J.)

Fermi level shift, manifests itself externally as a voltage at the photocell terminals. This voltage is the open-circuit voltage, V_{op}. If the terminals of the photocell are shorted, the current that flows is known as the short-circuit current, I_{sh}.

When a voltage is applied to a photocell in the dark state, the VI characteristic curve observed is similar to the curve of a conventional diode, as

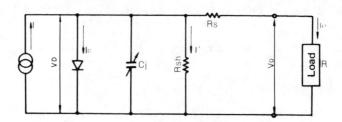

I_L	: current generated by the incident light
I_D	: diode current
C_j	: junction capacitance
R_{sh}	: shunt resistance
R_s	: series resistance
I'	: shunt resistance current
V_D	: voltage at the diode terminals
I_o	: output current
V_o	: output voltage

FIGURE 2.30
Photocell equivalent circuit. (Courtesy of Hamamatsu Corp., Middlesex, N.J.)

❶ AC signal FET pre-amplifier

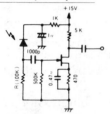

FET: 2SK19, 2SK43, 2N4392, etc.
R: Load resistance, selected to achieve desired frequency response and sensitivity.

❷ High sensitivity light detector using an OP-amp

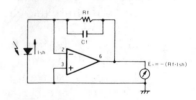

$E_o = -(Rf \cdot Ish)$

Rf: Feedback resistance
Cf: Feedback capacitance
Cutoff frequency $(-3dB) = \dfrac{1}{2\pi \cdot Cf \cdot Rf}$ (Hz)

❸ Light difference circuit

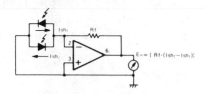

$E_o = | Rf \cdot (Ish_2 - Ish_1)|$

By placing two photocells back to back with opposing polarities, the resultant output Ish is the difference of the incident light striking the two devices. Equal light results in 0 output.

❹ Multiphotocell circuit using analog switches

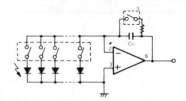

Analog switch: RCA CD4051B, Siliconix DG501 etc.
Cc: Hold capacitance

❺ Illumination meter

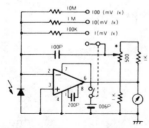

OP-AMP: CA3130
Silicon photocell: S1133 (0.5μA/100/x)
* Meter calibration adjustment V.R.

❻ High speed light detector using a video amplifier

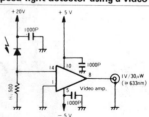

1V/30μW (@ 633nm)

Video AMP: LM733CH (bandwidth: 200MHz)
R: Load resistance, chosen to achieve desired frequency response and sensitivity.

Note: The OP-amp used in the above circuits (②,③,④) should be one with a very small bias current, FET input circuit.

Representative Listing of FET Input OP-amps

Manufacturer	Types	Manufacturer	Types
● Analog Devices	AD503, AD515, AD523	● RCA	CA3130, CA3140, CA3160
● Burr Brown	3521, 3522, 3523, 3527	● Signetics	NE536
● Intersil	8007, 8043, ICH8500	● Teledyne	1421, 1425, 1426, 1429, 1439
● National Semiconductors	LH0022, LH0042, LH0052, LF156, LF256, LF356, LF13741	● Toshiba	TA7505M
● NEC	μPC252A	● Tokyo Musen Kizai	LX706, LX7031, LX7032

FIGURE 2.31
Application examples of the photocell. (Courtesy of Hamamatsu Corp., Middlesex, N.J.)

shown in Figure 2.27(a). When light strikes the photocell, however, the curve shifts to position 1 in Figure 2.27(b). Increasing the amount of incident light shifts the characteristic curve still further to position 2 of the figure. Under such conditions an open circuit will exhibit a voltage V_{op} (or Vop'), while shorting the photocell terminals will result in the flow of the short circuit current I_{sh} (or Ish'). The open-circuit voltage is a forward voltage in the sense used with a conventional diode.

The usual method of achieving logarithm compression is to use a logarithmic diode as the feedback element in an op-amp circuit (see Figure 2.28).

When connecting with a finite load resistance R, care should be taken with the design since linearity is affected by R. Smaller values of R result in operation close to the I_{sh} curve and a wide linear operating range (see Figure 2.29). A photocell equivalent circuit is shown in Fig. 2.30.

Figure 2.31 shows application examples, which include the following circuitry:

1. AC signal FET preamplifier.
2. High-sensitivity light detector using an operational amplifier.
3. Light difference circuit.
4. Multiphotocell circuit using analog switches.
5. Illimination meter.
6. High-speed detector using a video amplifier.

2.14
PHOTOCONDUCTIVE CELL AS A
LIGHT DEVICE [f]

The photoconductive transducer cell is based on the phenomenon of *photoresistivity,* the decrease in resistance exhibited by certain semiconductors as a result of the application of light. Since all photosensitive semiconductor materials do not have a linear resistivity change when exposed to light, devices not exhibiting such a change are called *cells* rather than photoresistors. Some photojunction diodes and transistors exhibit the photoconductive effect over part of their characteristics.

For the visible range, photoconductive cells are usually made of cadmium sulfide or cadmium selenide. Lead sulfide cells are used as infrared detectors.

Photoconductive cells are constructed from a suitably processed powder of the compounds mentioned; in cadmium cells, the powder is

---

[f]The material in Section 2.14 is excerpted from Harry E. Thomas, *Handbook of Biomedical Instrumentation and Measurements,* Reston Publishing Co., Reston, Va. 1974.

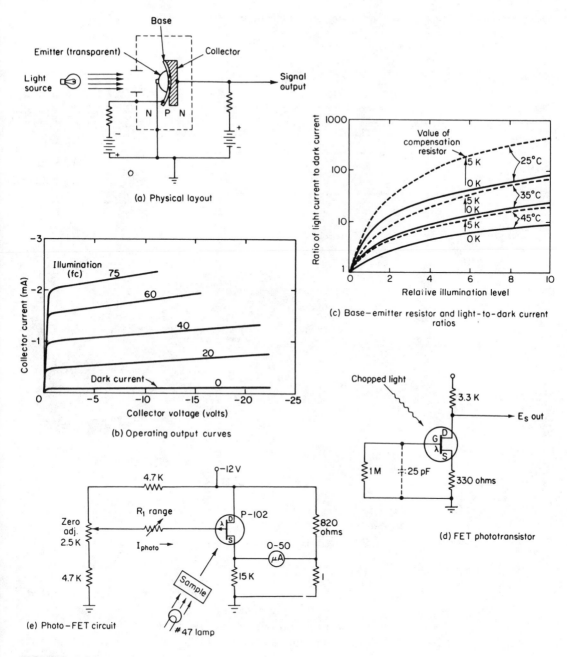

FIGURE 2.32

Phototransistor construction and operation: (a) physical layout; (b) collector voltage versus current curves; (c) base–emitter resistor variation; (d) FET phototransistor; (e) photo-FET circuit. (From Harry E. Thomas, *Handbook of Biomedical Instrumentation and Measurements,* Reston Publishing Co., Reston, Va., 1974)

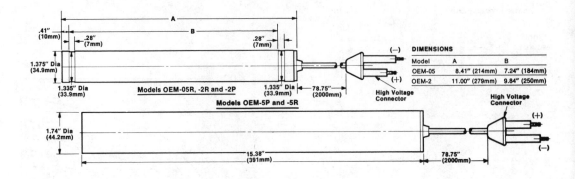

FIGURE 2.33
Helium neon cased laser heads. (Courtesy of
Aerotech Inc., Pittsburgh, Pa.)

pressed and sintered into a 1-mil-thick layer over a ceramic base in various
patterns to fit the light-collecting requirements of particular units.

As a transistor, the phototransistor amplifies the photoconductive current generated across the reverse-biased emitter–base junction. The light
applied to the base through a hole in the case is the input parameter [see
Figure 2.32(a)]. The phototransistor [Figure 2.32(a)] is a device consisting
of a base-to-emitter photodiode and a base-to-collector diode. The photodiode responds to light energy as a simple photodiode, but the transistor
arrangement of the base–collector junction provides a gain, causing the
phototransistor to have the propensity of increased sensitivity.

In the conventional manner, we can pick off from collector voltage versus current curves [see Figure 2.32(b)] any point corresponding to a given level of illumination; the amplification there is equal to beta times the input junction current. Unfortunately, the collector–base leakage currents now become the dark current, which is also amplified.

In the phototransistor the correct base–emitter resistor governs the ratio of light to dark current; the resistor must also be chosen so that danger of thermal runaway is minimized. Figure 2.32(c) shows a base–emitter resistor variation with temperature.

2.15
LASERS

The laser (light amplification of simulated by emission of radiation) is being used in industry and medicine. The laser is used industrially for machining and welding in electronic instrumentation and medically in the treatment of cancer, eye surgery, and many other applications.

The carbon dioxide laser used in industry can also be used as a surgical tool, which, among other uses, is suitable for primary treatment of small invasive carcinomas of the vocal cord.

The American National Standards Institute's ANSI Z136.1-1980 Laser Safety Standard discusses laser classifications, control measures, safety and surveillance programs, and associated nonbeam hazards.

Helium neon cased laser heads (Figure 2.33) are packaged in open-ended black anodized aluminum housing. All models are standard, with a rear-entry high-voltage cable and connector, with a 47-kΩ ballast register mounted inside the laser head. In addition, customers using OEM laser heads are responsible for compliance with applicable Bureau of Radiological Health of the Food and Drug Administration regulations.

2.16
FIBER OPTIC TRANSDUCERS

Fiber optics deals with the transmission or guidance of light rays along transparent fibers of glass or plastic material. It has found use in communications and medicine.

A multipurpose fiber optic pressure transducer takes the form of a catheter, with the pressure-sensing part at one end and the necessary bulky components at the other. A light guide of glass fibers inside the catheter transports light from a light source to the measuring tip. Here the pressure variations affect a movable membrane that reflects a variable amount of light into another light guide leading to a photodetector. The variations in the signal from the detector are thus proportional to the applied pressure.

In communication, fiber optic links can be optimized for greater than 10 Mbit/s operations.

2.17
REVIEW QUESTIONS

1. How does a technician select the proper transducer?
2. What is a medical sensor?
3. A strain gage has a gage factor of 170. It is 20 cm long and a resistance of 500 Ω in the unstressed condition. The gage is placed in a pressure vessel and stressed to 20.01 cm. What is the new resistance?
4. A certain LVDT has a 1-mV output when the core is moved up 20 μin. What will be the output if the core is moved 40 μin.?
5. List five displacement transducers. Discuss each transducer.
6. Describe force and torque transducers.
7. Describe motion transducers.
8. Describe velocity transducers.
9. Describe acceleration transducers.
10. What is a jerk?
11. What is a bioelectronic electrode?
12. List the types of strain gage transducers.
13. What is meant by gage factor?
14. Describe a strain gage calibration technique.
15. What is meant by strain gage temperature compensation?
16. What is a thermocouple? Discuss the digital thermometer.
17. List five uses of a thermistor.
18. Describe a potentiometric transducer.
19. Describe the use for a linear variable transformer transducer.
20. Describe a capacitor displacement transducer.
21. Describe a piezoelectric transducer.
22. Describe the operation of photosensitive transducer devices.
23. Describe the operation of radiation transducers and lasers.

2.18
REFERENCES

1. G. F. Harvey, ed. *ISA Transducer Compendium*, 2nd ed., Part 1, Plenum Publishing Co., New York, 1969.
2. P. Webb, "Bioastronautics Data Book," NASA SP-30006, National Aeronautics and Space Administration, Washington, D.C., 1964.

3. G. F. Harvey, ed. *ISA Transducer Compendium*, 2nd ed., Part 2, Plenum Publishing Co., New York, 1969.

4. *Omega Temperature Measurement Handbook*, OMEGA ENGINEERING INC., Stamford, Conn., 1982.

5. Clevite Brush Displacement Transducer Bulletin 639-5, Metripak Transducer Models 33 03, 33 04, 33 05, 33 06, July 1, 1967.

6. A. E. Robertson, *Microphones*, 2nd ed., Hayden Book Co., Rochelle Park, N.J., 1963.

7. F. Fraim and P. Murphy, "Miniature Electret Microphones," *Journal of the Audio Engineering Society*, vol. 18, no. 5, Oct. 1970.

8. "Survey of Transducers and Signal Converts," *Electronic Instrument Digest*, 31–40, May 1969.

9. YSI Precision Thermistors Products, Yellow Springs Instrument Co., Yellow Spring, Ohio, Jan. 1980.

10. A. D. Little, "Temperature Measurement with the Variable Capacitive Temperature Transducer," *IEEE Transactions on Industrial Electronics and Control Instrumentation*, vol. IECI-17, no. 2, April 1970.

11. L. McKay, The Development of an Intracranial Pressure Transducer, Master of Science in Bioengineering Thesis, Polytechnic Institute of Brooklyn, June 1968.

12. W. Welkowitz and S. Deutsch, *Biomedical Instruments Theory and Design*, Academic Press, New York, 1976.

13. R. S. C. Cobbold, *Transducers for Biomedical Measurements*, John Wiley & Sons, New York, 1974.

14. Edward E. Herceg, *Handbook of Measurement and Control*, Schaevitz Engineering Co., Pennsauken, N.J., 1978.

15. P. Kantrowitz, G. Kausourou and L. Zucker, *Electronic Measurements*, Prentice-Hall, Englewood Cliffs, N.J., 1979.

16. Thermistor Handbook, Cat. No. 181-A, Thermometrics, Edison, N.J., 1980.

3

Data-Acquisition
Systems[a]

3.1
INTRODUCTION

Data-acquisition systems (Figures 3.1 and 3.2) via analog signals are used in communication, electronic, and medical applications. Conversion to digitized systems is widely used today because complex circuits are low cost, accurate, and relatively simple to implement. In addition, there is rapid growth in use of microprocessors and microcomputers to perform difficult digital control and measurement functions.

The electronic devices [1–10] that perform the interfacing function between the analog and digital worlds are the analog-to-digital converter, abbreviated A/D converter or ADC, and the digital-to-analog converter, abbreviated D/A converter or DAC. ADCs or DACs are generally classified as data converters; they are discussed in detail in Chapters 4 through 6.

Some of the specific applications in which data converters are used include data telemetry systems, pulse code modulated communications,

[s] The material in this chapter is from Eugene L. Zuck, *Data Acquisition and Conversion Handbook*, pp. 3-25, 4th printing, Sept. 1981. © 1979, by Datel-Intersil, Inc. Reprinted with the permission of Datel-Intersil, Inc.

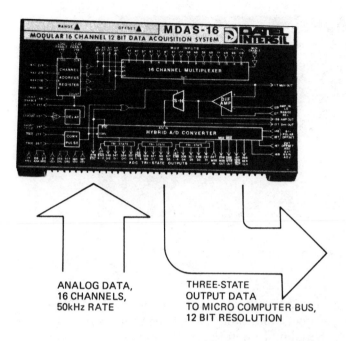

ANALOG DATA,
16 CHANNELS,
50kHz RATE

THREE-STATE
OUTPUT DATA
TO MICRO COMPUTER BUS,
12 BIT RESOLUTION

MECHANICAL DIMENSIONS - INCHES (MM)

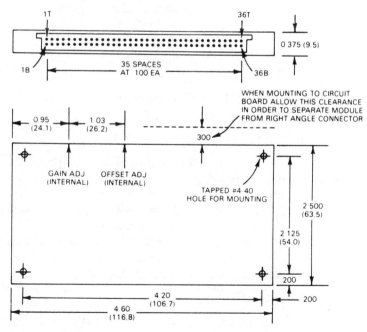

FIGURE 3.1
MDAS-16 data-acquisition system. (Courtesy of
Datel-Intersil, Inc.)

FIGURE 3.2
PDAS-250 data-acquisition system. (Courtesy of
Datel-Intersil, Inc.)

automatic test systems, computer display systems, video signal processing systems, data logging systems, and sampled-data control systems. In addition, every laboratory digital multimeter or digital panel meter contains an A/D converter.

Besides A/D and D/A converters, data-acquisition systems may employ one or more of the following circuit functions:

1. Transducers

2. Amplifiers

3. Filters

4. Nonlinear analog functions

5. Analog multiplexers

6. Sample-holds

The interconnection of these components is shown in the diagram of the data-acquisition portion of a computerized feedback control system in Figure 3.3.

The input to the system is a *physical parameter*, such as temperature, pressure, flow, acceleration, or position, which are analog quantities as discussed in Chapter 2. The parameter is first converted into an electrical signal by means of a transducer; once in electrical form, all further processing is done by electronic circuits.

Next an amplifier or signal conditioner boosts the amplitude of the transducer output signal to a useful level for further processing. Transducer outputs may be microvolt or millivolt level signals, which are then amplified to 1- to 10-V levels. Furthermore, the transducer output may be a high-impedance signal, a differential signal with common-mode noise, a current output, a signal superimposed on a high voltage, or a combination of these. The amplifier, in order to convert such signals into a high-level voltage, may be one of several specialized types.

The amplifier is frequently followed by a low-pass *active filter,* which reduces high-frequency signal components, unwanted electrical interference noise, or electronic noise from the signal. The amplifier is sometimes also followed by a special *nonlinear analog function* circuit that performs a nonlinear operation on the high-level signal. Such operations include squaring, multiplication, division, rms conversion, log conversion, or linearization.

The processed analog signal next goes to an *analog multiplexer,* which sequentially switches between a number of different analog input channels. Each input is in turn connected to the output of the multiplexer for a specified period of time by the multiplexer switch. During this connection time a *sample-hold* circuit acquires the signal voltage and then holds its value while an *analog-to-digital converter* converts the value into digital form. The resultant digital word goes to a computer data bus or to the input of a digital circuit.

Thus the analog multiplexer, together with the sample-hold, time shares the A/D converter with a number of analog input channels. The timing and control of the complete data-acquisition system is done by a digital circuit called a *programmer-sequencer,* which in turn is under control of the computer. In some cases the computer itself may control the entire data-acquisition system.

While this is perhaps the most commonly used data-acquisition system configuration, there are alternatives. Instead of multiplexing high-level signals, low-level multiplexing is sometimes used with the amplifier following the multiplexer. In such cases only one amplifier is required, but its gain may have to be changed from one channel to the next during multiplexing. Another method is to amplify and convert the signal into digital form at the transducer location and send the digital information in serial form to the computer. Here the digital data must be converted to parallel form and then multiplexed onto the computer data bus.

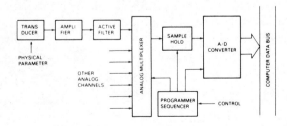

FIGURE 3.3
Block diagram of data-acquisition system.
(Courtesy of Datel-Intersil, Inc.)

3.2
CODING FOR DATA CONVERTERS

3.2.1 Natural Binary Code

A/D and D/A converters interface with digital systems by means of an appropriate digital code. While there are many possible codes to select, a few standard codes are almost exclusively used with data converters. The most popular code is *natural binary*, or *straight binary*, which is used in its fractional form to represent a number:

$$N = a_1 2^{-1} + a_2 2^{-2} + a_3 2^{-3} + \cdots + a_n 2^{-n} \tag{3.1}$$

where each coefficient a assumes a value of 0 or 1. N has a value between 0 and 1.

A binary fraction is normally written as 0.110101, but with data-converter codes the decimal point is omitted and the code word is written 110101. This code word represents a fraction of the full-scale value of the converter and has no other numerical significance.

The binary code word 110101 therefore represents the decimal fraction $(1 \times 0.5) + (1 \times 0.25) + (1 \times 0.125) + (1 \times 0.0625) + (0 \times 0.03125) + (1 \times 0.015625) = 0.828125$, or 82.8125% of full scale for the converter. If full scale is +10 V, then the code word represents +8.28125 V. The natural binary code belongs to a class of codes known as positive weighted codes since each coefficient has a specific weight, none of which is negative.

The leftmost bit has the most weight, 0.5 of full scale, and is called the *most significant bit*, or MSB; the rightmost bit has the least weight, 2^{-n} of full scale, and is therefore called the *least significant bit*, or LSB. The bits in a code word are numbered from left to right from 1 to n.

LSB (Analog Value) is given by the following equation:

$$\text{LSB (Analog Value)} = \frac{\text{FSR}}{2^n} \tag{3.2}$$

where FSR is full-scale range. Table 3.1 is a useful summary of the resolution, number of states, LSB weights, and dynamic range for data converters from 1 to 20 bits of resolution.

The *dynamic range* of a data converter in decibels (dB) is found as follows:

$$\text{DR (dB)} = 20 \log 2^n = 20n \log 2$$
$$= 20n\,(0.301) = 6.02n \tag{3.3}$$

where DR is dynamic range, n is the number of bits, and 2^n the number of states of the converter. Since 6.02 dB corresponds to a factor of 2, it is simply necessary to multiply the resolution of a converter in bits by 6.02. A 12-bit converter, for example, has a dynamic range of 72.2 dB.

An important point to notice is that the maximum value of the digital

TABLE 3.1
Resolution, number of states, LSB weight, and
dynamic range for data converters

RESOLUTION BITS n	NUMBER OF STATES 2^n	LSB WEIGHT 2^{-n}	DYNAMIC RANGE dB
0	1	1	0
1	2	0.5	6
2	4	0.25	12
3	8	0.125	18.1
4	16	0.0625	24.1
5	32	0.03125	30.1
6	64	0.015625	36.1
7	128	0.0078125	42.1
8	256	0.00390625	48.2
9	512	0.001953125	54.2
10	1 024	0.0009765625	60.2
11	2 048	0.00048828125	66.2
12	4 096	0.000244140625	72.2
13	8 192	0.0001220703125	78.3
14	16 384	0.00006103515625	84.3
15	32 768	0.000030517578125	90.3
16	65 536	0.0000152587890625	96.3
17	131 072	0.00000762939453125	102.3
18	262 144	0.000003814697265625	108.4
19	524 288	0.0000019073486328125	114.4
20	1 048 576	0.00000095367431640625	120.4

code, that is, all 1's, does not correspond with analog full scale, but rather with one LSB less than full scale, or $FS(1 - 2^{-n})$. Therefore, a 12-bit converter with a 0- to +10- V analog range has a maximum code of 1111 1111 1111 and a maximum analog value of +20 V $(1 - 2^{-12}) = +9.99756$ V. In other words, the maximum analog value of the converter, corresponding to all 1's in the code, never quite reaches the point defined as analog full scale.

3.2.2 Other Binary Codes

Several other binary codes are used with A/D and D/A converters in addition to straight binary. These codes are *offset binary, two's complement, binary coded decimal* (BCD), and their complemented versions. Each code has a specific advantage in certain applications. BCD coding, for example, is used where digital displays must be interfaced, such as in digital panel meters and digital multimeters. Two's complement coding is used for computer arithmetic logic operations, and offset binary coding is used with bipolar analog measurements.

Not only are the digital codes standardized with data converters, but so are the analog voltage ranges. Most converters use unipolar voltage ranges of 0 to +5 V and 0 to +10 V, although some devices use the negative ranges 0 to −5 V and 0 to −10 V. The standard bipolar voltage ranges are ±2.5V, ±5, and ±10 V. Many converters today are pin-programmable between these various ranges.

Table 3.2 shows straight binary and complementary binary codes for a

TABLE 3.2
Binary coding for 8-bit unipolar converters

Fraction of FS	+10 V FS	Straight Binary	Complementary Binary
+FS − 1 LSB	+9.961	1111 1111	0000 0000
+$\frac{3}{4}$ FS	+7.500	1100 0000	0011 1111
+$\frac{1}{2}$ FS	+5.000	1000 0000	0111 1111
+$\frac{1}{4}$ FS	+2.500	0100 0000	1011 1111
+$\frac{1}{8}$ FS	+1.250	0010 0000	1101 1111
+1 LSB	+0.039	0000 0001	1111 1110
0	0.000	0000 0000	1111 1111

unipolar 8-bit converter with a 0- to +10-V analog full-scale range. The maximum analog value of the converter is +9.961 V, or one LSB less than +10 V. Note that the LSB size is 0.039 V, as shown near the bottom of the table. The *complementary binary* coding used in some converters is simply the logic complement of straight binary.

When A/D and D/A converters are used in bipolar operation, the analog range is offset by half scale, or by the MSB value. The result is an analog shift of the converter transfer function as shown in Figure 3.4. Notice for this 3-bit A/D converter transfer function that the code 000 corresponds to −5 V, 100 with 0 V, and 111 with +3.75 V. Since the output coding is the same as before the analog shift, it is now appropriately called offset binary coding.

Table 3.3 shows the offset binary code together with *complementary*

TABLE 3.3
Popular bipolar codes used with data converters

FRACTION OF FS	±5V FS	OFFSET BINARY	COMP. OFF. BINARY	TWO'S COMPLEMENT	SIGN-MAG BINARY
+FS - 1 LSB	+4.9976	1111 1111	0000 0000	0111 1111	1111 1111
+¾ FS	+3.7500	1110 0000	0001 1111	0110 0000	1110 0000
+½ FS	+2.5000	1100 0000	0011 1111	0100 0000	1100 0000
+½ FS	+1.2500	1010 0000	0101 1111	0010 0000	1010 0000
0	0.0000	1000 0000	0111 1111	0000 0000	1000 0000
-¼ FS	-1.2500	0110 0000	1001 1111	1110 0000	0010 0000
-½ FS	-2.5000	0100 0000	1011 1111	1100 0000	0100 0000
-¾ FS	-3.7500	0010 0000	1101 1111	1010 0000	0110 0000
-FS +1 LSB	-4.9976	0000 0001	1111 1110	1000 0001	0111 1111
-FS	-5.0000	0000 0000	1111 1111	1000 0000	—

*NOTE: Sign Magnitude Binary has two code words for zero as shown here.

	SIGN-MAG BINARY
0+	1000 0000 0000
0-	0000 0000 0000

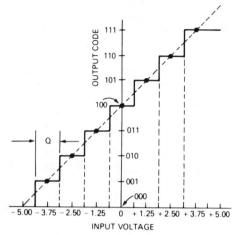

FIGURE 3.4
Transfer function for bipolar 3-bit A/D converter.
(Courtesy of Datel-Intersil, Inc.)

offset binary, two's complement, and *sign-magnitude binary* codes. These are the most popular codes employed in bipolar data converters.

The two's complement code has the characteristic that the sum of the positive and negative codes for the same analog magnitude always produces all zero's and a carry. This characteristic makes the two's complement code useful in arithmetic computations. Notice that the only difference between two's complement and offset binary is the complementing of the MSB. In bipolar coding, the MSB becomes the sign bit.

The sign-magnitude binary code, infrequently used, has identical code words for equal-magnitude analog values except that the sign bit is different. As shown in Table 3.3, this code has two possible code words for zero: 1000 0000 or 0000 0000. The two are usually distinguished as 0+ and 0−, respectively. Because of this characteristic, the code has maximum analog values of ± (FS − 1 LSB) and reaches neither analog +FS or −FS.

3.2.3 BCD Codes

Table 3.4 shows BCD and complementary BCD coding for a 3-decimal digit data converter. These are the codes used with integrating-type A/D converters employed in digital panel meters, digital multimeters, and other decimal display applications. Here four bits are used to represent each decimal digit. BCD is a positive weighted code but is relatively inefficient, since in each group of 4 bits, only 10 out of a possible 16 states are utilized.

The LSB analog value (or quantum, Q) for BCD is

$$\text{LSB (analog value)} = Q = \frac{\text{FSR}}{10^d} \qquad (3.4)$$

TABLE 3.4
BCD and complementary BCD coding

Fraction of FS	+10 V FS	Binary-coded Decimal			Complementary BCD		
+FS − 1 LSB	+9.99	1001	1001	1001	0110	0110	0110
+$\frac{3}{4}$ FS	+7.50	0111	0101	0000	1000	1010	1111
+$\frac{1}{2}$ FS	+5.00	0101	0000	0000	1010	1111	1111
+$\frac{1}{4}$ FS	+2.50	0010	0101	0000	1101	1010	1111
+$\frac{1}{8}$ FS	+1.25	0001	0010	0101	1110	1101	1010
+1 LSB	+0.01	0000	0000	0001	1111	1111	1110
0	0.00	0000	0000	0000	1111	1111	1111

where FSR is the full-scale range and d is the number of decimal digits. For example, if there are 3 digits and the full-scale range is 10 V, the LSB value is

$$\text{LSB (analog value)} = \frac{10 \text{ V}}{10^3} = 0.01 \text{ V} = 10 \text{ mV}$$

BCD coding is frequently used with an additional overrange bit that has a weight equal to full scale and produces a 100% increase in range for the A/D converter. Thus, for a converter with a decimal full scale of 999, an overrange bit provides a new full scale of 1999, twice that of the previous scale. In this case, the maximum output code is 1 1001 1001 1001. The additional range is commonly referred to as $\frac{1}{2}$ digit, and the resolution of the A/D converter in this case is $3\frac{1}{2}$ digits.

Likewise, if this range is again expanded by 100%, a new full scale of 3999 results and is called $3\frac{3}{4}$ digits resolution. Here two overrange bits have been added and the full-scale output code is 11 1001 1001 1001. When BCD coding is used for bipolar measurements, another bit, a sign bit, is added to the code, and the result is sign-magnitude BCD coding.

3.3
OPERATIONAL AND INSTRUMENTATION AMPLIFIERS

The front end of a data-acquisition system extracts the desired analog signal from a physical parameter by means of a transducer and then amplifies and filters it. An amplifier and filter are critical components in this initial signal processing.

The amplifier must perform one or more of the following functions: boost the signal amplitude, buffer the signal, convert a signal current into a voltage, or extract a differential signal from common-mode noise. To accomplish these functions requires a variety of different amplifier types.

The most popular type of amplifier is an *operational amplifier,* which is a general-purpose gain block with differential inputs. The op-amp may be connected in many different closed-loop configurations, of which a few are shown in Figures 3.5 and 3.6. The gain and bandwidth of the circuits shown depend on the external resistors connected around the amplifier. An operational amplifier is a good choice in general where a single-ended signal is to be amplified, buffered, or converted from current to voltage.

In the case of differential signal processing, the *instrumentation amplifier* is a better choice since it maintains high impedance at both of its differential inputs, and the gain is set by a resistor located elsewhere in the amplifier circuit. One type of instrumentation amplifier circuit is shown in Figure 3.7. Notice that no gain-setting resistors are connected to either of the input terminals. Instrumentation amplifiers have the following important characteristics:

1. High-impedance differential inputs.

2. Low input offset voltage drift.

3. Low input bias currents.

4. Gain easily set by means of one or two external resistors.

5. High common-mode rejection ratio.

Strain gage bridge transducers, RTD bridges, and many other transducers discussed in Chapter 2, have outputs using signal-conditioning circuitry with differential input amplifiers. Today these transducers can be a part of a data-acquisition system. Operational amplifiers, because of their requirement for feedback to the input circuitry, are unable to provide a high-impedance balanced input.

Instrumentation amplifiers eliminate the feedback problem by using a

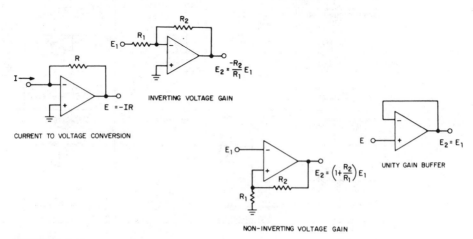

FIGURE 3.5
Operational amplifier configurations. (Courtesy of Datel-Intersil, Inc.)

separate set of terminals for gain setting by a resistor. This allows the inputs to float within a range ultimately restricted by the amplifiers power-supply voltage. Manufacturers now have available new integrated-circuit instrumentation amplifiers to serve transducer output applications via data-conversion systems.

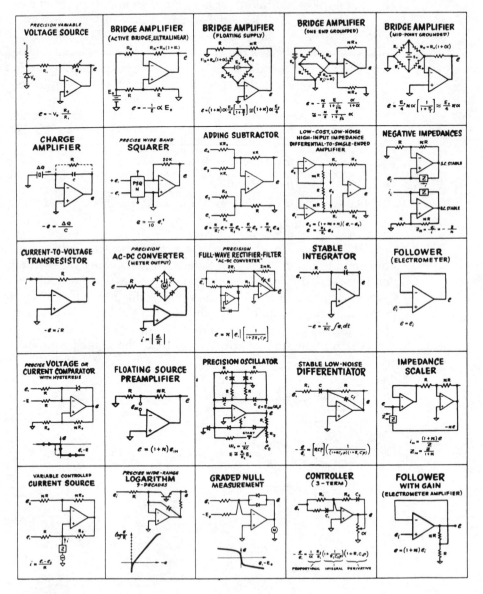

FIGURE 3.6
Operational amplifier applications. (Courtesy of Teledyne Philbrick)

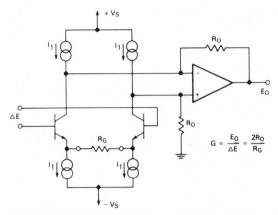

FIGURE 3.7
Simplified instrumentation amplifier circuit.
(Courtesy of Datel-Intersil, Inc.)

3.3.1 Common-mode Rejection

Common-mode rejection ratio is an important parameter of differential amplifiers. An ideal differential input amplifier responds only to the voltage difference between its input terminals and does not respond at all to any voltage that is common to both input terminals (common-mode voltage). In nonideal amplifiers, however, the common-mode input signal causes some output response, even though it is small compared to the response to a differential input signal.

The ratio of differential and common-mode responses is defined as the common-mode rejection ratio. *The common-mode rejection ratio (CMRR) of an amplifier is the ratio of differential voltage gain to common-mode voltage gain and is generally expressed in decibels:*

$$\text{CMRR} = 20 \log_{10} \frac{A_D}{A_{CM}} \tag{3.5}$$

where A_D is the differential voltage gain and A_{CM} is the common-mode voltage gain. CMRR is a function of frequency and therefore also a function of the impedance balance between the two amplifier input terminals. At even moderate frequencies, CMRR can be significantly degraded by small unbalances in the source series resistance and shunt capacitance.

3.3.2 Other Amplifier Types

Several other special amplifiers are useful in conditioning the input signal in a data-acquisition system. An *isolation amplifier* is used to amplify a differential signal, which is superimposed on a very high common-mode voltage, perhaps several hundred or even several thousand volts. The isolation amplifier has the characteristics of an instrumentation amplifier with a very high common-mode input voltage capability.

Another special amplifier, the *chopper stabilized amplifier*, is used to accurately amplify microvolt-level signals to the required amplitude. This amplifier employs a special switching stabilizer that gives extremely low input offset voltage drift. Another useful device, the *electrometer amplifier*, has ultralow input bias currents, generally less than 1 picoampere (pA) and is used to convert extremely small signal currents into a high-level voltage.

3.4
FILTERS

A *low-pass filter* frequently follows the signal-processing amplifier to reduce signal noise. Low-pass filters are used for the following reasons: to reduce man-made electrical interference noise, to reduce electronic noise, and to limit the bandwidth of the analog signal to less than half the sampling frequency in order to eliminate frequency folding. When used for the last reason, the filter is called a *presampling filter* or *antialiasing filter*.

Man-made electrical noise is generally periodic, as for example in power line interference, and is sometimes reduced by means of a special filter such as a *notch filter*. Electronic noise, on the other hand, is random noise with noise power proportional to bandwidth and is present to transducer resistances, circuit resistances, and in amplifiers themselves. It is reduced by limiting the bandwidth of the system to the minimum required to pass desired signal components.

No filter does a perfect job of eliminating noise or other undesirable frequency components, and therefore the choice of a filter is always a

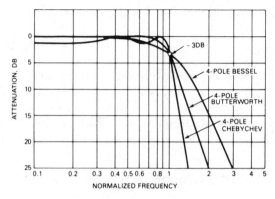

FIGURE 3.8
Some practical low-pass filter characteristics.
(Courtesy of Datel-Intersil, Inc.)

compromise. Ideal filters, frequently used as analysis examples, have flat passband response with infinite attenuation at the cutoff frequency, but are mathematical filters only and not physically realizable.

In practice, the systems engineer has a choice of cutoff frequency and attenuation rate. The attenuation rate and resultant phase response depend on the particular filter characteristic and the number of poles in the filter functions. Some of the more popular filter characteristics include Butterworth, Chebychev, Bessel, and elliptic. In making this choice, the effect of overshoot and nonuniform phase delay must be carefully considered. Figure 3.8 illustrates some practical low-pass filter response characteristics.

Passive *RLC* filters are seldom used in signal-processing applications today owing chiefly to the undesirable characteristics of inductors. Active filters are generally used now since they permit the filter characteristics to be accurately set by precision, stable resistors and capacitors. Inductors, with their undesirable saturation and temperature drift characteristics, are thereby eliminated. Also, because active filters use operational amplifiers, the problems of insertion loss and output loading are eliminated.

3.5
SETTLING TIME

A parameter that is specified frequently in data acquisition and distribution systems is *settling time*. The term *settling time* originates in control theory, but it is now commonly applied to amplifiers, multiplexers, and D/A converters.

Settling time is defined as the time elapsed from the application of a full-scale step input to a circuit to the time when the output has entered and remained within a specified error band around its final value. The method of application of the input step may vary depending on the type of circuit, but the definition still holds. In the case of a D/A converter, for example, the step is applied by changing the digital input code, whereas in the case of an amplifier the input signal itself is a step change.

The importance of settling time in a data-acquisition system is that certain analog operations must be performed in sequence, and one operation may have to be accurately settled before the next operation can be initiated. Thus a buffer amplifier preceding an A/D converter must have accurately settled before the conversion can be initiated.

Settling time for an amplifier is illustrated in Figure 3.9. After application of a full-scale step input, there is a small delay time, following which the amplifier output slews, or changes at its maximum rate. *Slew rate* is determined by internal amplifier currents that must charge internal capacitances. As the amplifier output approaches final value, it may first overshoot and then reverse and undershoot this value before finally entering and remaining within the specific error band. Note that settling time is measured to the point at which the amplifier output *enters* and *remains* within

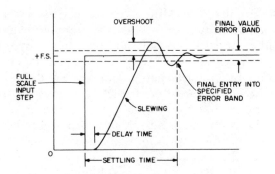

FIGURE 3.9
Amplifier settling time. (Courtesy of Datel-Intersil, Inc.)

the error band. This error band in most devices is specified to either ±0.1% or ±0.01% of the full-scale transition.

3.5.1 Amplifier Characteristics

Settling time, unfortunately, is not readily predictable from other amplifier parameters such as bandwidth, slew rate, or overload recovery time, although it depends on all of these. It is also dependent on the shape of the amplifier open-loop gain characteristics, its input and output capacitance, and the dielectric absorption of any internal capacitances. An amplifier must be specifically designed for optimized settling time, and settling time is a parameter that must be determined by testing.

One of the important requirements of a fast-settling amplifier is that it have a single-pole open-loop gain characteristic, that is, one that has a smooth 6-dB/octave gain roll-off characteristic to beyond the unity gain crossover frequency. Such a desirable characteristic is shown in Figure 3.10.

It is important to note that an amplifier with a single-pole response can never settle faster than the time indicated by the number of closed-loop time constants to the given accuracy. Figure 3.11 shows the output error as a function of the number of time constants τ, where

$$\tau = \frac{1}{2\pi f} \tag{3.6}$$

and f is the closed-loop 3-dB bandwidth of the amplifier.

Actual settling time for a good-quality amplifier may be significantly longer than that indicated by the number of closed-loop time constants due to slew-rate limitation and overload recovery time. For example, an amplifier with a closed-loop bandwidth of 1 MHz has a time constant of 160 ns,

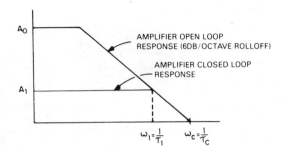

FIGURE 3.10
Amplifier single-pole loop gain characteristic.
(Courtesy of Datel-Intersil, Inc.)

which indicates a settling time of 1.44 μs (9 time constants) to 0.01% of final value. If the slew rate of this amplifier is 1 V/μs, it will take more than 10 μs to settle to 0.01% for a 10-V change.

If the amplifier has a nonuniform gain roll-off characteristic, then its settling time may have one of two undesirable qualities. First, the output may reach the vicinity of the error band quickly, but then take a long time to actually enter it. Second, it may overshoot the error band and then oscillate back and forth through it before entering and remaining inside it.

Modern fast-settling operational amplifiers come in many different types, including modular, hybrid, and monolithic amplifiers. Such amplifiers have settling times of 0.1% or 0.01% of 2 μs down to 100 ns and are useful in data acquisition and conversion applications. Figure 3.12 is an example of an ultrafast settling operational amplifier of the hybrid type.

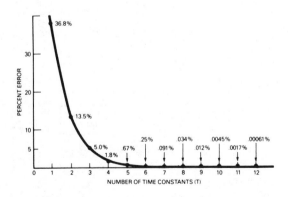

FIGURE 3.11
Output settling error as a function of number of time constants. (Courtesy of Datel-Intersil, Inc.)

FIGURE 3.12
Ultrafast settling hybrid operational amplifier.
(Courtesy of Datel-Intersil, Inc.)

3.6
ANALOG MULTIPLEXER OPERATION

Analog multiplexers are the circuits that time share an A/D converter among a number of different analog channels. Since the A/D converter in many cases is the most expensive component in a data-acquisition system, multiplexing analog inputs to the A/D is an economical approach. Usually, the analog multiplexer operates into a sample-hold circuit that holds the required analog voltage long enough for A/D conversion.

As shown in Figure 3.13, an analog multiplexer consists of an array of

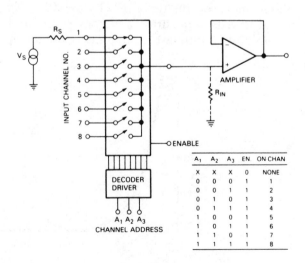

FIGURE 3.13
Analog multiplexer circuit. (Courtesy of Datel-Intersil, Inc.)

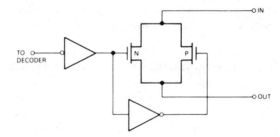

FIGURE 3.14
CMOS analog switch circuit. (Courtesy of Datel-
Intersil, Inc.)

parallel electronic switches connected to a common output line. Only one switch is turned on at a time. Popular switch configurations include 4, 8, and 16 channels, which are connected in single (single-ended) or dual (differential) configurations.

The multiplexer also contains a decoder-driver circuit that decodes a binary input word and turns on the appropriate switch. This circuit interfaces with standard transistor-transistor-logic (TTL) inputs and drives the multiplexer switches with the proper control voltages. For the 8-channel analog multiplexer shown, a one-of-eight decoder circuit is used.

Most analog multiplexers today employ the CMOS switch circuit shown in Figure 3.14. A CMOS driver controls the gates of parallel-connected *P*-channel and *N*-channel MOSFETs. Both switches turn on together, with the parallel connection giving relatively uniform on-resistance over the required analog input voltage range. The resulting on-resistance may vary from about 50 Ω to 2 kΩ, depending on the multiplexer; this resistance

FIGURE 3.15
A group of monolithic CMOS analog multiplexers.
(Courtesy of Datel-Intersil, Inc.)

increases with temperature. A representative group of monolithic CMOS analog multiplexers is shown in Fig. 3.15.

3.6.1 Analog Multiplexer Characteristics

Because of the series resistance, it is common practice to operate an analog multiplexer into a very high load resistance such as the input of a unity-gain buffer amplifier. The load impedance must be large compared with the switch on-resistance and any series source resistance in order to maintain high transfer accuracy. *Transfer error* is the input-to-output error of the multiplexer with the source and load connected; error is expressed as a percent of input voltage.

Transfer errors of 0.1% to 0.01% or less are required in most data-acquisition systems. This is readily achieved by using operational amplifier buffers with typical input impedances from 10^8 to 19^{12} Ω. Many sample-hold circuits also have very high input impedances.

Another important characteristic of analog multiplexers is *break-before-make* switching. There is a small time delay between disconnection from the previous channel and connection to the next channel, which assures that two adjacent input channels are never instantaneously connected together.

Settling time is another important specification for analog multiplexers; it has the same definition previously given for amplifiers except that it is measured from the time the channel is switched on.

Throughput rate is the highest rate at which a multiplexer can switch from channel to channel with the output settling to its specified accuracy.

Crosstalk is the ratio of output voltage to input voltage with all channels connected in parallel and off; it is generally expressed as an input-to-output attenuation ratio in decibels.

As shown in the representative equivalent circuit of Figure 3.16, analog multiplexer switches have a number of leakage currents and capacitances

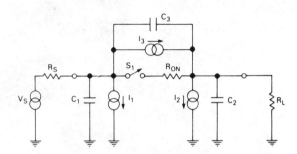

FIGURE 3.16

Equivalent circuit of analog multiplexer switch.
(Courtesy of Datel-Intersil, Inc.)

associated with their operation. These parameters are specified on data sheets and must be considered in the operation of the devices. Leakage currents, generally in picoamperes at room temperature, become trouble-some only at high temperatures. Capacitances affect crosstalk and settling time of the multiplexer.

3.6.2 Analog Multiplexer Applications

Analog multiplexers are employed in two basic types of operation: high level and low level. In *high-level multiplexing*, the most popular type, the analog signal is amplified to the 1- to 10-V range ahead of the multiplexer. This has the advantage of reducing the effects of noise on the signal during the remaining analog processing. In *low-level multiplexing*, the signal is amplified after multiplexing; therefore, great care must be exercised in handling the low-level signal up to the multiplexer. Low-level multiplexers generally use two-wire differential switches in order to minimize noise pickup. Reed relays, because of essentially zero series resistance and the absence of switching spikes, are frequently employed in low-level multi-plexing systems. They are also useful for high common-mode voltages.

A useful specialized analog multiplexer is the *flying-capacitor* type. This circuit, shown as a single channel in Figure 3.17 has differential inputs and is particularly useful with high common-mode voltages. The capacitor connects first to the differential analog input, charging up to the input voltage, and is then switched to the differential output, which goes to a high-input-impedance instrumentation amplifier. The differential signal is therefore transferred to the amplifier input without the common-mode volt-age and is then further processed up to A/D conversion.

To realize large numbers of multiplexed channels, you can connect analog multiplexers in parallel using the enable input to control each de-vice. This is called *single-level multiplexing*. You can also connect the output of several multiplexers to the inputs of another to expand the number of channels; this method is *double-level multiplexing*.

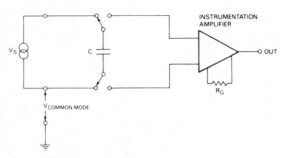

FIGURE 3.17
Flying-capacitor multiplexer switch. (Courtesy of Datel-Intersil, Inc.)

3.7
OPERATION OF SAMPLE-HOLD CIRCUITS

Sample-hold circuits, discussed earlier, are the devices that store analog information and reduce the aperture time of an A/D converter. A sample-hold is simply a voltage-memory device in which an input voltage is acquired and then stored on a high-quality capacitor. A popular circuit is shown in Figure 3.18.

A_1 is an input buffer amplifier with a high input impedance so that the source, which may be an analog multiplexer, is not loaded. The output of A_1 must be capable of driving the hold capacitor with stability and enough drive current to charge it rapidly. S_1 is an electronic switch, generally an FET, which is rapidly switched on or off by a driver circuit that interfaces with TTL inputs.

C is a capacitor with low leakage and low dielectric absorption characteristics; it is a polystyrene, polycarbonate, polypropylene, or Teflon type. In the case of hybrid sample-holds, the MOS-type capacitor is frequently used.

A_2 is the output amplifier that buffers the voltage on the hold capacitor. It must therefore have extremely low input bias current, and for this reason an FET input amplifier is required.

There are two modes of operation for a sample-hold: *sample* (or tracking) *mode*, when the switch is closed; and *hold mode*, when the switch is open. Sample-holds are usually operated in one of two basic ways. The device can continuously track the input signal and be switched into the hold mode only at certain specified times, spending most of the time in tracking mode. This is the case for a sample-hold employed as a deglitcher at the output of a D/A converter, for example.

Alternatively, the device can stay in the hold mode most of the time and go to the sample mode just to acquire a new input signal level. This is the case for a sample-hold used in a data-acquisition system following the multiplexer.

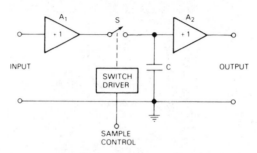

FIGURE 3.18
Popular sample-hold circuit. (Courtesy of Datel-Intersil, Inc.)

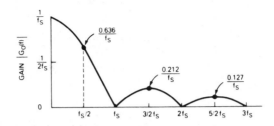

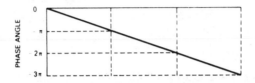

FIGURE 3.19
Gain and phase components of zero-order hold
transfer function. (Courtesy of Datel-Intersil, Inc.)

3.7.1 Sample-hold as a Data-Recovery Filter

A common application for sample-hold circuits is *data-recovery, or signal-reconstruction, filters.* The problem is to reconstruct a train of analog samples into the original signal; when used as a recovery filter, the sample-hold is known as a *zero-order hold.* It is a useful filter because it fills in the space between samples, providing data smoothing.

As with other filter circuits, the gain and phase components of the transfer function are of interest. By an analysis based on the impulse response of a sample-hold and use of the Laplace transform, the transfer function is found to be

$$G_o(f) = \frac{1}{f_s}\left[\frac{\sin \pi \, (f/f_s)}{\pi \, (f/f_s)}\right] \epsilon^{-j\pi f/f_s} \tag{3.7}$$

where f_s is the sampling frequency. This function contains the familiar $(\sin x)/x$ term plus a phase term, both of which are plotted in Figure 3.19.

The sample-hold is therefore a low-pass filter with a cutoff frequency slightly less than $f_s/2$ and a linear phase response that results in a constant delay time of $T/2$, where T is the time between samples. Notice that the gain function also has significant response lobes beyond f_s. For this reason, a sample-hold reconstruction filter is frequently followed by another conventional low-pass filter.

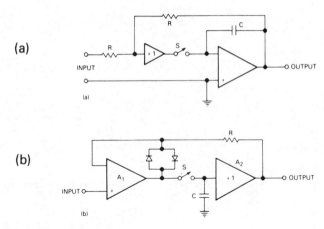

(a)

(b)

FIGURE 3.20
Two closed-loop sample-hold circuits. (Courtesy of Datel-Intersil, Inc.)

3.7.2 Other Sample-Hold Circuits

In addition to the basic circuit of Figure 3.18, several other sample-hold circuit configurations are frequently used. Figure 3.20 shows two such circuits, which are closed-loop circuits, as contrasted with the open-loop circuit of Figure 3.18. Figure 3.20(a) uses an operational integrator and another amplifier to make a fast, accurate inverting sample-hold. A buffer amplifier is sometimes added in front of this circuit to give high input impedance. Figure 3.20(b) shows a high input impedance noninverting sample-hold circuit.

FIGURE 3.21
Ultrafast sample-hold modules that employ diode-bridge switches. (Courtesy of Datel-Intersil, Inc.)

The circuit in Figure 3.18, although generally not as accurate as those in Figure 3.20, can be used with a diode-bridge switch to realize ultrafast acquisition sample-holds, such as those shown in Figure 3.21.

3.7.3 Sample-Hold Characteristics

A number of parameters are important in characterizing sample-hold performance. Probably most important of these is *acquisition time*. The definition is similar to that of settling time for an amplifier. It is the time required, after the sample-command is given, for the hold capacitor to change to a full-scale voltage charge and remain within a specified error band around final value.

Several hold-mode specifications are also important. *Hold-mode droop* is the output voltage change per unit time when the sample switch is open. This droop is caused by the leakage currents of the capacitor and switch and the output amplifier bias current. *Hold-mode feedthrough* is the percentage of input signal transferred to the output when the sample switch is open. It is measured with a sinusoidal input signal and caused by capacitive coupling.

The most critical phase of sample-hold operation is the transition from the sample mode to the hold mode. Several important parameters characterize this transition. *Sample-to-hold offset* (or step) *error* is the change in output voltage from the sample mode to the hold mode, with a constant input voltage. It is caused by the switch transferring charge onto the hold capacitor as it turns off.

Aperture delay is the time elapsed from the hold command to when the switch actually opens; it is generally much less than a microsecond. *Aperture uncertainty* (or *aperture jitter*) is the time variation, from sample to sample, of the aperture delay. It is the limit on how precise is the point in time of opening the switch. Aperture uncertainty is the time used to determine the aperture error due to rate of change of the input signal.

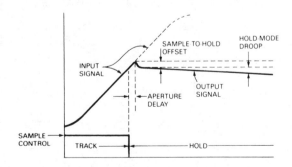

FIGURE 3.22
Some sample-hold characteristics. (Courtesy of Datel-Intersil, Inc.)

Several of these specifications are illustrated in the diagram of Figure 3.22.

Sample-hold circuits are simple in concept, but generally difficult to fully understand and apply. Their operation is full of subtleties, and they must therefore be carefully selected and then tested in a given application.

3.8
REVIEW QUESTIONS

1. Draw a block diagram of a data-acquisition system. Discuss briefly the main function of each box.

2. Discuss the meaning of binary coded decimal and least significant bit.

3. Discuss the uses of an operational and instrumentation amplifier. Draw an operational amplifier voltage comparator and an instrumentation amplifier.

4. Given:

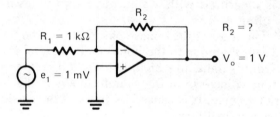

Question: Find R_2.

5. Given:

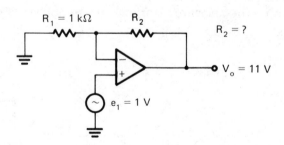

Find R_2.

6. Discuss the use of an isolation amplifier, chopper-stabilized amplifier, and electrometer amplifier. Draw an operational amplifier to represent these amplifiers.

7. Discuss the purpose of a low-pass filter for data-acquisition systems.

8. Discuss the purpose of setting time.

9. Discuss the analog multiplexer operation with a diagram.

10. Discuss the purpose of a CMOS switch circuit.

11. Discuss the analog multiplexer characteristics.

12. Discuss the analog multiplexer applications with a diagram.

13. Discuss the operation of sample-hold circuits.

14. Discuss the sample hold as a recovery filter.

15. Discuss the characteristics of sample-hold circuits.

3.9
REFERENCES

1. Eugene L. Zuch, *Data Acquisition and Conversion Handbook*, 4th printing, Datel-Intersil, Inc., Mansfield, Mass., Sept. 1981.

2. Daniel J. Dooley, *Data Conversion Integrated Circuits*, IEEE Press Books, New York, 1980.

3. Walter G. Jung, *IC Op-Amp Cookbook*, Howard W. Sams and Co., Indianapolis, 1980.

4. Daniel H. Sheingold, *Transducer Interfacing Handbook*, Analog Devices, Inc., Norwood, Mass., 1981.

5. Charles Wojslaw, *Electronic Concepts, Principles, and Circuits*, Reston Publishing Co., Reston, Va., 1980.

6. Daniel H. Sheingold, Analog-Digital Conversion Notes, Analog Devices, Inc., Norwood, Mass., 1977, 1980.

7. Robert W. Glines, Specifying and Testing Sample-Hold Amplifiers, Bulletin AN-30, Teledyne-Philbrick, Dedham, Mass., Nov. 1980.

8. Dennis Santucci, Designing High Speed Data Acquisition Systems, Bulletin AN-21, Teledyne-Philbrick, Dedham, Mass., April 1978.

9. Robert W. Glines, Specifying and Testing Multiplexers, Bulletin AN-31, Teledyne-Philbrick, Dedham, Mass., May 1977.

10. F. Goodenough, Operational Amplifier Parameter Definition and Measurement Guide, Bulletin AN-23, June 1976, Teledyne-Philbrick, Dedham, Mass., June 1976.

11. Bernard M. Gordon, *The Analogic Data-Conversion Systems Digest*, 4th ed., Analogic Corp., Wakefield, Mass., 1981.

12. Data Conversion/Acquisition Databook, National Semiconductor Corp., Santa Clara, Calif., 1980.

4

Analog-to-Digital Converters

4.1
INTRODUCTION

Conversion of the analog signals to digital words can be accomplished in several ways, including integration, counting, successive approximation, and parallel conversion. While each technique has its advantages, successive approximation and dual-slope integration are the most popular A/D conversion methods. Other important types include servo loop tracking, synchro-to-digital (S/D), and voltage-to-frequency (V/F) converters.

A/D construction methods range from integrated circuits, both monolithic and hybrid, to high-performance modules. Conversion methods can vary from manufacturer to manufacturer and are being improved rapidly. One or more comparators are key elements used in analog-to-digital converter design. Important characteristics of A/D converters include resolution in bits, linearity at 25°C maximum, input options, output code options, maximum conversion time, and temperature coefficient.

Noise is of prime importance in high-resolution A/D converters equal to or greater than 12 bits, where the noise can easily exceed the least significant bit (LSB) value, thereby reducing the useful resolution.

A specification of $\pm\frac{1}{2}$ LSB linearity requires that the sum of either positive or negative errors of the individual bits do not exceed $\frac{1}{2}$ LSB.

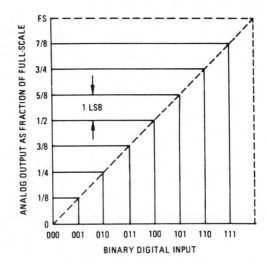

FIGURE 4.1
Transfer characteristic of 3-bit A/D converter.
(Courtesy of ILC Data Device Corp.)

4.2
A/D CONVERTER TERMINOLOGY AND PARAMETERS[a]

Although A/D and D/A converters perform inverse operations, their input–output or transfer characteristics, shown in Figures 4.1 and 4.2, are not exactly inverse. In the D/A converter each digital input produces a single value of the analog output, and all the input–output points fall on a straight line; but in the A/D converter any analog value within $\pm\frac{1}{2}$ LSB of the ideal value will give the same digital output, so the digital output is a step approximation to the ideal straight-line relationship. A plot of quantizing error versus analog input is shown at the top of Figure 4.2. This plot shows that the quantizing error varies cyclically in a sawtooth fashion from $+\frac{1}{2}$ LSB through 0 to $-\frac{1}{2}$ LSB. The rms value of that sawtooth is

$$\text{rms error} = \frac{1/2 \text{ LSB}}{2\sqrt{3}} = \frac{\text{LSB}}{4\sqrt{3}} \tag{4.1}$$

The maximum quantizing error of $\pm\frac{1}{2}$ LSB (called the *quantizing uncertainty*) is inherent in the discrete conversion process, and can only be reduced by reducing the size of the LSB. Since LSB $= \frac{1}{2}^N$, the size of the

*[a] Sections 4.2 through 4.4 and 4.6 through 4.9 are used courtesy of the ILC Data Device Corp., Bohemia, N.Y. and is reprinted with their permission.

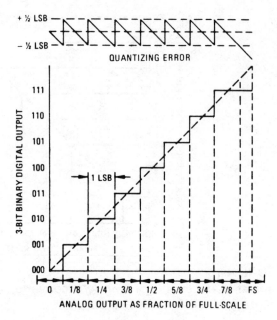

FIGURE 4.2
Transfer characteristic of 3-bit A/D converter.
(Courtesy of ILC Data Device Corp.)

LSB is reduced by increasing N, the number of bits. The theoretical resolution of an N-bit converter is 2^N, which means that, ideally, a D/A converter will be able to generate 2^N discrete levels of analog output voltage, and an A/D converter will be able to respond with separate digital code numbers to 2^N discrete levels of analog input voltage. If internal noise and/or drift in a converter exceeds quantizing step levels, missing codes will occur, and the actual resolution will be less than the theoretical resolution.

Monotonicity in a converter means that the output continuously increases with the input. A monotonic A/D converter will generate successively increasing codes as the analog input increases, and a monotonic D/A converter will generate a continuously increasing analog output voltage with increasing values of the digital input code.

The linearity of a converter is defined as the maximum deviation of the transfer function from a best straight line through that transfer function. The relative accuracy of a converter is the maximum deviation of the analog input or output voltage from a straight line referred to the source of reference voltage, V_{REF}. Absolute accuracy is the maximum deviation of the analog input or output voltage from a straight line referred to the NBS absolute volt.

The major types of A/D converters employ (in order of increasing speed) counter or ramp, dual-slope integration, successive approximation, serial–parallel, and parallel techniques. The successive-approximation technique has the highest performance–cost ratio in the present state of the art, and permits a wider range of trade-offs between speed and resolution than any other type. Each A/D converter described in this section is a successive-approximation type, offering a high resolution–speed product at low cost. Video types are parallel and two step (serial–parallel).

Figure 4.3 is a simplified block diagram of a successive-approximation A/D converter. The analog input voltage, V_{in}, to be digitized and the output voltage, V_{da}, of a D/A converter are fed to a comparator. The comparator output is fed to a control logic circuit that drives a register, which in turn controls the D/A converter and thus determines the voltage V_{da}. The closed loop thus formed operates by successive approximations to reduce the difference between the D/A output voltage V_{da} and the analog input voltage V_{in} to less than the value of the LSB. In this condition, a number represented by the digital output of the register is equal (within the value of the LSB) to the analog input to the comparator.

The system operates as follows: a start conversion command clears the register and converter. Bit 1 (the MSB) is set to 1, and its weight, $\frac{1}{2}$ full scale (FS), is compared to the analog input V_{in}. If V_{in} is greater than the MSB, the MSB remains at 1, and bit 2 ($\frac{1}{4}$ FS) is set to 1. If the combined weights of bits 1 and 2 ($\frac{1}{2}$ FS + $\frac{1}{4}$ FS = $\frac{3}{4}$ FS) now exceed V_{in}, bit 2 is set to 0, and bit 3 ($\frac{1}{8}$ FS) is set to 1. Any succeeding bit that does not cause the combined bit weight to exceed V_{in}, remains set at 1, while any succeeding bit that causes the combined bit weight to exceed V_{in} is set to 0. When the LSB has been tried,

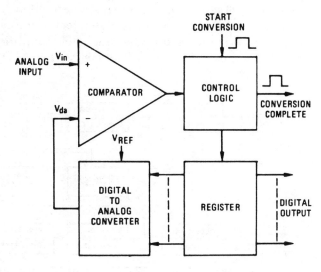

FIGURE 4.3
Simplified block diagram of successive approximation A/D converter. (Courtesy of ILC Data Device Corp.)

the control logic transmits a conversion complete signal. The output of the register at that time is a digital number representing the quantized analog voltage V_{in} as a fraction of the D/A reference voltage V_{REF}. There is one value of V_{REF} for which the digital output represents V_{in} exactly in volts, so the A/D converter is a digital voltmeter. However, if V_{REF} is a separate input variable, the A/D converter becomes a ratiometric converter, the digital output of which represents the fraction V_{in}/V_{REF}.

Applications of successive approximation A/D converters include the following:

1. High-speed data-acquisition systems.
2. Pulse-code modulation systems.
3. Waveform sampling and digitizing.
4. Automatic test systems for productions.
5. Digital process control systems.

The speed of a successive approximation converter is limited by the need to wait for each bit to settle for each trial. At the cost of increased complexity, it is possible to use a two-step technique, which can achieve more than twice the speed. In a two-step process, the bits are tried in two groups, called the coarse bits and the fine bits. Each group of bits is tried with its own A/D converter, and each A/D requires less resolution (fewer bits). A two-step conversion process is faster than single-stage conversion, not only because a new conversion can be initiated while the fine bits are being tried, but also because the settling times for each bit are reduced when an A/D converter requires less accuracy.

Figure 4.4 shows how the two-step technique can be implemented. The analog input is sampled by the track-hold amplifier and held for the first part of the conversion cycle. The coarse bits A/D converter digitizes the

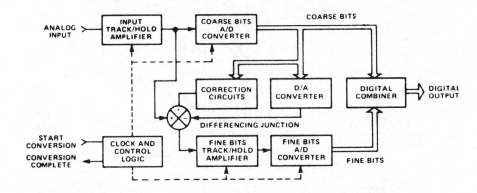

FIGURE 4.4
Two-step A/D conversion. (Courtesy of ILC Data Device Corp.)

most significant bits at a relatively fast rate and the coarse bits are stored in the digital combiner. The analog voltage corresponding to the coarse bits which is formed by the D/A converter is subtracted from the analog input signal being held by the input track-hold amplifier. Correction circuits are usually required at the differencing junction to compensate for nonlinearities in the A/D and D/A, track-hold errors, and temperature drifts. With a PROM these errors, which are highly repeatable, can be uniquely corrected for each converter.

As soon as the summed voltage from the differencing junction has been acquired by the fine bits track-hold amplifier, the input track-hold can acquire a fresh analog input voltage to commence a new conversion cycle. Meanwhile, the fine bits A/D converter digitizes the fine bits, and the digital combiner accepts them to complete the first conversion cycle. The entire two-step conversion cycle is regulated by a clock with control logic, and a start-conversion pulse initiates each conversion. An important advantage of two-step conversion is that the ultimate accuracy of conversion is determined by the accuracy of the subtraction process, which involves the D/A and track and hold amplifier. These circuits need to settle only once during each conversion cycle. The internal coarse and fine bit A/D converters may be successive approximation or flash converters.

The flash or parallel quantization method, which is illustrated in Figure 4.5, is the most rapid A/D conversion technique. A string of comparators is used to compare the analog input voltage directly with voltage steps on a resistance ladder. In Figure 4.5, for instance, if the analog input voltage lies

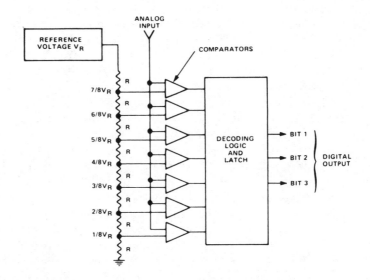

FIGURE 4.5
Three-bit flash conversion. (Courtesy of ILC Data Device Corp.)

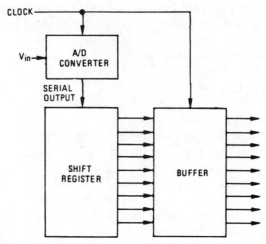

FIGURE 4.6
Conversion of A/D serial output to parallel output. (Courtesy of ILC Data Device Corp.)

between $\frac{3}{8}V_R$ and $\frac{4}{8}V_R$, the lower three comparators would respond and the upper four would not. The decoding logic translates the comparator responses into appropriately coded digital output.

Flash quantization is rapid because signals propagate very quickly through the parallel-connected comparators and the decoding logic. In contrast, the throughput delay in a successive approximation converter with n bits includes n D/A settling times and n comparator response times. On the other hand, the complexity of a flash quantization converter is greater because the number of comparators required and the complexity of the decoding logic depend on the number of bits. A resolution of n bits will require 2^n resistors and $2^n - 1$ comparators. The 3-bit converter shown in Figure 4.5 requires $2^3 = 8$ resistors and $2^3 - 1 = 7$ comparators. A converter with 6-bit resolution requires $2^6 = 64$ resistors and $2^6 - 1 = 63$ comparators.

Since the A/D conversion process takes time, the digital output will be in error if the analog input voltage changes during the conversion. A sample-hold circuit (S/H) is therefore often employed at the analog input to sample V_{in} before conversion and to hold the A/D input at the value of V_{in} that existed at the instant of the start-conversion command, until conversion is complete. The conversion-complete signal may then be used to return the S/H to its sampling mode. If the converter is a ratiometric type with variable V_{REF}, it will be necessary to apply a S/H to that input, too.

The digital output of an A/D converter may appear simultaneously on N lines in parallel, or serially, on one line. Where it is necessary to convert such serial data to equivalent parallel data, the arrangement of Figure 4.6 may be used. Note that serial data may be transmitted in either of two ways. In return-to-zero (RZ) transmission, the levels return to ground between successive bits; in non-return-to-zero (NRZ), the levels change only when the leading edge of a clock pulse is present.

4.3
GENERAL APPLICATIONS OF A/D CONVERTERS

The fundamental application of A/D converters is the conversion of analog signals to digital form (i.e., digitizing), and this application is included in all the following applications of A/D converters.

Figure 4.7 shows how an A/D converter with an internal reference, V_{REF}, and a multiplying D/A converter can be used to multiply two analog voltages with excellent precision. If both are 12-bit devices, the overall accuracy will be on the order of 0.1%. Input V_1 is fed to the A/D converter, the digital output of which corresponds to V_1/V_{REF}.

This digital number and input V_2 are fed to the multiplying D/A converter, which generates the analog output voltage V_1V_2/V_{REF}.

Figure 4.8 shows a similar arrangement, employing a ratiometric A/D converter dividing an analog input V_1 by a second input V_2.

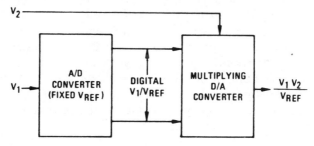

FIGURE 4.7
Precision multiplication of analog voltages.
(Courtesy of ILC Data Device Corp.)

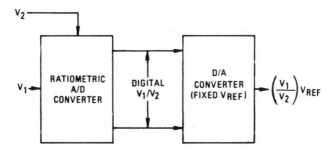

FIGURE 4.8
Precision division of analog voltages. (Courtesy of ILC Data Device Corp.)

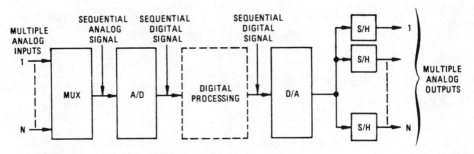

FIGURE 4.9
Use of MUXs for multiplexing and distribution.
(Courtesy of ILC Data Device Corp.)

4.4
MULTIPLEXERS (MUX)

When it is required to perform A/D conversion on multiple analog inputs, a separate A/D converter may be employed for each, or an analog MUX (discussed in Chapter 3) may be used to sample the analog inputs and feed the sampled analog data sequentially to a single A/D converter, which develops a corresponding sequential digital output. Conversely, an analog MUX may be used to distribute the sequential analog output of a D/A converter to multiple analog output channels. Figure 4.9 shows both processes. Although not shown, means must be provided for relating each segment of the sequential data to the corresponding analog channel so that data from each analog input channel is always distributed to the same-numbered analog output channel.

4.4.1 Multiplexer Terminology and Parameters

The transfer accuracy of any channel of a MUX is a measure of the difference between the input voltage, e_{in}, in the output voltage e_o:

$$\text{Transfer accuracy} = \frac{100\,(e_{in} - e_o)}{e_{in}}\% \tag{4.2}$$

Throughput rate is the highest rate at which the MUX will switch from channel to channel while maintaining a specified transfer accuracy. Throughput rate is the inverse of the sum of switching and settling times, as follows:

$$\text{Throughput rate} = \frac{1}{\text{switching time} + \text{settling time}} \tag{4.3}$$

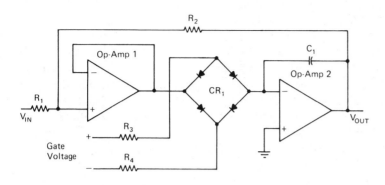

$$t \approx \cfrac{C_1}{\left(\cfrac{I_R + I_B}{\Delta_{OUT}}\right)} \qquad C \approx \frac{t(I_R + I_B)}{\Delta_{OUT}}$$

$I_R \approx CR_1$ Diode Leakage

$I_B \approx$ Op-Amp 2 Input Bias

$\Delta_{OUT} \approx$ Difference in Output Voltage

Gate Voltage $\approx$ 1.5 V + Max V_{IN}

$R_1 \geqslant \dfrac{V_{IN} \times 0.1}{\text{Input Bias of Op-Amp 1}}$

$R_2 = R_1 \times$ Gain $\qquad R_3 = R_4 = 1\ k\Omega$

FIGURE 4.10

Track-hold (sample-hold) amplifier using op-amp.
(From John D. Lenk, *Manual for Operational Amplifier Users*, Reston Publishing Co., Reston, Va., 1976)

Crosstalk attenuation is a measure of the MUX's ability to prevent signals at the "off" inputs from appearing at the output. When a test voltage e_{test} of specified amplitude and frequency is applied to all the "off" channel inputs, and the output voltage e_o is measured, the crosstalk attenuation in decibels is given by

$$\text{Crosstalk attenuation (dB)} = 20 \log_{10}\left(\frac{e_{test}}{e_o}\right) \tag{4.4}$$

4.5
SAMPLE-HOLD AND TRACK-HOLD
MODULES [b]

Where analog inputs vary significantly during multiplexing or during A/D conversion, the multiplexed or digitized outputs may contain large errors. Such errors may be avoided by use of sample-hold or track-hold techniques.

Figure 4.10 is the working schematic of two op-amps used as a track and hold (T/H) amplifier, also known as a sample and hold circuit. With this circuit, when two voltages are applied to the gate inputs, the diode bridge conducts and the output voltage tracks the input voltage. In this case, V_{out} equals V_{in} and follows the variation of V_{in}. However, when the polarity of V_{out} is reversed from that of V_{in}, an op-amp connected as a unity gain amplifier at the output is used.

A sample or track command on the control input causes the S/H to

[b] John D. Lenk, *Manual for Operations Amplifier Users*, Reston Publishing Co., Reston, Va., 1976.

acquire the analog input rapidly and to follow it closely; so in the sample or track mode the S/H output tracks the varying analog input. When a hold command appears at the control input, the S/H stores the analog input value that existed at the instant of the hold command, and the S/H output remains constant at the stored value until the next sample or track command. If the S/H is held in the sample or track mode most of the time, the module may be designated a track-hold, or T/H module. S/H and T/H modules usually have unity gain without inversion. The control inputs are usually designed to accept standard transistor-transistor logic (TTL) levels, where logic 1 is one command and logic 0 is the other command.

4.5.1 Sequential versus Simultaneous Sampling

Where analog data are to be sampled-held, multiplexed, and converted to digital data, the S/H process can be performed either before or after multiplexing. One S/H can be used at the output of the MUX, to sample the data after multiplexing, as shown in Figure 4.11, or one S/H can be used in each input channel of the MUX to sample the data before multiplexing (Figure 4.12).

The process shown in Figure 4.11 is inherently sequential in that the MUX output represents successive inputs at successive times. The sampling process of Figure 4.12 can be either sequential or simultaneous, depending on how the sample commands are applied to the S/H amplifier. If the sample commands are applied simultaneously to all the S/H units,

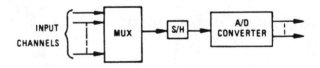

FIGURE 4.11
Sequential sampling after multiplexing. (Courtesy of ILC Data Device Corp.)

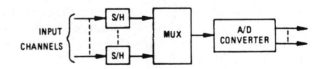

FIGURE 4.12
Simultaneous sampling before multiplexing. (Courtesy of ILC Data Device Corp.)

then the multiplexed output will represent the inputs as of the same instant. Such *time coherence* of multiplexed data is important in certain applications. Note that simultaneous sampling permits an inherently greater conversion rate than sequential sampling, because the S/H settling time in the simultaneous system is only that of any one S/H, whereas the total S/H settling time in the sequential system is the S/H settling multiplied by the number of channels.

4.6
VIDEO A/D CONVERTERS

The speed of analog-to-digital conversion can be greatly increased by employing $(2^N - 1)$ comparators, which simultaneously measure the analog input voltage, V_{in}. Successive comparators are biased in increments of $V_{REF}/2^N$ (= 1 LSB).

Figure 4.13 is a simplified diagram of a 4-bit parallel A/D converter employing 15 (= $2^4 - 1$) comparators. The negative inputs are fed from a resistance divider connected to reference voltage V_{REF} so that the reference voltage for comparator 1 is $\frac{1}{16} V_{REF}$ (= 1 LSB), the reference voltage for comparator 2 is $\frac{2}{16} V_{REF}$ (= 2 LSB), . . . , and the reference voltage for comparator 15 is $\frac{15}{16} V_{REF}$ (= FS − LSB). The input voltage, V_{in}, to be quantized, is fed simultaneously to the (+) inputs of all the comparators. The output of any comparator is 0 when its (+) input is less than its reference input, and is 1 when its (+) input exceeds its reference input. Thus, if V_{in} is less than $\frac{1}{16} V_{REF}$, the outputs of all the comparators are 0s, and if V_{in} is greater than $\frac{15}{16} V_{REF}$, the outputs of all the comparators are 1s. At any intermediate value

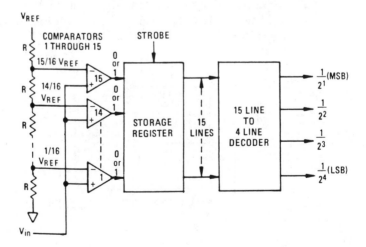

FIGURE 4.13
Four-bit parallel A/D converter. (Courtesy of ILC Data Device Corp.)

of V_{in}, the outputs of all comparators with reference voltages less than V_{in} will be 1s, and the outputs of all comparators with reference voltages greater than V_{in} will be 0s. The 15 comparator outputs, which change progressively from 0s to 1s, constitute an unweighted binary code, which can be represented with equal resolution by a 4-bit 2^N-weighted code. The 15-line data are therefore fed to a storage register, which, when strobed, feeds a 15-line to 4-line decoder, the outputs of which represent the bits $2^{-1} (= \text{MSB})$, 2^{-2}, 2^{-3}, and $2^{-4} (= \text{LSB})$.

Since N-bit parallel A/D conversion requires $2^N - 1$ comparators, the number of comparators and the decoder complexity both increase geometrically with resolution. Thus 6-bit resolution would require 63 comparators, 8 bits would require 255 comparators, and 10 bits would require 1,023 comparators. Even if such large numbers of comparators were economically feasible, the problem of driving so many in parallel would remain.

The number of bits can be increased with only a linear increase in the number of comparators by employing parallel A/D converters in tandem. Figure 4.14 is a simplified diagram of an 8-bit serial–parallel A/D converter employing two 4-bit parallel A/D converters of the type shown in Figure 4.13. The analog input voltage, V_{in}, is fed through a sample-hold circuit to the first 4-bit parallel A/D converter, which determines the four most significant bits (2^{-1}, 2^{-2}, 2^{-3}, and 2^{-4}) of the held voltage. These four bits are fed to four of the eight parallel digital output lines and also to a subtracting D/A converter. This subtracting D/A converter generates the analog equivalent of the four most significant bits, and subtracts that analog equivalent from the held value of the analog input voltage, V_{in}. Since the first 4-bit parallel A/D converter quantizes the analog input voltage V_{in} within $\frac{1}{16}V_{REF}$, the output of the subtracting D/A converter will not exceed $\frac{1}{16}V_{REF}$. This output is fed to the second 4-bit parallel A/D converter, the reference voltage of which is $\frac{1}{16}V_{REF}$, and which therefore determines the four least significant bits, 2^{-5}, 2^{-6}, 2^{-7}, and 2^{-8}. A second S/H, before the subtracting

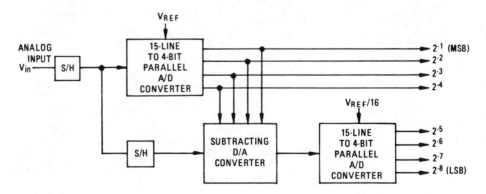

FIGURE 4.14
Eight-bit serial–parallel (two-step) A/D converter.
(Courtesy of ILC Data Device Corp.)

D/A converter, holds the analog input voltage for the second 4-bit conversion, permitting the first S/H and first 4-bit A/D converter to start a new conversion as soon as they complete the previous conversion. This two-step serial–parallel conversion technique employs a total of only 30 comparators to achieve 8-bit resolution, whereas a one-step parallel A/D converter would require 255 comparators.

4.7
VIDEO FREQUENCY MULTIPLEXER

The video-frequency multiplexers (VMUXs) described in this section employ field-effect transistors (FETs) as switches, as shown in the simplified VMUX schematic of Figure 4.15. These FET switches are designed for break-before-make action to eliminate the possibility of introducing transients back into the input signal channels. Each FET switch is controlled by a separate gate input driven by external high-speed TTL circuits and is turned on when the gate is driven to logic 0. Only one input signal channel should be turned on at a time, because if two or more input signal channels are turned on simultaneously, those channels will be connected together

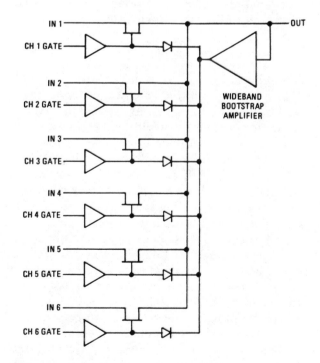

FIGURE 4.15
Simplified schematic of video-frequency MUX.
(Courtesy of ILC Data Device Corp.)

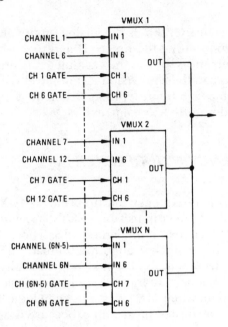

FIGURE 4.16
One-level multiplexing of video-frequency MUXs.
(Courtesy of ILC Data Device Corp.)

through the low ON resistance of the FET switches. Note that, since the FET switches are turned on by logic 0 gate levels, loss of power to the MUX will connect all the signal inputs together. If it is vital to maintain isolation between channels even with power down, some means of maintaining gate voltage must be provided during such down times.

Note that the MUX output is fed back through a wideband amplifier to the gates of the FET switches. This "bootstrap" arrangement forces the gate voltage of the conducting FET to track the drain-source voltage. This maintains V_{GS} at 0, ensuring that the FET ON resistance remains constant and does not vary with the input signal.

Since all FET devices have drain-to-source leakage when OFF, and leakage from gate to both drain and source when ON, excessive settling times will occur unless a path is provided for such leakage currents. This can be accomplished by making sure that one channel is on at all times; if that is not feasible, a 100-MΩ resistor should be connected from output to ground to provide the required return path for the FET leakage currents.

VMUX modules are easily combined in parallel to obtain multiples of six channels by connecting their outputs together, as shown in Figure 4.16. Channels 1 through $6N$ each appear at the common output as gates 1 through $6N$ are each applied. This one-level arrangement is thus equivalent to a single N-channel MUX. The number (N) of MUXs that can be paralleled is limited by leakage current and by output capacitance. The first limitation occurs because leakage currents from the output sides of all $(6N - 1)$ OFF FETs flow through the single ON FET and through the source resistance of the ON channel. The voltage drop across the source resistance of the ON channel caused by the sum of $(6N - 1)$ leakage currents flowing through that source resistance constitutes a static voltage error at the input from the ON channel. Since this voltage error is proportional to the number of channels, there will be some number of MUXs for which the error will exceed acceptable levels. The second limitation of simple paralleling of MUXs is that the output capacitance is the sum of the individual FET output capacitances, and the settling time thus increases with the number of channels.

The limitations described are removed by the two-level multiplexing technique shown in Figure 4.17, which isolates the outputs of the MUXs from each other, and thus limits leakage currents and output capacitance. In the examples, channels 1 to 36 are fed to VMUX's 1 to 6, and their multiplexed outputs are fed to VMUX 7. The leakage current in any ON channel is limited to the sum of the output currents of the five OFF FET switches in the same MUX. This is a sevenfold reduction from the leakage current error of six VMUXs in parallel. Similarly, the output capacitance of each VMUX is only one-sixth that of six VMUXs in parallel.

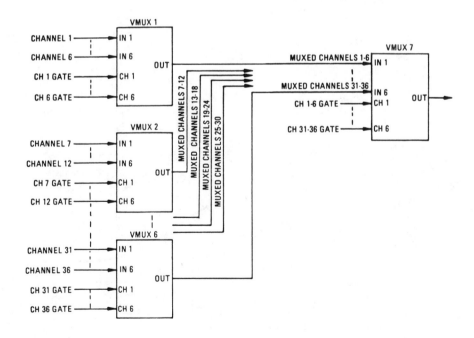

FIGURE 4.17

Two-level multiplexing of video-frequency multiplexers. (Courtesy of ILC Data Device Corp.)

4.7.1 Video Sample-Hold and Track-Hold Amplifiers

Although the circuitry of video S/H and T/H amplifiers may be different from that of the lower-speed units, the basic block diagrams are identical.

In a S/H or T/H amplifier, a sample command pulse is applied to the control input whenever it is desired to sample the analog signal voltage at the analog input. The switching amplifier then closes the switch, connecting the output of the first buffer amplifier to the holding capacitor for the duration of the sample command pulse. The sample command pulse duration must be at least equal to the *acquisition time*, which is the time necessary to charge the capacitor to some specified percentage of the analog input voltage. A commonly used criterion for the acquisition time is $9\,RC$ time constants, which permits the holding capacitor to charge to 99.99% of the analog input voltage.

As the end of the sample command pulse, the switching amplifier opens the switch, disconnecting the holding capacitor from the input buffer amplifier. The uncertainty of jitter of the actual time at which the switch opens, called the *aperture time* t_a, is a measure of the repeatability of the switch. If the analog input varies during the aperture time, there will be an *aperture-time error* in the sampled voltage. As shown in Figure 4.18, the aperture-time error is

$$e_a = \frac{dV}{dt}\,t_a \tag{4.5}$$

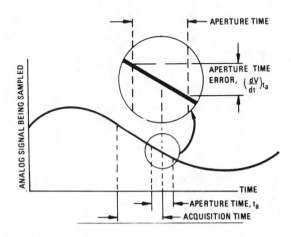

FIGURE 4.18
Aperture time error. (Courtesy of ILC Data Device Corp.)

where

$$e_a = \text{aperture-time error, volts}$$

$$dV/dt = \text{rate of change of analog input, volts/second}$$

$$t_a = \text{aperture time, seconds}$$

If the sampled voltage is a sine wave, $V = 2\pi ft$, then the aperture-time error is

$$e_a = \frac{dV}{dt}t_a = (2\pi f \cos 2\pi ft)t_a$$

The worst-case condition occurs at the maximum dV/dt value, when $\cos 2\pi ft = 1$, so that

$$e_a \text{ max} = (2\pi f)t_a - 6.28ft_a$$

and

$$\%e_a \text{ max} = 100 \times 6.28ft_a = 628ft_a\%$$

For a 1-MHz signal and an aperture time of 100 ps, the maximum percent of error will be

$$\%e_a \text{ max} = 628(10^6)(100 \times 10^{-12}) = 0.0628\%$$

The calculation of percent of aperture-time error for sinusoids of given frequencies and aperture times is facilitated by the nomograph of Figure 4.19. This assumes full-scale input voltage at the frequency f, and thus gives

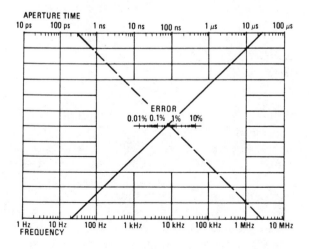

FIGURE 4.19
Aperture-time error nomograph. (Courtesy of ILC Data Device Corp.)

maximum aperture-time error. (Input voltages that are less than full scale will have proportionately smaller aperture-time errors.)

As an example of the use of the nomograph, assume a maximum allowable aperture-time error of 0.5% and an upper frequency limit of 2.6 MHz. Draw a straight line from 2.6 MHz on the lower scale, through 0.5% on the center scale, to the intersection with the aperture-time scale. This gives the maximum required aperture time as 300 ps. This is about three times the 100-ps minimum aperture of contemporary high-accuracy (0.1% linearity) S/H amplifiers.

The nomograph can also be used to compute the aperture-time error of an A/D converter with varying or low-frequency AC input by treating the conversion time as aperture time. Assume an A/D converter with 22-μs conversion time and a dc input varying at a maximum rate of 20 Hz. A straight line from 20 Hz on the lower scale to 22 μs on the upper scale intersects the middle scale at a maximum aperture-time error of 0.3%. This can easily be reduced to less than 0.003% by using S/H amplifiers at the inputs of the A/D converter.

At the end of the sample command, when the sampling switch opens, the holding capacitor is disconnected from the output of the input buffer amplifier. The output buffer amplifier has very high input resistance, so the voltage across the holding capacitor decays very slowly. This held voltage appears at the output of the output buffer amplifier. The voltage decay with time after sampling is seen as the output *drift*, which is expressed as the voltage change per hold-time interval. In slow- and medium-speed S/H amplifiers, leakage currents are the primary cause of capacitor voltage decay and thus of output voltage drift, but in high-speed S/H and T/H amplifiers, capacitor voltage decay and the corresponding output drift are caused mainly by dielectric absorption within the capacitor.

4.8
SYNCHRO AND RESOLVER CONVERTERS

Synchros and resolvers are shaft-angle transducers used in measurement transmission and control of angular position, velocity, and acceleration. These devices are essentially rotary transformers resembling small ac motors. The rotors and stators are wound so that the magnetic coupling between any rotor winding and any stator winding varies sinusoidally with the spatial angle between the magnetic axes of the windings. Synchro transducers and related devices are used in a vast number of applications where shaft angles and machine tool positions must be determined.

Standard synchro and resolver designators, given in Table 4.1, comprise two or three letters. The first letter gives the type: C (control) or T (torque). The remaining letters give the function: X (transmitter), R (receiver), T (transformer), DX (differential transmitter), or DR (differential receiver).

TABLE 4.1
Synchro and Resolver Designations

Type	Designation	Windings	Terminals
TORQUE SYNCHROS			
Torque transmitter	TX	1 R/3S	2 R/3S
Torque receiver	TR	1 R/3S	2 R/3S
Torque differential transmitter	TDX	3R/3S	3 R/3S
Torque differential receiver	TDR	3 R/3S	3 R/3S
CONTROL SYNCHROS			
Control transmitter	CX	1 R/3S	2 R/3S
Control transformer	CT	1 R/3S	2 R/3S
Control differential transmitter	CDX	3 R/3S	3 R/3S
RESOLVERS			
Resolver transmitter	RX	2 R/2S	4 R/4S
Resolver differential	RD	2 R/2S	4 R/4S
Resolver control transformer	RC	2 R/2S	4 R/4S
TRANSOLVERS	TY	2 R/3S or 3 R/2S	4 R/3S or 3 R/4S

Note that the transformer (T) exists only in control (C) types, as CT, and that there are no torque transformers.

Resolver designations comprise two letters, the first of which is R (resolver), and the second of which gives the function: X (transmitter), C (control transformer), or D (differential). Where the specific function of the resolver is not known, the general designation RS is often used.

Synchros have three stator windings and either one or three rotor windings. The stator windings are distributed so that their magnetic axes are 120° apart, and they are wye-connected to three leads designated $S1$, $S2$, and $S3$. Single-winding rotors have only one magnetic axis and are connected to leads designated $R1$ and $R2$. The schematic diagram at the left in Figure 4.20 and the equivalent symbolic diagram at the right are used to represent synchros with one rotor winding. Since there are several types of synchros with one rotor winding, the type designations (TX, TR, CX, or CT) are usually included in the drawing. All types of a given size are externally quite similar; they differ only in their internal construction, which depends on intended use. Synchro torque transmitters (TX) and control transmitters (CX) usually have their rotors energized from a sine-wave reference source (e.g., 115 V ac for 60-Hz synchros and 16 V ac for 400-Hz synchros), and generate three-wire voltages at their stator terminals. Torque receivers (TR)

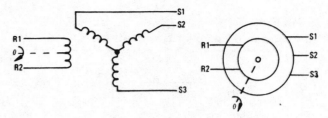

FIGURE 4.20
Synchro transmitter receiver and transformer diagrams. (Courtesy of ILC Data Device Corp.)

and control transformers (CT) usually have their stators energized by three-wire voltages from transmitters.

If a TX or CX rotor is energized by an ac reference voltage, $E \sin (2\pi ft)$, the induced voltages appearing at the stator terminals will be proportional to the sine of the spatial angle, θ, between the magnetic axis of the rotor winding and the magnetic axes of the stator windings:

$$V_{1-3} = K_1 E \sin \theta \sin (2\pi ft + a_1) \tag{4.6}$$

$$V_{2-3} = K_2 E \sin (\theta + 120°) \sin (2\pi ft + a_2) \tag{4.7}$$

$$V_{2-1} = K_3 E \sin (\theta - 120°) \sin (2\pi ft + a_3) \tag{4.8}$$

where f is the carrier or reference frequency of excitation; K_1, K_2, K_3 are the (ideally equal) rotor-stator transfer functions; and a_1, a_2, a_3 are the (ideally equal or zero) rotor-stator phase shifts at the carrier frequency.

If the induced voltages appearing at the stator terminals of a TX are applied to the corresponding stator terminals of a TR, the TR stator will develop a magnetic field at the same spatial angle θ with respect to the TR stator that the TX rotor has with respect to the TX stator. If the unconnected TR rotor is rotated to the angle θ, it will be aligned with the stator field, and maximum voltage will be induced in it. If the TR rotor is rotated to $\theta \pm 90°$, the rotor magnetic axis will be perpendicular to the stator magnetic axis, and

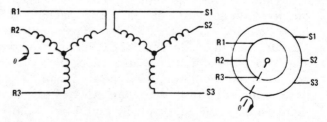

FIGURE 4.21
Synchro differential transmitter and differential receiver diagrams. (Courtesy of ILC Data Device Corp.)

minimum (ideally 0) voltage will be induced in the TR rotor. (In the example, the TX and TR may be replaced by a CX and CT, respectively.)

Three-winding rotors are wound like the stators and are wye-connected to leads designated $R1, R2$, and $R3$, as shown in Figure 4.21. Synchros with three-winding rotors are called *differentials* and are designated TDX, TDR, or CDX.

If the three output voltages from a synchro transmitter are applied to the three stator terminals of a synchro differential, they will establish a stator magnetic field at the same spatial angle θ with respect to the stator that the transmitter rotor has with respect to its stator. If the differential rotor is at a spatial angle θ with respect to its stator, the voltages induced at the differential rotor terminals will be similar to those given by Eqs. (4.6) through (4.8), but with θ replaced by $(\theta + \phi)$.

Resolvers have two stator windings and two rotor windings. The stator windings are distributed so that their magnetic axes are mutually perpendicular (90° apart). One stator winding is connected to leads designated $S1$ and $S3$, and the other stator winding is connected to leads designated $S2$ and $S4$, as shown in Figure 4.22. The rotor windings are similarly distributed so that their magnetic axes are 90° apart. One rotor winding is connected to leads designated $R1$ and $R3$, and the other rotor winding is connected to leads designated $R2$ and $R4$. Resolvers are designated RX (resolver transmitter), RC (resolver control transformer), or RD (resolver differential).

The resolver is so named because it can resolve an input voltage, E, into two output voltages $KE \sin \theta$ and $KE \cos \theta$. If one stator winding (e.g., $S1$–$S3$) of a resolver is energized by an ac voltage $E \sin (2\pi ft)$, the voltages induced in the two rotor windings will vary as the sine or cosine of the rotor-stator rotation angle, θ:

$$V_{1-3} = K_1 E \sin \theta \sin (2\pi ft + a_1) \tag{4.9}$$

$$V_{2-4} = K_2 E \cos \theta \sin (2\pi ft + a_2) \tag{4.10}$$

where K_1, K_2 are the (ideally equal) rotor-stator transfer functions, and a_1, a_2 are the (ideally equal or zero) rotor-stator phase shifts.

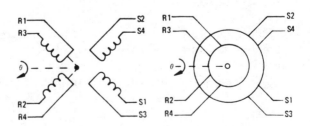

FIGURE 4.22
Resolver diagram. (Courtesy of ILC Data Device Corp.)

The preceding is an example of conversion from polar coordinates (E, θ) to Cartesian or rectangular coordinates:

$$E_x = V_{1-3}, \qquad E_y = V_{2-4}$$

Conversely, if a voltage $E_1 \sin (2\pi ft)$ is applied to winding $S1$–$S3$, and a voltage $E_2 \sin (2\pi ft)$ is applied to winding $S2$–$S4$, the resultant stator field will be at the angle

$$\phi = \tan^{-1} \frac{E_1}{E_2}$$

and it will be proportional to the resultant,

$$E_R = \sqrt{E_1^2 + E_2^2}$$

If the rotor is at the spatial angle θ with respect to the stator, the voltages induced in the two rotor windings will be

$$V_{1-3} = KE_R \sin (\theta + \phi) \sin (2\pi ft) \qquad (4.11)$$

$$V_{2-4} = KE_R \cos (\theta + \phi) \sin (2\pi ft)$$

This is an example of rotation of rectangular coordinates; that is, the coordinates E_1 and E_2 along one set of axes have been converted into coordinates V_{1-3} and V_{2-4} along a new set of axes rotated through the angle ϕ from E_1, E_2 axes.

Transolvers are crosses between synchros and resolvers that can convert between three-wire synchro data and four-wire resolver data, while simultaneously adding or subtracting a rotation angle to the converted data. They may have three-winding synchro rotors and two-winding resolver stators, as in Figure 4.23(a), or resolver rotors and synchro stators, as in Figure 4.23(b). In the transolver designation TY, the T does not stand for torque, but represents the two mutually perpendicular windings of a resolver. Similarly, the Y represents the wye-connected windings of a synchro.

4.8.1 Data Transmission Using Torque Synchros

The oldest application of synchros is for transmitting angular position data to remote indicators, as shown in the schematic and symbolic diagrams of Figure 4.24.

Rotor leads R_1 of a local synchro torque transmitter (TX) and of a remote torque receiver (TR) are connected to one side of an ac reference line, and the rotor leads $R2$ are connected to the other side of the reference line. Transmitter stator leads $S1$, $S2$, and $S3$ are connected to the corresponding receiver stator leads. Since the TX rotor is energized from the ac reference line, it will induce in each of the three stator windings a voltage propor-

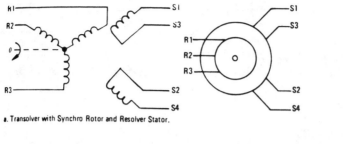

a. Transolver with Synchro Rotor and Resolver Stator.

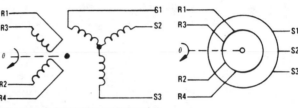

b. Transolver with Resolver Rotor and Synchro Stator.

FIGURE 4.23
Transolver diagrams. (Courtesy of ILC Data Device Corp.)

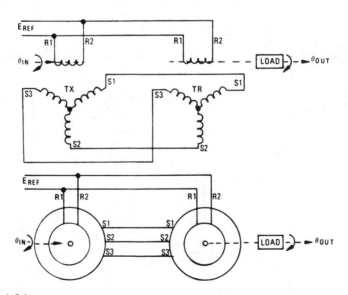

FIGURE 4.24
Data transmission using torque synchros.
(Courtesy of ILC Data Device Corp.)

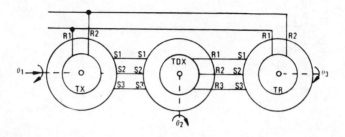

FIGURE 4.25
Data transmission using synchro torque differentials. (Courtesy of ILC Data Device Corp.)

tional to the sine of the angle between the magnetic axis of the rotor and the magnetic axes of the stator windings, so the terminal voltages V_{1-3}, V_{2-3}, and V_{2-1} will be as represented by Eqs. (4.6) through (4.8). Since corresponding TX and TR stator windings are connected together, the voltages induced in the TX stator windings by the energized TX rotor are applied to the TR stator windings, which set up a magnetic field at the angle θ_{out} (with respect to the TR stator) equal to the angle θ_{in} of the TX rotor with respect to the TX stator). Since the TR rotor is energized from the same reference lines as the TX rotor, it develops a magnetic field that induces various voltages in the TR stator. If the TR rotor is at the angle θ_{out}, the voltage it induces in the TR stator windings will be equal to the voltages applied to those windings, and

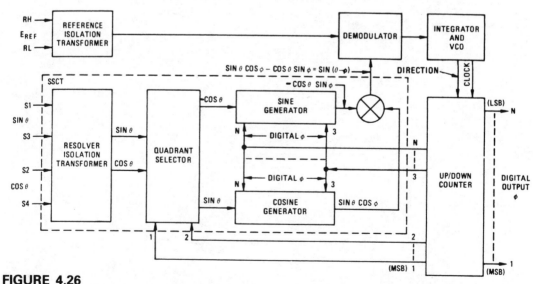

FIGURE 4.26
Tracking-type resolver-to-digital converter block diagram. (Courtesy of ILC Data Device Corp.)

minimum current will flow between the two stators. If the TR rotor is not at the angle θ_{out}, the voltages it induces in the TR stator windings will not be equal to the applied voltages; currents will then flow between the TX and TR stator windings, developing a torque that tends to align the TR rotor at the angle θ_{out}. Depending on the ratings, several torque receivers can be operated from one torque transmitter, although the accuracy of all the TRs is reduced by each additional TR and by increased mechanical loads.

The basic synchro data-transmission system of Figure 4.24 can be expanded to permit addition or subtraction of two angular inputs, as shown in Figure 4.25. The torque transmitter and torque receiver are connected as before, but a torque differential transmitter is inserted in the three-wire line between the TX and the TR. The TR output ($\theta_3 = \theta_1 - \theta_2$) is the difference between the TX input angle (θ_1) and the TDX input angle (θ_2). Note that if the TR is replaced by a TX, and the TDX by a TDR, the rotor angles (θ_1 and θ_3) of the two TXs are the inputs, and the output is the TDR rotor angle, $\theta_2 = \theta_1 - \theta_3$. Reversal of two wires between the TDX and the TR, so that $R1$ connects to $S2$ and $R3$ connects to $S1$, will cause the output angle to be the sum of the two input angles.

4.9
SYNCHRO-TO-DIGITAL AND RESOLVER-TO-DIGITAL CONVERTERS

Figure 4.26 is the block diagram of a tracking-type resolver-to-digital (R/D) converter employing an SSCT. Four-wire resolver data ($\sin \theta$ and $\cos \theta$) are fed through an isolation transformer to a quadrant selector. The quadrant selector determines the quadrant in which the angle θ falls, and feeds sine θ or cos θ (of appropriate polarity, in accordance with Table 3.3) as reference voltages to the function generators. The transfer functions of the function generators are controlled by a digital input angle, ϕ, generated by an up–down counter. The outputs $\cos \theta \sin \phi$ of the sine generator and $\sin \theta \cos \phi$ of the cosine generator are fed to a differencing junction. The output of the differencing junction,

$$\sin \theta \cos \phi - \cos \theta \sin \phi = \sin (\theta - \phi)$$

is fed to a demodulator along with the reference voltage E_{REF}. The demodulator develops a dc analog voltage proportional to $\sin (\theta - \phi)$, where $(\theta - \phi)$ is the error or difference between the digital output angle, ϕ, and the synchro input angle θ. Note that, for small errors, $\sin (\theta - \phi) = (\theta - \phi)$. The dc analog error voltage is fed to an analog integrator. The amplitude of the integrator output determines the frequency of a voltage-controlled oscillator (VCO), clock pulses from which are fed to an up–down counter. The polarity of the integrator output determines whether the counter counts up or down. If ϕ is less than θ, the error $(\theta - \phi)$ is positive, and an up-count control signal is fed to the counter. If ϕ is greater than θ, the error $(\theta - \phi)$

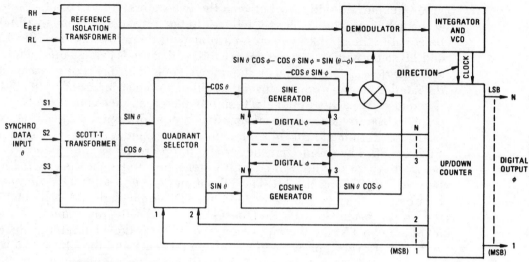

FIGURE 4.27
Tracking-type synchro-to-digital converter block diagram. (Courtesy of ILC Data Device Corp.)

is negative, and a down-count control signal is fed to the counter. The closed loop thus operates like a servomechanism to reduce the error $(\theta - \phi)$ to zero, by making the digital output angle ϕ equal to the resolver-format input angle θ. Note that the quadrant selector is controlled by the two most-significant bits of the digital angle ϕ, just as in the D/R and D/S counters previously described.

Figure 4.27 shows a tracking-type synchro-to-digital (S/D) converter. This differs from the R/D converter only in employing a Scott-T input transformer to change the synchro-format input into resolver-format data.

Figure 4.28 shows the block diagram of a two-speed S/D converter employing two solid-state control transformers.

In addition to tracking-type converters, there are a number of sampling types, some of which employ successive approximation techniques, as shown in Figure 4.29, and some of which employ harmonic oscillator techniques, as shown in Figure 4.30. The successive approximation type of S/D converter operates like the successive approximation A/D converter described previously. The harmonic oscillator type of S/D converter employs two integrating operational amplifiers and one inverting operational amplifier connected in a closed loop that oscillates at the frequency $f = 1/(2\pi \sqrt{R_1 C_1 R_2 C_2})$. The output voltages V_1 and V_2 of the integrators are the X and Y components of a phasor R that rotates with the constant angular velocity $\omega = 2\pi f$. The time required for this phasor to rotate from its initial position (determined by the input synchro or resolver angle θ) to the position where the Y component, V_2, passes through zero in a positive direction is directly proportional to the input angle θ. The instantaneous

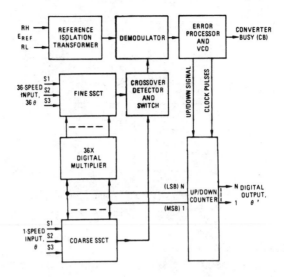

FIGURE 4.28
Two-speed synchro-to-digital converter block diagram. (Courtesy of ILC Data Device Corp.)

values of sin θ and cos θ, respectively, are set into the integrators as initial conditions when switches $S3A$ and $S3B$ are closed and switches $S1$ and $S2$ are connected to the Scott-T transformer. When all the switches are simultaneously set to the positions shown in the diagram, the phasor R represented by the voltages V_1 and V_2 will begin to rotate at the uniform angular

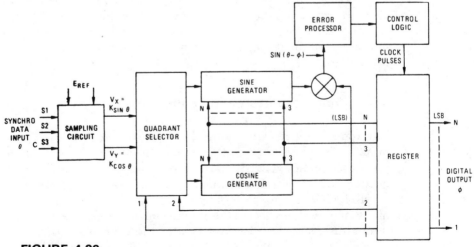

FIGURE 4.29
Successive approximation synchro-to-digital converter block diagram. (Courtesy of ILC Data Device Corp.)

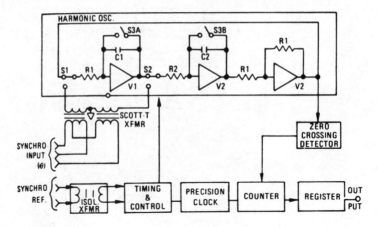

FIGURE 4.30
Harmonic oscillator synchro-to-digital converter
block diagram. (Courtesy of ILC Data Device
Corp.)

velocity $\omega = 2\pi f$. The initial-condition angle θ is determined from the time required for the voltage V_2 to pass through zero in a positive direction. This is done by counting clock pulses from a precision clock-pulse generator that is started at the same time as the oscillator is started, and stopping the count when a zero-crossing detector senses that the voltage V_2 has crossed zero in a positive direction. The clock-pulse counter drives a register that develops the digital output.

The harmonic oscillator S/D converter can be made into an R/D converter by replacing the Scott-T transformer with two isolation transformers. In either case, the digital output can be scaled to represent, in engineering units, any parameter (e.g., lb, psi, mpg, ft/s, ft.³) that can be linearly related to shaft angle.

4.9.1 Inductosyn-to-digital converters

The rotary Inductosyn (R) is functionally similar to a resolver with a large number (N) of poles (e.g., 108, 144, 360) that operates at a high reference carrier frequency (e.g., 10 kHz). When the rotor is excited by an ac reference voltage, the stator output voltages are of the form sin $P\theta$ and cos $P\theta$, where P is the number of pole pairs ($P = N/2$) and θ is the angle of rotation of the rotor with respect to the stator. The envelope of each output thus goes through P cycles per 360° rotation of the input shaft, just like a P-speed resolver (i.e., a resolver with 1:P step-up gearing). Angular accuracies are on the order of 0.1 and 1.0 s of arc. The basic design concept makes possible a linear form of Inductosyn, capable of measuring displacement or linear motion as small as 2 to 10 μ in. (0.05 to 0.25 μm).

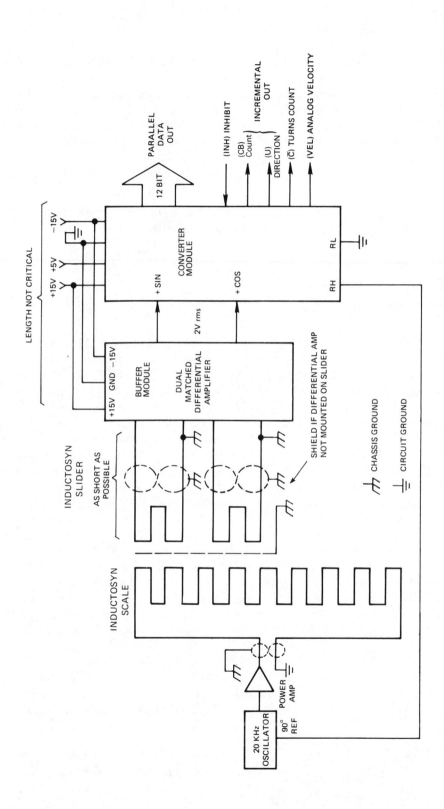

FIGURE 4.31
Typical Inductosyn connection. (Courtesy of ILC Data Device Corp.)

Figure 4.31 shows how a linear or rotary Inductosyn is used in conjunction with an Inductosyn-to-digital (I/D) converter and a buffer module. The clock and up–down signals that drive the internal up–down counter are available as external *toggle* and *direction* signals for driving an external BCD up–down counter, driver, and display, which can have any number of decades depending on the Inductosyn length (in cycles). Toggle outputs of 1000, 2000, or 4000 counts per cycle can be programmed to permit use of the I/D converter with linear Inductosyns having pitches of 0.1, 0.2, or 0.4 in. (1, 2, or 4 mm).

NOTE: IDC 35300 has a CB pulse that comes out every LSB change. Thus for a 12-bit converter, 4096 pulses are available.

4.10
SPECIALIZED APPLICATIONS OF THE
A/D CONVERTER [c]

Figure 4.32 shows the block diagram of a 12-bit hybrid A/D converter with 5-μs conversion time and ±0.012% full-scale range linearity error used in electronic instrumentation. The ADH-8585 and ADH-8586 are complete in a 32-pin triple DIP hermetically sealed metal package and are pin compatible with other generic ADC-85 12-bit A/D converters. An advanced design using thin-film and MSI technologies results in conversion times of 10 μs for the ADH-8585 and 5 μs for the ADH-8586, corresponding to word rates of nearly 100 and 200 kHz, respectively. Conversion times may be reduced to less than this by short cycling and increasing the internal clock rate if less resolution and accuracy are acceptable. Gain and offset errors can be trimmed to zero using two external potentiometers, making accuracy equal to the ±$\frac{1}{2}$ LSB linearity. The internal reference and internal clock are externally accessible, and an external clock may be used.

Because of their high reliability, hermetically sealed metal cases, and wide operating temperature ranges, the ADH-8585 and ADH-8586 will meet the most demanding military and industrial requirements. Typical applications include data-acquisition systems, automatic test equipment, and electronic countermeasures systems. Standard processing at no added cost is based on MIL-STD-883, except for burn-in, which is an option. These converters are excellent for remotely located and hard-to-access equipment where small size and high MTBF are critical.

The main elements of the ADH-8585 and ADH-8586 as shown in the block diagram, Figure 4.32, are a voltage comparator, a 12-bit successive approximation register (SAR), and a 12-bit digital-to-analog converter con-

---

[c] The material in Section 4.10 is used courtesy of ILC Data Device Corp.

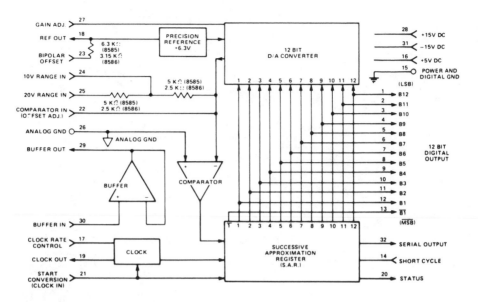

FIGURE 4.32
ADH-8585 and ADH-8586 block diagram.
(Courtesy of ILC Data Device Corp.)

nected in a closed loop. The analog input and the D/A converter output are superimposed at the sum point of the comparator. Successive approximation is used to make the D/A output equal to the analog input. Conversions are initiated by an external START CONVERSION pulse, and a STATUS logic output indicates when a conversion has been completed and output data are available.

The 10-V RANGE INPUT or 20-V RANGE INPUT is connected to the analog input (depending on scaling), and the BIPOLAR OFFSET is used to select unipolar or bipolar operation. An input buffer amplifier is provided for higher input impedance, but its use has been made optional. The buffer amplifier must be allowed to settle before a conversion can be initiated.

The rate of the internal clock is determined by the voltage provided at the CLOCK RATE CONTROL input. When this input is connected to power ground, the internal clock rate is appropriate for the full-accuracy 12-bit converter. However, it may be necessary to adjust the voltage to obtain the optimum clock rate.

The conversion rate can be increased by truncating the conversion at 10 or 8 bits using the SHORT CYCLE pin. Short cycling allows the clock to be speeded up. The internal clock rate can be increased by applying a more positive voltage to the CLOCK RATE CONTROL, thereby increasing the conversion time still further. It is also possible to use an external clock.

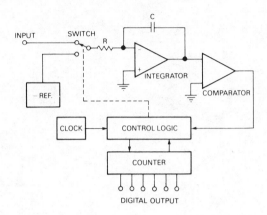

FIGURE 4.33
Dual slope A/D converter. (Courtesy of ILC Data
Device Corp.)

4.10.1   Integrating-type A/D Converters[d]

A class of A/D converters known as integrating type operates by an indirect
conversion method. The unknown input voltage is converted into a time
period, which is then measured by a clock and counter. A number of variations exist on the basic principle, such as *single-slope, dual-slope,* and
triple-slope methods. In addition, there is another completely different
technique known as the *charge-balancing* or *quantized feedback* method.

An integrating A/D converter can have exceptional noise-rejection
capability if you merely adjust its measurement cycle to equal the period (or
a multiple) of the noise frequency to be rejected.

The most popular integrating A/D converter methods are dual-slope
and charge-balancing; although both are slow, they have excellent linearity
characteristics with the capability of rejecting input noise. Because of these
characteristics, integrating-type A/D converters are almost exclusively
used in digital panel meters, digital multimeters, and other slow measurements applications.

The dual-slope technique, shown in Figure 4.33, is perhaps best
known. Conversion begins when the unknown input voltage is switched to
the integrator input; at the same time the counter begins to count clock
pulses and counts up to overflow. At this point the control circuit switches
the integrator to the negative reference voltage, which is integrated until
the output is back to zero. Clock pulses are counted during this time until
the comparator detects the zero crossing and turns them off.

[d] The material in Section 4.10.1 is from pages 18 and 19 of E. Zuch, *Data Acquisition and
Conversion Handbook,* 4th printing, Sept. 1981, and is used courtesy of Datel-Intersil, Inc.

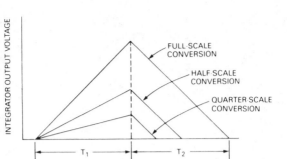

FIGURE 4.34
Integrator output waveform for dual slope A/D
converter. (Courtesy of ILC Datel-Intersil, Inc.)

The counter output is then the converted digital word. Figure 4.34 shows the integrator output waveform where T_1 is a fixed time and T_2 is a time proportional to the input voltage. The times are related as follows:

$$T_2 = T_1 \frac{E_{\text{IN}}}{V_{\text{REF}}} \tag{4.12}$$

The digital output word therefore represents the ratio of the input voltage to the reference.

Dual-slope conversion has several important features. First, conversion accuracy is independent of the stability of the clock and integrating capacitor as long as they are constant during the conversion period. Accuracy depends only on the reference accuracy and the integrator circuit linearity. Second, the noise rejection of the converter can be infinite if T_1 is set to equal the period of the noise. To reject 60-Hz power noise therefore requires that T_1 be 16.667 ms. Figure 4.35 shows digital panel meters that employ dual slope A/D converters.

FIGURE 4.35
Digital panel meters that employ dual slope A/D
converters. (Courtesy of Datel-Intersil, Inc.)

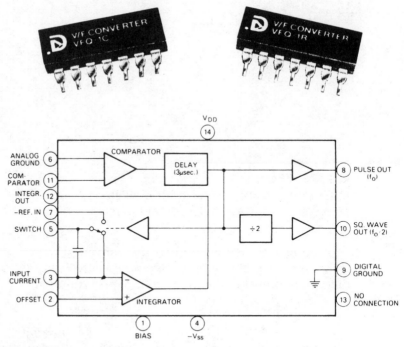

FIGURE 4.36
Low-cost monolithic voltage-to-frequency con-
verters, models VFQ-1C and VFQ-1R. (Courtesy of
Datel-Intersil, Inc.)

4.10.2   Voltage-to-Frequency Converters [e]

The charge-balancing, or quantized-feedback, method of conversion is
based on the principle of generating a pulse train with frequency proportional to the input voltage and then counting the pulses for a fixed period
of time. This circuit is shown in Figure 4.36. This monolithic voltage-to-
frequency converter, uses combined bipolar and CMOS technologies.

The VFQ-1 accepts a positive analog input current and produces an
output pulse train with a frequency linearly proportional to an input current. In addition to the pulse output, there is also a square-wave output at
half the pulse frequency. The full-scale output pulse rate can be set from 10
to 100 kHz by means of two external capacitors. Linearities are typically
0.01% for 10 kHz full scale and 0.1% for 100 kHz full scale; linearity holds
all the way down to zero.

The VFQ-1 internal circuitry shown in Fig. 4.36 includes an operational
integrator, a comparator, a digital delay circuit, a single-pole double-throw

---

[e] The material in Section 4.10.2 is used courtesy of Datel-Intersil, Inc.

MECHANICAL DIMENSIONS
INCHES (MM)

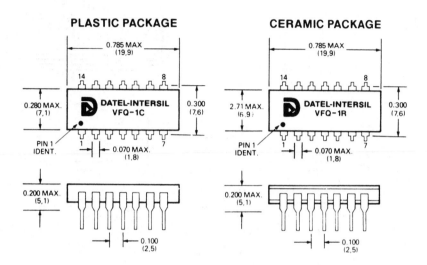

FIGURE 4.36 (Continued)

electronic switch, a start circuit, a divide-by-two circuits, and two output driver circuits. It operates on the well-known charge-balancing integrator principle. The two outputs are open collector *NPN*, which can sink up to 10 mA and give a logic **HI** output up to +18 V.

In normal operation this converter requires only five external components and a reference. If the zeroing adjustment is used, a trimming potentiometer and two more resistors are required. The VFQ-1 can be operated from dual ±4- to ±7.5-V supplies or from a single + 10- V to + 15-V supply. Current drain is 4 mA maximum. The device can also be operated as a frequency-to-voltage converter.

Voltage-to-frequency converters are commonly used in the following applications:

1. Digital multimeters

2. Digital panel meters

3. Remote data transmission

4. Totalizing measurements

5. Measurements in high noise measurements

6. High-voltage isolation measurements

7. Ratiometric measurements

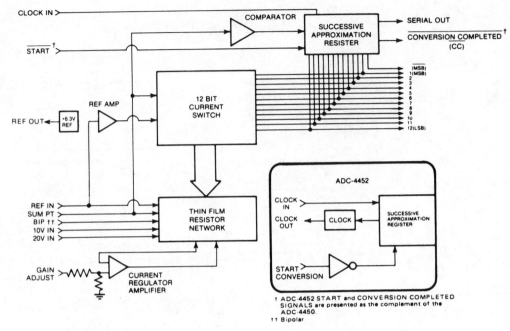

FIGURE 4.37
Block diagram of the ADC-4450. (Courtesy of ILC
Data Device Corp.)

4.10.3 A/D Converters Used to Digitize Signals [f]

The A/D converter is the largest single contributor of errors in a digital signal analyzer. To change that situation, the ILC Data Device Corporation has designed a new converter, the ADC-4450, specifically for digital instruments such as spectrum analyzers and fast Fourier (FFT) equipment. With 12-bit resolution and a 1.5-μs conversion, the 4450 specifications are well suited to instrument applications.

The ADC-4450 (Figure 4.37) series frequency domain, analog-to-digital converters are successive approximation devices, capable of operating with six unipolar and bipolar input ranges. Digital output codes include complementary offset binary, complementary binary, and complementary two's complement, when using MSB (see Figure 4.37). Output data are available in 12-bit parallel and serial formats, and the converter has a drive capability of five TTL loads.

In certain applications the converters may require offset and gain adjustments. The procedure, set forth in Figure 4.38, is to be used to make the

---

[f]The material in Section 4.10.3 is used courtesy of ILC Data Device Corp.

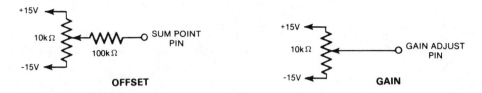

FIGURE 4.38
Offset and gain circuits. (Courtesy of ILC Data Device Corp.)

necessary adjustments for unipolar and bipolar input configurations (complementary offset binary codes shown).

Offset adjustment in unipolar configuration is as follows:

1. Set the input level to zero.

2. Adjust the potentiometer of the offset circuit (Figure 4.38) to achieve 50% dither between digital output code 111---10 and 111---11.

In bipolar configurations:

1. Set the input level to equivalent minus full scale plus 1 LSB.

2. Adjust the potentiometer to achieve 50% dither between digital output codes 111---10 and 111---11.

Gain adjustments are made for both input configurations with input levels set to plus full scale minus 1 LSB. Adjustment of the potentiometer

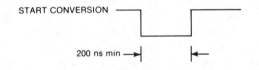

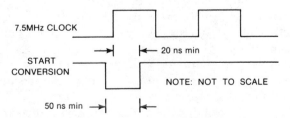

FIGURE 4.39
Start pulse timing. (Courtesy of ILC Data Device Corp.)

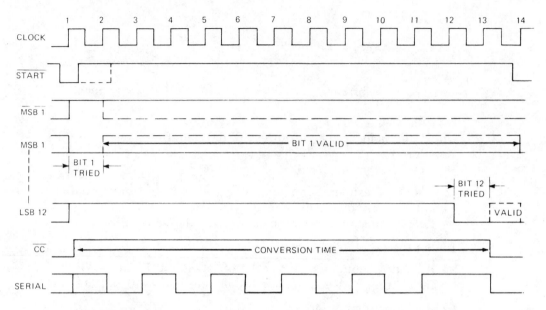

FIGURE 4.40
Converter timing. (Courtesy of ILC Data Device Corp.)

in the gain circuit (Figure 4.38) must achieve a 50% dither between digital output codes 000–––00 and 000–––01.

The ADC-4450 series converters may be operated synchronously or asynchronously with respect to start conversion input signal timing. The start conversion pulse must be 200 ns (1.5 clock pulses minimum) when operating in an asynchronous mode at a 7.5-MHz clock rate (Figure 4.39). Synchronous conversions are started on the 0 level of a negative 50-ns (minimum) start conversion pulse, applied to the START pin (ADC-4452 requires a positive pulse). The start conversion pulse specifications supplied in Figure 4.39 are intended to ensure sufficient time to reset internal logic, while the START line is low. Start must be low during at least one low-to-high clock transition. Allowing START to go high, after the reset period, will begin the conversion on the next low-to-high clock transition. A converter busy signal is provided on the CC pin. Logic 1 indicates that a conversion is in progress (logic 0 indicates data available). The ADC-4452 conversion completed (CC) is presented as the complement of the ADC-4450 output signal. Transfer of data may occur 20 ns, typically, after the trailing edge of the converter busy signal.

All timing specifications are based on a 7.5-MHz clock (guaranteed accuracy). Maximum clock frequency is 15 MHz, which produces a conversion time of 800 ns. (Clock frequencies above 7.5 MHz could result in linearity degradation.)

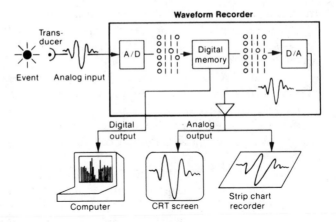

FIGURE 4.41
How the waveform digitizer fits into a data-acquisition system. (Courtesy of Gould Inc., Instrument Division, Biomation Operation, Santa Clara, Calif.)

ADC-4450 series units reset internal logic on the first low-to-high clock transition after START CONVERSION goes low. Conversion begins on the first low-to-high transition after START CONVERSION is allowed to go high (see Figure 4.40). The converter will remain in a reset mode as long as START CONVERSION is held low.

4.10.4 Waveform Digitizers Using A/D Converters[g]

Waveform digitizers represent a new-generation technology for capturing and recording electrical events of very short duration, as short as 10 ns. There are several reasons why systems utilizing these devices are replacing older capture systems such as oscilloscopes with cameras, storage oscilloscopes, chart recorders, FM tape recorders, and light beam oscillographs.

Systems using waveform digitizers are faster. They have better resolution. They provide digital output that can be directly accepted by computers, as well as analog output for making hard copy on chart recorders. Finally, they permit the researcher to study different time relationships of trigger events to signal of interest, a feature that none of the other systems offer. These special capabilities will all be described in detail.

A waveform digitizer captures a single-shot or infrequently recurring *analog* signal and converts it into *digital* code. After analog-to-digital (A/D) conversion, the information is stored in a semiconductor memory. The data

---

[g] The material in Section 4.10.4 is used courtesy of Gould Inc., Instrument Division, Biomation Operation, 4600 Old Ironside Drive, Santa Clara, Calif.

in memory can be output directly to a computer or can pass through a digital-to-analog (D/A) converter for output to a CRT screen or strip-chart recorder. Figure 4.41 shows how a waveform digitizer fits into typical data-acquisition systems.

As with all A/D converters, waveform digitizers sample the input signal at regular time intervals and then store that information in the form of digital code. The sampling rate, or frequency, can be controlled by adjusting the recorder's internal clock circuit (a crystal-controlled oscillator) or by applying the desired clock frequency from an external source such as a function generator.

The *slowest* Biomation waveform digitizer has a maximum clock rate of 5 MHz, permitting capture of events as fast as 800 ns in duration. The fastest model, with a clock rate of 500 MHz, can capture events down to 10 ns in duration. All models filter the input to prevent erroneous data due to aliasing of high-frequency signal components.

With the improvements of integrated-circuit technology, newer A/D converters will find newer uses in electronic instrumentation. Precision resistor networks are important to the performance of A/D converters. By means of state-of-the-art, thin-film technology, it is now possible to produce high precision stable resistive networks economically, for use in high-performance hybrid converters that combine the advantages of both discrete-component and monolithic circuits.

4.11
REVIEW QUESTIONS

1. Discuss four methods of achieving A/D conversion. Which type is the most popular? Why?

2. Discuss the uses of the voltage-to-frequency converters.

3. Draw a modified block diagram of successive approximation A/D converters.

4. Discuss the necessity for a two-step A/D conversion and the 3-bit flash conversion.

5. Draw a video A/D converter. State where it is used.

6. Draw a simplified schematic of a video-frequency MUX.

7. Draw a successive approximation synchro-to-digital converter and a two-speed synchro-to-digital converter.

8. Discuss an A/D converter to be used for spectrum analyzers.

9. Discuss the uses of synchro-to-digital converters.

10. Draw a block diagram of a waveform digitizer with a digitizer readout.

4.12
REFERENCES

1. Eugene L. Zuch: *Data Acquisition and Conversion Handbook*, 4th printing, Datel-Intersil, Inc., Mansfield, Mass., Sept. 1981.

2. Smed Ruth: "Electronic Signal Converters: An Overview," *Instruments and Control Systems*, Aug. 1981.

3. Data Converters, DDC Product Catalog, ILC Data Device Corporation, Bohemia, N.Y., 1981.

4. Synchro Conversion Handbook, 2nd printing, ILC Data Device Corporation, Bohemia, N.Y., 1979.

5. Op Amps and Data Conversion Products, Teledyne Philbrick, Dedham, Mass., 1982.

6. Dennis Santucci: Designing High Speed Acquisition Systems, Bulletin AN-21, Teledyne Philbrick, Dedham, Mass., April 1976.

7. R. W. Jacobs: Repetitive Mode Operation for Models 4109/4111 A/D Converters, Application Bulletin AN-28, Teledyne Philbrick, Dedham, Mass., Jan. 1977.

8. R. W. Jacobs: Specifying and Testing Analog to Digital Converters, Bulletin AN-24, Teledyne Philbrick, Dedham, Mass., July 1976.

9. Barry Blesser: Implementation of the Analog-to-Digital Conversion, *Digital Audio, dB*, Nov. 1980.

10. *Instruments and Systems Handbook*, Datel-Intersil, Inc., Mansfield, Mass., 1981 (published yearly).

5

Digital Instrumentation Systems

5.1
INTRODUCTION

Examples of digital instrumentation systems comprise this chapter. We will start this chapter with concepts of integrated-circuit (IC) logic devices. This is followed by descriptions and circuitry used in logic probes, logic monitors, digital pulsers, logic analyzers, and IC digital testers, digital multimeter theory, digital panel meters, and an application of a data-acquisition system using computers. It is the intent of the authors to introduce the readers to the complexity of this new digital technology of the 1980s. Wherever possible, the system approach will be stressed. Since most instruments are digitized, we will discuss some of the useful types today.

5.2
IC LOGIC DEVICES [a]

Resistor-transistor logic, abbreviated RTL, was the first integrated-circuit logic form introduced about 20 years ago. The basic circuit is a direct translation from the discrete form into integrated format.

------

[a] The material in Section 5.2 is excerpted from John D. Lenk, *Logic Designers Manual,* Reston Publishing Co., Reston, Va., 1977.

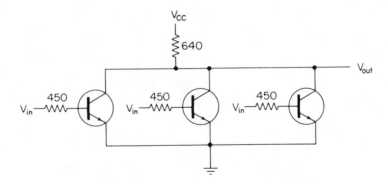

FIGURE 5.1
Basic RTL gate. (From John D. Lenk, *Logic De-signers Manual,* Reston Publishing Co., Reston, Va., 1977)

The basic RTL gate circuit shown in Figure 5.1 is presented here to illustrate the basic building block type of logic. The most complex elements are constructed simply by the proper interconnection of this basic circuit. The small resistor added in the base circuit increases the input impedance to assure proper operation when driving more than one load. Without this resistor, when several base–emitter junctions are driven from the same output, the input with the lowest base–emitter junction forward bias could severely limit the drive current to the other transistor bases. To turn the three transistors on, a voltage greater than the base-to-emitter voltage must be applied. In the case of RTL, the 1 state is approximately 0.9 V and the 0 state is approximately 0.2 V. With 0 V all the resistors are on. By placing a capacitor across the resistors in Figure 5.1, the slow response of the circuit can be improved; this is called resistor-capacitor-transistor logic (RCTL).

Diode-transistor logic, or DTL, is another logic form that was translated from discrete design into IC elements. DTL was very familiar to the discrete component logic designer in the form that used diodes and transistors as the main components (plus a minimum number of resistors). The diodes provided higher signal thresholds than could be obtained with RTL.

Figure 5.2 shows the basic DTL gate. Note that it requires two power supplies (to improve turn-off time of the transistor inverter).

High-threshold logic (HTL) is used where a high logic swing (13 V) and high noise immunity are needed. The HTL is the slowest of all IC logic families and requires the most power of all logic families. HTL is generally not being used for design of current logic systems (although HTL is found in many existing systems, particularly where logic must be connected to industrial and other heavy-duty equipment).

Figure 5.3 shows a typical HTL gate and transfer characteristics. The gate is identified as an MHTL, or Motorola HTL. The HTL is essentially the same as the DTL, except that a zener diode is used for D_1 in the HTL. The use of a zener for D_1 (with a conduction point of about 6 or 7 V) and the

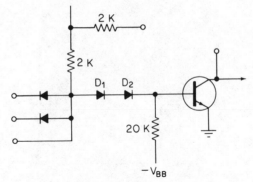

FIGURE 5.2
Basic DTL gate. (From John D. Lenk, *Logic Designers Manual,* Reston Publishing Co., Reston, Va., 1977)

higher supply voltage (a V_{cc} of about 15 V) produce the wide noise margins shown in the transfer characteristics of Figure 5.3. The HTL gate will not operate unless the input voltage swing is large. Noise voltages below about 5 V will have no effect on the gate.

Transistor-transistor logic, TTL or T^2L, has become one of the most popular logic families available in IC form. Most IC manufacturers produce at least one line of TTL, and often several lines. This fact gives TTL the widest range of logic functions. Typically, TTL has a 4.5-V logic swing.

Figure 5.4 shows an MTTL or Motorola TTL gate.

The emitter-coupled logic, ECL, shown in Figure 5.5, operates at very

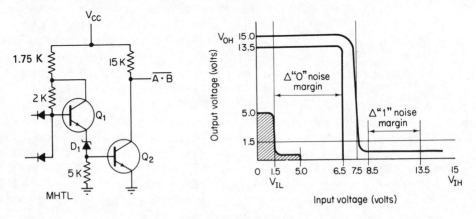

FIGURE 5.3
Motorola MHTL gate and transfer characteristics. (From John D. Lenk, *Logic Designers Manual,* Reston Publishing Co., Reston, Va., 1977)

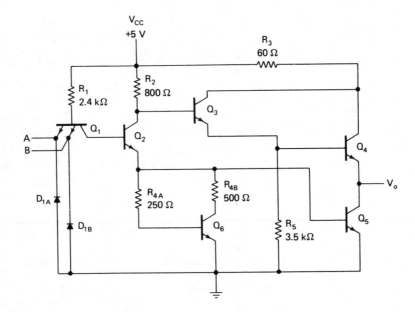

FIGURE 5.4
Improved Motorola MTTL gate. (Courtesy of
Motorola. From John D. Lenk, *Logic Designers
Manual,* Reston Publishing Co., Reston, Va., 1977)

high speeds (compared to TTL). Another advantage of ECL is that *both a
true and complementary* output are produced. Thus both OR and NOR
functions are available at the output. Note that when the NOR functions of
two ECL gates are connected in parallel, the outputs are ANDed, thus
extending the number of inputs.

The high operating speed is obtained since ECL uses transistors in the
nonsaturating mode. That is, the transistors do not switch full on or full off,
but swing above and below a given bias voltage. Delay times range from
about 2 to 10 nanoseconds (ns). ECL generates a minimum of noise and has
considerable noise immunity. However, as a trade-off for the nonsaturating
mode (which produces high speed and low noise), ECL is the least efficient. The ECL dissipates the most power for the least output voltage.

The ECL is still the fastest of all logic families. The disadvantages of
ECL are high power consumption (the highest next to HTL) and a low logic
swing (usually less than 1 V, but some ECLs will provide nearly 2 Vs).

Although MOS logic is not limited to the complementary inverter or to
the complementary fabrication technique (known as CMOS), the complementary principle forms the backbone of most current MOS logic ICs, both
standard and custom built. The basic complementary inverter circuit is
shown in Figure 5.6. Note that this circuit is formed using an *N*-channel
MOSFET and a *P*-channel MOSFET on a single semiconductor chip. As

discussed in later chapters, some MOS logic devices are formed using only *P*-channel (PMOS) or *N*-channel (NMOS) fabrication techniques.

The complementary inverter of Figure 5.6 has the unique advantage of dissipating almost no power in either stable state. *Power is dissipated only during the switching interval.* And, since MOSFETs are involved, the capacitive input lends itself to direct-coupled circuitry. (The input gate of a MOSFET acts as a capacitor.) No capacitors are required between circuits; this results in a savings in component count and in circuit wiring.

When large-scale integrated circuits are used, CMOS is the best choice. TTL and ECL are used in medium-scale integrated circuitry.

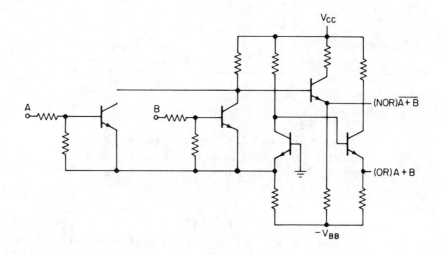

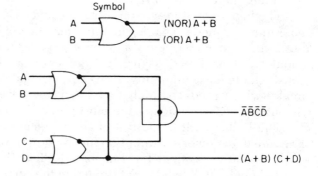

FIGURE 5.5
Emitter-coupled logic (ECL). (From John D. Lenk,
Logic Designers Manual, Reston Publishing Co.,
Reston, Va., 1977)

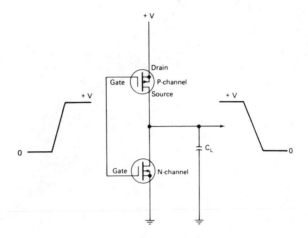

FIGURE 5.6
Basic MOS device complementary (CMOS) inverter circuit. (From John D. Lenk, *Logic Designers Manual,* Reston Publishing Co., Reston,
Va., 1977)

5.3
LOGIC PROBES [b]

Integrated circuits are very private devices. When something goes wrong, they don't cough . . . they don't turn red . . . they don't make funny little noises . . . they just don't work.

It is bad enough working on one IC when part or all of that IC goes bad. But in most cases, when an IC goes bad the result is a large, complex system that will not work the way it is supposed to.

Up until now it has not been very easy to tell just what was happening at any one point in a logic system. You could use a voltmeter and translate its readings into logic states in your head, but that is a clumsy process.

Oscilloscopes are another way of trying to look inside ICs. Indeed, for certain complex timing measurements, they have no equal. But oscillo-scopes are large and expensive. And it can take a very long time to set up an oscilloscope to measure exactly the phenomenon you are looking for. Logic probes are portable, circuit-powered digital instruments designed to simplify the chore of diagnosing state-oriented logic.

This ability to rapidly diagnose logic states and sense and optionally

---

[b] The material in Sections 5.3 through 5.5 is used courtesy of the Global Specialties Corp., headquarters New Haven, Conn. It is reprinted with their permission and is copyrighted by Global Specialties Corp.

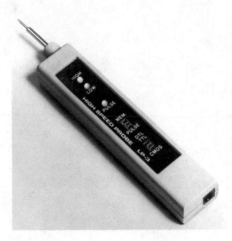

FIGURE 5.7
High-speed LP-3 logic probe. (Courtesy of Global
Specialties Corp.)

store pulses has led to almost universal acceptance of logic probes as the
proper diagnostic instrument for all but the most complex digital service
and maintenance problems. Logic probes offer speeds and storage capabilities only equaled by the very fastest oscilloscopes.

Logic probes are applied to digital troubleshooting tasks in much the
same way as signal tracers are used in audio and RF circuits; often, in
conjunction with a digital pulser, they are used in much the same way as a
signal injector. Together, these two instruments permit methodical
stimulus–response evaluation of specific sections of circuitry. Since entire
logic trees can be followed and predicted results can be compared with
observations, faults are quickly isolated and repairs can be effected.

Combining logic monitors, whether clip-on or benchtop, with a digital
pulser greatly simplifies the testing of sequential circuits. Now entire ICs
or logic trees can be simultaneously monitored while the circuit is exercised. All output and input states are conveniently visible for immediate
detection of improper operation.

The logic probe, LP-3 (Figure 5.7), detects, memorizes, and displays
logic levels, pulses, and voltage transients in mixed and single logic family
systems.

The LP-3 detects out-of-tolerance logic signals and open circuit nodes,
as well as transient events down to 10 ns, while providing the user with an
instant easily interpreted high-intensity light-emitting diode (LED) readout. LEDs are discussed in detail in Chapter 9.

The probe tip of the LP-3 is connected to a dual threshold window
comparator and a bipolar edge detector. The window comparator bias network establishes the LOGIC 1 and LOGIC 0 threshold levels. The levels are

fixed in the DTL/TTL mode (2.25 and 0.8 V); in the CMOS/HTL mode, the thresholds are determined by the applied V_{cc} voltage: LOGIC $1 > 70\%$ V_{cc}, LOGIC $0 < 30\%$ V_{cc}.

The LP-3's pulse-detecting system consists of a level adjusted high-speed amplifier that drives a bipolar monostable pulse stretching circuit. The pulse stretcher converts level transitions as well as narrow pulses to one-tenth of a second pulses that drive one of the three readout LEDs. In the memory mode, the output of the pulse catcher is fed to a latching flip-flop. Table 5.1 shows the interpretation of the LEDs of the logic probe LP-3.

Testing the LP-3 for logic level tests and pulse tests are now given. The logic level test equipment required includes the following (see Figs. 5.8 through 5.10):

1. 5-V dc regulated power supply.

2. $3\frac{1}{2}$ digit, DVM, input impedance greater than 1 MΩ.

3. 1000-Ω linear potentiometer.

TABLE 5.1
Interpreting the LEDs of LP-3 logic probe

	LED STATES			INPUT SIGNAL	
	HIGH	LO	PULSE		
● LED ON	○	●	○	o————	LOGIC "0" NO PULSE ACTIVITY
○ LED OFF	●	○	○	o————	LOGIC "1" NO PULSE ACTIVITY
* BLINKING LED	○	○	○	————	ALL LEDS OFF 1. TEST POINT IS AN OPEN CIRCUIT. 2. OUT OF TOLERANCE SIGNAL. 3. PROBE NOT CONNECTED TO POWER. 4. NODE OR CIRCUIT NOT POWERED.
	●	●	*	⊓⊓⊓	THE SHARED BRIGHTNESS OF THE HI AND LO LEDS INDICATE A 50% DUTY CYCLE AT THE TEST POINT. (<1.5MHz)
	○	○	*	⊓⊓⊓⊓⊓	HIGH FREQUENCY SQUARE WAVE (>1.5MHz) AT TEST NODE. AS THE HIGH FREQUENCY SIGNALS DUTY CYCLE SHIFTS FROM A SQUARE WAVE TO EITHER A HIGH OR LOW DUTY CYCLE PULSE TRAIN EITHER THE LO OR HI LED WILL BECOME ACTIVATED.
	○	●	*	⊔⊔⊔⊔	LOGIC "0" PULSE ACTIVITY PRESENT POSITIVE GOING PULSES SINCE HI LED NOT "ON" PULSE TRAIN DUTY CYCLE IS LOW RE <15%. IF THE DUTY CYCLE WERE INCREASED ABOVE 15% HI LED WOULD START TO TURN ON.
	●	○	*	⊓⊓⊓	LOGIC "1" PULSE ACTIVITY PRESENT NEGATIVE GOING PULSES, SINCE LO LED NOT "ON" PULSE TRAIN DUTY CYCLE IS HIGH RE >85% IF THE DUTY CYCLE WERE REDUCED TO <85% "LO" LED WOULD START TO TURN ON.

Courtesy of Global Specialties Corp.

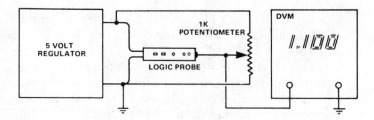

FIGURE 5.8
Logic level test circuit. (Courtesy of Global Specialties Corp.)

DTL/TTL test includes the following:

1. DTL/TTL: cmos switch in the DTL/TTL position.
2. PULSE/MEM switch in the pulse position.
3. Adjust the 1-kΩ potentiometer until the low LED goes on.
    - Maximum logic 0 voltage, 0.9 V
4. Adjust the 1-kΩ potentiometer until the HI LED goes on.
    - Maximum logic 1 voltage, 2.55 V
    - Minimum logic 1 voltage, 2.10 V

The test equipment for the pulse test of the LP-3 logic probe includes
the following:

1. 5-V dc regulated power supply.

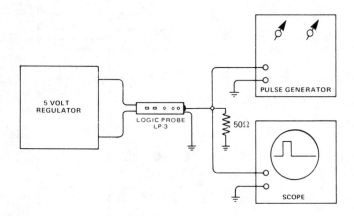

FIGURE 5.9
Pulse test circuit of the logic probe LP-3.
(Courtesy of Global Specialties Corp.)

FIGURE 5.10
Pulse generator testing. (Courtesy of Global Specialties Corp.)

2. Pulse generator model 4001 or equivalent.

3. Oscilloscope 250-MHz bandwidth (HP 1720 or equivalent).

DTL/TTL test includes the following:

1. DTL/TTL: CMOS switch in the DTL/TTL position.

2. PULSE/MEM switch in the pulse position.

3. Pulse generator set (see Figure 5.10).

4. Pulse LED will flash.

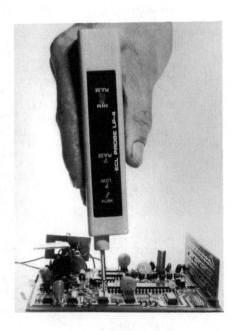

FIGURE 5.11
LP-4 ECL probe. (Courtesy of Global Specialties Corp.)

Memory test includes the following:

1. PULSE/MEM switch in the MEM position.
2. Test in the pulse test circuit.
4. Pulse LED will stay on.

NOTE: the pulse detection circuitry is independent of the DTL/TTL switch position.

Specifications for the LP-3 are shown in Table 5.2. Figure 5.11 shows the LP-4 ECL logic probe.

Emitter-coupled logic (ECL), discussed in Section 5.1, has long been used in mainframe computers wherein the fastest possible switching rates translate almost directly into system performance. More recently, ECL usage has expanded into a broader range of applications, including those listed in Table 5.3.

The LP-4 detects and indicates valid ECL levels using LED indicators for HI (logic 1) and LO (logic 0) levels using positive true logic (see Figure

TABLE 5.2
Specifications of the LP-3

	DTL/TTL	HTL/CMOS
IMPUT IMPEDANCE..... 500,000 Ohms		
THRESHOLDS .... Switch selectable		
Logic 1 (HI)	2.25V ± .015V	70% V_{cc}
Logic 0 (LO)	0.80V ± .010V	30% V_{cc}
MIN. DETECTABLE PULSE WIDTH....... 10 nanoseconds		
MAX. INPUT SIGNAL FREQUENCY ...... 50 MHz		
PULSE DETECTOR ..... High speed pulse train or single events (+ or − transitions) activate 1/10 second pulse stretcher, light PULSE LED switch selectable.		
PULSE MEMORY ...... Pulse or Level transition detected and stored until reset, keeping PULSE LED lighted.		
INPUT OVERLOAD PROTECTION .... ± 40V continuous 117VAC for less than 15 seconds.		
POWER REQUIREMENTS . 5 Volt V_{cc} @ 30 mA 15 Volt V_{cc} @ 40 mA 36 Volts max. with power lead reversal protection.		
OPERATING TEMPERATURE ..... 0° to 50° C		

Courtesy of Global Specialties Corp.

TABLE 5.3
Applications of emitter-coupled logic

Communications	Instrumentation	Peripherals
Error detection and correction	Signal generators	Add-on memories
High-speed modems	Correlators	Memory control systems
Signal processors	Frequency meters	Fast Fourier processors
Data compression	Counting–timing circuits	Dedicated computers
Digital filters	Frequency synthesizers	Preprocessors
Phase-locked loops	A/D converters	

Courtesy of Global Specialties Corp.

5.12). A third LED, labeled PULSE, indicates single shot or multiple pulses. It is capable of detecting a single-shot occurrence down to 3 ns in duration and pulse trains with repetition rates up to 100 MHz minimum (150 MHz typical at 50% duty cycle). With the PULSE LED flashing (pulse train) the HI or LO LEDs will indicate a positive or negative pulse polarity or duty cycle by their relative brightness (see Figure 5.13).

The LP-4 includes a two-position slide switch for the selection of PULSE or MEMORY (latch) modes. When in the PULSE position, the PULSE LED will flash a single 0.3-s (stretched) pulse indication by the application of a single pulse occurrence. With the presence of a pulse train, it will flash continuously at approximately a 3-Hz rate. With the switch in the MEMORY position, a single shot pulse will be stored and the PULSE LED will remain *on* until it is *reset* by toggling the switch. This is a useful feature for detecting spurious glitches or transients.

The LP-4 will also detect valid static ECL levels (see Figure 5.12). A logic 1 = -0.810 to -1.100 V dc; logic 0 = -1.50 to -1.850 V dc. The absence of pulses and a particular logic state (all LEDs off) results when the input of an ECL line receiver is biased to V_{BB} (-1.30 V) with no input signal. With the probe input floating, all LED indicators will be off since the probe input is biased at V_{BB} (-1.30 ± 0.05 V dc), the nominal ECL reference level.

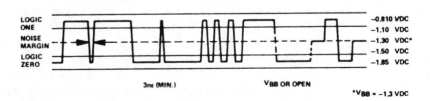

FIGURE 5.12
Valid ECL levels and waveform. (Courtesy of Global Specialties Corp.)

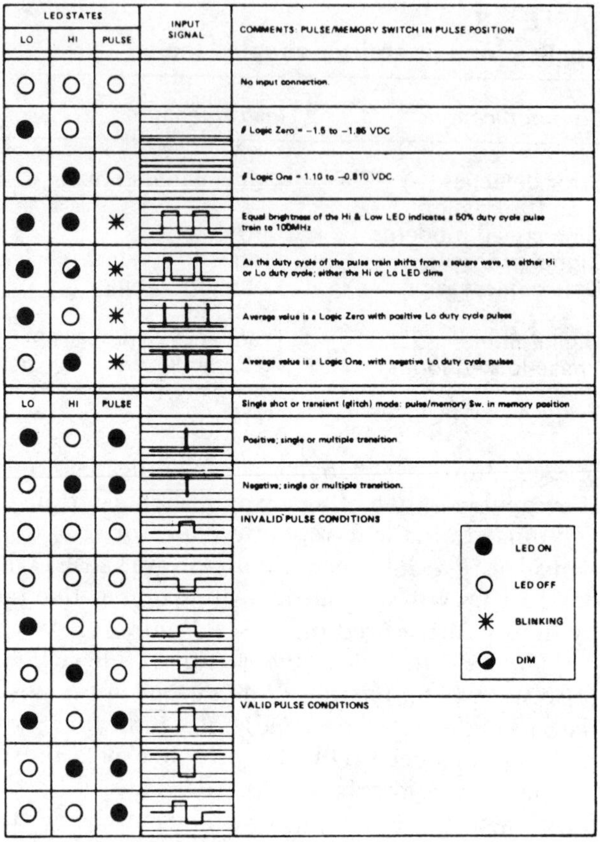

| LED STATES | | | INPUT SIGNAL | COMMENTS: PULSE/MEMORY SWITCH IN PULSE POSITION |
LO	HI	PULSE		
O	O	O		No input connection.
●	O	O	▭	∅ Logic Zero = -1.5 to -1.86 VDC.
O	●	O	▭	∅ Logic One = -1.10 to -0.810 VDC.
●	●	✳	⊓⊓⊓	Equal brightness of the Hi & Low LED indicates a 50% duty cycle pulse train to 100MHz
●	◑	✳	⊔⊔⊔	As the duty cycle of the pulse train shifts from a square wave, to either Hi or Lo duty cycle; either the Hi or Lo LED dims
●	O	✳	⌷⌷⌷	Average value is a Logic Zero with positive Lo duty cycle pulses
O	●	✳	⊓⊓⊓	Average value is a Logic One, with negative Lo duty cycle pulses
LO	HI	PULSE		Single shot or transient (glitch) mode: pulse/memory Sw. in memory position
●	O	●	⊥	Positive; single or multiple transition
O	●	●	⊤	Negative; single or multiple transition.
O	O	O	⊓	INVALID PULSE CONDITIONS
O	O	O	⊔	
●	O	O	⊓	
O	●	O	⊔	
●	O	●	⊓	VALID PULSE CONDITIONS
O	●	●	⊔	
O	O	●	⊓⌐	

Legend:
● LED ON
O LED OFF
✳ BLINKING
◑ DIM

FIGURE 5.13
LED states, input signal, and comments for pulse memory switch in pulse switch. (Courtesy of Global Specialties Corp.)

Figure 5.13 shows the various static and dynamic conditions as displayed by the LP-4 LED indicators.

Other features of the LP-4 logic probe include an input impedance greater than 10 kΩ to avoid circuit loading, input overload protection of ±100 V dc continuous, ±220 V dc transients, and 120 V dc for 30 s (to 1 kHz). Power requirements are −5.2 V dc (V_{EE}) at 100 mA with supply overload protection of −12 V to +200 V dc.

The LP-4 is basically comprised of the eight functional blocks shown in Figure 5.14, which includes the input protection circuitry, input buffer, high-speed logic 1 and 0 detectors, valid pulse selector, trigger and latch circuitry, 3-Hz (0.3 s) astable, ± voltage reference, and a power-supply protection circuit. The specifications for the ECL logic probe (Figure 5.11) are shown in Table 5.4.

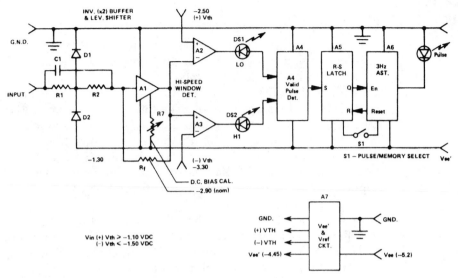

FIGURE 5.14
ECL-logic probe (simplified functional block diagram). (Courtesy of Global Specialties Corp.)

TABLE 5.4
Specifications for ECL logic probe model LP-4

THRESHOLD VOLTAGES, POSITIVE LOGIC:
Logic one (1): $-1.10 \pm .05V$; Logic zero (0): $-1.50 \pm .05V$;
$V_{EE} = -5.20$; $T_A = 25°C$. Tol. ± 0.100 @ $V_{EE} \pm 5\%$

INPUT IMPEDANCE: Greater than 10K Ohms

INPUT PULSE WIDTH (MINIMUM): 3 nanoseconds

INPUT PULSE REPETITION RATE:
100 MHz minimum (150 MHz typical at 50% duty cycle).

INPUT OVERLOAD PROTECTION: ± 100 VDC continuous;
± 220 VDC transients; 120 VAC for 30 seconds (to 1 KHz).

POWER REQUIREMENTS: -5.2 VDC $\pm 5\%$ at 100 milliAmpers.
Supply Voltage protection: -12 to $+200$ VDC.

OPERATING TEMPERATURE RANGE: $+10 - 40°C$.

5.4
LOGIC MONITORS

The logic monitor, LM-1, shown in Figure 5.15, simultaneously displays the static and dynamic logic states of DTL, TTL, HTL or a CMOS 14-pin and 16-pin digital DIP IC. The voltage at each IC lead is measured by one of 16 independent binary-optical voltmeters. When one of the input voltages exceeds the 2-V threshold, the LED corresponding to the activated input pin is turned on. Inputs below the threshold or uncommitted (floating) do not activate their corresponding LEDs. A built-in power-seeking gating network locates the most positive and negative voltages applied to the IC under test. It then feeds them to the internal buffered amplifiers and LED drivers.

During the design, breadboarding, and testing phases of a new logic system, the designer usually has full control of the system variables (clock, power supplies, input–output transducers, etc.) and can easily isolate ICs for detailed investigation with the logic monitor.

When a logic block needs an additional gate, inverter, flip-flop, register, or the like, the logic monitor can quickly "show" the designer where unused logic elements are located within the system. Nonfunctioning components can easily be located and replaced. Long-term testing of individual modules can be implemented by merely clipping the LM-1 onto the questionable IC.

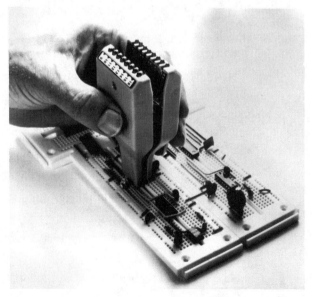

FIGURE 5.15
LM-1 logic monitor. (Courtesy of Global Specialties Corp.)

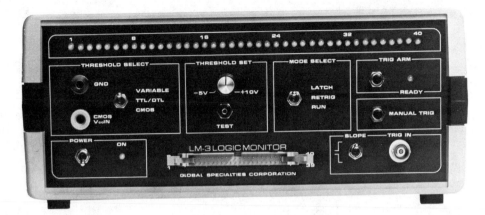

FIGURE 5.16
LM-3 logic monitor. (Courtesy of Global Specialties Corp.)

Since the entire IC can be monitored simultaneously, direct fast visual correlation of IC inputs and outputs simplifies and expedites signal-tracing data transfer and system fault-finding operations.

Mixed logic design, DTL, TTL, HTL, CMOS, where designers take advantage of individual logic family characteristics (i.e., DTL input, CMOS signal processing, and TTL or HTL outputs) can be used with the LM-1.

When dealing with multiple PC board systems, the LM-1 demonstrates its utility. One LM-1 on the inputs or outputs of the driving–receiving board and one on the board under test enable the designer to visually observe the results of any modification or simulation on one board while full attention can be directed to the focal point of the investigation.

The specifications for the LM-1 are shown in Table 5.5.

Figure 5.16 shows the Global Specialties Model LM-3 triggerable 40-channel logic monitor, an innovative logic test instrument in one convenient benchtop package. It has 40 precision threshold, high-speed, high-impedance logic state indicators, combined with a unique and highly flexible triggerable latching circuit. The turn-on thresholds adjust for any logic family and supply voltage. Forty easy-clips connect to the IC pins, test points, or bus line combination selected. The LM-3's 40-LED display then accurately follows the data or holds (freezes) the display in any number of ways.

In the RETRIG mode, the display follows the data until a rising or falling edge (switch selected) appears at the TRIG input; this latches display. The next selected edge updates latched display each time it appears. Also, MANUAL TRIG push button permits manual triggering.

In the LATCH mode, display follows data and ignores TRIG input until TRIG ARM button is pushed and READY LED lights. Upon the occurrence of a

TABLE 5.5
Specifications for LM-1

Input Threshold	2.0 ± 0.2 V dc
Input Impedance	100,000 Ω
Input Voltage Range	4 V minimum, 15 V maximum across any two or more input leads
Maximum Current Drain	200 mA at 10 V
Operating Temperature Range	5° to 45°C
Weight	3 ounces (85 g)
Maximum Dimensions	$L \times W \times D$: 4.0 × 2.0 × 1.5 in. 102 × 51 × 38 mm

Courtesy of Global Specialties Corp.

TRIGGER input the display latches. Any further TRIGGER inputs will not change the display until the TRIG ARM button is pushed. MANUAL TRIG push button also functions in this mode. In the RUN mode, display LEDs follow input data.

The LM-3 offers unprecedented flexibility in logic family compatibility through a unique threshold-selection scheme. Three methods of determining logic threshold levels are provided:

1. A fixed +2.2-V dc threshold is provided for logic operating at standard TTL/DTL levels.

2. A variable threshold between −5 V dc and +10 V dc is controlled by a front panel adjustment.

3. A supply-dependent threshold, determined at 70% of the V_{cc} of the circuit under test, is sampled through front panel dual banana jack connectors.

A front panel test point permits precise monitoring of logic threshold with a voltmeter regardless of which of the three methods is selected.

The 40 data inputs of Figure 5.17 are organized in the following manner: The input signal enters and passes through a passive voltage divider before entering the comparator. Across the input resistor is a variable capacitor that is used to speed up the response and to compensate for any stray capacitance of the comparator's input. Two diodes protect the comparator's inputs.

The data that appear at the comparator's input are compared to V_r (reference) and will produce a low output if greater than V_r. The output of the comparator is held in the load mode, which allows it to follow the input data. When a high condition appears at the data input, the comparator's output will be low, causing the 74193's output to go low and turn LED on.

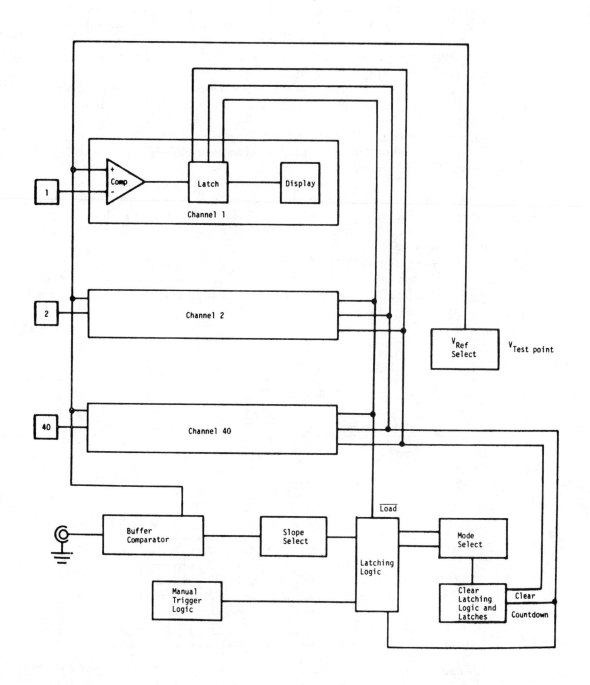

FIGURE 5.17
LM-3 logic monitor block diagram. (Courtesy of
Global Specialties Corp.)

The trigger circuit operates in the following manner: The input signal comes in and goes through a voltage-divider network. A variable capacitor is across the input resistor to speed up the response and compensate for stray capacitance. This signal is fed into a FET, which has a high input impedance and is not affected by the voltage divider. The source of the FET has a constant current source, which draws 3 mA. The FET output signal is fed into a high-speed comparator, whose output is in turn taken off the inverting port and fed into an exor, which sets up an invert–noninvert trigger input signal. The output of this exor is fed into a positive pulse detecting one shot. This one shot clocks two D flip-flops shifting the D input data to the Q, $\bar{Q}$ output.

The data at the input of the D flip-flops is selected by the MODE SELECT switch. Upon the transition of the MODE SELECT switch, a one shot is triggered, which clears the data latches (74193). This signal is delayed and inverted to have the latches count down. This will have all the LEDs show a low output.

The logic monitor LM-3 (Figure 5.16) can be used to troubleshoot and locate glitches of a microprocessor system. Glitches are discussed in Chapter 6. The glitch has to be at least 100 ns long. One technique is shown in Figure 5.18.

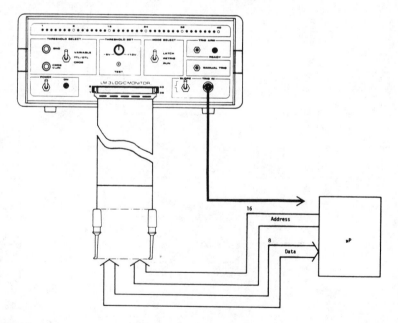

FIGURE 5.18
LM-3 used to troubleshoot and locate glitches on a microprocessor. (Courtesy of Global Specialties Corp.)

The LM-3 data inputs monitor the address and data lines while the trigger input is probed around the circuit latching the data when a transition occurs.

The LM-3 can perform some of the basic functions of a signature analyzer. This is accomplished by monitoring a known good board under a specific set of conditions, with data recorded at different stages of the circuit. This data will be compiled in either a troubleshooting book or on a schematic stating the exact set of conditions when the data was recorded. When one of these boards fails, the LM-3 can be reconnected to the PC board at the data points indicated on the schematic, reproducing the set of conditions stated on the schematic. The user can trace the circuit through, comparing the new data with the old until a discrepancy in the new reading is located. This will demonstrate that the previous stage was working, but this stage is defective.

5.5
DIGITAL PULSER

The digital pulser shown in Figure 5.19 fulfills a long neglected need in the digital service industry, a completely automatic pulse generator in a pencil-sized probe, tailored for both the laboratory and field service. The DP-1 allows the user to conveniently pulse any family of digital circuits. By obtaining its power from the circuit it is testing, the DP-1 self-adjusts the amplitude of its output pulse to the input requirements of the circuit under test.

When the pulser tip is connected to the circuit node to be tested, the DP-1's *autopolarity sensing system* selects the sink or source pulse required to activate the test point. Simply depress the push button once to produce a clean, bounce-free pulse. When the push button is held down for more than 1 s, the DP-1 produces a pulse train at a 100 pps rep rate.

The DP-1 is capable of sinking or sourcing 100 mA or 60 TTL loads at a 1.5-microsecond (μs) pulse width. If a wider pulse is required, simply move the TTL/CMOS slide switch to the CMOS position to increase the pulse width to 10 μs. This allows reliable triggering of even the slowest CMOS devices at the lowest V_{cc}'s.

The DP-1 in combination with the LP-1 logic probe or the LM-1 logic monitor produces an extremely effective method of troubleshooting, and in many cases more useful than an oscilloscope.

Figure 5.20 shows the hookup for checking out a 7490 decade counter using a pulser and a logic monitor. The pulser CMOS/TTL slide switch is set to TTL, and the pulser tip is connected to the R(0) input of the 7490. The logic monitor is clamped onto the 7490 to display all the logic states of the counter simultaneously.

Depressing the DP-1 push button once puts a zero pulse into the 7490 and clears all the outputs to zero. The pulser can now be applied to the clock

FIGURE 5.19
Digital pulser DP-1. (Courtesy of Global Specialties Corp.)

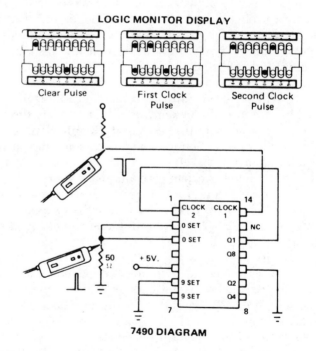

FIGURE 5.20
Hookup for checking a decade counter. A logic monitor display is also shown. (Courtesy of Global Specialties Corp.)

input and single step or jog the 7490 through its decade cycle. When the counter is pulsed, all four outputs can be seen changing state, while simultaneously monitoring the power supply input, clock inputs, and clear lines. This shows the great advantage of the logic monitor pulser approach over the oscilloscope, which at best can only monitor one or two points at a time. Figure 5.21 shows this method of troubleshooting.

With troubleshooting gates and decoders, the use of a logic probe is recommended. Such desirable probes incorporate pulse stretchers to aid in viewing even the fastest pulses of the DP-1. In addition, the digital pulser can be used to indicate logic states and pulse polarity, as well as estimate duty cycles.

In Figure 5.21(a) a two-gate circuit is being tested. $G1$'s output is held high, causing $G2$'s output to be low. By applying the pulser to the output of $G1$, the pulser overrides the out state of $G1$ and puts a train of "zero" pulses into the gate of $G2$. The logic probe connected to the output of $G2$ has its low LED on, but now the pulse LED starts flashing. This shows the gate is passing the input pulses in proper polarity.

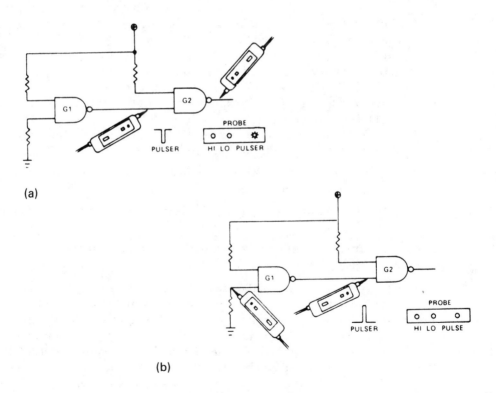

(a)

(b)

FIGURE 5.21
Troubleshooting gates. (Courtesy of Global Specialties Corp.)

In Fig. 5.21(b) the probe is moved to the output of $G1$ and the pulser is applied to the low gate input. The pulser now produces a series of (one) pulses when the push-button is held down. However, the probe's pulse LED does not respond, indicating a defective gate.

5.6
LOGIC ANALYZERS [c]

Logic systems are built using hardware and software. A malfunction at the point of integrating hardware and software could send the engineer back to check the control lines or to verify that the data lines and controlled circuits are transmitting the state flow properly. Different analysis instruments are optimal for different phases in the development of a logic system. Digital multimeters and oscilloscopes are required to create the base of the hardware used to assure that the circuits do indeed function properly and within the limits specified.

As systems become more sophisticated, analysis tools as logic analyzers are needed to obtain a better overview of the system being diagnosed. The logic timing analyzer is frequently called the hardware designer's tool. A primary function of the timing analyzer is to provide a display of the timing relations between signals on the logic circuits discussed in Section 5.1. The logic monitors discussed in Section 5.5 are examples of logic analyzers.

The Tektronix 7D02/Option 01 (Figure 5.22) is two complete logic analyzers in one. The state analyzer has unprecedented triggering power to search out malfunctions in complex, intelligent systems. A separate timing analyzer simultaneously monitors hardware to pinpoint timing problems. Features of the 7D02 include the following:

1. Unprecedented triggering power; trigger on the problem instead of around it; reduce system debug time.

2. Disassembled mnemonics for fast, simplified interpretation of state information.

3. Trigger on glitches and store them in a separate memory; lets you know exactly when glitches occurred.

The 7D02 logic analyzer can acquire and store up to 52 channels of state information in the synchronous mode, using the clock of the system under test. The basic instrument contains 28 channels, with an expansion option (03) increasing the number of stored channels to 44. A timing option (01) provides 8 additional channels of state information for a total of 52 synchronously stored channels. Alternatively, the timing option provides 7 channels of asynchronous timing simultaneously with 44 channels of state information. The state and timing sections can be operated independently or used as a trigger source for one another.

---

[c] The material in Section 5.6 is used courtesy of Tektronix, Inc.

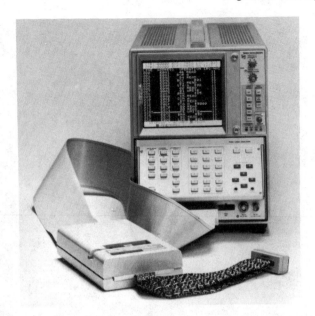

FIGURE 5.22
Tektronix Model 7D02 logic analyzer. (Courtesy of
Tektronix, Inc.)

All the 7D02's data-acquisition resources are under the control of a
powerful user language that allows the trigger menu to be user defined. The
trigger menu may be a simple word recognition of a complex nonsequential
algorithm including real run-time counter status. Through user programming, almost any combination of resources can be employed to construct
specific triggers or data qualifiers. The 7D02 can track nonsequential activities, make decisions to follow branches, monitor two real-time counters in
either time or events mode, and trigger when they reach some predefined
number.

5.7
DIGITAL IC TEST SYSTEM[d]

The GenRad 1731 linear test system (Figure 5.23) is controlled by two
microprocessors, which are discussed in detail in Chapter 16. The main
microprocessor, a Z80-based microcomputer, handles test calculations and
the display. A second processor controls the mass-storage functions, freeing
the Z80 for testing at high throughput.

Family boards are designed modularly to accommodate future testing
needs. Because the 1731 uses random-access memory (RAM), discussed in

---

[d]The material in Section 5.7 is used courtesy of GenRad, Inc.

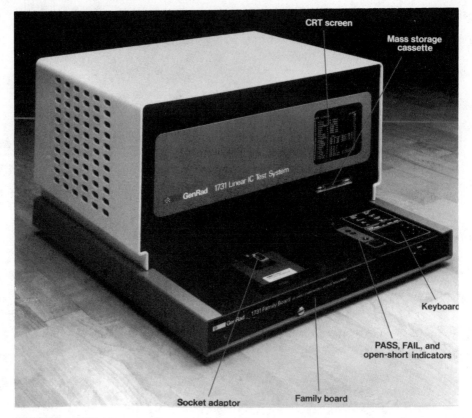

FIGURE 5.23

GenRad 1731 linear IC test system. (Courtesy of GenRad)

Chapter 16, the operating software to accommodate new tests and new linear device families can be easily modified. The capability needed to test the most popular classes of ICs (op-amps, fixed and adjustable voltage regulators, voltage comparators, voltage followers, and current mirror amperages) is available on one family board, and new boards are being developed for additional classes of ICs.

The 1731's specifications equal or exceed those of large IC test systems. For example, when op-amps are tested, bias current, positive–negative input current, and offset current can all be measured with 1.5-pA resolution. Op-amp slew rates are tested to 1000 V/μs, a specification that no other system, large or small, can top. With performance like this, you can rely on the 1731 for rapid, repeatable, and accurate tests.

Data cartridges provide a permanent record of the operating system and subsequent test plans written from the user keyboard. System output modes are flexible and easy to use. A 5-in. CRT permits easy program entry; you

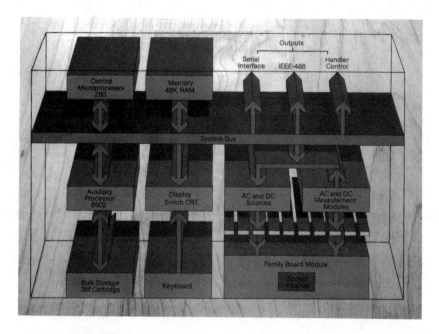

FIGURE 5.23 (Continued)

simply fill in the blanks on the displayed program guide. Both PASS and FAIL test results are clearly shown.

An IEEE-488 standard bus interfaces with calculators, computers, and their data libraries. A 20-mA dc current loop and an optional RS-232 link offer direct data output to printers and computers.

The 1731 solves the problems commonly associated with automatic testing and handling. Testing at the handler interface is as simple and precise as testing at the socket adaptor. The unique architecture of the detachable family module maintains the accuracy of the ac tests and low-current measurements at the handler interface. There is no need for fine tuning the compensation networks for cable capacitance or for guard-banding because of cable length.

In brief, the 1731 gives you all the big-system capability you need, but with benchtop convenience. The extensive range of tests includes:

1. Operational amplifiers

2. Voltage regulators

3. Voltage comparators

4. Voltage followers

5. Current mirror amperages

IC digital testers vary from manufacturer to manufacturer due to the update in microprocessor electronic instrumentation technology.

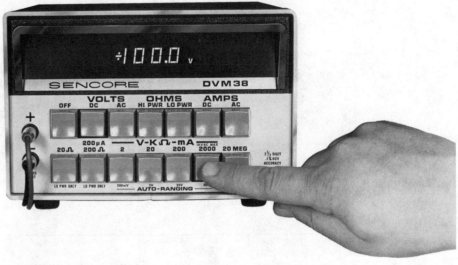

FIGURE 5.24
Sencore DVM38 digital multimeter. (Courtesy of
Sencore, Inc.)

5.8
DIGITAL MULTIMETER
OPERATION THEORY[e]

5.8.1 Introduction

This section describes the theory of operation of the DVM38 shown in
Figure 5.24. Simplified diagrams and descriptions are included for a basic
understanding of the functions and operations of each circuit. These will be
helpful in learning how the DVM38 operates.

5.8.2 Theory of Operation

Three main circuit functions must be performed by the DVM38 on each
measurement for it to be viewed as the digital display (see Figure 5.25). The
signal-processor section converts the input signal to a dc voltage that is
suitable to drive the analog-to-digital (A/D) converter. The converter produces the digital coding that operates the display section to produce the
digital readout.

---

[e]The material in Section 5.8 is used courtesy of Sencore, Inc.

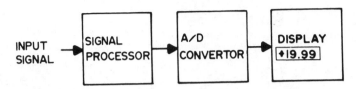

FIGURE 5.25
Simplified block diagram of DVM38. (Courtesy of
Sencore, Inc.)

The signal processor includes the function and range switches to select the operation to be performed on the input signal. The dc inputs are dropped to their proper level by range voltage dividers and current shunts. An ac-to-dc converter rectifies and filters all ac signal inputs. The HI and LO power ohms circuits produce a dc output voltage proportionate to the unknown resistance. The processor output is a ±0 to −2 V dc analog input used by the A/D converter.

The heart of the DVM38 is the MN2301 LSI IC chip, which converts the ±0 to −2 V dc analog input voltage to the digitally coded outputs that will give the proper display readout of the measured signal. This A/D converter uses the dual-slope method of converting the analog input to the binary-coded data (BCD) for the display circuits. IC207 receives a 160-kHz signal from the clock circuit and divides these pulses by 8. The 20-kHz pulses are then counted to three 2000-count periods for one sample period (see Figure 5.26). The first count period integrates, or ramps, the analog input voltage. The slope of this ramp, and hence the peak point of the ramp, varies directly with the amount of the input voltage.

The averaging during this ramp time reduces noise and transient interference. During the second count period, the integrated input is reduced to zero at a constant discharge rate. The time (number of clock counts) required for this to occur is proportionate to the input voltage. The analog-to-digital converter produces the number of the counts in the discharge period

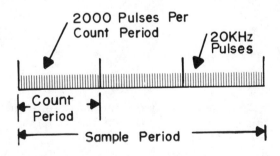

FIGURE 5.26
Pulse of sample period. (Courtesy of Sencore, Inc.)

POWER SUPPLY

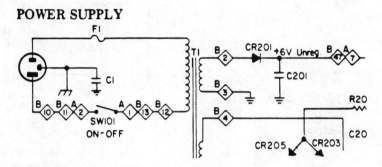

FIGURE 5.27
Count period of A/D converter. (Courtesy of Sencore, Inc.)

as the coded data output for the display. The third count period has no signals applied and is used to set the autozeroing circuitry within IC207 (see Figure 5.27). The A/D converter IC also provides outputs for the automatic polarity and autoranging features.

The display section converts the BCD output to a decimal output and presents the decoded information on 7-segment LED (light-emitting diode) readouts, discussed in Chapter 9. The display, polarity, and decimal driver circuits accomplish the decoding process.

The preceding results in an extremely accurate and clearly readable presentation of the measured signal.

5.8.3   Circuits of the Digital Multimeter

Useful circuits in the digital multimeters include the following:

1. Power supply
2. AC voltage converter
3. Clock
4. Ramp
5. Digital output
6. Autorange

The power supply of Figure 5.28 provides four regulated and one unregulated output to power all DVM38 circuitry. The ac power line is transformer isolated and fuse protected from the stepped down secondary circuits. A bridge rectifier circuit ($CR202$–205) and filter networks ($R201$, $C202$, $R202$, $C203$) feed the regulator ICs with an unregulated positive and negative 15 V. Circuit ground is referenced to a center tap on the transformer secondary.

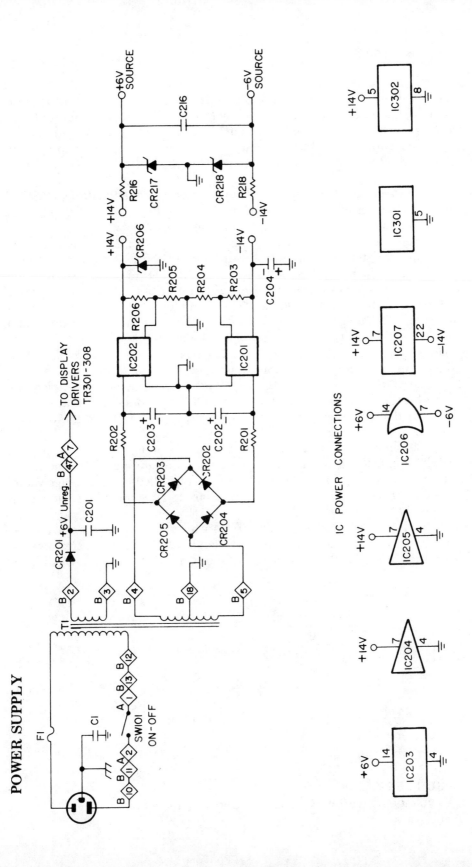

FIGURE 5.28
Power supply and IC power connections.
(Courtesy of Sencore, Inc.)

Sources of positive and negative 14 V power the ICs and also serve as the B+ line for some of the transistor circuits. The +14-V output is regulated by $IC202$. The divider formed by $R205$ and $R206$ at the output forms a control voltage at the junction, which is fed back to $IC202$ to determine the amplifier gain. The -14-V regulator, $IC201$, and the divider formed by $R203$ and $R204$ operate in a like manner. $CR206$ and $C204$ protect and filter the sources.

The 14-V sources are stepped down and zener regulated through $CR217$ and $CR218$ to provide the positive and negative 6-V sources. $C216$ further filters the output against circuit interaction.

A separate secondary winding and rectifier ($CR201$) are used for the +6-V unregulated line that supplies the B+ for the LED readout drivers.

The ac voltage input from the ac voltage divider or current shunt is fed to the ac voltage converter (Figure 5.29) to be changed to the dc voltage required by the A/D converter. The input enters $IC204$ after passing through a dc blocking network ($R248$, $C214$) and the protection circuit ($CR210$, $CR207$). $IC204$ is an op-amp that amplifies the signal to approximately 13 V p–p. This level is required to overcome the losses to the output signal due to rectification and averaging of the input ac signal.

The CMOS op-amp is extremely linear, but does not have adequate drive for the detector circuit. Therefore, $TR203$ is connected as an emitter follower that provides a constant output impedance to drive the detector circuit.

$C207$ blocks the dc component, but passes ac to the detector circuit. $CR209$ shunts the ac negative half-cycle to ground through $R210$ and $C208$. $CR208$ passes the positive half-cycle that charges $C210$ to the average value. The output is filtered through $R212$ and $C209$ and becomes the analog input for the A/D converter, $IC207$.

Precision resistors $R211$ and $R207$ form an ac voltage divider that determines the ac feedback level to $IC204$ through $C229$. This feedback, plus the dc return through $R208$, determines the op-amp gain and improves linear-

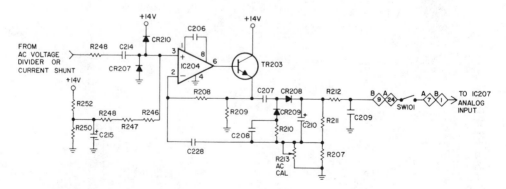

FIGURE 5.29
AC voltage converter. (Courtesy of Sencore, Inc.)

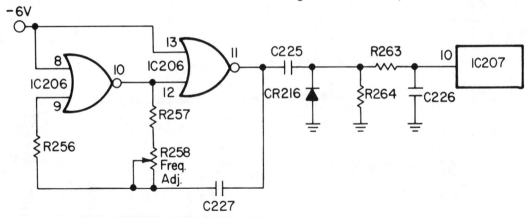

FIGURE 5.30
Clock. (Courtesy of Sencore, Inc.)

ity. $R213$ serves as the ac calibration control by paralleling $R207$ in the divider.

Since the $IC204$ negative supply is at ground potential, the divider formed by $R252$ and $R250$ between the $+14$-V source and ground offsets the (+) input to pin 3 to $+7$ V for proper dc balance at the output. High-resistance isolation from B+ is maintained by $R246, R247,$ and $R248$. $C215$ filters this line.

The clock, shown in Figure 5.30, generates a 160-kHz pulse output to be used by the A/D converter. The clock uses $IC206$ in a NOR gate flip-flop configuration. The feedback path through $C227, R258,$ and $R257$ determines the oscillator frequency, with $R258$ calibrating the frequency. $R256$ resets the first gate at the end of each cycle.

The ac output from the gates passes through the dc blocking capacitor, $C225$, and is clipped by $CR216$. $R263$ and $C226$ form an integrator that rounds the leading edge of the pulse so that $IC207$ will accept it. $R264$ restores the dc path.

The key operation performed by the analog-to-digital converter $IC207$ consists of charging the integrating capacitor $C221$ through $R296$ to a value proportional to the analog input voltage. This ramp is produced during the first count period (see Figure 5.31).

This integrated input voltage on $C221$ is then discharged to zero at a constant rate through $R222$ and $R220$, if positive, or $R221$ and $R219$, if negative. The conversion to coded digital output is made when the voltage reaches zero on $C221$.

The third, or zeroing, count period is used to establish the zero reference for the charge–discharge cycle. Zero shift and errors are eliminated with this comparison each sample period.

Extremely high stability and accuracy is achieved by making the read, count, and zero operations within a single sample period (300 ms).

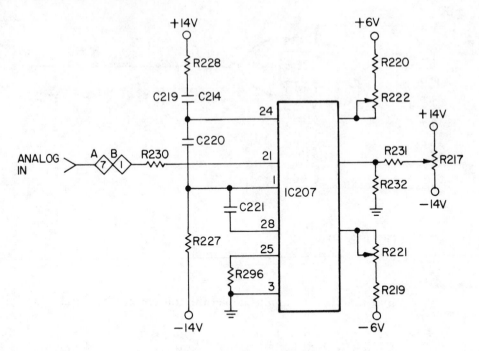

FIGURE 5.31
Ramp. (Courtesy of Sencore, Inc.)

The 000 readout is determined by the $R217$ control, and the divider $R231$ and $R232$. $C220$ improves stability during the auto zero interval. $C219$ and $R228$ stabilize the feedback circuit to prevent oscillation on the overrange display. $R227$ improves the ramp linearity. Calibration is made by $R222$ on positive analog input voltages and by $R221$ on negative inputs.

When the integrating capacitor in the ramp circuit is discharged to zero during the count period, the number of clock pulses at that instant is latched, converted, and transferred to the binary-coded decimal (BCD) circuits in $IC207$ (see Figure 5.32).

These four digits are stored and then displayed in sequence at a 5-kHz rate. The BCD output selects which cathode elements are to be lit while the anode multiplex output individually strobes the display digits. Only the display segments that have both the cathode and anode selected will complete the current path and illuminate.

$IC302$ accepts the BCD output and converts it to 7-segment information for the display drivers in $IC301$.

The autorange circuitry shown in Figure 5.33 performs the two functions of (1) changing the sensitivity of $IC207$ for either high or low scale measurement, and (2) providing the correct display readings for the decimal and v, mv indicators.

*IC*206 is a latch that maintains a HI or LO state until changed by an input pulse. The output voltage is the U-LINE (underscale) state.

When the ramp pulse counter in *IC*207 determines a count, or reading, below 180 (approximately one-tenth of a full scale), it produces a positive pulse at pin 4. *IC*206 then latches the U-LINE at pin 3 to a HI state. If the

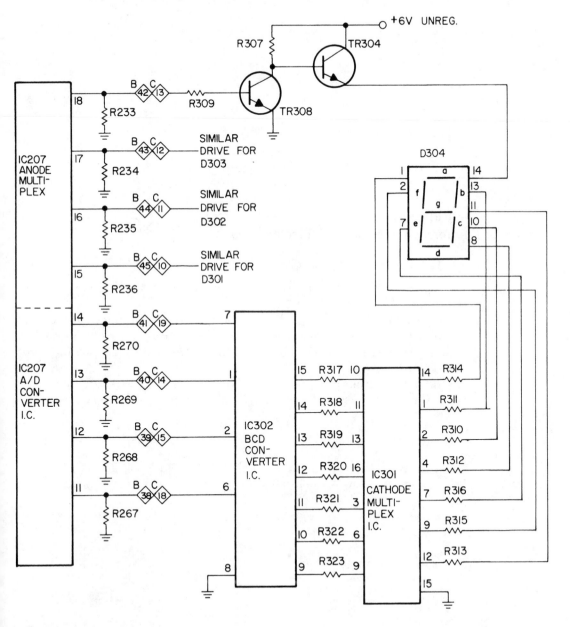

FIGURE 5.32
Digital output. (Courtesy of Sencore, Inc.)

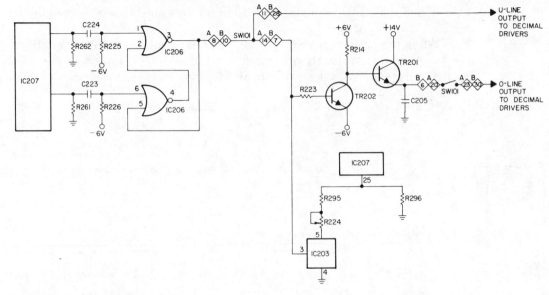

FIGURE 5.33
Autorange. (Courtesy of Sencore, Inc.)

count goes above 1999, a pulse is produced at pin 6, causing the U-LINE to be reset to the LO state by $IC\,206$.

Transistor $TR\,202$ inverts the U-LINE state to produce the O-LINE (overscale) state. The U-LINE and O-LINE are always in opposite states. $TR\,201$ serves as an emitter follower on the O-LINE to the V and MV indicator driver transistors.

$R\,296$ determines the gain of $IC\,207$ by providing a 1-MΩ path to ground. When the U-LINE goes to the HI state, $IC\,203$ completes the ground return for

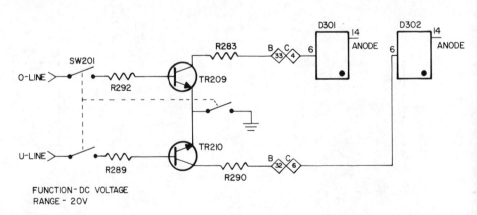

FIGURE 5.34
Decimal drivers. (Courtesy of Sencore, Inc.)

$R295$ and $R294$. These now parallel $R296$ for a total resistance of 100 kΩ, increasing the $IC207$ gain 10 times.

The proper decimal display is controlled by the function and range switches and the O-LINE and U-LINE states. A simplified example of the three decimal positions is shown in Figure 5.34. The correct pair of driver transistors is selected by grounding both emitters with the range switch. The O-LINE and U-LINEs will then control which transistor will be driven ON with the HI state on the base. This HI state saturation will complete the ground return for display decimal, lighting the proper indicator of either the auto-range high or low scale.

Figure 5.35 shows the complete schematic diagram of the DVM38 digital multimeter by Sencore.

5.9
DIGITAL PANEL METERS

The DM-3100B, shown in Figure 5.36, is a very low cost, dual ac-powered digital panel meter. Analog voltages over the range of ± 1.999 V dc are displayed with $3\frac{1}{2}$ digits of resolution. The DM-3100B is powered from the ac line; 115 or 230 V ac are pin-selectable. The unit can provide +5 V and −5 V dc (at 100 and 5 mA, respectively) to power to customer-supplied external circuitry.

The DM-3100B uses a self-illuminated red LED display with 0.56-in.-high numerals. It is clearly visible from many feet away in normal or dim light.

Inputs to the DM-3100B are balanced differential (80-dB common-mode rejection), so the meter will accurately display small signals even in electrically noisy industrial environments. CMOS circuitry results in an extremely high input impedance (1000 MΩ, typically) and a very low bias current of 5 pA; inputs with a source impedance as high as 100 kΩ can be displayed with accuracy. The input circuitry will also safely tolerate over-voltages up to ± 250 V dc (155 V rms). Inputs are sampled and displayed about four times per second.

Autozeroing and a ratiometric reference in–out loop, as shown in Figure 5.37, permit the DM-3100B to be used for drift correction in bridge-type measurement systems. Meter accuracy is adjustable to $\pm\frac{1}{10}$% (± 1 count). Temperature drift of zero is ± 1 count from 0° to 50°C, while temperature drift of gain runs typically ± 50 ppm of reading/°C.

The ac-powered DM-3100B was designed for installations where existing dc supplies are noisy, inaccessible, or overloaded. This meter may be used wherever a voltage, or a unit that can be made proportional to voltage, must be displayed with accuracy and clarity. The basic input range of ± 1.999 V dc can be expanded with a simple voltage divider to display voltages up to ± 1 kV or up to ± 2 A using current shunts. Blank pads on the meter's circuit board can accept user-supplied voltage attenuator resistors current shunts or digital ohmmeter components.

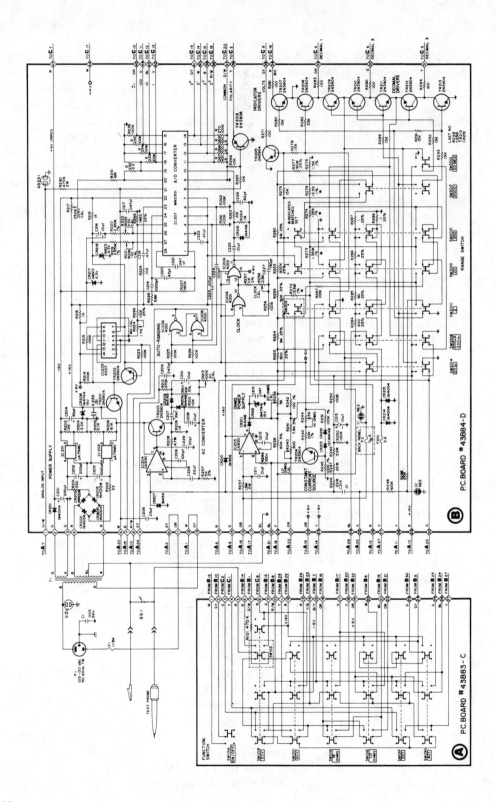

192

FUNCTION BOARD

SCHEMATIC REFERENCE	PART NO.	DESCRIPTION
SW101	25A223	Switch
SW102	25A225	Switch

RANGE BOARD

SCHEMATIC REFERENCE	PART NO.	DESCRIPTION
IC201	69G16	IC, Neg Volt Reg. uA79MG
IC202	69G15	IC, Pos Volt Reg. uA78MG
IC203	69G12	IC, Array, CMOS, 4007
IC204, 205	69G21	IC, Op Amp, CA3130
IC206	69G7	IC, Quad NOR Gate, 4001
IC207	69G17	IC, A-D Converter, MN2301
TR201–205, 207–213	19A33	Transistor, NPN, 2N3904
TR206	19A34	Transistor, PNP, 2N3906
CR201, 202, 204, 214	16S10	Diode, 1N4004, 1A
215, 203, 205	50C4-11	Diode, Zener, 15V, 1N4744
CR206, 212	50C5-1	Diode, 1N456
CR207, 210	50C5-2	Diode, 1N4148
CR208, 209, 211, 213, 216, 219	50C12-1	Diode, Zener, 6.3V, 1N823
CR217, 218	15C7-13	Control, 100K
R213	15C7-25	Control, 250K
R217	15A19-1	Control, 500K, Multiturn
R221, 222	15C7-10	Control, 1.7K
R224	15C7-14	Control, 5K
R239, 244	15C7-3	Control, 500
R240	15C7-31	Control, 2.5K
R258	25A224B	
SW201	25A225	
SW202	26G1	Switch
J201, 202	44G12	Switch
F201		Fuse, 2A, 3AG

Resistors

R207	14C30-5003A	5K, 2%, ½ W
R211	14C28-1004A	10K, .5%, ½ W
R219, 220	14C29-2856A	2.85 Meg, 1%, ½ W
R230	14C31-105	470K, .10%, ½ W
R238	14C28-8004A	80K, .5%, ½ W
R241	14C28-7903A	7.9K, .5%, ½ W
R242	14C32-1002A	100, 1%, 1 W
R243	14C29-2004A	20K, 1%, ½ W
R245, 251	14C29-1005A	100K, 1%, ½ W
R253, 284	14C33-9002A	900, 25%, 1 W
R254	14C27-9003A	9K, .25%, ½ W
R255	14C27-9004A	90K, .25%, ½ W
R257	14C30-1104A	11K, 2%, ½ W
R259	14C27-9005A	900K, .25%, ½ W
R260	14C27-9006A	9 Meg, .25%, ½ W
R272	14C25-1357A*	13.5 Meg, .1%, 1 W
R273	14C27-906A	9 Meg, .1%, 1 W
R274	14C25-1356A*	1.35 Meg, .1%, 1 W
R276	14C26-1354A*	13.5K, .1%, ½ W
R277	14C27-9007A	90 Meg, .25%, ½ W
R279	14C26-1603A	1.5K, .1%, ½ W
R285	14C27-9001A	90, .25%, ½ W
R286	14C27-9000A	9, .25%, ½ W

*Indicates matched sets

SCHEMATIC REFERENCE	PART NO.	DESCRIPTION
R287	14C27-1000A	1.0, .25%, ½ W
R288	43E11	Resistance Wire, 4¼"
R295	14C27-1105A	110K, .25%, ½ W
R296	14C27-1006A	1 Meg, .25%, ½ W

Capacitors

C201, 202	24G252	200 mF, 50V, Lytic
C203	24G272	1000 mF, 25V, Lytic
C204	24G271	50 mF, 35V, Lytic
C205	24G31	.03 mF, 600V, Disc
C206, 223, 224, 229	24G172	100 pF, 500V, Disc
C207, 210	24G335	10 mF, 16V, Tantalum
C208, 215, 216	24G120	10 mF, 15V, Lytic
C209, 213, 217	24G212	.047 mF, 250V, Mylar
C211	24G303	.01 mF, 100V, Mylar
C212, 225, 230	24G207	.01 mF, 100V, Mylar
C214	24G331	.1 mF, 1000V, Mylar
C218, 222	24G289	.47 mF, 100V, Film
C219	24G214	1000 pF, 500V, Disc
C220	24G192	47 pF, 100V, Disc
C221	24G237	.047 mF, 33V, Film
C226	24G95	82 pF, 100V, Disc
C227	24G230	130 pF, 100V, Film
C228	24G309	.22 mF, 100V, Mylar

DISPLAY BOARD

IC301	69G19	IC, CA3081
IC302	69G18	IC, MC14558
TR301–308	19A33	Transistor, 2N3904
D301, 302	20G15	Lamp, 10ES
D301	23G58	Display, L.E.D., ±1
D302, 303, 304	23G57	Display, L.E.D., 7 Segment

CHASSIS COMPONENTS

C1	24G305	Capacitor, .005 uF, 3KV
	28B65	Transformer
T1	44G5	Fuse, .2A, 3AG
F1	44G13	Fuse, 1/16A, 3AG
F2	64G28	Fuse Holder
	27G16	Line Cord, SPT-3, 6 ft.
	21A58	Red Filter
	63K16	Pushbutton
	37G25	Rubber Foot
	39G91	Probe and Lead Assembly
GD2	25A132	Probe Switch
GD201	33G245	1.5 KV Surge Arrester
	35G248	230V Gas Discharge Device

P.C. BOARD #43B82-A

NOTES:
1. ALL CAPACITANCES GREATER THAN ONE ARE pf, LESS THAN ONE ARE uf.
2. UNLESS OTHERWISE NOTED.
3. DENOTES CIRCUIT GROUND, DENOTES CHASSIS GROUND.
4. DASHED LINES (- - - -) REPRESENT MECHANICAL PARTS AND CONNECTIONS.
5. FUNCTION SWITCH SHOWN IN OFF POSITION.
6. ALL VOLTAGES MEASURED WITH RESPECT TO CIRCUIT GROUND USING MEGOHM INPUT DIGITAL MULTIMETER
7. FRONT PANEL CONTROL NAMES ARE ENCLOSED IN A BOX.
8. PC BOARD CONNECTIONS CORRESPOND TO SCHEMATIC REFERENCE NUMBERS.

FIGURE 5.35
DVM38 digital multimeter complete schematic diagram and parts list. (Courtesy of Sencore, Inc.)

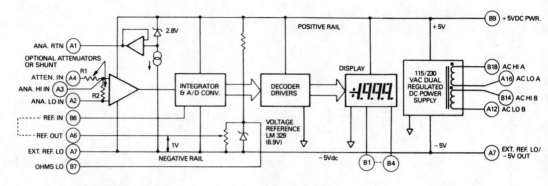

FIGURE 5.36
Datel-Intersil models DM-3100B simplified block
diagram. (Courtesy of Datel-Intersil, Inc.)

Figure 5.38 shows the digital panel meter model 3100LM, which uses
a dual slope A/D converter. This panel meter is useful in the design of
modern electronic instruments.

5.10
DATA-ACQUISITION APPLICATION TO
COMPUTERS

Transducers have been discussed in Chapter 2 in detail. Voltage-to-
frequency converters have been discussed in Chapter 4. These two devices
can be combined to form a data-acquisition system that will be very useful
in the electronic instrumentation in the next five to ten years. The data-
acquisition system is shown in Figure 5.39.

The Intel 2920 is a signal processor chip used for the following:

1. Telecommunication

2. Process control

3. Digital signal processing

4. Modem applications, discussed in Chapter 10

5. Guidance and control

6. Speech processing

7. Industrial automation

Following an analog signal and an antialiasing filter, the Intel 2920
digital-signal-processor, in a single integrated-circuit chip, forms the following blocks:

1. Sample and hold circuit

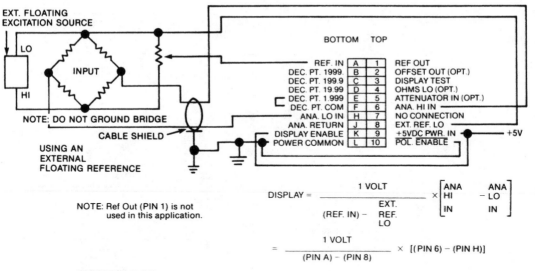

FIGURE 5.37
Ratiometric connections with bridge inputs. DPM
is an abbreviation for the digital panel meter DM-
3100B. (Courtesy of Datel-Intersil, Inc.)

2. A/D converter

3. Digital processor

4. D/A converter

5. Sample and hold circuit.

The output signal (analog) is fed to a reconstruction filter and readout
device.

FIGURE 5.38
Digital panel meter model 3100LM. (Courtesy of
Datel-Intersil, Inc.)

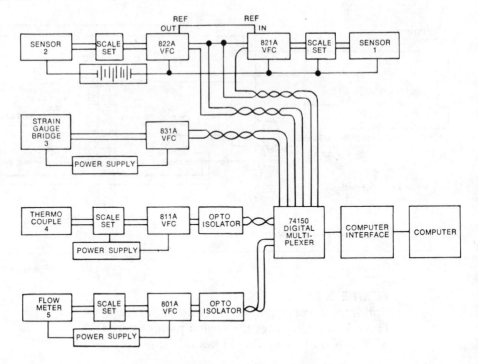

Voltage-to-frequency converter = V/F = VFC

FIGURE 5.39
Data-acquisition system. (Courtesy of Dynamic
Measurement Corp., Winchester, Mass.)

Today, digital-signal-processing integrated circuit chips are a reality
and are approaching the art of microprocessors.

The data-acquisition system using voltage-to-frequency converters
(VFCs) (Figure 5.39) shows driving opto-isolators to eliminate common-
mode and ground-loop concerns. The voltage-to-frequency converter,
model 831A, incorporates its own optoisolator for this purpose and also
features a high-impedance differential input, which does not load the strain
gage bridge. The voltage-to-frequency converter 811A is chosen for its
100-kHz full-scale rate, which gives the highest resolution with low-level
thermocouple signal levels. The low-power-drain VFC models 821A and
822A are used with a battery supply that also provides isolation. The 821A
internal reference will supply up to three other low-power VFCs for minimum power drain. In such a system many transducers using the proper VFC
are fed to a digital multiplexer and to an interface device and finally a
computer. The reader may call this complex, but today's technology provides us with the opportunity to study analog signals that were impossible
to accurately analyze a few years ago.

5.11
REVIEW QUESTIONS

1. Briefly discuss and draw the following IC logic devices: RTL, DTL, HTL, TTL, ECL, and CMOS. Are there advantages of one over the others?

2. Discuss the uses of a logic probe and how it works.

3. Discuss the uses of a logic monitor and how it works. Discuss its specifications.

4. Discuss the uses of a logic monitor.

5. Discuss the uses of a digital IC test system.

6. Discuss the uses and operation of a digital multimeter.

7. Draw a digital multimeter clock, ramp, and digital output circuit and discuss how they operate.

8. Discuss the uses of a digital panel meter.

9. Describe the ratiometric principle in digital instruments.

10. Given three transducers (strain gage, thermocouple, and flow), draw a data-acquisition and conversion-to-digital system.

5.12
REFERENCES

1. John D. Lenk, *Logic Designer's Manual,* Reston Publishing Co., Reston, Va., 1977.

2. New Techniques of Digital Troubleshooting Application, Bulletin AN 163-2, Hewlett-Packard Co., Palo Alto, Calif., 1981.

3. Eugene L. Zuck, *Data Acquisition and Conversion Handbook,* 4th printing, Datel-Intersil, Inc., Sept. 1981.

4. Electronic Instruments and Systems, Handbook and Catalogue on Measurement/Computation, Hewlett-Packard, Palo Alto, Calif., 1982.

5. James Coffron, *Getting Started in Digital Troubleshooting,* Reston Publishing Co., Reston, Va., 1979.

6. 2920 Analog Signal Processor Design Handbook, Intel Corporation, Santa Clara, Calif., 1980.

6

Digital-to-Analog Converters

6.1
INTRODUCTION

In Chapter 4, the authors have shown how analog signals have been digitized. Digital instruments have been investigated in detail in Chapter 5. In this chapter, we will analyze how digitized outputs can be translated to an analog waveform. The digital-to-analog data converters must use thin-film resistor technology for combining both discrete-component and monolithic circuits. Important characteristics for D/A converters include resolution (bits), nonlinearity at 25°C maximum, input code options (integrating circuit logic), analog output options, and settling time to percentage of full-scale maximum, and temperature coefficients of gain and offset.

6.2
DIGITAL CONVERTER CODES [a]

The inputs to D/A converters and the outputs from A/D converters are digital numbers in various binary codes, so called because they consist of binary bits, each of which has two possible states designated on and off, true

---

[a]The material in Sections 6.2 through 6.8 is used courtesy of the ILC Data Device Corp., Bohemia, N. Y. Reprinted with permission.

and false, or 1 and 0. Each state is represented by a discrete voltage level with respect to ground. The off, false, or 0 state is represented by the logic 0 level; and the on, true, or 1 state is represented by the logic 1 level.

D/A converters accept inputs of digital numbers in one or more of the following binary codes: straight or natural binary (BIN), offset binary (OBN), inverted OBN, one's complement (1SC), two's complement (2SC), and binary-coded decimal (BCD). The input circuits are generally DTL/TTL compatible, positive true, which means that they will interface with DTL and/or TTL (logic 0 = 0 to +0.8 V and logic 1 = +2.0 to +5.5 V).

A/D converters generate digital-number outputs in one or more of the following binary codes: BIN, inverted BIN, OBN, 2SC, and BCD. The outputs are generally DTL/TTL compatible, positive true, which means that each logic 0 output bit is at a level between 0 and +0.4 V, and each logic 1 bit is at a level between +2.4 and +5.5 V.

Straight BIN and BCD are designated unipolar codes, because they represent analog quantities having only one polarity. However, since the two polarities ($+$) and ($-$) can be represented by 1 bit (e.g., logic 0 = minus and logic 1 = plus), BIN and BCD can be made bipolar by adding a sign bit, usually in front of the digital number. They are then called sign-magnitude BIN or BIN plus sign and sign-magnitude BCD or BCD plus sign. OBN, inverted OBN, 1SC, and 2SC are designated bipolar codes because they can represent both positive and negative analog quantities. All bipolar codes, including BIN + sign and BCD + sign, are 1 bit longer than equivalent unipolar codes having the same resolution, but have twice the range.

Although binary digital codes are capable of representing both integers and fractions, the codes employed in D/A and A/D converters, whether unipolar or bipolar, are fractional codes consisting of binary bits having the weighted values $A_n/2^n$ (in straight BIN) or $A_n/2^{n-1}$ (in many other binary codes). The presence or absence of any bit in a digital number depends on the code relationship between that digital number and the particular analog quantity it represents. If the code calls for the presence of the nth bit, then $A_n = 1$, and the nth bit has the value $1/2^n$ (or $1/2^{n-1}$); if the code calls for the absence of the nth bit, then $A_n = 0$, and the value of the nth bit is 0. An N-bit fractional number in straight binary, for example, is the sum of N terms of the form $A_n/2^n$, as follows:

$$\frac{A_1}{2^1} + \frac{A_2}{2^2} + \frac{A_3}{2^3} + \cdots + \frac{A_n}{2^n} + \cdots + \frac{A_N}{2^N} = \frac{A_1}{2} + \frac{A_2}{4} + \frac{A_3}{8} + \cdots + \frac{A_n}{2^n} + \cdots + \frac{A_N}{2^N}$$

The first term, $A_1/2$, is the most-significant bit or MSB; the term $A_n/2^n$ is the nth bit; and the last term, $A_N/2^N$, is the least-significant bit or LSB. Since the LSB, $A_N/2^N$, is the smallest increment by which the digital number can change, it is equal to the resolution. Since each bit is half the weight of the previous bit, the resolution doubles for each additional bit. Table 6.1 lists the weights and resolutions of 1 to 20 binary bits.

The value of a digital number is the sum of the weights of all terms having coefficients $A - 1$. The maximum value of a digital number is the

TABLE 6.1
Calculation of a 10-bit digital number 1001100110

Bit	Weight	Code	Value
1	0.5	1	0.5
2	0.25	0	0.0
3	0.125	0	0.0
4	0.0625	1	0.0625
5	0.03125	1	0.03125
6	0.015625	0	0.0
7	0.007812	0	0.0
8	0.003906	1	0.003906
9	0.001953	1	0.001953
10	0.0009766	0	0.0
			0.599609

Courtesy of ILC Data Device Corp.

TABLE 6.2
Binary bit weights and resolution

Bit No. n	Fraction $1/2^n$	Decimal $(0.5)^n$	Percent Resolution	PPM Resolution
1	1/2	0.5	50.0	500000
2	1/4	0.25	25.0	250000
3	1/8	0.125	12.5	125000
4	1/16	0.0625	6.25	62500
5	1/32	0.03125	3.125	31250
6	1/64	0.015625	1.5625	15625
7	1/128	0.007812	0.7812	7812
8	1/256	0.003906	0.3906	3906
9	1/512	0.001953	0.1953	1953
10	1/1024	0.0009766	0.0977	977
11	1/2048	0.00048828	0.0488	488
12	1/4096	0.00024414	0.0244	244
13	1/8192	0.00012207	0.0122	122
14	1/16384	0.000061035	0.0061	61
15	1/32768	0.0000305176	0.0031	30.5
16	1/65536	0.0000152588	0.0015	15.3
17	1/131072	0.00000762939	0.00076	7.6
18	1/262144	0.000003814697	0.00038	3.8
19	1/524288	0.000001907349	0.00019	1.9
20	1/1048576	0.0000009536743	0.000095	0.95

Courtesy of ILC Data Device Corp.

sum of all the terms when all coefficients $A_n = 1$, and is less than unity by the weight of the LSB, as follows:

$$\sum_{n=1}^{N} \frac{A_n}{2^n} = 1 - \frac{1}{2^N} = 1 - LSB$$

Thus the larger the number of bits N, the closer the value of a fractional digital number can approach unity. It is also noteworthy that the sum of all the terms to the right of any term can never exceed the value of that term. Since the weight of each bit in any binary code is related only to its position, it is not necessary to state the denominators, but only the value of the coefficient, 0 or 1. Thus, in straight BIN, the 10-bit digital number 1001100110 has the value of 0.599609, calculated as follows and as shown in Table 6.1.

Table 6.2 gives the binary bit weights and resolution. Tables 6.3 and 6.4 give examples of 12-bit unipolar codes representing analog voltage ranges of 0 to +5 V FS and 0 to +10 V FS. Table 6.5. gives examples of 12-bit bipolar codes representing analog voltage ranges of 16 V FS. Note that 1SC and BIN + sign each have two words for zero, the upper of which is called 0+, and the lower of which is called 0−. Because 1SC and BIN + sign have two words for zero, their range is one LSB smaller than that of OBN and 2SC.

6.3
TRANSLATION BETWEEN CODES

When digital input data are not in a code acceptable to the input circuits of a particular D/A converter, or when digital output data from an A/D converter are not in the code required by subsequent equipment, it becomes necessary to translate the data to the required code. Some D/A converters can be externally connected so as to accept either of two codes. The following list tells how to translate from any one to any other of the four following bipolar codes: BIN magnitude + sign, OBN, 1SC, and 2SC.

1. To convert from BIN magnitude + sign to:
   - 1SC: If MSB = 1, complement other bits.
   - 2SC: If MSB = 1, complement other bits, add 00 ... 01.
   - OBN: Complement MSB. If new MSB = 1, complement other bits; add 00 ... 01.

2. To convert from 1SC to:
   - 2 SC: If MSB = 1, add 00 ... 01.
   - OBN: Complement MSB. If new MSB = 0, add 00 ... 01.
   - BIN magnitude + sign: If MSB = 1, complement other bits.

3. To convert from 2SC to:
   - 1SC: If MSB = 1, add 11 ... 11.

TABLE 6.3
Twelve-bit unipolar binary codes

Scale	+10 V FS	+5 V FS	Straight Binary			Complementary Binary		
+FS − 1 LSB	+9.9976	+4.9988	1111	1111	1111	0000	0000	0000
+7/8 FS	+8.7500	+4.3750	1110	0000	0000	0001	1111	1111
+3/4 FS	+7.5000	+3.7500	1100	0000	0000	0011	1111	1111
+5/8 FS	+6.2500	+3.1250	1010	0000	0000	0101	1111	1111
+1/2 FS	+5.0000	+2.5000	1000	0000	0000	0111	1111	1111
+3/8 FS	+3.7500	+1.8750	0110	0000	0000	1001	1111	1111
+1/4 FS	+2.5000	+1.2500	0100	0000	0000	1011	1111	1111
+1/8 FS	+1.2500	+0.6250	0010	0000	0000	1101	1111	1111
0 +1 LSB	+0.0024	+0.0012	0000	0000	0001	1111	1111	1110
0	0.0000	0.0000	0000	0000	0000	1111	1111	1111

Courtesy of ILC Data Device Corp.

TABLE 6.4
Twelve-bit unipolar BCD codes

Scale	+10 V FS	+5 V FS	Binary-Coded Decimal			Complementary BCP		
+FS − 1 LSB	+9.99	+4.95	1001	1001	1001	0110	0110	0110
+7/8 FS	+8.75	+4.37	1110	0000	0000	0001	1111	1111
+3/4 FS	+7.50	+3.75	1100	0000	0000	0011	1111	1111
+5/8 FS	+6.25	+3.12	1010	0000	0000	0101	1111	1111
+1/2 FS	+5.00	+2.50	1000	0000	0000	0111	1111	1111
+3/8 FS	+3.75	+1.87	0110	0000	0000	1001	1111	1111
+1/4 FS	+2.50	+1.25	0100	0000	0000	1011	1111	1111
+1/8 FS	+1.25	+0.62	0010	0000	0000	1101	1111	1111
0 +1 LSB	+0.01	+0.00	0000	0000	0001	1111	1111	1110
0	+0.00	0.00	0000	0000	0000	1111	1111	1111

Courtesy of ILC Data Device Corp.

TABLE 6.5
Twelve-bit bipolar codes

Scale	5 V FS	Offset Binary	Two's Complement	One's Complement	Magnitude + Sign
+FS − 1 LSB	+4.9976	1111 1111 1111	0111 1111 1111	0111 1111 1111	1111 1111 1111
+3/4 FS	+3.7500	1110 0000 0000	0110 0000 0000	0110 0000 0000	1110 0000 0000
+1/2 FS	+2.5000	1100 0000 0000	0100 0000 0000	0100 0000 0000	1100 0000 0000
+1/4 FS	+1.2500	1010 0000 0000	0010 0000 0000	0010 0000 0000	1010 0000 0000
0	0.0000	1000 0000 0000	0000 0000 0000	0000 0000 0000*	1000 0000 0000*
				1111 1111 1111*	0000 0000 0000*
−1/4 FS	−1.2500	0110 0000 0000	1110 0000 0000	1101 1111 1111	0010 0000 0000
−1/2 FS	−2.5000	0100 0000 0000	1100 0000 0000	1011 1111 1111	0100 0000 0000
−3/4 FS	−3.7500	0010 0000 0000	1010 0000 0000	1001 1111 1111	0110 0000 0000
−FS +1 LSB	−4.9976	0000 0000 0001	1000 0000 0001	1000 0000 0000	0111 1111 1111
−FS	−5.0000	0000 0000 0000	1000 0000 0000	—	—

*1SC and Magnitude + Sign each have two words for zero, the upper of which is called 0+ and the lower of which is called 0−.

Courtesy of ILC Data Device Corp.

- OBN: Complement MSB.
- BIN magnitude + sign: If MSB = 1, complement other bits, add 00 ... 01.

4. To convert from OBN to:
- 1SC: Complement MSB. If new MSB = 1, add 11 ... 11.
- 2SC: Complement MSB.
- BIN magnitude + sign: Complement MSB. If new MSB = 1, complement other bits, add 00 ... 01.

6.4
TRANSFER CHARACTERISTICS AND ACCURACY

The transfer characteristics of D/A converters will be discussed in terms of a 3-bit converter with straight binary and offset binary coding. The difference between these two codings is just an analog offset equal to one-half of full scale. This offset is generally pin programmable in converters that offer both unipolar and bipolar coding.

For a D/A converter, analog output values such as 1 LSB, 2 LSB, and so on, should coincide with the input bit codes. This is shown by the ideal transfer characteristic for a 3-bit D/A converter illustrated in Figure 6.1. For straight binary coding, for instance, the 0 LSB output level coincides with 000 and the 1 LSB level coincides with 001. Intermediate values of analog output should not exist except as brief transitions while a code change occurs.

It can be seen from Figure 6.1 that the largest analog output obtained in both straight and offset binary coding is FS − 1 LSB. In some converters with offset binary coding, the output is offset by 1 LSB so that the largest positive value is FS, and the largest negative value is −FS + 1 LSB. The bit table equivalent to the transfer characteristics in Figure 6.1 is shown in Figure 6.2.

The accuracy of a D/A converter is measured by determining the analog output for each input digital code. The total error is the sum of the offset error, the gain error, and the linearity error.

For a unipolar converter the offset is usually specified to be at the code for zero analog output, but it could be designated by the manufacturer to be at the code at either end of the range. For a bipolar converter, the offset could also be assigned to the code for zero output near the center of the range. The offset error is the difference between the measured and the theoretical output value at the code designated for offset.

The gain error is measured at the codes at opposite ends of the range. The gain is equal to the difference between the measured outputs at these two codes (keeping negative signs for negative output), divided by the

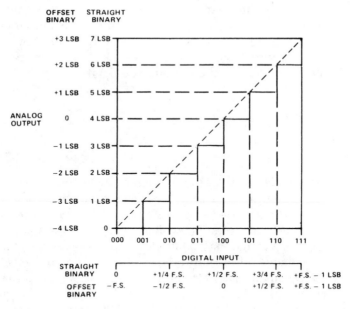

FIGURE 6.1
Ideal transfer characteristic for a 3-bit D/A converter. (Courtesy of ILC Data Device Corp.)

theoretical output span between them. For the 3-bit converter just discussed, the input span is FS − 1 LSB for unipolar coding, and 1 FS − 1 LSB for bipolar coding.

The linearity is obtained by determining the difference between the measured and theoretical output at all other codes. Before subtracting the theoretical value, the measured value at each point must be corrected by first subtracting the offset error and then multiplying by the gain correction.

ANALOG OUTPUT		DIGITAL CODE		
Unipolar Straight Binary	Bipolar Offset Binary	(MSB) Bit 1	Bit 2	(LSB) Bit 3
+ F.S. − 1 LSB	+ F.S. − LSB	1	1	1
+ 3/4 F.S.	+ 1/2 LSB	1	1	0
+ 1/2 F.S. + 1/2 LSB	+ 1 LSB	1	0	1
+ 1/2 F.S.	0	1	0	0
+ 1/2 F.S. − 1 LSB	− LSB	0	1	1
+ 1/4 F.S.	− 1/2 F.S.	0	1	0
+ 1 LSB	− F.S. + 1 LSB	0	0	1
0	− F.S.	0	0	0

FIGURE 6.2
Bit table for 3-bit D/A converter. (Courtesy of ILC Data Device Corp.)

The calculations are greatly simplified if the converter offset and gain errors are both trimmed to zero before linearity is measured.

Monotonicity is usually also required. If the converter is monotonic, the analog output will never decrease if the digital code increases, and will never increase if the digital code decreases.

6.5
DIGITAL-TO-ANALOG GENERAL CONCEPTS

D/A converters or DACs generate analog output voltages that are digitally controlled, binary-weighted, discrete fractions of some reference voltage V_{REF}. Unless otherwise designated, D/A converters usually have a regulated internal reference voltage source so that V_{REF} is fixed, and the analog output voltage has a single value for each digital-input code. When designated ratio, multiplying, external-reference, or universal-reference D/A converters, the reference voltage V_{REF} is externally supplied as a separate input. The analog output voltage is then the product of the variable analog voltage V_{REF} and the discrete fraction represented by the digital input code.

Contemporary D/A converters employ networks of precision resistors controlled by bipolar transistors or FET switches turned on and off in accordance with the digital input code. Figure 6.3 shows a D/A converter employing a network of summing resistors, having the relative weightings R, $2R$, $4R$, $8R$, ..., $2^{(N-1)} R$, connected to the $(-)$ summing point of an operational amplifier. The input end of resistor R is connected to the output of the amplifier, and the input ends of resistors $2R$, $4R$, $8R$, ..., $2^N R$ are all

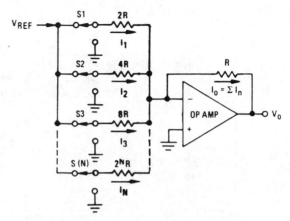

FIGURE 6.3
D/A converter employing $2^N R$ summing network.
(Courtesy of ILC Data Device Corp.)

connected to a source of reference voltage V_{REF}. The high amplifier gain and the large negative feedback operate to maintain the summing point of virtual ground, so if switches $S1, S2, S3, \ldots, S(N)$ are closed, the currents

$$I_1 = V_{REF}/2R \quad (= \text{MSB})$$

$$I_2 = V_{REF}/4R$$

$$I_3 = V_{REF}/8R \tag{6.1}$$

$$\vdots \qquad \vdots$$

$$I_N = V_{REF}/2^N R \quad (= \text{LSB})$$

flow through the resistors $2R, 4R, 8R, \ldots, 2^N R$ to the summing point. The amplifier has high input resistance, so the current into the $(-)$ input is negligible, and the sum of the input currents is equal to the feedback current I_0 out of the summing point. The output voltage is therefore

$$V_{\bar{0}} = -I_0 R = -V_{REF} \left(\frac{A_1}{2} + \frac{A_2}{4} + \frac{A_3}{8} + \ldots + \frac{A_N}{2^N} \right) \tag{6.2}$$

$$= V_{REF} \left[\sum_{n-1}^{N} \left(\frac{A_n}{2^n} \right) \right]$$

where

$$A_n = 1 \text{ if switch } n \text{ is at position } 1$$

$$A_n = 0 \text{ if switch } n \text{ is at position } 0$$

The configuration in Figure 6.3 therefore generates analog output voltages equal to the binary-weighted values of the switch positions.

D/A conversion by means of $2^N R$ summing networks is limited to 12 or 13 bits because more bits would require impractically high resistances. Resistance R_{12}, for example, must be 4096 times R, and R_{20} must be 1,048,576 times R. Also, since the network resistance varies greatly with the combinations of switch settings, the amplifier design cannot be optimized to achieve fastest settling time.

These limitations are overcome by means of the $R-2R$ ladder network shown in Figure 6.4, for which it can be shown that switches $S1, S2, S3, \ldots,$ $S(N)$ cause output increments $-V_{REF}/2, -V_{REF}/4, -V_{REF}/8, \ldots, -V_{REF}/2^N$. This network is economically realizable because the resistor values are either R or $2R$.

Also, the network resistance seen from the summing point is constant at the low value of R, so the operational amplifier may be designed for fastest settling time. Most D/A converters therefore employ $R-2R$ ladder networks, with low-resistance FET switches that turn on and off in nanoseconds.

D/A converters include a resistance network and switches that together provide a current output. The simplest concept of a D/A converter is shown in Figure 6.5. The resistance network produces a series of binary-weighted currents. These are connected to the output by switches controlled by the

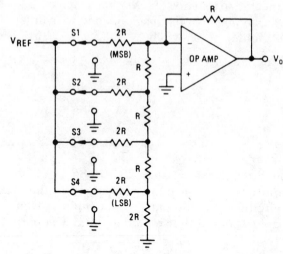

FIGURE 6.4
D/A converter employing *R–2R* ladder network.
(Courtesy of ILC Data Device Corp.)

digital input. The summed current is available directly as an output current I_{OUT} or is transformed by an op-amp into a voltage output V_{OUT}. The chief requirements for the converter are linearity (which depends on the bit current ratios), scale factor accuracy, and low offset drift.

A difficulty with the arrangement in Figure 6.5 is that turning the bit currents off and on changes voltages in the resistance network and so may affect stability. This problem is eliminated by the arrangement shown in Figure 6.6, where the resistor currents are constant because they are switched either to actual ground or to the virtual ground of an op-amp.

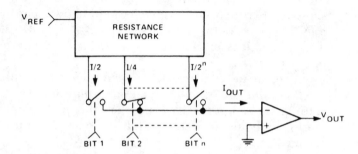

FIGURE 6.5
Basic D/A diagram. (Courtesy of ILC Data Device Corp.)

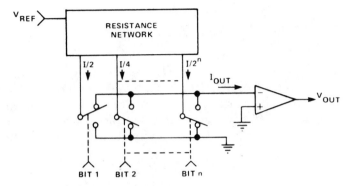

FIGURE 6.6
Arrangement with constant resistor currents.
(Courtesy of ILC Data Device Corp.)

There are many other possible types of resistance networks and switching arrangements. Currents, for instance, can be switched as equal currents before they are given binary weighting by the network, as shown in Figure 6.7. The method chosen for a particular D/A converter depends on the best compromise of characteristics and cost.

For a multiplying D/A converter, the output current must remain proportional to the reference voltage without saturation effects over an acceptable range of bipolar input voltages (usually ±10 V) and frequencies. Otherwise, multiplying and fixed reference D/A converters are quite similar in basic design concepts.

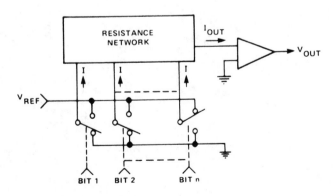

FIGURE 6.7
Another D/A converter arrangement using binary weighting by the network. (Courtesy of ILC Data Device Corp.)

6.6
DIGITAL-TO-ANALOG SYSTEMS

An important system of D/A converters is shown in Figure 6.8, where multiplexed digital data are fed to several D/A converters, and it is desired that each converter respond only to its channel. The j-channel multiplexed data are fed to a register that is strobed when the ith channel data appear and holds these data until the next strobe. The output of the register is fed

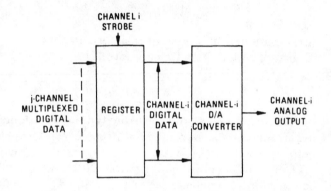

FIGURE 6.8
Digital data-distribution systems. (Courtesy of ILC Data Device Corp.)

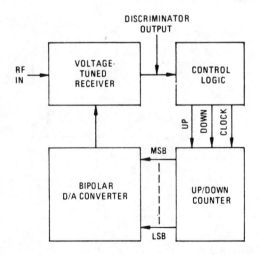

FIGURE 6.9
Digital AFC employing D/A converter. (Courtesy of ILC Data Device Corp.)

to the D/A converter, which generates the ith channel analog output. This output is held constant until updated by the next strobe. This system is equivalent to a long-term sample-hold circuit with digital input and analog output.

Figure 6.9 shows a bipolar D/A converter employed in a digital servo for automatic frequency control of a voltage-tuned receiver. The bipolar discriminator output (e.g., negative when the receiver is tuned below the signal, positive when the receiver is tuned above the signal) is fed to a control logic circuit that sends clock pulses and either an up-count or a down-count command to an up–down counter, the digital output of which drives the D/A converter. The analog output voltage is proportional to the frequency error, and its polarity is such as to reduce that error.

6.7
MULTIPLYING D/A CONVERTERS AND APPLICATIONS

In the D/A converters described previously, the output voltage is the product of the reference voltage V_{REF} and the binary-weighted fraction represented by the switch settings. If the circuits are designed to operate with wide-ranging values of V_{REF}, as in Figure 6.10, they are designated multiplying, ratio, external-reference, or universal-reference D/A converters. If both the analog input V_{REF} and the digital input are unipolar, the output is in one quadrant; if either input is unipolar and the other is bipolar, the output is in two quadrants; if both digital and analog inputs are bipolar, the output is in four quadrants.

Applications of multiplying D/A converters include the following:

1. Multiplying analog and digital inputs

2. Digitally programmable attenuator

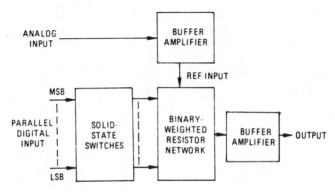

FIGURE 6.10
Simplified block diagram of multiplying D/A converter. (Courtesy of ILC Data Device Corp.)

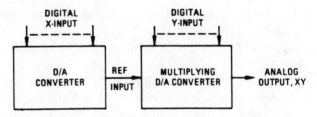

FIGURE 6.11
Multiplying two digital inputs. (Courtesy of ILC
Data Device Corp.)

3. Multiplying two digital inputs

4. Polar-to-rectangular conversion

5. PPI sweep

Refer to Figures 6.11 through 6.13.

Multiplying-type D/A converters are also useful in audio signal processing [4]. The basic block diagram of Figure 6.10 shows that a multiplying D/A converter can be used to generate an analog output voltage $X V_{REF}$, where X is a digital input and V_{REF} is an analog input.

The D/A converter in the preceding application can be considered a digitally programmable attenuator, the output of which ($X V_{REF}$) is a digitally controlled fraction (X) of the analog input voltage (V_{REF}). If the converter accepts bipolar values of V_{REF}, it will accept ac and will thus function as a digitally programmable ac attenuator.

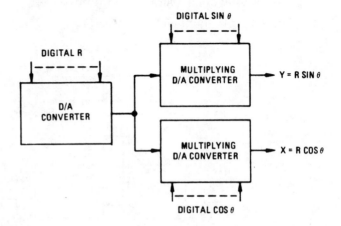

FIGURE 6.12
Polar-to-rectangular conversion. (Courtesy of ILC
Data Device Corp.)

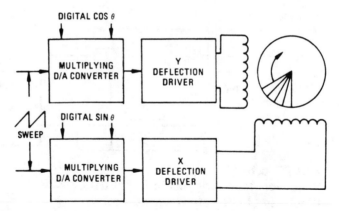

FIGURE 6.13
PPI sweep generator. (Courtesy of ILC Data Device Corp.)

Figure 6.11 shows how an analog output voltage XY can be obtained from the digital inputs X and Y by use of an internal-reference D/A converter and an external-reference D/A converter.

Figure 6.12 shows how one internal-reference D/A converter and two multiplying D/A converters can be used to obtain the analog rectangular coordinates $Y = R \sin 0$, and $X = R \cos 0$ from the polar coordinates R and 0. Digital R is fed to the D/A converter, and digital sin 0 and digital cos 0 are fed to the multiplying D/A converters.

Figure 6.13 shows how a rotating radial sweep, as used in a PPI radar, can be generated from an analog linear sweep and digital sin 0 and digital cos 0 inputs.

6.8
VIDEO D/A CONVERTERS

The basic block diagrams of video D/A converters are the same as for the lower-speed units, but the components have been selected and the circuitry designed for operation at very high speeds. Section 6.8.1 discusses a basic problem called data skew that occurs in very fast D/A converters.

6.8.1 Reduction of Skew in Video D/A Converters

Charge-storage effects in TTL circuits cause the 0 to 1 transition delay to be greater than the 1 to 0 transition delay. The difference between the two delay times is called *skew*. When skewed data are fed into a high-speed D/A converter (one that can perform conversions in times of the same order

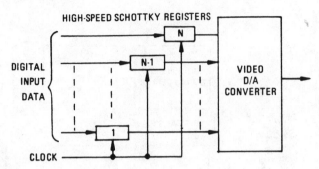

FIGURE 6.14
Reduction of slew by reclocking of data.
(Courtesy of ILC Data Device Corp.)

of magnitude as the TTL transition time), the unequal delays during the transition times may be sensed as false codes, some of which may even call for full-scale output. In such cases the D/A converter output will slew from the value corresponding to the previous word toward the full-scale value corresponding to the false word that appears during the transition, and then back to the value corresponding to the next word. The D/A output during the transition may thus be a spike of appreciable magnitude. Since the amplitudes of such spikes are proportional to the slew rate, those amplitudes can be reduced by limiting the slew rate. This is practicable in systems where the D/A output changes only by small increments between conversions.

When large, rapid changes occur, the input data should be reclocked, as shown in Figure 6.14, through very high speed Schottky registers, which have transition times as short as 4 ns. Although this technique reduces the skew of the input data, it cannot reduce the internal skew of the D/A converter input circuits, which will therefore cause some spiking.

6.9
DIGITAL-TO-SYNCHRO AND
DIGITAL-TO-RESOLVER CONVERTERS

A synchro transmitter (TX or CX), when excited by an ac reference voltage, converts the input angle θ of a mechanical shaft rotation into equivalent three-wire synchro data. Where the input angle θ is not in the form of a mechanical shaft rotation but is in the form of parallel binary-coded digital data, a digital-to-synchro (D/S) converter may be used to convert the binary-coded digital data into corresponding three-wire synchro data. Figure 6.15 shows the functional equivalence between a D/S converter and a TX or CX.

Similarly, a digital-to-resolver (D/R) converter may be used to convert parallel binary-coded digital angle data into corresponding four-wire re-

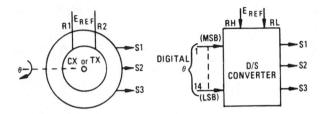

FIGURE 6.15
Synchro transmitter and D/S converter compared. (Courtesy of ILC Data Device Corp.)

solver data. Figure 6.16 shows the functional equivalence between a D/R converter and a resolver with one rotor winding.

Figure 6.17 is the block diagram of a theoretical D/R converter. An ac reference voltage, $E_{REF} \sin 2\pi ft$, is fed to the reference inputs of sine and cosine function generators; and a digital input corresponding to the angle θ is fed to the function inputs of the generators. The outputs of the generators are $E_{REF} \sin 2\pi ft \sin \theta$ and $E_{REF} \sin 2\pi ft \cos \theta$, respectively. These voltages are fed through buffer amplifiers to the output terminals. The configuration of Figure 6.17 requires the function generators to operate over all four quadrants (0° to 360°).

Figure 6.18 shows a more practical configuration in which the sine and cosine function generators are required to operate only in the first quadrant (0° to 90°). In this arrangement the ac reference voltage, $E_{REF} \sin 2\pi ft$, is fed to a center-tapped transformer that provides isolation and develops two equal and opposite reference voltages, $\pm E_{REF} \sin 2\pi ft$. These voltages are fed through a quadrant sector to the sine and cosine generators. The quadrant selector is controlled by the two most-significant bits of the digital θ input. The sine and cosine generators are directly controlled by bits 3

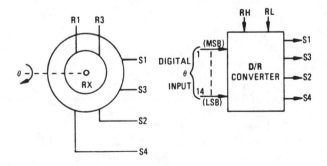

FIGURE 6.16
Resolver transmitter and D/R converter compared. (Courtesy of ILC Data Device Corp.)

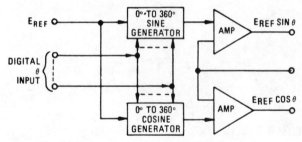

FIGURE 6.17
Block diagram of a theoretical D/R converter.
(Courtesy of ILC Data Device Corp.)

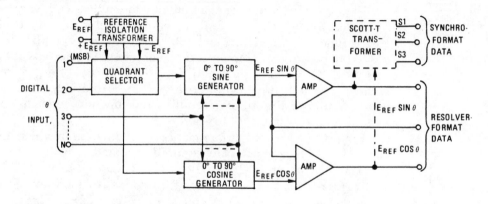

FIGURE 6.18
Block diagram of practical D/R and D/S converters. (Courtesy of ILC Data Device Corp.)

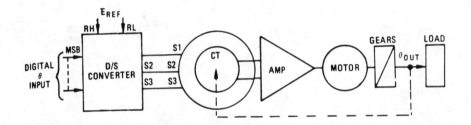

FIGURE 6.19
Hybrid synchro follow-up servo using D/S converter. (Courtesy of ILC Data Device Corp.)

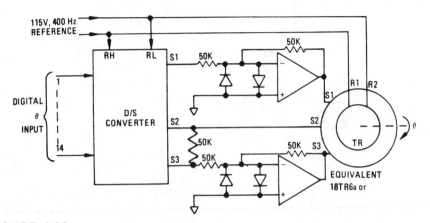

FIGURE 6.20
Driving torque receiver from D/S converter.
(Courtesy of ILC Data Device Corp.)

through N of the digital θ input. The outputs of the generators are $E_{REF} \sin 2\pi ft \sin \theta$ and $E_{REF} \sin 2\pi ft \cos \theta$, respectively, for $0° \leq \theta \leq (360° - \text{LSB})$.

The D/R converter is changed into a D/S converter by changing its resolver-format outputs into three-wire synchroformat outputs by means of a Scott-T transformer, as shown, or by means of a transolver set to 0°. The D/S converters described in this section are generally D/R converters with Scott-T transformers.

Figures 6.19 through 6.21 show some typical applications, in which D/S converters fed digital angle data (e.g., from a navigation or simulation computer) are used to position a synchro follow-up servo or a torque receiver (TR).

Since most D/S and D/R converters operate at low power levels, buffer amplifiers must be used between them and loads such as torque synchros. Figure 6.20 shows a D/S converter driving a TR through buffer amplifiers. Figure 6.21 shows an alternate arrangement in which a D/R converter

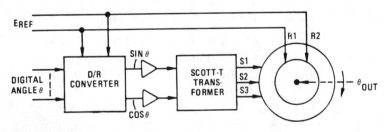

FIGURE 6.21
Using D/R and R/S converters to drive torque receivers. (Courtesy of ILC Data Device Corp.)

feeds buffer amplifiers that raise the power level of the D/R output, and a Scott-T transformer converts the receiver-format output of the buffers into the synchroformat required to drive the TR.

6.10
SPECIALIZED APPLICATIONS OF THE
D/A CONVERTER [b]

The Datel-Intersil DAC-7520 and DAC-7521 are monolithic high-accuracy, low-cost 10-bit and 12-bit resolution, multiplying digital-to-analog converters. Datel-Intersil's thin film on CMOS process enables 10-bit accuracy with DTL/TTL/CMOS compatible operation. Digital inputs are fully protected against static discharge by compensating diodes to ground and positive supply. The functional diagram and chip topography are shown in Figure 6.22.

Typical applications for the DAC-7520 and DAC-7521 include digital–analog interfacing, multiplication, and division; programmable power supplies; CRT character generation; digitally controlled gain circuits, integrators and attenuators; and so on.

These D/A converters feature the following:

1. DAC-7520: 10-bit resolution; 10-bit linearity.

2. DAC-7521: 12-bit resolution; 10-bit linearity.

3. Low power dissipation: 20 mW (max).

4. Low nonlinearity temperature coefficient: 2 ppm or FSR/°C (max).

5. Current settling time: 500 nS to 0.05% of FSR.

6. Supply voltage range: +5 to +15 V.

7. DTL/TTL/CMOS compatible.

8. Full input static protection.

The test circuits shown in Figures 6.23 through 6.28 apply for the DAC-7520. Similar circuits can be used for the DAC-7521.

The definition of terms required in specifying the D/A converters shown in Figures 6.23 through 6.28 include the following:

1. **Nonlinearity:** Error contributed by deviation of the DAC transfer function from a best straight-line function. Normally expressed as a percentage of full-scale range. For a multiplying DAC, this should hold true over the entire V_{REF} range.

2. **Resolution:** Value of the LSB. For example, a unipolar converter with n bits has a resolution of $(2^{-n})(V_{REF})$. A bipolar converter of n bits has a resolution of $[2^{-(n-1)}][V_{REF}]$. Resolution in no way implies linearity.

---

[b] The material in Sections 6.10 and 6.10.1 is used courtesy of Datel-Intersil, Inc.

CHIP TOPOGRAPHY

FUNCTIONAL DIAGRAM

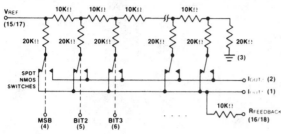

(Switches shown for Digital Inputs "High")

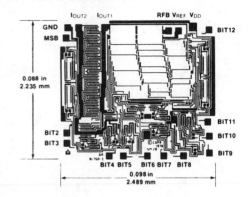

FIGURE 6.22
Functional diagram and chip topography of Datel-Intersil DAC-7520 and DAC-7521 monolithic D/A converters. (Courtesy of Datel-Intersil, Inc.)

3. **Settling time:** Time required for the output function of the DAC to settle to within $\frac{1}{2}$ LSB for a given digital input stimulus (i.e., 0 to full scale).

4. **Gain:** Ratio of the DAC's operational amplifier output voltage to the nominal input voltage value.

5. **Feed-through error:** Error caused by capacitive coupling from V_{REF} to output with all switches OFF.

6. **Output capacitance:** Capacity from $I_{OUT\ 1}$ and $I_{OUT\ 2}$ terminals to ground.

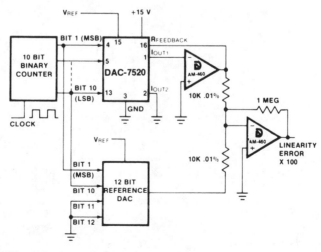

FIGURE 6.23
Nonlinearity. (Courtesy of Datel-Intersil, Inc.)

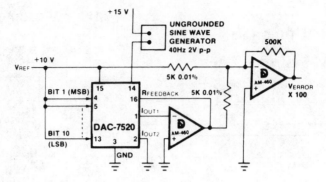

FIGURE 6.24
Power supply rejection. (Courtesy of Datel-Intersil, Inc.)

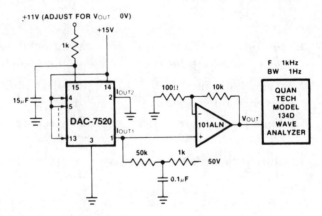

FIGURE 6.25
Noise. (Courtesy of Datel-Intersil, Inc.)

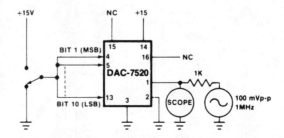

FIGURE 6.26
Output capacitance. (Courtesy of Datel-Intersil, Inc.)

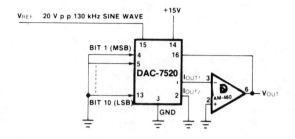

FIGURE 6.27
Feed-through error. (Courtesy of Datel-Intersil,
Inc.)

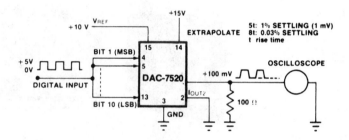

FIGURE 6.28
Output current settling time. (Courtesy of Datel-
Intersil, Inc.)

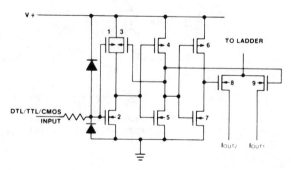

FIGURE 6.29
CMOS switch. (Courtesy of Datel-Intersil, Inc.)

7. **Output leakage current:** Current that appears on $I_{OUT\ 1}$ terminal with all digital inputs LOW or on $I_{OUT\ 2}$ terminal when all inputs are HIGH.

A simplified equivalent circuit of the DAC is shown in Figure 6.22. The NMOS DPDT switches steer the ladder leg currents between $I_{OUT\ 1}$ and $I_{OUT\ 2}$ busses, which must be held at ground potential or virtual ground potential. This configuration maintains a constant current in each ladder leg independent of the input code.

Converter errors are further eliminated by using separate metal interconnections between the major bits and the outputs. Use of high-threshold switches reduces the offset (leakage) errors to a negligible level.

The level shifter circuits are comprised of three inverters with a positive feedback from the output of the second to the first (Figure 6.29). This configuration results in DTL/TTL/CMOS compatible operation over the full military temperature range. With the ladder DPDT switches driven by

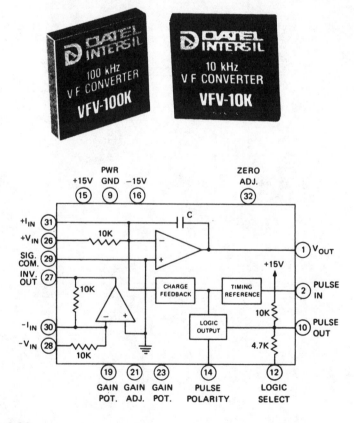

FIGURE 6.30
Frequency-to-voltage converter. (Courtesy of Datel-Intersil, Inc.)

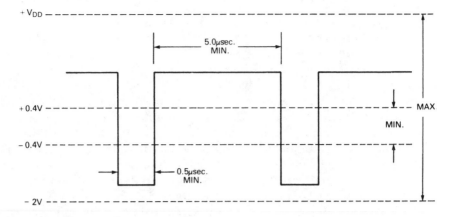

FIGURE 6.31
Input waveform limits of a frequency-to-voltage
converter. (Courtesy of Datel-Intersil, Inc.)

the level shifter, each switch is binarily weighted for an ON resistance
proportional to the respective ladder leg current. This assures a constant
voltage drop across each switch, creating equipotential terminations for the
2R ladder resistors, resulting in accurate leg currents.

6.10.1 Frequency-to-Voltage Converter

In a frequency-to-voltage (F/V) converter (Figure 6.30), either 0 to +10 V or
0 to −10 V outputs can be chosen by pin connection. Output pulses can be
selected to be positive or negative going, with DTL/TTL, CMOS, or high-
level logic interfacing. The output is short circuit proof to common or either
supply voltage. The results of these universal pin-connectable operating
characteristics is wide flexibility in applications.

Figure 6.31 gives the input waveform limits of a frequency-to-voltage
converter. When used as an F/V converter, an external capacitor can be
used to reduce output ripple to a specified level.

6.10.2 Glitch Problem in the D/A
Converter[c]

Glitches (voltage spikes) in the output of conventional D/A converters are
primarily caused by data skew and by switches that cause faster turn-on
than turn-off times. Thus, whenever a code change occurs, there will be a
short period of time (measured in nanoseconds) when some spurious code
will exist. The faster D/As will attempt to follow this code, resulting in a

---

[c]The material in Section 6.10.2 is used courtesy of ILC Data Device Corp.

transient known as a glitch. The worst case occurs at the major carry point when the input code is transitioning from 1000 ... 0 to 0111 ... 1. The spurious code may then be 1111 ... 1, and the D/A output will momentarily slew full scale (or close to it).

The DDAC 13-bit deglitched D/A converter overcomes these glitches by using a carefully designed low-transient track-and-hold amplifier (deglitcher) after the D/A. Thus the deglitcher can be gated into the hold mode during the D/A output transient and released into the track mode after the transients have settled out. The resultant output then makes a very clean transition from one value to another. Using this technique, glitches are not only greatly reduced but are only a function of deglitcher design, and hence are the same for any transition and can be effectively filtered out.

As illustrated in Figure 6.32, the user's strobe pulse activates the timing circuit in the deglitcher. The timing circuit simultaneously opens the T/H amplifier switch so that its output remains constant and activates the latch so that the input data bits are converted to an analog current by the D/A converter. After the analog current conversion has been completed, the T/H amplifier switch is closed again. The T/H amplifier then tracks the D/A converter, converting the D/A current into a voltage. The system output becomes stable after the T/H settles out.

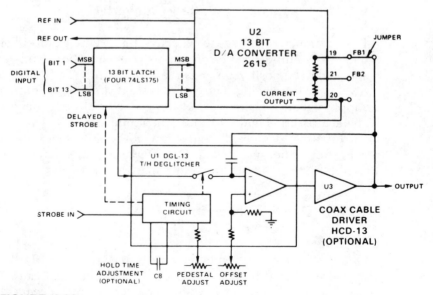

FIGURE 6.32
DDAC card assembly schematic (with connections for ±10 V output). (Courtesy of ILC Data Device Corp.)

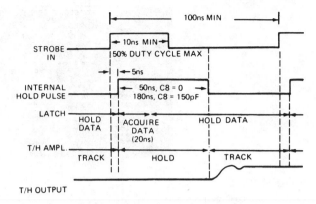

FIGURE 6.33
Card assembly timing diagram. (Courtesy of ILC
Data Device Corp.)

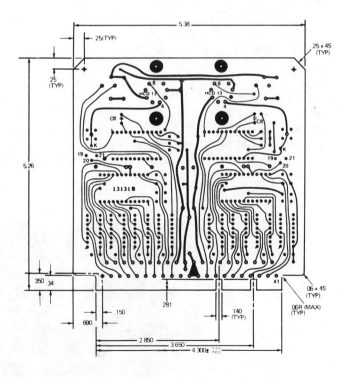

FIGURE 6.34
Printed circuit board layout. (Courtesy of ILC Data
Device Corp.)

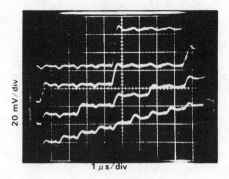

FIGURE 6.35
Glitch output waveforms. (Courtesy of ILC Data
Device Corp.)

Figure 6.33 illustrates the timing of the PC card assembly DDAC. The leading edge of the input strobe triggers a one shot in the deglitcher, which, after a delay of 5 ns, generates a hold pulse. The leading edge of the hold pulse causes the latch to acquire the input data bits. This acquisition process takes up to 20 ns, and the latch bit values then remain constant. Note that the data bits are acquired during an interval of time between 5 and 25 ns after the leading edge of the strobe pulse, and the input digital data should be held constant during this time interval. The leading edge of the hold pulse also opens the switch to the T/H amplifier, so the T/H retains the previous output while conversion is taking place.

Conversion takes place during the hold cycle pulse. The minimum hold pulse duration is 50 ns (typical), but the user may increase this interval to 180 ns or more by adding a capacitor to the printed circuit board. A printed circuit board layout is shown in Figure 6.34.

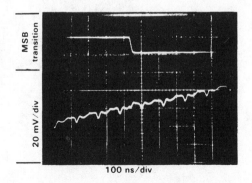

FIGURE 6.36
Glitch at MSB transition. (Courtesy of ILC Data
Device Corp.)

At the end of the hold pulse, the T/H amplifier switch closes, and the T/H begins to track the D/A converter output again. The strobe pulse interval should be long enough to allow time for the T/H amplifier output to settle and for the analog signal to be read out.

The glitch characteristics of the DDAC of Figure 6.32 are a function of interconnection layout and deglitcher design. Figures 6.35 and 6.36 illustrate the waveform of the DDAC on the printed circuit card assembly.

Figure 6.35 illustrates the output waveform for incrementing at the 9th, 10th, 11th, and 12th bit levels into a 5-MHz filter. Update rate is 1 MHz, 20 mV/div vertical and 1 μs/div horizontal.

Figure 6.36 illustrates the waveform around the MSB transition (upper trace). Conditions are 8-MHz word rate, 100 ns/div horizontal, 20 mV/div (lower trace) vertical, and into a 20-MHz filter. Note that while the output is incremented at the 12-bit level, there is no noticeable increase in the glitch during the MSB transition.

6.11
REVIEW QUESTIONS

1. Discuss the ideal transfer characteristic for a 3-bit D/A converter with a 3-bit table.

2. Draw a D/A converter employing $2^N R$ summing elements and discuss the usage of such a D/A converter.

3. Draw a D/A converter using a R–$2R$ network and discuss the usage of such a D/A converter.

4. Discuss a digital data-distribution system.

5. Discuss a digital AFC employing a D/A converter.

6. Discuss briefly a multiplying D/A converter.

7. List five applications of multiplying D/A converters.

8. Discuss briefly basic problems of data skew in video D/A converters.

9. Draw a digital-to-synchro converter and a digital-to-resolver converter. Discuss the usages.

10. Draw the functional diagram of a monolithic D/A converter with 12-bit resolution.

11. What are the features of a monolithic D/A converter with 12-bit resolution?

12. Discuss the definitions of nonlinearity, resolution settling time, gain, feedback error, output capacitance and output leakage current in the specification of a monolithic D/A converter with 12-bit resolution.

13. Discuss the usage of a frequency-to-voltage converter.

14. Define the problem of glitch in a D/A converter.

15. Discuss briefly how the deglitched D/A converter works.

6.12
REFERENCES

1. Eugene L. Zuck, *Data Acquisition and Conversion Handbook*, 4th printing, Datel-Intersil, Inc., Mansfield, Mass., Sept. 1981.

2. Data Converters Product Catalog, 1981, ILC Data Device Corp., Bohemia, N.Y., 1981.

3. Data Conversion/Acquisition A/D Advisor, vol. 1, no. 5, National Semiconductor, Santa Clara, Calif., Oct. 1981.

4. Walter G. Jung, "Application Considerations for IC Data Converters Useful in Audio Signal Processing," *JAES*, vol. 25, no. 12, Dec. 1977.

5. R. W. Jacobs: Specifying and Testing Digital to Analog Converters, Bulletin AN-26, Teledyne Philbrick, Dedham, Mass., July 1976.

6. Stuart B. Michaels, "D-a glitch—causes abound but solutions are rare," *Electronic Design*, vol. 29, no. 8, April 16, 1981.

7. *Data Acquisition Component Handbook*, Datel-Intersil Co., Mansfield, Mass., 1981 (published yearly).

8. Ed Maddox, Current Steering Chip Upgrades Performance of D/A Converter, Application Notes AN-17, Teledyne Philbrick, Dedham, Mass., April 1974.

9. Op Amps and Data Conversion Products, Teledyne Philbrick, Dedham, Mass. 1982.

10. Jim Williams, Exploit D/A Converters in Unusual Controller Designs, Electronic Design News (EDN), p. 111–118, Nov. 25, 1981.

7

Electronic Instrumentation Waveform Generation

7.1
INTRODUCTION

In this chapter electronic instrumentation waveform generation is discussed with the use of practical instrument examples. In general, electronic instruments vary from manufacturer to manufacturer on waveform generation. Electronic instruments can generate sine, cosine, square, triangle, and ramp waveforms, as well as other nonlinear waveform shapes. We will start our analysis of waveform generation using the function generator, followed by a discussion of a programmable microprocessor oscillator. We will then analyze a pulse generator. This will be followed by describing the function of signal analyzers. Emphasis in this area will include spectrum and modulation analyzers.

A function generator, test oscillator, audio oscillator, or signal generator, as we call this device, is an indispensable tool widely used in electronic instrumentation in industry, communication, medicine, production, automation, research, and many other applications. The function generator oscillator performs a wide range of operations, including the frequency response of amplifiers, bandpass characteristics of filters, radio or TV receiver alignment, or simply tracing faults in many electronic instruments, many discussed in previous chapters. Function generators have a range from less than 1 Hz to 100 kHz and higher. High-frequency oscillators from 100 kHz to 500 MHz or higher generally use some variation of an inductance–capacitance (LC) tank circuit.

Pulse generators range from simple digital units to microprocessor instruments. The output impedance of pulse generators delivers clean signals to integrated-circuit families, resistive loads, and the end of undetermined cables. Variable clocks may have speeds to 1 GHz and amplitude from 20 to 100 V.

Signal analyzers are used generally in electronics to identify and measure signals from nonlinear effects in the process of amplification; filtering and mixing; observing purity of electronic signals; and analyzing modulated communication signals, complex networks, and integrated circuits.

7.2
THE FUNCTION GENERATOR[a]

The Global Specialties Corporation Model 2001 sweepable function generator of Figure 7.1 is a multiple waveform signal generator, capable of producing precision sine, square, or triangle waveforms. The 2001 offers advanced features that include the following:

1. Separate TTL output
2. Variable dc offset control
3. DC-coupled output
4. 600-Ω output impedance
5. VCO and sweep capabilities
6. Separate −40-dB output

Switch-selectable sine, square, and triangle waveforms are continuously adjustable over a 10 : 1 calibrated vernier in five decade ranges, from 1 Hz to 0.1 mHz. In addition, a ±10-V sweep input can vary the frequency over a 100 : 1 range with a surprisingly linear characteristic over the range. A switch-selected variable dc offset can shift the waveform at the main outputs up to ±10 V.

The 2001 offers three main outputs. Two are amplitude variable: The HI output provides 0.1 to 10 V peak-to-peak into an open circuit and .05.5 V into a matching 600-Ω load; the LO output delivers 1 to 100 mV into an open circuit, and 0.5 to 50 mV into 600 Ω. The 40-dB variable amplitude control is flat within ±0.5 dB over its range. A third output, the TTL output, provides a constant-amplitude TTL square wave, frequency, and phase identical to the main outputs, buffered to drive up to TTL loads.

---

[a] The material in Section 7.2 is used courtesy of Global Specialties Corp.

FIGURE 7.1
Model 2001 sweepable function generator.
(Courtesy of Global Specialties Corp.)

Refer to Figure 7.2 for the location of all front panel features. These front panel numbers will be referred to throughout Section 7.2. All push-button switches are push-to-activate, push-to-release switches. The RANGE and MODE groups of switches are interlocked, allowing only one switch within each group of switches to be activated at any one time.

The control and connection locations are numbered from 1 to 12 and are described as follows:

1. **Power switch.** Power is supplied to the 2001 when the power switch is pressed in.

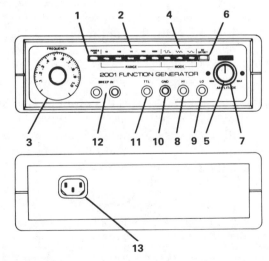

FIGURE 7.2
Control and connection locations. (Courtesy of Global Specialties Corp.)

2. **Range switch.** The range switches act as multipliers for the frequency vernier. As an example, if the frequency dial is set to 0.5 with the 1K range switch depressed, the frequency at every output is 500 Hz.

3. **Frequency vernier.** The frequency vernier acts as a continuous fine tuning adjustment for setting the output frequency. The vernier is marked with nine major divisions, 0.1 to 1.0, subdivided into five minor divisions. For example, to set the 2001 to 50 kHz,

   Set the frequency vernier to 0.5.

   Depress the 100K range push button.

   The output frequency should now be 50 kHz.

4. **Mode switches.** Whichever of the $\sqcap$, $\diagup$; or $\sqcap$ push-button switches is pressed in, the associated waveform appears at the output terminals.

5. **Amplitude control.** The amplitude control is a linear potentiometer that varies the dc-coupled level up to 10 V (with no output load), and up to 5 V with a 600-Ω load. This control is the inner shaft of a dual concentric control; the outer ring is the OFFSET control.

6. **DC offset switch.** When pressed in, the dc offset push button activates the dc offset feature. The centerline of the waveform at the HI output can then be offset from its calibrated zero to ±5 V at no load and ±2.5 V at 600 Ω. When the dc offset switch is released, dc offset is removed from the output waveform, leaving it symmetrically centered with reference to ground.

7. **DC offset control.** The dc offset control is enabled when the dc offset pushbutton is pressed in. This control is coupled to the outer ring of the dual concentric pair (the amplitude control being coupled to the inner shaft). With the control fully counterclockwise, −5 V dc offset is added to the output signal (no load). As the control is turned clockwise, the amount of dc offset varies linearly through zero (as the midpoint) to +5 V dc offset in its fully clockwise setting.

8. **High output jack.** The HI output jack is intended as the main signal output of the 2001; its output impedance is a constant 600 Ω at any amplitude, to a maximum 5 V p-p (driving a 600-Ω load, maximum 10 V p-p unloaded). The output signal characteristics are determined by the settings of the function switch, frequency vernier, range switch and dc offset controls.

9. **Low output jack.** The LO output is −40 dB below the HI output, features a constant 600-Ω output impedance, and is capable of a maximum 100-mV p-p amplitude (unloaded, maximum 50 mV p-p driving 600Ω).

10. **Ground jack.** Connection of the ground jack to the ground on your equipment is strongly recommended.

11. **TTL output jack.** The TTL output jack produces a TTL-compatible

square wave output capable of driving 10 TTL loads. The TTL output wave frequency and phase are identical to those of the square-wave signal at the HI and LO outputs.

12. **Sweep input jack.** An external voltage applied to the sweep input jacks will linearly vary the frequency at all the outputs. A positive voltage will cause an increase in frequency, a negative voltage a decrease. The change in frequency from that indicated on the dial is given by the formula:

$$\Delta f = 0.093 \times V_i \times R \tag{7.1}$$

where

Δf = change of frequency, Hz

V_i = input voltage, Volts

R = multiplier indicated by RANGE switch setting

Note that while it is possible to obtain very low frequencies with the application of negative sweep voltages, it is inevitable that accuracy will deteriorate at these frequencies. The 2001 determines frequency within a range switch (2) setting by summing the setting of the frequency vernier (3), which divides a carefully calibrated internal voltage, with the voltage applied at the sweep input jack (12). Maximum linear sweep range is achieved with the frequency vernier (3) near 0.9; but while linearity suffers when the sum of these voltages reaches the end of the usable sweep range, no damage to the 2001 will occur as long as the sweep input (12) does not exceed ±12 V.

In checking out the 2001 sweepable function generator, use Figure 7.3. To acquaint the reader with all the functions of the 2001, the following checkout procedure is recommended:

1. **Equipment:** oscilloscope, Tektronix model 450 or equivalent; and 10 : 1 attenuator probe with shunt capacity of 15 pF or less.

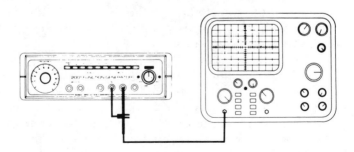

FIGURE 7.3
Checkout configuration. (Courtesy of Global Specialties Corp.)

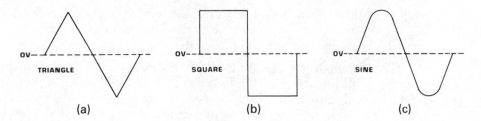

FIGURE 7.4
Checking waveforms. (Courtesy of Global Specialties Corp.)

2. Set the oscilloscope as follows:

CONTROL	POSITION
Sweep time	0.1 ms/cm
Vertical display	2 V/cm (centered on screen)
Trigger mode	Normal
Trigger slope	+
Input	Normal, dc

3. Set the controls on the 2001 as follows:

PUSH BUTTON	POSITION
Power	on (in)
Range	1K
Mode	⊓⌐
DC offset	off (out)

4. Set the frequency vernier to 0.3.

5. Set the amplitude control to MAX. Connect the oscilloscope to the 2001 as shown in Figure 7.3.

Figure 7.4 shows the various waveforms as they will be observed. Part (a) shows the waveform with the ∧ push button in. Part (b) is the waveform with the ⊓ push button in. Part (c) is the waveform with the ∿ push button in.

All waveforms are 10 V p-p with no dc component in the waveform.

Figure 7.5 shows Figure 7.4(a) with dc offset added. The dc offset push button (6) must first be pushed in to activate the dc offset feature; then the dc offset control will vary the vertical position of the waveform trace on the oscilloscope. With the amplitude control (5) set at MAX, maximum dc offset will result in waveform clipping, as seen in Figure 7.5(a) and (d). Figure 7.5(b) and (c) shows maximum allowable dc offset before clipping.

Connect the scope probe to the TTL output (11). Figure 7.6 shows the waveform on the scope. The TTL output (11) produces a fast rise time 0- to 5-V square wave capable of driving 10 TTL loads. The TTL output is independent of the mode switches (4), dc offset (6), and amplitude (5) controls. Its frequency and phase are the same as that of the main outputs.

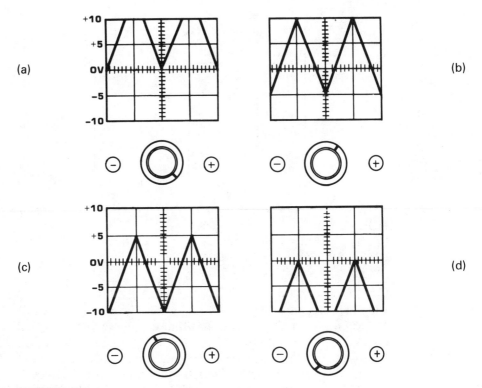

FIGURE 7.5
Operation of dc offset. (Courtesy of Global Specialties Corp.)

The 2001 sweepable function generator with its large voltage range, low distortion, and five-decade frequency response is ideal for audio testing. The 600-Ω output impedance of the 2001 concurs with the dBm scale, standard for audio testing.

Figure 7.7 shows the relationship between dBm, voltage and milliwatts of power. (The power relationships holds only for 600 Ω.)

The volume unit (VU) system and the dBm system are directly related, with zero VU equal to +4 dBm. Figure 7.8 shows a simple circuit for calibrating volume indicator (VI) meters. The amplitude of the 2001 should be set for 0 dBm or 0.7746 V rms at 1 kHz (approximately 2.2 V p-p) into a

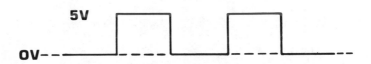

FIGURE 7.6
TTL output. (Courtesy of Global Specialties Corp.)

| dB down | | Level | dB up | |
Volts	MilliWatts	dbm	Volts	MilliWatts
0.7746	1.000	—0+	.7746	1.000
0.6905	.7943	1	.8691	1.259
0.6167	.6310	2	.9752	1.585
0.5484	.5012	3	1.094	1.995
0.3882	.2512	6	1.546	3.981
0.2449	.1000	10	2.449	10.000
0.07746	.0100	20	7.746	100.000
0.02448	.0010	30	24.49	1W
0.007746	.0001	40	77.46	10W
0.000774	1.00 × 10-6	60	774.6	100W

FIGURE 7.7
Charts of decibel relationships. (Courtesy of
Global Specialties Corp.)

load of 600 Ω. The VU meter should then read 0 VU. If the meter reads —4
VU, this indicates that an attenuator has been included in the meter case.

Figure 7.9 shows a basic amplifier frequency response test circuit. Most
power amplifiers require an input level between —20 and 0 dB. At these
levels, the 2001 HI output (8) should be used. The cable connecting the 2001
to the amplifier input should be shielded and, if necessary, the load
matched with a 600-Ω resistor at its input terminals. The amplifier's output
terminals should be monitored by an oscilloscope, and both input and
output terminals should be monitored by an ac voltmeter with a decibel

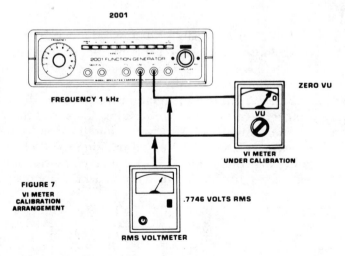

FIGURE 7.8
Volume indicator meter calibration arrangement.
(Courtesy of Global Specialties Corp.)

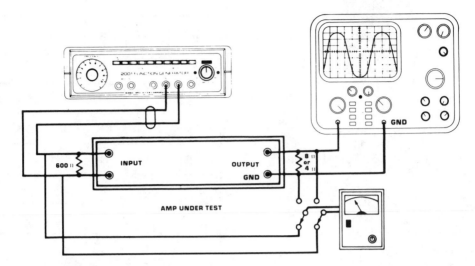

FIGURE 7.9
Amplifier frequency response test. (Courtesy of
Global Specialties Corp.)

scale. The oscilloscope will then display any hum, noise, or oscillation
appearing on the signal. Care should be taken with the input and output
leads; they should be as short as possible and kept far enough apart to
prevent coupling.

To perform the frequency response test, first set the 2001 to its 10K range,
frequency vernier   to   0.1,   and   the   mode   switch   to $\sim$. Next,
attach the ac voltmeter across the load at the amplifier output terminals.
Then adjust the 2001 amplitude control (5) to a level −10 dB below the
maximum rated output of the amplifier. This establishes a 0-dB reference
level, against which other readings will be compared.

Power Ratings (rms)			Output (rms)	
8 Ω (W)	4 Ω (W)	E^2 (V)	Full Output (V)	10 dB below Max. Output (V)
1	2	8	2.8	2.3
5	10	40	6.3	5.2
10	20	80	8.9	7.3
20	40	160	12.6	10.3
50	100	400	20.0	16.3
100	200	800	28.0	23.1

FIGURE 7.10
Power–output relationships. (Courtesy of Global
Specialties Corp.)

L = Low
LL = Very Low
H = High
OK = Suitable, Proper

✳Sharp
Cutoff
or Peaked

Waveform	LF Gain	LF Delay	HF Gain	HF Delay	Damping
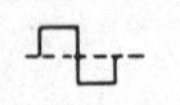	OK	OK	OK	OK	OK
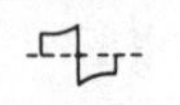	L	OK	H	OK	H
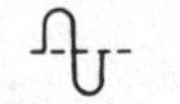	H	OK	L	OK	H
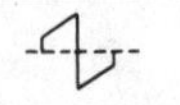	OK	H	OK	L	H
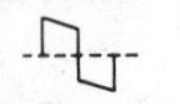	OK	L	OK	H	H
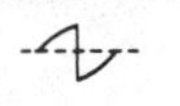	H	H	L	L	H
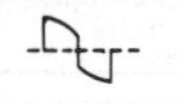	H	L	L	H	H
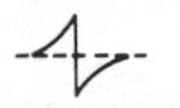	L	H	H	L	H
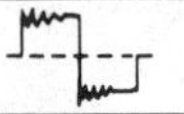	OK	OK	H	OK	OK
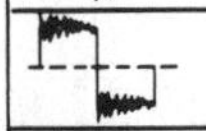	OK	OK	H	OK	L
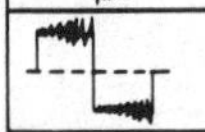	OK	OK	H	OK	LL
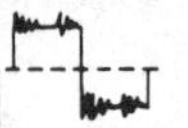	OK	OK	✳	OK	L

FIGURE 7.11
Square wave patterns and interpretations of each
shape. (Courtesy of Global Specialties Corp.)

Figure 7.10 shows a convenient table of power and voltage relationships for 4- and 8-Ω speakers. It also lists the voltage level −10 dB below
the maximum output levels at different power ratings. Choose the rating
closest to the amplifier under test. Testing at −10 dB below maximum
output ensures that the amplifier output will not saturate. This level is
approximately two-thirds of the full power output.

Starting from the amplifier's lowest specified frequency (in most cases
20 Hz), increase the 2001 output frequency, noting changes in the decibel
reading on the ac voltmeter and monitoring waveform changes on the oscilloscope. These readings, when drawn on semilog graph paper, plot the
amplifier's frequency response curve.

Square waves are used in audio testing to display a wide range of
frequencies simultaneously. Mathematically, a square wave can be considered as a fundamental frequency summed with a series of its odd harmonics, which serve to square off the waveshape. For an amplifier to repro-

duce a square wave, it must have a flat frequency response from 0.1f to 10f, where f represents the fundamental frequency of the square wave.

The traditionally used frequencies are 50 Hz for the low-frequency test and 10 kHz for the high-frequency test. Figure 7.11 shows the typical square wave patterns and an interpretation of these shapes.

The basic circuit in Figure 7.9 can also be used for testing preamps. Before testing, adjust all tone controls for a flat response. The following are the tests for the various types of preamplifiers.

Set the 2001 LO output (9) to 10 mV (unloaded) by setting its HI output (8) to 1 V (unloaded). Connect the preamp input to the LO output (9). Figure 7.12 shows the RIAA equalization curve, along with the correct

STANDARD RIAA REPRODUCING CHARACTERISTIC			
Hz	dB	Hz	dB
30	+ 18.61	5,000	− 8.23
50	+ 16.96	6,000	− 9.62
70	+ 15.31	7,000	− 10.85
100	+ 13.11	8,000	− 11.91
200	+ 8.22	9,000	− 12.88
300	+ 5.53	10,000	− 13.75
400	+ 3.81	11,000	− 14.55
700	+ 1.23	12,000	− 15.28
1,000	0.00 ✳	13,000	− 15.95
2,000	− 2.61	14,000	− 16.64
3,000	− 4.76	15,000	− 17.17
4,000	− 6.64		

✳ based on standard 1000 Hz tone.

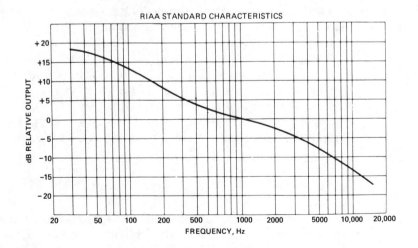

FIGURE 7.12
RIAA standard characteristics. (Courtesy of Global Specialties Corp.)

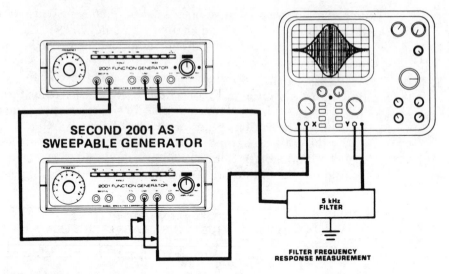

FIGURE 7.13
Audio sweep circuit for testing filter characteristics. (Courtesy of Global Specialties Corp.)

decibel values at 23 different frequencies. The actual measurements should fall within ±2 dB of RIAA equalization. Other equalization curves may similarly be matched.

Since microphone output levels are generally −70 dB, extra care in connecting the input is recommended. To set the 2001 for a −70-dB output, first set the HI output (8) to 0 dB. Next, connect a 20-Ω resistor across the LO output (9). This will bring the LO output down to −70 dB and provide a source impedance of 20 Ω. Before making the frequency run, set all tone controls for a flat response.

Figure 7.13 shows an audio sweep circuit for testing filter characteristics. The sweep input of the 2001 is driven by a triangle waveform. This waveform also drives the X axis of the oscilloscope. Since the triangular waveform exhibits a linear change of voltage with time, the frequency of the 2001 will increase at a linear rate as the oscilloscope trace sweeps from left to right and back again. This allows calibration of the oscilloscope face in linear increments.

First set the sweep voltage to zero and adjust the scope trace so that it is at center screen. This allows the X axis to display both + and − sweep voltages.

Next, to set the 2001 to the middle of the sweep range, press in the 10K range push button (2), and set the frequency vernier (3) to 0.55; then adjust the sweep voltage to 9.7 V p-p. This will sweep the frequency of the 2001 from 1 to 10 kHz. Set the X axis gain of the oscilloscope to 1.0 V/cm. The oscilloscope will then display the graphic bandpass characteristic of the filter.

The circuit diagram of the 2001 is shown in Figure 7.14. The heart of the 2001 is waveform generator IC A3 (ICL-8038). This IC produces all three waveforms. The range switches (2) select tuning capacitors C5 through C9. These five capacitors set the decade frequency ranges from 1 Hz to 100 kHz.

The frequency vernier R12 provides continuous frequency adjustment within a preselected decade. High and low end frequencies are calibrated by trimmers R10 and R13. The voltage at R12 is buffered by A1-C, then fed to A1-A, where it is summed with the sweep input. This combined voltage is then fed to the VCO input pin of A3.

The 2001 requires three regulated supplies: +12 V is created by a 12-V IC regulator (A4); A1-B and Q3 form a −12-V tracking regulator, slaved to the regulated +12-V output; a +5-V regulator is formed by voltage divider R1, R2, A1-D, and Q1. This regulator powers the TTL circuitry.

The square wave output of A3 drives Q2, which in turn drives A2, the TTL output buffer. The selected output of A3 is first buffered by Q4 and Q5, and then fed through the amplitude control to the main amplifier input. Q6 and Q7 are a differential input amplifier for the output drivers. This differential amplifier drives the complimentary push–pull output pair through Q8, a common-emitter driver.

Feedback resistors R37 and R38 establish the overall voltage gain of the amplifier at $V0/V1 = 11$. The summing junction at the base of Q7 is also fed current from the dc offset control through R34. When the dc offset switch is off, the amplifier output is zeroed by potentiometer R30.

Since the output impedance at the junction of R42 and R41 is very low, a series resistor matches the HI output to 600 Ω. This resistor also isolates the main amplifier and its feedback loop from the load. An attenuator network, R45 and R46, provides a low output level −40 dB below the high output.

Specifications for the 2001 follow:

1. **Frequency range:** 1 Hz to 100 kHz in 5-decade range, push button selectable; 50-increment 10 : 1 dial calibrated ±5% of setting at 10 Hz, 100 Hz, 1 kHz, 10 kHz.

2. **Waveforms:**

Sine:	Less than 2% THD
Triangle:	Linearity better than 1%
Square:	Rise and fall times under 100 ns (600 Ω, 20 pF) for ±2% time symmetry
TTL square:	Rise and fall times under 25 ns

3. **Sweep range:** 100 : 1 max; 10 : 1 linear range; 0- to ±10-V sweep input; input impedance 30,000 Ω.

4. **Outputs:**

HI:	0.1 to 10 V p-p open circuit, 0.05 to 5 V p-p into 600 Ω
LO:	1 to 100 mV open circuit, 5 to 50 mV into 600 Ω; 40-dB amplitude control ±0.5 dB flat.
TTL square:	Drives 10 TTL loads

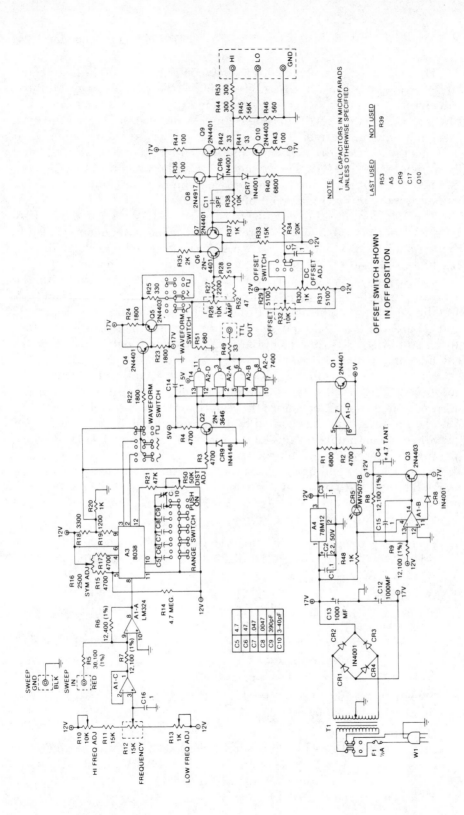

FIGURE 7.14

Circuit diagram of the 2001 sweepable function generator. (Courtesy of Global Specialties Corp.)

5. **DC offset:** Switch selectable; ±5 V into an open circuit; max offset (ac + dc before clipping) ±10 V at HI output and ±1 V at LO into open circuit; ±5 V at HI and ±0.5 V at LO into 600-Ω load

6. **Power:** 6 W at 105 to 125 V ac 50/60 Hz; 220 to 240 V ac 50/60 Hz optional

7. **Operating temperature:** 0 to 50°C (calibrated at 25°C ± 5%)

8. **Size (W × H × D):** 10 × 3 × 7 in. (254 × 76 × 178 mm)

9. **Weight:** 2.2 lb (1 kg)

7.3
MICROPROCESSOR-BASED FULLY PROGRAMMABLE SWEEP OSCILLATOR[b]

Oscillators vary from manufacturer to manufacturer. The first micro-processor-based fully programmable sweep oscillator mainframe is shown in Figure 7.15. This new HP8350A sweep oscillator family comprises a microprocessor-based mainframe and an extensive selection of RF plug-ins, which gives the microwave engineer a state-of-the-art, completely program-mable, and versatile swept signal source. The HP8350A mainframe operates with 24 different plug-ins. Nineteen plug-ins operating from 10 mHz to 22 GHz come from the existing HP 86200 series plug-in line and are com-patible with the 8350A via a low-cost adapter. The remaining six plug-ins are from the new HP 83500 series, covering a 10-mHz to 26.5-GHz fre-quency range. One plug-in, the 83592A, covers the entire 10-mHz to 20-GHz range, with 10 mW of output power. In addition, all front panel controls on the 8350A and 83500 series plug-ins are computer controllable via the HP-IB. The 8350A may act as either a talker or a listener. With all these new features, the 8350A sweep oscillator provides an excellent source for both bench and automatic applications.

The significant features of the HP8350A sweep oscillator include the following:

1. All function values can be set using the precise data entry keyboard, step keys, or one of the four knobs that provide the traditional analog feel. Parameter values are shown on digital LED displays.

2. An innovative marker section generates up to five independent fre-quency markers, measures the difference between any two markers, sweeps between two marker settings, and permits marker frequency to be counted while sweeping using the HP5343A frequency counter.

3. Up to nine complete front panels setups can be saved in memory and later recalled when measurements are repeated. For what are effec-

[b]The material in Section 7.3 is reprinted from *Measurement Computation News*, July/Aug. 1981. Courtesy of Hewlett-Packard Co., Palo Alto, Calif.

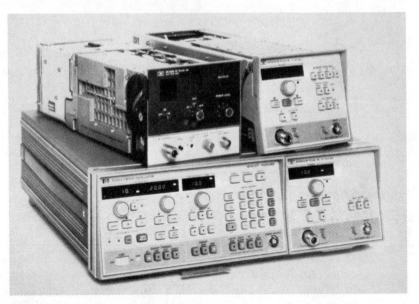

FIGURE 7.15
The first microprocessor-based, fully programma-
ble sweep oscillator mainframe. (Courtesy of
Hewlett-Packard, Palo Alto, Calif.)

tively simultaneous measurements over different frequency ranges or
power levels, the 8350A can even alternate on successive sweeps between any two saved front panel settings.

4. Automatic self-check and diagnostic capabilities are built in for greater
   confidence and easier calibration.

5. The 8350A is fully compatible with both scalar and vector network
   analyzers.

6. Six plug-ins are available with the widest frequency coverage (10 MHz
   to 20 GHz in one unit) and the highest solid-state output power of 10 to
   50 mW available.

7.4
PULSE GENERATORS [c]

The 4001 of Figure 7.16 is a compact, flexible, wide range dc to 5 MHz,
reliable, low-cost pulse generator. Its uncomplicated rugged design and
high-quality components ensure long and dependable service. Designed
primarily as a pulse or clock source, the 4001 is compatible with IC and

---

[c] The material in Section 7.4 is used courtesy of Global Specialties Corp.

FIGURE 7.16
Model 4001 ultravariable pulse generator.
(Courtesy of Global Specialties Corp.)

discrete component circuits. It also finds application as a system stepper (one-shot mode), gated oscillator (gate mode), or pulse stretcher (trigger mode). With a minimum of adjustments, the 4001 can also serve as a missing pulse detector or a frequency discriminator. The output can be complemented or converted to a square wave with the push of a button.

Features such as two simultaneous independent outputs (TTL and variable) with rise and fall times of less than 30 ns; 20-ns TTL compatible leading-edge sync pulse output; independent pulse width and pulse spac-

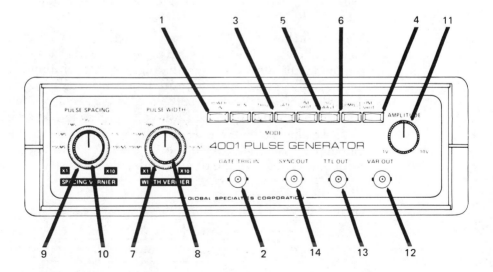

FIGURE 7.17
Front panel features. (Courtesy of Global Specialties Corp.)

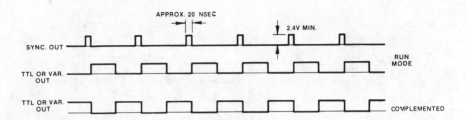

FIGURE 7.18
Run mode. (Courtesy of Global Specialties Corp.)

ing controls (both selectable from 100 ns to 1 s), eliminating incompatible frequency and pulse width settings; four front panel push-button-selectable operating modes (run, trigger, gate, and one-shot); and an uncomplicated front panel layout, all add up to an effective solution to a wide range of pulse source problems.

Refer to Figure 7.17 for the location of all front panel features. These numbers will be referred to throughout Section 7.4.

The pushbuttons are numbered from (1) to (14) and are described as follows:

1. **Power switch:** When the push button is depressed, the 4001 is turned on.

2. **Gate/trig input:** The gate/trig input terminals are dc coupled to the 4001 internal circuitry. The input signal can be a sine wave greater than 1.7 rms or a positive pulse greater than 2.4-V amplitude must not exceed ±10 V.

3. **Mode switches:** *Note:* The mode switches are mechanically inter-locked, allowing only one switch to be activated at a time. RUN: In the run mode, the 4001 is self-oscillating. All external inputs are discon-nected, and all timing controls are functional (see Figure 7.18). TRIG: In the trig mode, the 4001 outputs produce a synchronous positive-going output pulse for each positive input trigger. The output pulse width is adjusted by the pulse width and width vernier controls. The output pulse is initiated by the positive edge of the input trigger. When the mode switch is in the trigger mode, the pulse spacing and spacing

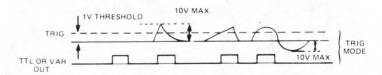

FIGURE 7.19
Trig mode. (Courtesy of Global Specialties Corp.)

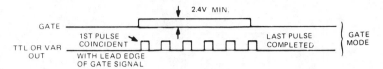

FIGURE 7.20
Gate mode. (Courtesy of Global Specialties Corp.)

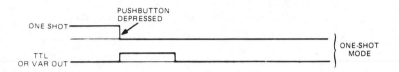

FIGURE 7.21
One-shot mode. (Courtesy of Global Specialties
Corp.)

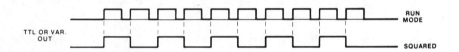

FIGURE 7.22
Conversion to square wave. (Courtesy of Global
Specialties Corp.)

vernier controls are inactive (see Figure 7.19). GATE: In the gate mode, the 4001 output produces trains of pulses for the duration of the applied gating signal. The leading edge of the gating signal starts the output pulse train. The first pulse in the train is synchronized with the lead edge of the gating signal. Both the Pulse Spacing and Pulse Width controls program the pulse train parameters. If the gating pulse ends while an output is present, the last pulse will be completed (see Figure 7.20).

4. **One-shot:** In the one-shot mode, push button 4 initiates an output pulse; the output pulse occurs as the push button is depressed. Pulse parameters are set by the pulse width and width vernier controls. Pulse spacing and spacing vernier controls are not active (see Figure 7.21).

5. **Square wave:** When this button is depressed, the output is converted to a square wave. The output now changes state with every positive edge of the original "programmed" waveform. This divides the frequency of the signal by 2. All inputs and controls are still functional (see Figure 7.22).

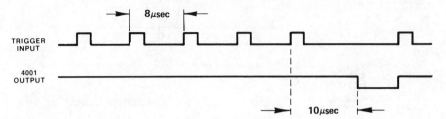

FIGURE 7.23
Use of pulse generator as a missing pulse detector. (Courtesy of Global Specialties Corp.)

6. **Complement:** When this button is depressed, the TTL and VAR outputs are automatically complemented (Figure 7.18). All inputs and controls are still functional.

7. **Pulse width switch:** The pulse width switch is used to select output pulse widths from 100 ns to 1 s in seven ranges. When used in conjunction with the width vernier, continuous adjustment over the entire instrument range is achieved.

8. **Width vernier:** The width vernier is used for continuous adjustment of pulse width between the limits of the range set on the pulse width switch. A slight overlap at both ends of the vernier range ensures continuous adjustment over the entire seven decades of pulse width adjustment.

9. **Pulse spacing switch:** The pulse width switch is used to select output pulse spacing from 100 ns to 1 s in seven ranges. When used in conjunction with the spacing vernier, continuous adjustment over the entire instrument range is achieved.

10. **Spacing vernier:** The space vernier is used for continuous adjustment of pulse spacing between the limits of the range set on the pulse spacing switch. A slight overlap at both ends of the vernier range ensures continuous adjustment over the entire seven decades of pulse spacing adjustment.

11. **Amplitude control:** The amplitude control adjusts the amplitude of the voltage at the VAR OUT BNC connector from 0.1 to 10 V.

12. **Var Out:** The VAR OUT BNC connector provides a convenient means for interconnecting the generator output to its destination. The VAR OUT signal has a rise and fall time of 30 ns and output impedance of 60 Ω.

13. **TTL out:** The TTL output BNC connector is fed by four TTL gates in parallel, providing a TTL fanout of 40. Rise and fall times are less than 20 ns. Both TTL and VAR OUT pulses are derived from the same internal source and are synchronous.

14. **Sync out:** The SNYC OUT BNC connector produces an output pulse 20 ns wide and 20 ns in advance of the main output pulses. The sync pulse amplitude is a minimum of 2.4 V and can drive 10 TTL loads.

The 4001 pulse generator with its many features and ease of operation is a welcome addition to any laboratory. The following is just a sampling of the varied uses the 4001 offers.

Program the 4001 pulse width for 10 μs. Set the input trigger pulse repetition period (PRP) to 8 μs. Each time the trigger pulse goes positive, the 4001 is reset and must time out to its full 10 μs.

The outputs of the 4001 remain in the high state. If one of the trigger pulses is not present (missing), the 4001 output will time out to 10 μs and then return to its low state until the next trigger pulse occurs (see Figure 7.23).

The 4001 and a digital logic probe such as Global Specialties LP-3 make a superior troubleshooting team. Use the 4001 as a signal injector to inject either a pulse train, a single one-shot, or the complement of either. Then trace through the circuits with our LP-3 and quickly find the defective component.

By substituting your 4001 for the microprocessor system clock, you can give your microprocessor the capability of stepping through its microprogram either a step at a time (in one-shot mode) or at much reduced speed, by using long timing periods in RUN mode.

Note, however, that some microprocessors have a minimum clock speed below which correct operation is not assured. If in doubt, check the data sheet for the microprocessor that you are using.

Proportional radio control is usually implemented by sending a variable mark-space ratio low frequency modulation on a radio-frequency signal. Your 4001 may be used to simulate the radio transmitter when testing the low-frequency stages of your receiver. The 4001 is also ideal for simulating the joystick input to the transmitter.

The 4001 again shows its versatility in testing audio amplifiers. Square waves are used in audio testing to display a wide range of frequencies simultaneously. Square waves consist of a fundamental frequency and a series of odd harmonics to square off the wave shape.

For an amplifier to reproduce a square wave, it must have a flat frequency response from $0.1F$ to $10F$, where F is the fundamental frequency of the square wave. The traditional test frequencies are 50 Hz for the low-frequency test and 10 kHz for the high-frequency end. Figure 7.24 shows the typical square wave patterns and an interpretation of these shapes.

Connect the 4001 to the amplifier under test as shown in Figure 7.25 and observe output on the scope. Figure 7.26 shows a convenient table of power and voltage relationships for 4- and 8-Ω speakers. It also lists the output voltage level for 10 dB below the maximum output of amplifiers with different power ratings. Choose the rating closest to the amplifier under

Waveform	LF Gain	LF Delay	HF Gain	HF Delay	Damping
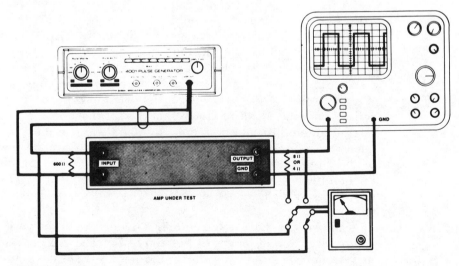	OK	OK	OK	OK	OK
	L	OK	H	OK	H
	H	OK	L	OK	H
	OK	H	OK	L	H
	OK	L	OK	H	H
	H	H	L	L	H
	H	L	L	H	H
	L	H	H	L	H
	OK	OK	H	OK	OK
	OK	OK	H	OK	L
	OK	OK	H	OK	LL
	OK	OK	*	OK	L

L = Low
LL = Very low
H = High
* = Sharp Cutoff or peaked

FIGURE 7.24
Typical square-wave patterns and interpretations
of these shapes. (Courtesy of Global Specialties
Corp.)

FIGURE 7.25
Amplifier frequency response test. (Courtesy of
Global Specialties Corp.)

POWER RATINGS (RMS)			OUTPUT (RMS)	
8 Ω	4 Ω	E²	FULL OUTPUT	10 dB BELOW MAX. OUTPUT
1W	.5W	8V	2.8V	2.3V
5W	2.5W	40V	6.3V	5.2V
10W	5.0W	80V	8.9V	7.3V
20W	10.0W	160V	12.6V	10.3V
50W	25.0W	400V	20.0V	16.3V
100W	50.0W	800V	28.0V	23.1V

FIGURE 7.26
Power–output relationships. (Courtesy of Global Specialties Corp.)

test. Testing at 10 dB below maximum output ensures that the amplifier will not be in saturation. This level is approximately two-thirds of the full power output.

If a transmission line is not terminated at the far end by its characteristic impedance, reflections will occur. This phenomenon can be used to find faults on transmission lines. Using your 4001, you can find out if the cable under test is open, or short circuited, and with some simple calculations, you can find the length of the cable. The equipment includes:

- 4001 pulse generator
- 50-Ω coaxial cable to be tested (approximately 10 m long)
- Oscilloscope
- Passive probe 10 : 1
- 50-Ω termination
- Adaptor banana female, BNC male
- BNC T, connector

Set up the equipment as shown in Figure 7.27. Set the 4001 as follows:

1. Amplitude, 6.5 V
2. Pulse width time, 5 μs
3. Pulse space time, 5 μs
4. Square wave, engaged

Testing Transmission Lines

If a transmission line is not terminated at the far end by its characteristic impedance, reflections will occur. This phenomenon can be used to find faults on transmission lines. Using your 4001 you can find out if the cable under test is open, or short circuited, and with some simple calculations, you can find the length of the cable.

Equipment

1 — 4001 pulse generator
1 — 50Ω coaxial cable to be tested ($\cong$10m long)
1 — oscilloscope
1 — passive probe 10:1
1 — 50Ω termination
1 — adaptor banana female — BNC male
1 — BNC T — connector

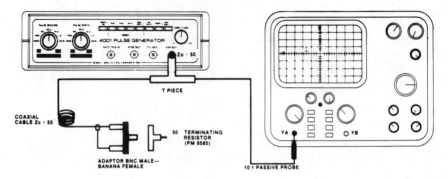

FIGURE 7.27
Transmission line test. (Courtesy of Global Specialties Corp.)

(use VAR output connector). Set the scope as follows:

1. Vertical display, 1 V/cm

2. Input coupling, dc

3. Trig source, int

4. Trig mode, auto

5. Trig level, adjusted

6. Sweep time, 2 μs/cm

With the far end open, the scope should display a signal as shown in Figure 7.28(a). The final amplitude is reached in two steps. At the moment the 4001 meets the 50 Ω of the cable itself, the output is at nominal value [midscale dots B on Figure 7.28(a)] and a reflection takes place at the open end. When this reflection feeds back to the pulse generator output, it tells the 4001 that the far end is open, and the open-circuit voltage of the 4001 appears.

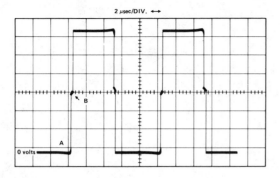

(a) Output of pulse generator with the transmission line.

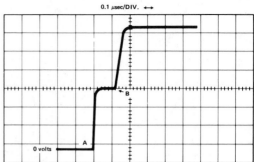

(b) Change time/div of scope to 0.1 μs and observe waveform.

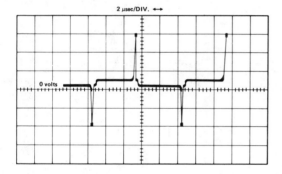

(c) Short circuiting of cable.

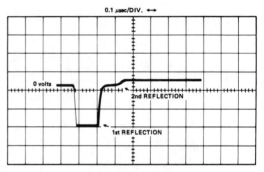

(d) Short circuiting of cable.

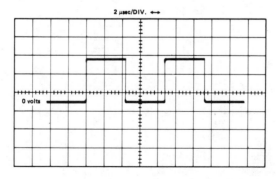

(e) Terminating the transmission line far end with 50-Ω termination results in the illustrated waveform.

FIGURE 7.28
(Courtesy of Global Specialties Corp.)

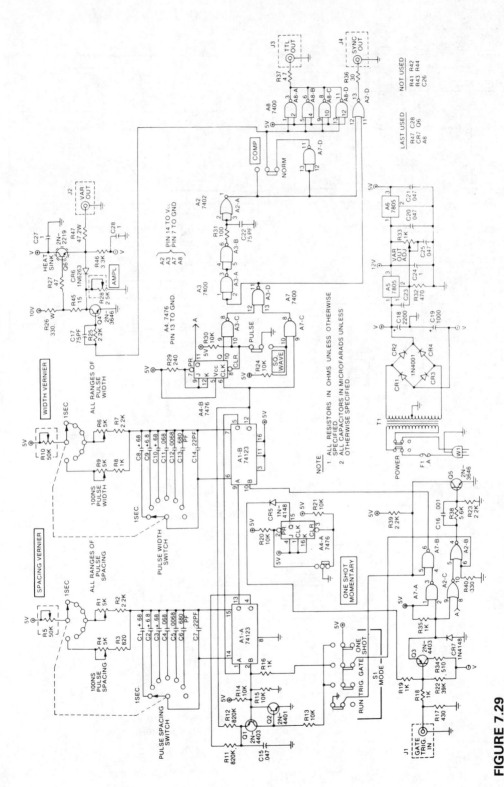

FIGURE 7.29

Schematic diagram of the 4001 pulse generator. (Courtesy of Global Specialties Corp.)

Now change the time/div of the scope to 0.1 μs and observe the signal in Figure 7.28(b). The time between points A and B is the time that it takes for the missing signal to reach the open end and return. For example, assume that, as in Figure 7.28(b), the reflection takes 120 ns. It is known that the velocity of a signal in a coaxial cable is about 0.7C ($C = 3 \times 10^8$ m/s). If the cable has length L, it will take $2L/0.7C$ before the signal returns as a reflection. Using the time observed on the scope, the cable length is calculated as follows:

$$T = \frac{2L}{0.7C} \tag{7.2}$$

where T is time for return reflection. Solving for L, we get

$$L = \frac{0.7CT}{2}$$

$$= \frac{(0.7) \times (3 \times 10^8 \text{ m/s}) \times (120 \times 10^{-9} \text{ s})}{2}$$

$$= 12.6 \text{ m}$$

Remember that the accuracy of this result is determined by the time-base accuracy.

Short circuiting the end of the cable results in the waveform shown in Figure 7.28(c) and (d). After 120 ns the 4001 "knows" that its output is short circuited and the voltage drops to zero. The cable load influences this ideal behavior, and zero means "almost" zero, which can be noticed from the offset level with respect to the start.

Terminating the far end with 50 Ω results in the waveform shown in Figure 7.28(e). Matching the far end of the cable with a resistor equal to the characteristic impedance completely eliminates reflections from the far end, resulting in a perfect square wave at the generator end of the cable.

Figure 7.29 shows the schematic diagram of the 4001 pulse generator. The heart of the pulse generator is the integrated circuit $A1$ dual monostable multivibrator 74123. The two monostable multivibrators are cross-coupled from the Q output of the first to the A input of the second, and vice versa. The cross-coupled monostable multivibrator circuits oscillate as long as their B inputs are high. However, since the coupling between the two circuits is purely dc, it is possible for the oscillator to latch up. The problem is overcome by a special triggering circuit that is gated on if $A1$ fails to oscillate. Transistors $Q1$ and $Q2$ form a hook oscillator. $R14$ and $R15$ bias the base of $Q1$ to $\frac{1}{2}V_{cc}$ while the emitter of $Q1$ is controlled by $R11$ and $R12$. These resistors are connected to the $\overline{Q}$ output of the A1-A and A1-B. As long as $A1$ oscillates, one of the $\overline{Q}$ outputs will be high. The average voltage at the emitter of $Q1$ will be half the $\overline{Q}$ high voltage. This holds the emitter of $Q1$ more negative than its base, and $Q1$ is cut off.

If the oscillator stops, both $\overline{Q}$s go high. $Q1$'s emitter voltage rises above its base voltage and starts to conduct. $Q1$ and $Q2$ turn on and latch, discharging C 15. As Q 1 and Q 2 recover from the latch condition, the B input of A1-A goes high, forcing Q of A1-A low, turning off the hook circuit and restarting the oscillator.

Specifications for the 4001 include the following:

1. **Frequency range:** 0.5 Hz to 5 MHz.

2. **Pulse width and spacing controls:** 100 ns to 1 s in seven overlapping ranges. Independent variable width and spacing controls. Two concentric, single-turn verniers provide continuous adjustments between ranges.

3. **Duty cycle:** 10^7 to 1 range, continuously adjustable.

4. **Accuracy:** ±5% of control settings calibrated at minimum and maximum vernier settings.

5. **Jitter:** Less than 0.1% + 50 ps.

6. **Operating modes (push button selectable):**
   - RUN: 0.5 Hz to 5 MHz, frequency settable through pulse width and spacing controls.
   - TRIG: DC to approximately 10 MHz from external source.
   - GATE: Generator starts synchronously with leading edge of gate signal, one-shot push button can manually activate gate in this mode.
   - ONE SHOT: Enables manual one-shot push button.
   - SQ WAVE: Square wave may be obtained at the outputs by depressing square-wave push button.
   - COMPL: Outputs may be inverted by depressing the COMPL push button, without losing sync time reference. (*Note:* Pulse spacing controls not active during trigger and one-shot modes.)

7. **Triggergate input requirements:** TTL compatible input, dc-coupled logic input: pulses > 2.4 V peak > 40 ns wide. Sine wave input: > 1.7 V rms < 10 MHz. Input impedance: 400-Ω maximum input ±10 V.

8. **Outputs:**
   - VAR OUT: Amplitude, 0.1 to 10 V, adjustable via single-turn vernier; rise and fall time, 30 ns; impedance, 50 Ω.
   - TTL OUT: Fan out, 40 TTL loads; sink, 64 mA at 0.8 V maximum; rise and fall time less than 20 s.
   - SYNC OUT: Amplitude 2.4 V minimum; fan out, 10 TTL loads; sink, 16 mA at 0.8 V maximum; rise and fall time less than 20 ns; sync pulse lead time, greater than 20 ns.

9. **Power requirements:** 105 to 125 50/60 Hz, 6 W; 220 to 240 V ac, 50/60 Hz optional.

10. **Operating temperature:** 0° to 50°C (calibrated at 25°C ± 5°C).

7.5
GENERAL SPECIALIZED APPLICATIONS FOR PULSE GENERATORS [d]

The pulse generator finds many applications because of the great flexibility of its output. Some of these are as follows:

1. Transient analysis of linear systems, such as overload recovery of amplifiers and damping of servo systems.

2. Radar system testing where the pulse generator is used to simulate the video of a radar return. In these applications the variable delay of the pulser is used for range calibration, and the amplitude is varied to test the sensitivity of the video circuits.

3. Component testing, such as transistor recovery time and photomultiplier switching speeds.

4. Communication system testing. This application area is very varied. The pulse generator can be used to perform sensitivity tests like those of the radar example, plus worst-case data generation and bit error rate measurements.

5. Integrated-circuit testing. This is the largest of all the application areas so we will take a more detailed look at the various requirements of this area.

Each individual IC family has certain unique characteristics that separate it from the other logic families. Because of this, manufacturers produce a pulse generator for each logic family. This family of pulse generators is capable of the < 1-ns transition time requirements of ECL, large 16-V amplitudes required by MOS-related logic families, as well as the many other unique features like offset, variable transition time, pulse delay, and so forth.

Integrated circuits are divided into three groups: small-scale, medium-scale, and large-scale integrated devices. For simple functions like quad NAND gates of hex inverters, or any simple function that requires only a few components on a chip, this chip of IC is considered to be a *small-scale* integrated device (SSI). More complicated devices such as divide-by-10 counters of BCD to decimal decoder–drivers are considered *medium-scale* integrated devices (MSI). Many complex functions such as RAMs, ROMs, and things like a 16-bit multiple port file with 3-state outputs would be considered as *large-scale* integrated devices (LSI).

Any single logic family can be used to make up an SSI, MSI, or LSI device; however, as the device becomes more complex, the test circuits for that device become more complex and therefore more expensive.

The logic family itself will determine which pulse generator is required

---

[d] The material in Section 7.5 is used courtesy of Hewlett-Packard Co., Palo Alto, Calif.

TABLE 7.1
Integrated-circuit families and parameters

A. *IC Families*
1. Resistor-transistor logic (RTL) results in the smallest die size (minimum space on a silicon wafer) for standard bipolar functions. It is easy to process and has low to medium power dissipation. RTL is known primarily for its economy.
2. Modified diode-transistor logic (DTL) is low in cost, has logic familiar to most designers, is available from many sources, and can be used for most general-purpose designs.
3. High-threshold logic (HTL) is designed for noise immunity and finds application in industrial environments and locations likely to have high electrical noise levels. It is noted for its ability to interface easily with discrete devices and electromechanical components.
4. Transistor-transistor logic (TTL) has characteristics that are similar to DTL and is noted for many complex functions and the highest available speed of any saturated logic. Moreover, many sources are available with an excellent rate of complex-function introductions.
5. Emitter-coupled logic (ECL) is known for performance and logic flexibility. It has the highest speed of any of the logic forms.
6. Metal oxide semiconductors (MOS) devices are noted for small die and large complex repetitive circuits, such as memories and shift registers. Power dissipation is low to moderate, with low cost per gate.
7. Complementary metal oxide semiconductor (CMOS) devices employ both *p*- and *n*-channel MOS components and are known for very low power dissipation with moderate delay times.

B. *Parameters of Concern*
1. **Speed**: operating frequency, clock rate, toggle rate, and repetition rate.
2. **Propagation delay**: The time between an input logic level (or transition) change and the output level (or transition) change.
3. **Transition time**: Minimum or maximum switching time that device will accept or produce.
4. **Threshold**: the input voltage level required to transfer a device from one state to another.
5. **Offset**: amount of dc that pulse reference level is shifted from ground, + or −.

for IC test. The degree of device integration will determine how many pulse generators are required and how complex the test circuitry will become. Typically, an SSI device, whether it be TTL, DTL, or whatever, will require one or two pulse generators and one or two power supplies. An MSI or LSI device might require a PRBS generator or a word generator in addition to a number of pulse generators and power supplies.

Integrated-circuit logic has been discussed in detail in Chapter 5. The IC logic family and parameters are given in Table 7.1.

Logic family, typical operating speed, typical propagation delay, typical threshold, and typical logic levels are given in Table 7.2.

The tests that engineers and technicians can perform are as follows:

1. **Toggle rate:** How fast can this IC run? Usually, this test is performed with a sine-wave input signal.

TABLE 7.2
Integrated-circuit logic characteristics

Logic Family	Typical Operating Speed (MHz)	Typical Propagation Delay (ns)	Typical Threshold (V)	Typical Logic Levels	
				High (V)	Low (V)
RTL	8–15	12	0.7–1.2	1–2	0–0
DTL	12–30	30	1.3–1.7	2–5	0–0
HTL	4	90	7.5	15	1.5
TTL	15–60	6–12	1.5	3–5	0.2
ECL	60–400	1–4	−0.8−−1.2	0.5–0.9	−1.5−−1.8
MOS	2	300	−1.5−−10	+5	−15
CMOS	5	70	3–9	5–15	0

2. **Propagation delay:** The time for an input signal to propagate through the IC to the output. The most important test parameter is the transition time of input signal.

3. **Setup and hold time:** Measuring the time that data must be held on the input to a flip-flop before they are clocked into the circuit.

4. **Noise sensitivity:** In this test the offset of the pulse generator is adjusted to just below the threshold level of the gate, and then a narrow pulse is adjusted in amplitude until the gate changes state.

7.6
BASIC SIGNAL ANALYZER CONCEPTS

Basic analyzers for frequency response signal analysis include the following systems:

1. Spectrum analyzers

2. Digital Fourier analyzers

3. Wave analyzers

4. Distortion analyzers

5. Audio analyzers

6. Modulation analyzers

The spectrum analyzer is a swept receiver that provides a visual display of amplitude versus frequency. It illustrates on a single display how energy is distributed in the system as a function of frequency, illustrating the absolute value of Fourier components of a given waveform. The spectrum analyzer is discussed in detail in Section 7.7.

The Fourier analyzer uses digital sampling and transformation techniques to form a Fourier spectrum display that has phase as well as amplitude distortion.

The wave analyzer is a tuned voltmeter, showing on a meter the magnitude of the energy in a specific frequency window, which is tunable over a specific frequency range.

The distortion analyzer collectively measures the energy outside a specific bandwidth, tuning out the fundamental signal and displaying the energy of the harmonics and other distortion products on the meter. The total harmonic distortion analyzer is used in communication equipment and audio and ultrasonic sound systems. (See Section 7.7.1 for a discussion of harmonic distortion, intermodulation distortion, and transient intermodulation distortion.)

The audio analyzer performs the same measurements as the distortion analyzer but includes measurements of the signal-to-noise ratio, frequency count, true rms dvm and dc dvm.

The modulation analyzer tunes to a desired signal and then recovers the entire envelope of AM, FM, and phase modulation by processing and displays.

The preceding analyzers are manufactured by Hewlett-Packard and others.

7.7
THE SPECTRUM ANALYZER[e]

A spectrum is a relationship usually represented by a plot of the absolute or relative *value* of some parameter against *frequency*. Every physical phenomenon, in whatever system (electromagnetic, mechanical, thermal, hydraulic) has a unique spectrum associated with it.

In electronics, we deal with phenomena in the form of *signals:* fixed or varying electrical quantities, such as voltage, current, or power. We can describe such a signal by making a *time-domain* plot of its *amplitude* against *time,* as in the left-hand graphs of Figure 7.30.

Regardless of the time-domain properties of the signal—fixed or varying, periodic, aperiodic (or even a single transient, never repeated)—there is a spectrum (a *frequency-domain* plot) that uniquely corresponds to it. In Figure 7.30, the right-hand plots are spectra corresponding to the left-hand time-domain plots opposite them.

A spectrum analyzer draws a picture analyzing the signal spectra in question. The graphic representation has two axes: frequency and amplitude (Figure 7.30), much as an oscilloscope graphs time and amplitude. Various display modes are available such as log and linear amplitude, and various spans can be selected.

---

[e]From *Spectrum Analysis—Theory, Implementation, and Applications*, 3rd ed., Wavetek® Rockland, Rockleigh, N.J., 1977.

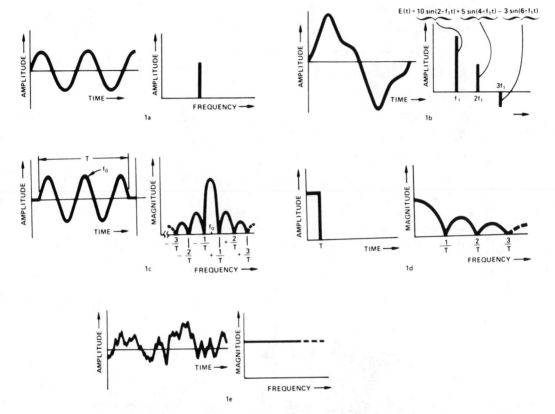

FIGURE 7.30
Signals and their spectra. (Courtesy of Wavetek ®
Rockland, Rockleigh, N.J.)

Basically, a spectrum analyzer is a tuned receiver with selectable frequency range and spans, selectable IF bandwidths, and a linear or log detector, all coupled to an oscilloscope display. A commercial spectrum analyzer is shown in Figure 7.31.

Many measurements now performed by more cumbersome and time-consuming means can be best performed with the spectrum analyzer. Measurements of waveforms, distortion, signal to noise, and amplitudes of complex components are easy with spectrum analysis.

In present-day technology, Tektronix has developed the 492P, a compact spectrum analyzer for frequency ranges of 50 kHz to 220 GHz with microprocessor-aided controls, which makes analysis automatic.

Measurements possible in high fidelity and AM transmitter work with the spectrum analyzer include the following:

1. Power output

2. Harmonic distortion

3. Intermodulation distortion

4. Frequency response

5. Signal-to-noise measurements

6. Distortion versus output

7. Power bandwidth

8. Damping factor

9. Square-wave response

10. Transient intermodulation distortion

Other spectra measurements using spectrum analyzers include the following: measuring filter–network characteristics by impulse excitation, ultrasonic waveform analysis and impedance measurement, real-time analysis, automated phase-noise measurement, biomedical waveform analysis; vibration analysis, acoustic resonance and impedance measurements, and others.

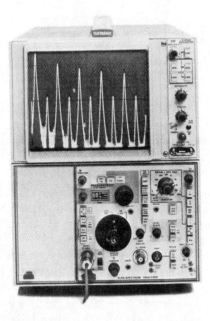

5L4N

20 Hz to 100 kHz
Selectable Impedance
Calibrated Appropriate to Impedance Selected
Single-Ended Input
Differential (Balanced) Input
On Screen Dynamic Range 80 dB (Full 8 div)
Intermode >70 dB Down
Resolution Bandwidth 10 Hz to 3 kHz
Auto Resolution
Built-in Tracking Generator
20 Hz to 20 kHz Log Sweep*

FIGURE 7.31
Tektronix 5L4N spectrum analyzer with frequency range of 20 Hz to 100 kHz. The analyzer features selectable input impedances, 80 dB of dynamic range, and a built-in tracking generator. (Courtesy of Tektronix, Inc., Beaverton, Ore.)

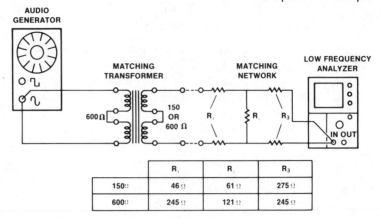

	R₁	R₂	R₃
150Ω	46 Ω	61 Ω	275 Ω
600Ω	245 Ω	121 Ω	245 Ω

FIGURE 7.32
Setup to check the test equipment. (Courtesy of
Tektronix, Inc.)

7.7.1 Audio Modulation Measurements and Other Applications Using the Spectrum Analyzer[f]

Audio modulation using the spectrum analyzer can be made using the
following recommended test equipment:

1. Use of the 5L4N (see Figure 7.28).

2. Audio generator such as the SG502 or equivalent.

3. Matching transformer such as WE 111C or the SG502 or equivalent.

4. Matching pad for the 5L4N input.

5. Miscellaneous patch cables.

We will concentrate on distortion measurements, including harmonic distortion, intermodulation distortion, and transient intermodulation distortion.

It should be emphasized that frequency response, percent of carrier
shift, carrier hum and noise, spurious and harmonics, modulation and other
maintenance operation can be made. Other spectrum analyzers with frequency ranges in television and radar can be used to perform similar functions.

To ensure a correct calibration procedure, hook up the circuit as shown
in Figure 7.32. The audio generator must be connected through a matching
transformer to the 5L4N to ensure noise and hum-free measurements.

---

[f]The material in Section 7.7.1 is excerpted from *AM Broadcast Measurements Using the
Spectrum Analyzer*, Tektronix, Inc., Beaverton, Ore., 1976. Reproduced by courtesy of Tektronix, Inc.

Use the following instructions:

1. Set the 5L4N to −10 dBM REF level and select the 600 Ω internal terminal ON. DBm refers to decibels (dB) with respect to a power level of 1 mW. The dBm scale is used extensively in broadcasting and transmission lines. Mathematically, the power gain, G_p, of an amplifier is

$$G_p = 10 \log \frac{P_2}{P_1} \, \text{dB} \qquad (7.3)$$

Therefore,

$$0 \text{ dBm:} \quad P_1 = P_2 = 1 \times 10^{-3} \text{ W}$$

$$-10 \text{ dBm:} \quad P_2 = 1 \times 10^{-4} \text{ W}$$

$$P_1 = 1 \times 10^{-3} \text{ W}$$

$$+10 \text{ dBm:} \quad P_2 = 1 \times 10^{-2} \text{ W}$$

$$P_1 = 1 \times 10^{-3} \text{ W}$$

For −10 dBm, $P_2 = 1 \times 10^{-4}$ W and a termination resistance provides an output voltage of 600×10^{-2} V or 0.2449 V or 244.9 mV.

2. Check the output level of the audio generator in the MAX position with all attenuation push buttons OUT. The analyzer should indicate a full screen signal (+10 dBm through the 20-dB pad with the analyzer input attenuator set at −10 dBm).

3. Check for harmonic content of the generator by looking at a 0- to 20-kHz span on the analyzer. Tune the generator frequency from 20 Hz to 10 kHz and note that the harmonics (as shown in Figure 7.33) are more than 70 dB down.

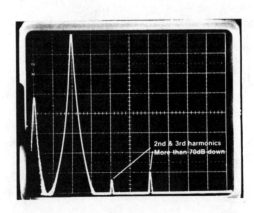

FIGURE 7.33
Audio generator test for harmonic distortion.
(Courtesy of Tektronix, Inc.)

4. Check the flatness of the matching transformer by connecting the tracking generator output of the 5L4N instead of the audio generator. Ideally, it should be virtually flat from 20 Hz to 10 kHz. If it is not, note the error for future reference.

5. Finally, the generator should be reconnected, and the output level set to minimum with all attenuation IN. The baseline of the analyzer display should be clean and free of noise and spurious signals from 20 Hz to 20 kHz. This test should also be performed on location to ensure that the transmitter is not interfering with the test equipment.

Harmonic distortion is the arithmetic sum of the amplitudes of all the separate harmonic components. This is the most common distortion test performed on transmitters, loudspeakers, and amplifiers and is measured by passing a pure sine-wave tone (at 1000 Hz for example) through a transmitter and measuring the sum of all the components (2000, 3000, 4000 Hz, etc.).

The low-frequency spectrum analyzer presents a graphic display for analysis of the harmonics of the audio signal. This technique also makes it possible to observe noise and hum separately from distortion. The 5L4N, in combination with a good audio generator such as the SG502 or equivalent, can make 70-dB (0.034%) harmonic distortion measurements.

Harmonic distortion can also be measured directly at radio frequency with a high-quality spectrum analyzer such as the 7L12 or 7L5. This direct measurement eliminates errors that might occur in the traditional amplitude-modulated detector or monitor that must be used for audio baseband measurements.

The procedure for harmonic measurements in audio work includes the following:

1. Connect an audio generator (such as the SG502) to the microphone input of the transmittter (A) (see Figure 7.34). Set the generator to 1000

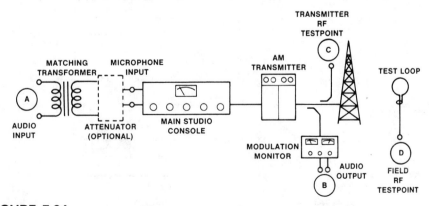

FIGURE 7.34
Standard test points for equipment performance tests. (Courtesy of Tektronix Inc.)

Hz and, with the input fader set to midrange, increase the generator output until 0 VU is indicated on the console.

The test points shown in Figure 7.34 should be located at the station. The 5L4N audio spectrum analyzer should be connected to the output of the modulation monitor or detector (B). The microphone input (A) to the main control room should also be accessible, preferably from the same point. The test equipment should be well grounded. Shielded cable should be used for all connections. If it is not practical to locate the generator and analyzer together (as in the case of remote transmitters), place the analyzer at the site of the modulation monitor and the generator in the main control room.

Signal-processing equipment should be defeated, although every attempt should be made to perform measurements with the equipment left in the processing chain. Most AGC-type equipment have built-in defeat switches that turn off the gain control loop without disturbing the signal path. Some engineers may patch out or bypass processing equipment. While this is not prohibited, it would be illegal to patch around any devices that purposely alter the frequency response (i.e., equalizers). This applies only to frequency-altering devices normally used between the board output and transmitters.

2. Connect the low-frequency spectrum analyzer to the monitor or diode detector test point (B). Set the analyzer controls to 10 dB/DIV, 1 kHz/DIV, and select a center frequency of 5 kHz. A display of the 1-kHz tone should be observed on the second graticule line from the left edge of the screen.

dB DIFFERENCE	ADD TO HIGHER LEVEL
Same (0dB)	3.01
1 dB	2.54
2	2.13
3	1.76
4	1.46
5	1.19
6	.97
7	.79
8	.64
9	.51

FIGURE 7.35
Correction factors for addition of components.
(Courtesy of Tektronix, Inc.)

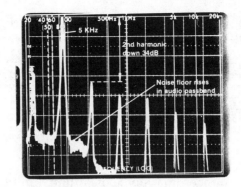

FIGURE 7.36
Harmonic distortion display of 5-kHz tone.
(Courtesy of Tektronix, Inc.)

3. Set the generator output for an indication of 25% on the modulation monitor.

4. Harmonic distortion may be measured by measuring the difference between the fundamental tone and the *sum* of the harmonics (second, third, fourth, etc.).

5. The sum of the harmonics may be added using the chart in Figure 7.35. Harmonics that are 10 dB or more down from the second do not have to be computed.

6. The difference between the fundamental and the harmonics may be measured in decibels as shown in Figure 7.36. These numbers may be converted to percentage of distortion using the chart in Figure 7.37. Record the results on a chart such as the one shown in Figure 7.38.

RATIO in dB	% of READING	RATIO in dB	% of READING
20. (40:60)	10% (1% .1%)	30 (50.70)	3.16% (.31, .031%)
21	8.9	31	2.87
22	7.94	32	2.51
23	7.08	33	2.24
24	6.31	34	2.00
25	5.62	35	1.78
26	5.01	36	1.59
27	4.47	37	1.41
28	3.98	38	1.26
29	3.55	39	1.12

FIGURE 7.37
Decibel to percent of distortion chart. (Courtesy of Tektronix, Inc.)

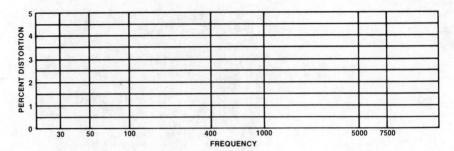

FIGURE 7.38
Chart to record harmonic distortion. (Courtesy of
Tektronix, Inc.)

7. Repeat steps 3 to 6 for modulation levels of 50%, 85%, and 100%.

8. Repeat steps 3 to 7 for frequencies of 30, 50, 100, 400, 5000, and 7500 Hz.

   Intermodulation distortion is determined by putting two or more pure tones through an amplifier and measuring the amount of each tone that is transferred (cross modulated) onto the others.

   While the FCC makes no references or recommendations concerning intermodulation in amplitude-modulated radio, it is recognized by many that harmonic distortion tests do not tell the entire story. Music is composed of complex tones. Harmonic distortion tests are performed using single tones. It is a very useful low-frequency test for low frequencies from 15 Hz or less to about 4kHz.

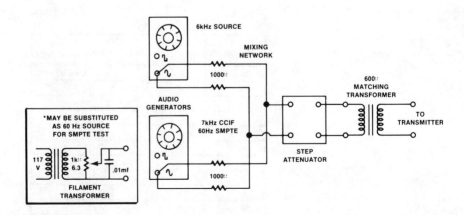

FIGURE 7.39
Intermodulation distortion equipment setup.
(Courtesy of Tektronix, Inc.)

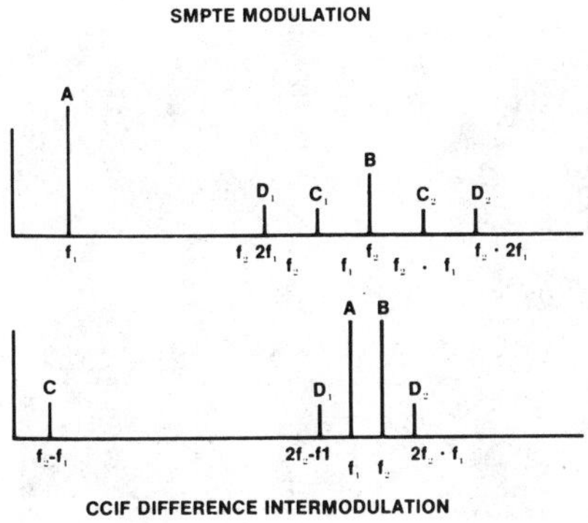

DISTORTION		COMMONLY USED FREQUENCIES
2nd	3rd	
$\frac{C_1 + C_2}{B}$	$\frac{D_1 + D_2}{B}$	60 Hz-5000 Hz 60 Hz-6000 Hz 60 Hz-7000 Hz (IHFM)
$\frac{C}{A + B}$	$\frac{D_1}{A + B}$	4000 Hz-5000 Hz 6000 Hz-7000 Hz 1200 Hz-1400 Hz

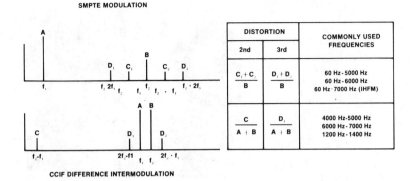

FIGURE 7.40
Two common intermodulation test standards.
(Courtesy of Tektronix, Inc.)

Studies performed many years ago, by the Society of Motion Picture and Television Engineers (SMPTE), proved that a correlation exists between actual listening tests and intermodulation distortion. This correlation was much higher than for harmonic distortion tests. Hence, the SMPTE has used the intermodulation distortion tests for motion picture film for years.

Most broadcasters are in the music business where intermodulation distortion tests would prove valuable. It is a welcome fact, then, that the audio spectrum analyzer is capable of measuring both intermodulation and harmonic distortion.

The following procedures may be used to measure intermodulation distortion by either of two techniques, the SMPTE or the CCIF. The exact details and options for these two tests are shown in Figure 7.39. The CCIF intermodulation test gives useful information for frequencies from 4 to 20 kHz, the high end of the audio spectrum. The two intermodulation distortion test standards presently discussed are summarized in Figure 7.40.

Any studio equipment may be also tested for intermodulation distortion. While both tests are standardized around higher-frequency audio tones, the lower-frequency cutoff of amplitude-modulation transmitters makes it desirable to use lower-frequency tones. We recommend that 60 Hz and 5 kHz be used for the SMPTE method and 4 and 5 kHz be used for the CCIF method.

The procedure for intermodulation distortion tests includes the following:

1. Set up a two-tone source connected to the input of the console or the transmitter as shown in Figure 7.39. If the two audio generators are not available, a filament transformer can be used for a 60-Hz tone. If this is done, however, the SMPTE technique must be used.
2. Set the tone ratio. If the SMPTE method is used, the 60-Hz tone should

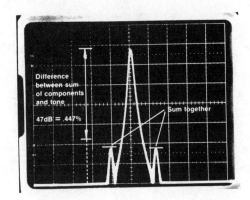

FIGURE 7.41
SMPTE test results. (Courtesy of Tektronix, Inc.)

be four times (12 dB) greater than the 5-, 6-, or 7-kHz tone. If the CCIF technique is used, the two tones should be equal in amplitude.

3. The console or transmitter input should be increased until 100% modulation (0 VU) is indicated.

4. The 5L4N should be connected to the modulation monitor or detector output and set for a span of 0 to 10 kHz.

5. SMPTE distortion is measured by noting the decibels down of the 60-Hz sidebands from the 6-kHz tone, as shown in Figure 7.41, and then converting to a percentage using the chart in Figure 7.37.

6. Intermodulation distortion measurement using the CCIF technique is performed by noting the decibels of the generated 1-kHz offset sidebands or 1-kHz tone (as shown in Figure 7.42), and then converting to a percentage using the chart shown in Figure 7.39. (Sum the components values to obtain true intermodulation distortion for the third-order distortion.)

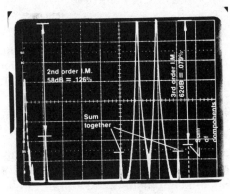

FIGURE 7.42
CCIF test results. (Courtesy of Tektronix, Inc.)

Transient intermodulation distortion is amplifier distortion that occurs principally during loud (high level), high-frequency passages. Most music contains some material that can cause transient intermodulation distortion. Amplifiers with large amounts of negative feedback are prone to transient intermodulation distortion because the amplifier loop, if improperly designed, requires too much time to respond to rapid transients.

Since the introduction of the transistor power amplifier, the *transistor sound* has been discussed. Even though, in many cases, a transistor amplifier tested better in terms of distortion than a tube counterpart, during a listening test the tube unit would unmistakably perform better. Transient intermodulation distortion is one explanation of these discrepancies. Transistor amplifiers perform excellently in tests using steady-state harmonic and intermodulation tests. However, music material generates amplifier distortion because of its transient nature.

A popular explanation for the source of transient intermodulation distortion is that the transients reach or exceed the skew rate of the amplifier, causing an instant, severe intermodulation condition until the time lapse of the negative feedback signal overcomes and corrects the distortion. Applying this explanation to the amplitude-modulation station, we have all heard an old transmitter that sounded a lot better than a more recent high-powered one, even though the new transmitter tested better on the harmonic distortion tests. The key factor is the use of high negative feedback. While transistors are still in the future for superpower transmitters, the broadcaster has been using large amounts of negative feedback normally associated with transistors to correct high-power tube-type modulation output stages. Many transmitters detect the radio frequency and apply feedback to the first audio stage.

Transient intermodulation distortion does not occur in the feedback loop, but when the signal is delayed in the forward direction. A number of factors contribute, such as mechanical length (often 8 or 10 ft through the transmitter cabinets) and bypassing, in the audio stages, to stop oscillation and radio frequency. Cases of 60% to 70% of transient intermodulation distortion have been found in transmitters that passed FCC harmonic distortion test standards.

Once again, the audio spectrum analyzer may be used to check transient intermodulation distortion along with *im* distortion and harmonic distortion. No measurement standard exists in any industry yet for transient intermodulation distortion; however, a low-frequency square wave with a high-frequency sine wave can produce easily interpreted results. Following is a technique that used a 6250-Hz sine wave and a 500-Hz square wave to yield a figure in percent of transient intermodulation distortion. Also included are some notes on possible cures for this distortion, as observed by the author.

To make transient intermodulation distortion use the following steps.

1. Connect the equipment as shown in Figure 7.43. Two SG502 audio generators must be combined as shown, one to produce 500-Hz square waves, the other to produce 6-kHz sine waves. An FG501 may be substi-

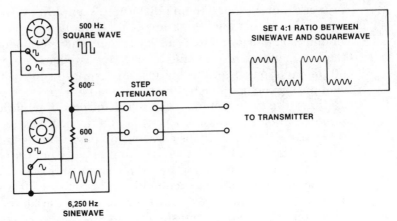

FIGURE 7.43
Transient intermodulation distortion equipment
setup. (Courtesy of Tektronix, Inc.)

tuted for the square-wave source. (The square wave should have excellent symmetry.) Connect the combined generators to the audio input of
the transmitter.

2. Temporarily connect the oscilloscope portion of the audio spectrum
   analyzer (5L4N) to the combined output of the two generators and set
   the ratio of the square wave to sine wave at 5:1. A pattern similar to
   Figure 7.44 should be obtained. This is the transient intermodulation
   distortion "stock" signal.

3. Reconnect the audio spectrum analyzer to the modulation monitor output (B) and display a span from 0 to 10 kHz.

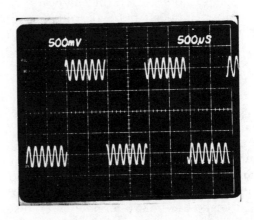

FIGURE 7.44
Transient intermodulation distortion stock signal.
(Courtesy of Tektronix, Inc.)

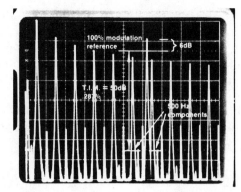

FIGURE 7.45
Transient intermodulation distortion test results.
(Courtesy of Tektronix, Inc.)

4. Look for 500-Hz sidebands on the 6.25-kHz tone in the positions noted
   in Figure 7.45. If these sidebands are 3 dB below the 6.25-kHz tone,
   100% *tim* distortion is indicated. To calculate actual transient intermodulation distortion, measure the sideband amplitude relative to the 100%
   modulation point (6 dB down from the 6.25-kHz tone) in decibels.
   Convert dB to percentage using the chart in Fig. 7.37.

NOTES: The largest factors contributing to transient intermodulation distortion problems in transmitters can be traced to the use of large amounts
of negative feedback. Since feedback corrects steady tone distortion, reducing feedback in the transmitter must be accompanied by other changes.
Sometimes the simplest cure is to temporarily remove the feedback connection and rebias the audio and output stages for better performance. This
will, of course, reduce the tube life on some stages, but once the feedback

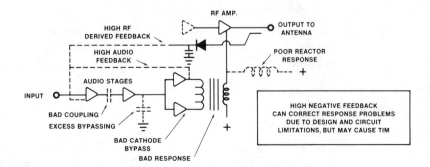

FIGURE 7.46
Common areas that contribute to transient intermodulation distortion problems. (Courtesy of Tektronix, Inc.)

(and less of it) is reconnected, the transmitter will sound better. Another thing to check carefully is that the signal path bypass capacitors are not pulling down the response in the audio range.

In rare instances, the entire feedback loop must be divided so that each stage has a small feedback loop. Figure 7.46 points up some common contributing factors of transient intermodulation distortion.

Transient intermodulation distortion is also useful today in the diagnosis of integrated logic circuits and microprocessor technology.

7.8
REVIEW QUESTIONS

1. Discuss briefly the uses and theory of a function generator.

2. Draw a block diagram to check the frequency response of a power amplifier using a function generator.

3. Discuss briefly the power amplifier square-wave response testing by use of a function generator.

4. Discuss briefly the microprocessor-based fully programmable sweep oscillator.

5. Discuss briefly the uses and theory of a pulse generator.

6. Discuss briefly the technique to find a fault on a microprocessor program and a digital logic flow system using a pulse generator.

7. Discuss the check of a transmission line with a pulse generator.

8. Discuss briefly the theory on the use of a spectrum analyzer; digital Fourier analyzer; distortion analyzer; audio analyzer; modulation analyzer.

9. Discuss briefly the procedure to make a harmonic distortion, intermodulation distortion, and transient intermodulation distortion measurement on an audio system.

10. (a) Discuss the meaning of dBm. (b) An amplifier has a power output of 100 W. Find the dBm value. (c) What is the voltage across a 600-Ω termination at 0 dBm?

7.9
REFERENCES

1. Clyde F. Coombs, Jr., *Basic Electronic Instrument Handbook*, McGraw-Hill Book Co., New York, 1972.

2. P. Kantrowitz, G. Kousourou, and L. Zucker, Electronic Measurements, Prentice-Hall, Inc., Englewood Cliffs, N.J., 1979.

3. Applications and Operation of the 8901A Modulation Analyzer, Application notes 286-1, Hewlett-Packard, Palo Alto, Calif., 1980.

4. Tektronix 1981 Catalogue, Tektronix, Inc., Beaverton, Ore., 1980. Published yearly.

5. Spectrum Analysis, Spectrum Analyzers Basics, Application Note 160, Hewlett-Packard, Palo Alto, Calif., 1974.

6. Morris Engelson, *Noise Measurements Using the Spectrum Analyzer, Part Two; Impulse Noise*, Tektronix Inc., Beaverton, Ore., 1975.

7. Clifford B. Shrock, *AM Broadcast Measurements Using the Spectrum Analyzer*, Tektronix Inc., Beaverton, Ore., 1976.

8. Clifford B. Shrock, *Standard Audio Tests*, Tektronix, Inc., Beaverton, Ore., 1975.

9. *Measurement/Computation Electronic Instruments and Systems*, Hewlett-Packard, Palo Alto, Calif., 1982. Published yearly.

10. *Signal Generator Seminar*, Hewlett-Packard, Palo Alto, Calif., revised Sept. 1974.

11. *Spectrum Analysis—Theory, Implementation and Analysis*, Wavetak® Rockland, Rockleigh, N.J., 1977.

12. Susumu Takahasi, and Susumu Tanaka, "A New Method of Measuring Transient Intermodulation Distortion: Comparison with the Conventional Method," *JAES*, vol. 30, no. 1/2, Jan./Feb. 1982.

8
Oscilloscopes

8.1
INTRODUCTION[a]

The cathode-ray oscilloscope is the most useful electronic instrument for the measurement of analog and digital waveforms. The oscilloscope examines the parameters of electrical events in real time. As the nature of these electrical events becomes more digital, the difficulty of triggering at the point in time of interest increases. The traditional scope trigger circuit is voltage level sensitive, usually over a wide bandwidth.

The oscilloscope can also present visual voltage operation of the transducers discussed in Chapter 2. This chapter discusses:

1. Analog dual trace oscilloscopes

2. Digital oscilloscopes

3. Special waveform programmable oscilloscopes

---

[a] The material in Section 8.1 to 8.4 is used courtesy of Hewlett-Packard Co., Palo Alto, Calif. From Bench Briefs, Service Information from Hewlett-Packard, pages 1–6, March–May 1981. Reprinted with permission of BENCH BRIEFS, a Hewlett-Packard Service organization.

In 1879, William Crookes demonstrated the ability to deflect cathode rays in a vacuum tube with a magnet. Cathode rays had earlier been shown capable of causing phosphorescence on the glass walls of a vacuum tube, but control of the area of phosphorescence was only possible by using shaping masks (solid structures in the tube incapable of producing dynamic deflection). The combination of focusing elements to produce a narrow electron beam (or cathode ray) aimed at a fluorescent target with dynamic electromagnetic-beam deflection became known as a *Crookes tube*. The Crookes tube, later to be commonly known as a *cathode-ray tube* (CRT) or more precisely as an *electron-beam tube*, offered much promise for displaying high-speed variations that could not be demonstrated on mechanical apparatus. By 1897, Carl F. Braun had constructed a "variable current apparatus" by using the Crookes tube, the first forerunner of the modern oscilloscope.

An oscilloscope is an instrument capable of presenting a luminous XY graph of any two related electrical parameters. One set of electrical signals is applied to the horizontal deflection (X-axis) system, and the other set of signals to the vertical deflection (Y-axis) system, whereupon the cathode-ray beam traces out the XY coordinate graph. The intensity of the beam can be controlled, or modulated, by an electrical signal (sometimes called the *video signal*) applied to the Z-axis system. Such a definition includes not only oscilloscopes and other related electronic instrumentation displays, but standard home television sets as well.

The major difference between an oscilloscope and a television is in the way the picture is formed on the screen of the CRT. In an oscilloscope, external signals control the vertical and horizontal movement of the electron beam. The path the beam traces on the phosphor of the CRT is a picture of those signals. In television, the beam is moved in a fixed pattern both vertically and horizontally; the picture is formed when the beam's intensity is modulated by the external video signal.

8.2
ELEMENTS AND TERMINOLOGY OF OSCILLOSCOPES[a]

In all XY display applications, the purpose is to draw an image on the screen for human interaction. However, before discussing the methods of drawing these images, it is important to discuss display-related terminology and some of the important elements of a picture. The important aspects of any picture are resolution, brightness and data density.

Two types of resolution, spot size and addressable, are important in discussing a display device. *Spot-size resolution* is simply the diameter of the electron beam spot divided into the screen dimensions. Since the light intensity of a spot is not uniform across its diameter, measurement of spot size is not an exact science.

To test spot-size resolution, a raster is displayed on the screen and reduced in size until the lines just become indistinguishable. At that point, the size of the raster is measured and divided by the number of lines in the raster to determine the resolution.

Addressable resolution, on the other hand, has nothing to do with spot size, but simply is the incremental accuracy in which the beam can be positioned. The smallest positional movement of the electron beam divided into the CRT dimensions then gives the addressable resolution. Unlike spot resolution, addressable resolution is an exact computable number. It is important to realize that addressable resolution holds no fixed relationship to spot resolution. Addressable resolution can be either greater or less than spot resolution. The relationship of addressable resolution to spot resolution does play a significant role in the perceived "quality" of a display or picture. More on this later.

Brightness is simply the light output of the picture being displayed. The relative light output of a displayed picture is significantly different from the maximum light output capability of the CRT, as will become apparent later.

The amount of data displayed (*data density*) is simply the area of the CRT screen that is lit versus the entire screen size (or the amount of lit area versus dark area). The lit portion of the screen consists of all the alphanumeric characters and vectors (lines) that comprise the picture being displayed.

The *raster* is a predetermined pattern of scanning lines that provides substantially uniform coverage of an area. In other words, the raster is the rectangular pattern of light that appears on the CRT screen when the screen is scanned by the electron beam.

The Z *axis* refers to that part of the display that controls CRT intensity. It includes the input circuit, an amplifier (called the Z-axis amplifier), and the circuit that drives the CRT control grid. The *control grid* controls the amount of beam current allowed to fall on the phosphor at the front of the tube, which varies the intensity of the picture.

8.3
CATHODE-RAY TUBE REFLECTION FUNDAMENTALS[a]

There are two methods of deflecting the beam inside the CRT; electromagnetic and electrostatic.

Electromagnetic deflection systems utilize an inductive coil (called a yoke) surrounding the neck of the CRT through which a current signal is passed to generate a magnetic field to deflect the electron beam (see Figure 8.1). At high frequencies, inductors with few turns are necessary to obtain fast current changes, and larger currents are required to obtain the required field strength. Consequently, above repetition rates of 20 kHz, large power

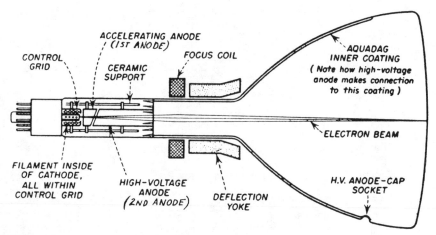

FIGURE 8.1
Electromagnetic deflection-type CRT. Commonly
used in television, computer display terminals,
and some inexpensive low-frequency (<20 kHz)
oscilloscopes. (Courtesy of Hewlett-Packard Co.,
Palo Alto, Calif.)

dissipations are required to obtain full-scan displays. Most television sets
and computer display terminals use the magnetic deflection scheme.

Electrostatic deflection systems consist of complex electron gun structures containing two sets of deflection plates, one set for horizontal deflection and one set for vertical deflection (see Figure 8.2). Electrostatic
deflection, involving voltage charging of capacitive plates to deflect the
electron beam, is capable of speeds several orders of magnitude higher than
electromagnetic for a comparable amplifier cost (but not CRT cost). Since
even inexpensive industrial oscilloscopes generally are capable of displaying 500 kHz or more on the vertical axis (and often on the horizontal axis as
well), it is not surprising that most oscilloscopes and *XY* displays use the
electrostatic deflection scheme.

The major differences between the two systems are light output and
frequency response. The magnetic system generally offers more light output, while the electrostatic system offers high-speed response at lower
power consumption. In addition, magnetic deflection allows a wider beam-
deflection angle than does electrostatic deflection. Moreover, when the
full-screen deflection bandwidth desired is less than 20 kHz, the electromagnetic deflection system (amplifier and CRT) has a substantial cost advantage over an electrostatic system. This is one of the reasons television
sets, many medical monitors, and some oscilloscopes rely upon CRTs with
electromagnetic deflection.

Brightness, which is a function of beam current, is governed by the
internal construction of the CRT. Given the same spot size, with all other

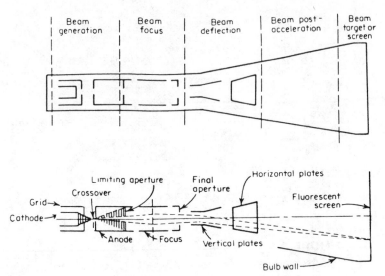

FIGURE 8.2
Electrostatic deflection-type CRT. Commonly used
in *XY* displays and oscilloscopes. (Courtesy of
Hewlett-Packard Co., Palo Alto, Calif.)

things equal, an electromagnetic CRT will be brighter than a comparable electrostatic CRT. This is mainly due to the fact that the electromagnetic CRT gun structure is very simple, allowing most of the beam current to pass through to the screen. The electrostatic gun has a large number of internal elements, as shown in Figure 8.2. Notice that the front-most element of the focus lens has a relatively small aperture, which strips away 70% to 90% of the beam current just before the beam enters the deflection plate region. The gun is constructed this way because the beam diameter must be smaller than the distance between the plates as it passes between them.

In either type of CRT, it is desirable to deflect the beam at as large a mechanical distance from the screen as feasible. This optimizes deflection sensitivity by providing the greatest deflection distance at the screen per volt, or gauss, of applied deflection field. In the electrostatic CRT, the plates must also be reasonably close together to achieve sufficient deflection field strength with an applied voltage in the range of 300 V. A greater swing would create significant *X* and *Y* amplifier design problems.

Electrostatic displays have faster deflection systems and use less current that electromagnetic displays. This is the result of physics; the load of the deflection plates is only the stray capacitance of a few picofarads. On the other hand, the inductance of the yoke is the load to the driving amplifiers. The higher the inductance (i.e., more turns and stronger magnetic field), the easier it is to deflect the beam; however, higher inductance requires high power. To lower the inductance reduces power requirements but also reduces deflection sensitivity. This then requires reduction in accelerating potentials in the CRT so that the beam can be deflected full screen, which also reduces the light output.

Some further points of comparison follow:

1. **Linearity and geometry.** To achieve a linear deflection in an electro-magnetic display, it is necessary to apply an S-shaped nonlinear voltage function to the deflection coils. To achieve good linearity in electro-magnetic displays, complex, nonlinear circuits must be designed which add to the expense and component count. For good geometry (minimum barrel and pincushion distortion) in electromagnetic displays, specially wound coils and additional compensation circuits, or both, are required. Also, the positioning of the yoke is very critical. Electrostatic CRTs, on the other hand, have inherently good linearity. Only a variable dc voltage needs to be applied to an element in the CRT for any minor geometric pattern correction.

2. **Z-axis delay:** Since deflection coils act as a delay line, it is necessary to introduce delay into the Z-axis signal path in high-speed electromagnetic displays. This coordinates the blanking and unblanking with the beam position; these delay lines are unnecessary in electrostatic displays.

3. **Resolution:** In either type of CRT, there is a trade-off of beam diameter versus brightness. Since an electromagnetic CRT passes so much more beam current, a much smaller spot size (higher resolution) can be obtained at the same level of brightness.

4. **Weight:** A high-speed electromagnetic display weights two to three times more than a comparable electrostatic display. The power trans-former is bigger and more heat sinks are required to meet the higher power demands of electromagnetic displays. The mass of the yoke and the increased number of components contribute to the added weight.

5. **Cooling:** Fast electromagnetic displays require forced-air cooling; electrostatic displays require only convective cooling.

8.4
CATHODE-RAY TUBE PICTURE DRAWING FUNDAMENTALS[a]

In addition to how the beam is deflected inside the tube is the actual method by which the picture is drawn. Again, two basic methods are used: *vector scan* and *raster scan*.

Other common names for vector drawing are dot writing, directed beam, and stroke writing. As the name implies, the beam is successively moved to each new location by signals applied to the X (horizontal) and Y (vertical) inputs. In an oscilloscope, the directed beam is linearly deflected to follow the signal pattern in the vertical direction, while often using a constant deflection rate in the horizontal direction to represent time. If the

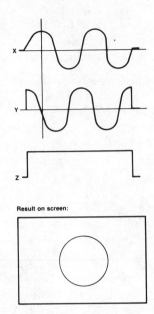

FIGURE 8.3
Example of vector or stroke beam deflection as
used in an *XY* display where the beam is on during movement. (Courtesy of Hewlett-Packard Co.,
Palo Alto, Calif.)

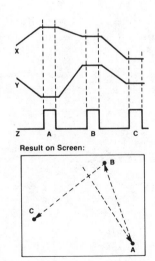

FIGURE 8.4
Example of dot writing (vector deflection) where
the beam is turned off during movement.
(Courtesy of Hewlett-Packard Co., Palo Alto, Calif.)

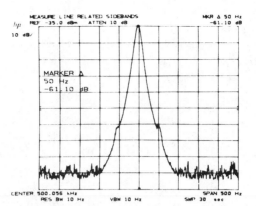

FIGURE 8.5
Application example of dot writing (electrostatic vector deflection). This type of display uses A/D converters and a large interactive memory. (Courtesy of Hewlett-Packard Co., Palo Alto, Calif.)

beam is left on during the writing cycle (as in a standard oscilloscope), the picture will be comprised of curves or straight-line segments on the CRT face (see Figure 8.3).

If the beam is blanked (turned off) during movement and turned on at rest, the picture will be comprised of a series of dots on the CRT face (see Figure 8.4). This type of dot writing lends itself well to the CRTs used in spectrum analyzers, air traffic control monitors, and other displays where annotation is displayed on the screen along with the main picture (see Figure 8.5).

In a raster system, the beam follows a fixed path covering the face of the tube in a series of interlaced parallel lines (usually horizontal), with blanked (turned off) retrace lines. In a raster system there are two methods of modulating the beam intensity: analog and digital.

Analog modulation is used in television. The principle of operation involves varying the intensity of the beam as it moves across the screen in accordance with the picture impulses of the video signals applied to the grid of the CRT.

The picture is traced line by line. The first odd-numbered line of the picture is traced by one sweep from left to right. When the beam reaches the right side of the screen, the horizontal blanking signal biases the tube beyond cutoff, and the beam is thus extinguished (blanked out). The beam remains blanked out long enough for it to move rapidly over to the left-hand side of the screen where it is in another position to start the trace of the second odd-numbered line of the picture. The horizontal blanking signal is now withdrawn and the beam moves across the screen as before, tracing another odd-numbered line of the picture.

As this process is repeated again and again, the vertical sweep signal is applied to the vertical deflection system of the CRT, which causes the beam to move gradually toward the bottom of the screen. Due to the downward pull of the slowly moving vertical sweep, each horizontal scanning line will be slightly below the preceding line.

When the beam reaches the bottom of the screen, the vertical blanking signal biases the CRT beyond cutoff, and the beam is blanked out for vertical retrace. The beam then moves rapidly (flyback) to the upper left-hand corner of the screen where it is in a position to start the traces of all the even-numbered lines to fill in the picture.

In the case of television, the composite picture is comprised of 525 lines scanned in an interlaced pattern at the rate of $\frac{1}{30}$ s. That is, the odd-line picture (first field) is scanned in the first $\frac{1}{60}$ s, and the even-line picture (second field) is scanned in the second $\frac{1}{60}$ s. This means that the horizontal lines are scanned at the rate of 525 times 30 or 15,750 lines/s (color TV is 15,734).

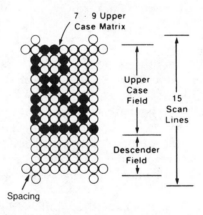

FIGURE 8.6

Example of digitally modulated raster scan method used to generate alphanumeric characters. This example shows a basic 7 pixel wide by 9 pixel high character cell as used in an HP CRT display terminal. The four additional scan lines beneath the 7 × 9 matrix are used for descender areas of lowercase letters and underlining. One pixel is used on either side for character-to-character spacing. In this particular example the display screen has a low aspect ratio of 5 × 10 in., for a capacity of 1920 character cells partitioned into 24 rows of 80 characters each. That equals 259,200 pixels (720 horizontal by 360 vertical). (Courtesy of Hewlett-Packard Co., Palo Alto, Calif.)

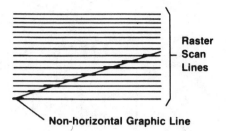

Non-horizontal Graphic Line

FIGURE 8.7
Exaggerated example of a nonhorizontal graphics
line drawn on a raster-type display. (Courtesy of
Hewlett-Packard Co., Palo Alto, Calif.)

All this scanning and modulation results in a picture of gray shades where the beam intensity ranges from full on to full off. In other words, when the video signal increases the beam intensity to full on, the phosphor on the CRT glows brightly; and when the video cuts off the beam intensity, the phosphor does not glow. Everything in between is the gray range, which forms the picture.

The other method of drawing the picture in a raster format is by using a digital signal to modulate the beam as it sweeps across the face of the CRT. The interlaced scanning procedure is the same. In this method, the face of the CRT is partitioned into rows and columns of dots, which are called *pixels* (refer to Figure 8.6). Pixels are defined as the smallest location on screen where the beam can be modulated on or off. The picture is created from line segments and dots produced by turning the beam intensity on and off at appropriate times. Special effects can be created when the beam intensity is modulated at points between full on and off.

For a low-cost alphanumeric display, the raster scan method offers several advantages over the directed beam. First, it does not require accurate, high-speed, and therefore expensive digital-to-analog converters. Instead, it uses counters on both axes to locate dots and characters. Second, the deflection circuitry need not have wideband, essentially identical channels; each axis can be optimized for its particular operating rate. Similarity with television techniques allows the use of low-cost television components in the deflection and high-voltage circuits. Finally, the raster display can minimize the power consumption of the monitor by means of an energy-conserving flyback-type horizontal deflection circuit.

However, "clean" graphics are difficult to display in a raster system. This is most obvious when lines are drawn that are close to, but not exactly horizontal; they will be displayed broken up into steps as shown in Figure 8.7.

In high-resolution line drawing situations, continuous lines have significant merit. It should be pointed out that vector systems have continuous lines at all addressable resolutions. Only the starting and stopping point

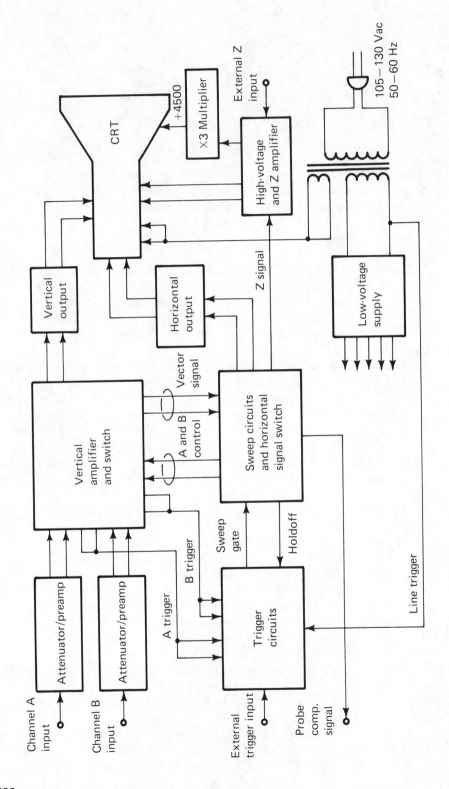

FIGURE 8.8
SC60 simplified block diagram. (Courtesy of Sen-core, Inc.)

286

accuracy of the vectors is sacrificed at lower resolutions. To achieve continuous lines in a raster system can place significant requirements upon it. Since, in a raster system, vectors are comprised of a serious of dots or pixels, to achieve the effect of a continuous line, the dots must be spaced so close as to merge. In fact, they need to overlap by approximately two-thirds to three-quarters of the spot diameter. This means that the addressable resolution (number of scan lines) should be three or four times greater than the spot resolution of the CRT. Therefore, a 1000-line spot resolution CRT requires 3000 or 4000 scan lines to achieve the same quality as a displayed line in a vector system. Of course, this has extremely adverse effects on light output, deflection systems, power, and cost. Therefore, in high-resolution line drawing applications, the quality of the continuous line may also become an issue.

Another example of resolution restriction can occur when trying to use an alphanumeric raster-type computer terminal to display a voltage versus time phenomenon. Remember that this type of CRT is the digital raster type with a fixed scan rate that modulates pixels on and off with a certain portion of time set aside for retrace.

8.5
THE DUAL-TRACE OSCILLOSCOPE[b]

The basic dual-trace oscilloscope (Figure 8.8) is usable in the design, manufacturing, or service of electrical or electronic equipment. Many such oscilloscopes can feature an extended bandwidth to 60 Hz (3 dB down) and is usable to 100 kHz. This feature is a must for servicing or designing with the newer, faster logic families; the high bandwidth is necessary to be able to see troublesome glitches, which can often completely alter a logic word. A second important feature is the fiddle-free trigger circuit that will provide solid sync up to 100 MHz. Solid sync plus a front panel that has only essential well-marked controls makes the SC60 extremely simple and easy to use by anyone, even those with no technical background.

This dual-trace oscilloscope can also be a video scope or a vector scope with just the push of a button. Preset sweep positions are offered for the vertical and horizontal components of a video signal, as well as special trigger processing through a sync separator if other sweep speeds are used for observing video signals. The vector scope uses the B channel for X deflection and maintains good phase relationship between X and Y deflection past 3.5 kHz.

Other more common features of channel A invert, channels A + B, dual alternate, dual chop, a beam finder, dc and ac input coupling, trace rotation, ×10 horizontal expand, line and external trigger, auto baseline, and a front

[b] The material in Section 8.5 is used courtesy of Sencore, Inc.

panel probe calibration signal make the SC60 truly a necessary instrument for servicing and designing today's modern circuits.

In Figure 8.8, the SC60 has straightforward circuitry and is rather basic in design. Input signals on channels A and B are each processed separately in a preamplifier stage. The signal first goes through a coarse step attenuator and later is more finely controlled with a vernier gain control in the amplifier. The output of each preamp is connected to the vertical amplifier, and each channel is further amplified individually before coming together at a diode electronic switch, where the A and/or B signals can be selected for display with the CRT display switch. The output of the diode switch is displayed as vertical, top to bottom, deflection on the CRT after first being amplified in the vertical output amplifier.

A sampling of signal from each channel is taken off prior to the diode switch in the vertical amplifier and is fed to the trigger circuits. Here either A, B line, or external triggers can be selected to control the start of the horizontal time base. The trigger circuits change any signal selected into a square wave with extremely fast transition times for solid jitter-free triggering. Once the horizontal trace has been started with a trigger, successive triggers are blocked out and remain so until the trace has swept completely across the screen and back before the trace can be started again with another trigger. In the auto baseline mode the retrace transition is used as a trigger in the absence of triggers from the trigger circuits to keep a trace displayed on the CRT.

The ramp waveform developed by the time base circuits is amplified in the horizontal output amplifier and applied to the horizontal deflection plates of the CRT, and is displayed as horizontal, left to right deflection.

In the vector operation, the channel B signal is routed through the horizontal amplifier and selected with the horizontal signal switch to take the place of the ramp waveform for horizontal deflection.

Operating voltages for the CRT are derived from a high-voltage oscillator circuit located in the Z amp and high-voltage supply block. Output from the oscillator is rectified to develop -1500 V dc for the CRT cathode and is tripled in a HV multiplier circuit to generate $+4500$ V dc on the CRT second anode. These high voltages are important for displaying low duty cycle signals with good intensity. The CRT filament power comes directly from the line power transformer.

The Z amp controls the CRT grid to blank the beam during retrace time and also during dual-chop transitions. An external Z input jack is provided for external control and signals.

A low-voltage supply provides all dc voltages needed to operate the various circuits used in the SC60. Many of the voltages provided are regulated for good, solid CRT displays that will not jump with line voltage variations.

Dual-trace analog oscilloscope applications include the following:

1. AC voltage measurements on one channel.
2. DC voltage measurements on one channel.

3. Measuring the dc level of an ac signal.

4. Comparison of voltages.

5. Comparison of ac signals.

6. Localization of intermittents.

7. Gain and distortion measurements.

8. Frequency measurements using calibrated time base.

9. Frequency measurements using Lissajous figures.

10. Frequency measurements using the modulated ring.

11. Phase-shift and time-delay measurements.

12. Time delay between signals.

13. Phase shift between signals.

14- Square wave testing.

15. Rise time measurements.

16. Coil and transformer testing.

17. Others.

The calibrated time base method of measuring frequency is the easiest to use.

1. Connect the signal to be measured to either the channel A or B input of the SC60 and adjust the controls of the SC60 to display one or two cycles of the signal. Use auto triggered horizontal sweep and be sure that the small center knob of the time base–frequency switch is in the full clockwise position.

2. Use the 10 × 10 grid to measure the horizontal distance between the two points that you wish to measure.

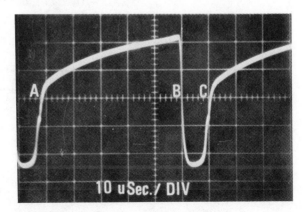

FIGURE 8.9
Time measurement with the SC60. (Courtesy of Sencore, Inc.)

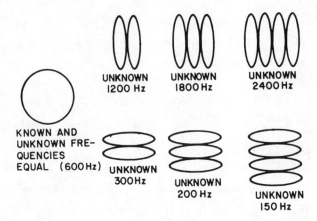

FIGURE 8.10
Typical Lissajous patterns. (Courtesy of Sencore, Inc.)

3. Multiply the distance measured by the setting of the time base–frequency switch.

Figure 8.9 shows the waveform at the grid of the horizontal output tube in a television receiver. The width of the negative-going pulse measured between points *B* and *C* is 1.1 divisions. The setting of the time base–frequency switch of 10 μs / DIV will result in a pulse 11 μs wide. The horizontal distance for one complete cycle of the drive signal, as measured between points *A* and *C* is 6.3 divisions, so one complete cycle is 62 μs long.

The formula for finding frequency if time is known is

$$\text{Frequency} = \frac{1}{\text{time in seconds for one cycle}} \qquad (8.1)$$

The time for one cycle in Figure 8.9 is 63 μs, or 0.000063 s. One divided by 0.000063 equals 15,850 Hz, or very close to the horizontal frequency of 15,750 Hz.

The sensitivity and bandwidth of the SC60 vector mode allow this type of measurement to be made from even the output of an RF generator up to 5 MHz.

1. Connect the known frequency to the channel B input and the unknown frequency to the channel A input.

2. Press the vector vertical input button, and adjust the controls so that the entire pattern is visible on the screen.

3. Adjust the known signal frequency so that the pattern holds steady, and count the number of vertical and horizontal loops.

4. Use the following formula to find the unknown frequency.

$$\text{Known frequency} \times \frac{\text{number of hor. loops}}{\text{number of vert. loops}} = \text{unknown frequency} \qquad (8.2)$$

The examples of Figure 8.10 show several patterns and the frequency that they represent. The known frequency used in the example is 600 Hz. As we see in Figure 8.10, frequency measurements can easily be made using the Lissajous figures.

The oscilloscope probe contains the signal-sensing circuitry. Such circuitry can contain the following:

1. Ten megohms shunted by a 5- to 15-pF capacitor.

2. Use of a FET source follower plus associated elements.

8.6
INTRODUCTION TO THE DIGITAL OSCILLOSCOPE[c]

As with its analog counterpart, the primary function of a digital oscilloscope is to graphically display the voltage and time characteristics of an analog waveform. The front panel controls are also very similar, including amplifier, triggering, sweep timing, and display. However, the digital oscilloscope not only acquires and displays analog signals, but also stores them in a solid-state memory and thereby offers many advantages.

The resolution and accuracy of an analog oscilloscope are restricted by screen size, phosphor characteristics, and the limitations of the human eye to a typical range of 2% to 3%. In contrast, a modern digital oscilloscope may offer an accuracy of 0.2% and resolution of 0.025%. Vertical resolution is a measure of the smallest voltage change distinguishable on a full-scale waveform. In digital oscilloscopes, it is specified by the number of bits in the analog-to-digital converter. The theoretical ramp voltage pictured in Figure 8.11 serves to clarify the meaning of "bits of resolution." A 2-bit digitizer distinguishes 1 in 4, a 4-bit digitizer 1 in 16, and so on. The actual ramp voltages shown in Figure 8.12 illustrate the amount of detail discernable with 12- and 15-bit digitizers using waveform expansion. Figure 8.13 shows an expansion of Figure 8.12, 64 times vertically, and shows detail possible with 12- and 15-bit digitizers. This digital representation of the analog signal is extremely accurate, with the voltage and time coordinates of each point precisely known. Thus, although the image is now reconverted to analog form for display, the reliance on CRT characteristics for measurement accuracy has been completely avoided.

Figure 8.14 shows an acquired shock waveform. Oscillations at the cursor junction seen only vaguely on the original waveform can be exam-

[c] The material in Section 8.6 is from Amelia Tegen and Jeremy Wright, *Oscilloscopes: The Digital Alternative*, Nicolet Instrument Co. This section is used courtesy of the Nicolet Instrument Corp., Oscilloscope Div., Madison, Wisc.

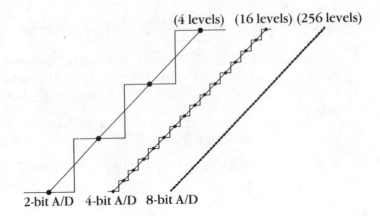

(4 levels) (16 levels) (256 levels)

2-bit A/D   4-bit A/D 8-bit A/D

FIGURE 8.11
Theoretical ramp voltage; vertical resolution
defined by the number of bits in the A/D converter. (Courtesy of Nicolet Instrument Corp.,
Oscilloscope Div., Madison, Wis.)

ined in great detail, using the cursors to select the area of interest. Figure
8.15 shows the expansion of the waveform in Figure 8.14 for detailed examination. The display of coordinate time and voltage also allows easy delta V,
delta T measurements, particularly on models having a relative numerics
feature.

The high horizontal resolution illustrated in Figures 8.14 and 8.15 is a
result of using a large memory, typically 4096 (4K) points. Some models

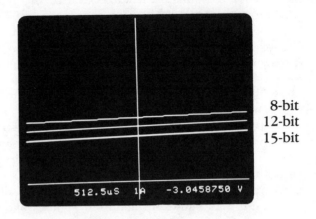

8-bit
12-bit
15-bit

512.5uS   1A   -3.0458750 V

FIGURE 8.12
Actual ramp voltages recorded from 8-, 12-, and
15-bit A/Ds. (Courtesy of Nicolet Instrument
Corp., Oscilloscope Div., Madison, Wis.)

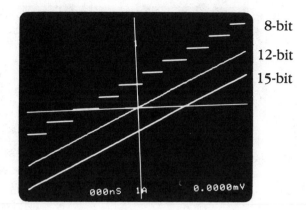

8-bit

12-bit

15-bit

FIGURE 8.13
Expanding Figure 8.12 by 64 vertically shows detail possible with 12- and 15-bit digitizers. (Courtesy of Nicolet Instrument Corp., Oscilloscope Div., Madison, Wis.)

even offer a 16K memory or a horizontal resolution of 1 part in 16,000! These large memories, coupled with the crystal-controlled time base, allow digital oscilloscopes to offer a much wider sweep range than analog oscilloscopes, typically from microseconds to several days! Thus the digital oscilloscope can also be used as a data logger or as a precision alternative to a strip-chart or light-beam recorder. As an example of such an application, Figure 8.16 shows an expanded five-day section of a nine-day recording of

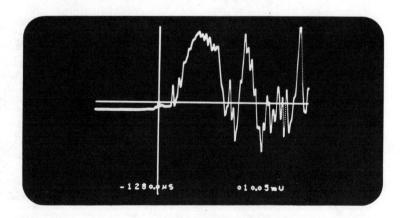

FIGURE 8.14
Shock waveform with cursor intersection on point of interest. (Courtesy of Nicolet Instrument Corp., Oscilloscope Div., Madison, Wis.)

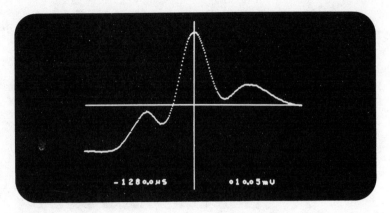

FIGURE 8.15
Expansion of waveform in Figure 8.14 for detailed
examination. (Courtesy of Nicolet Instrument
Corp., Oscilloscope Div., Madison, Wis.)

tidal movements in the earth's crust. Note the time numeric! Using a digital
oscilloscope to measure the output of a very sensitive *tiltmeter*, researchers
were able to distinguish changes in angle as small as a few nanoradians.

A further advantage of a digital oscilloscope, particularly when examin-
ing one-shot or transient phenomena, is its unique ability to "look back in
time." The use of a continuously updated input buffer memory allows the
capture and display of signals with both pretrigger and posttrigger information. For example, Figure 8.17 shows a transient on a power line captured
by triggering on the first sharp voltage spike. The trigger position, or zero

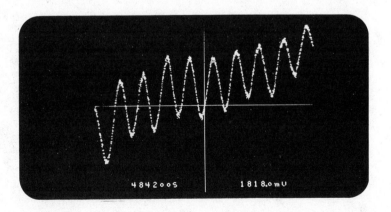

FIGURE 8.16
Expanded five-day section of a nine-day oscilloscope recording of tidal movements in the earth's
crust. (Courtesy of Sandia National Laboratories)

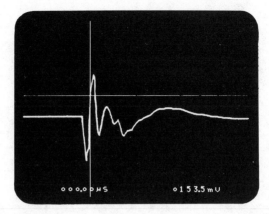

FIGURE 8.17
A power line transient, captured with both pre-
and posttrigger information. (Courtesy of Nicolet
Instrument Corp., Oscilloscope Div., Madison,
Wis.)

time, is indicated by the position of the vertical cursor. Data to the left of
this cursor are historical or in "negative" time; that is, they occurred before
the trigger. Both pre- and posttrigger parts of the waveform can now be
analyzed in detail using the display-expansion capabilities described earlier. This type of information is sometimes crucial when trying to discover
the cause of a transient or intermittent problem. Some digital oscilloscopes
offer a *cursor trigger* feature that allows the trigger point, or zero time, to
be set anywhere in the display area. Indeed, recent advances even enable
two-channel measurements, with each channel set at an independent pre-
or posttrigger delay.

Many digital oscilloscopes combine this transient capture ability with a
capture-and-store sequence. The Nicolet 2090, illustrated in Figure 8.18,
offers automatic transfer of data to a floppy disk, rearming the scope after
each data transfer to capture the next incoming transient. It can record and
store up to 32 sequential signals, and other models can store as many as 80
signals without operator intervention. When dealing with random phenomena such as intermittent problems, this is extremely useful, since it
frees the operator for more important tasks. The recorded data can be recalled into the oscilloscope at any time for further analysis or comparison
with other records. For example, Figure 8.19 shows eight signals captured
and stored two at a time and later recalled from disk for comparison.
However, the floppy disk is not limited to such automatic storage operations. It can be used for permanent storage of any type of waveform, complete with original time and voltage information. On some models the
floppy disk can even be used to complement the sweep range by allowing
the recording of up to 72 days of continuous information.

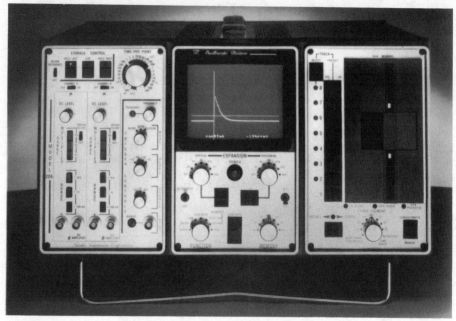

FIGURE 8.18
Nicolet 2090 oscilloscope. (Courtesy of Nicolet Instrument Corp., Oscilloscope Div., Madison, Wis.)

The concept of storing data for later analysis can be extended to storing known reference signals for real-time comparison with live data. Using the *retain reference* feature of some oscilloscopes, incoming signals can be visually compared with signals stored moments, or even years, earlier. This is particularly useful in quality-control applications where technicians have to adjust the signal output of a device to match a standard reference.

The comparison of waveforms can also be taken a step further using *XY* display of live or stored signals. In this mode, interdependent variables such as voltage and current, or stress and strain, can be displayed as functions of each other rather than time (see Figure 8.20). Using the cursor, the voltage coordinates of any point in the curve can be read directly. Since this is merely a change in display format, no data are lost: switching back to *YT* will restore the original time-related data and coordinates. By combining this feature with the reference-retention capability described previously, a live signal can be displayed as a function of a stored reference signal, particularly useful in the study of phase relationships.

These examples of display and storage flexibility illustrate some of the immediate benefits of a digital oscilloscope. Other benefits are perhaps less obvious but just as important. For example, obtaining a hard-copy record of displayed signals is no longer restricted to a camera. Most digital oscillo-

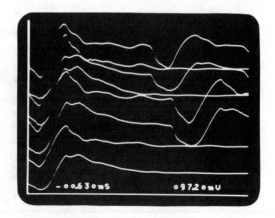

FIGURE 8.19
Eight waveforms are recalled from disk and
superimposed for comparison. (Courtesy of
Nicolet Instrument Corp., Oscilloscope Div.,
Madison, Wis.)

scopes offer *XY* and *YT* pen recorder outputs as standard, and some interface directly to high-speed digital plotters. The user merely adjusts the
display to his or her liking, then presses a button to obtain a clear permanent
record in a matter of seconds. Also, since waveforms are stored in digital
format, they are in an ideal form for further processing. Simple data-
manipulation functions, such as the subtraction of one signal from another,
baseline correction, and signal inversion, are commonly included as stan-

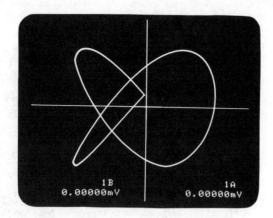

FIGURE 8.20
Related variables can be displayed as functions of
each other using *XY* mode. (Courtesy of Nicolet
Instrument Corp., Oscilloscope Div., Madison,
Wis.)

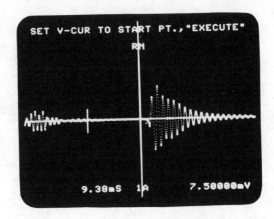

FIGURE 8.21
Push-button data manipulation via simple displayed instructions. (Courtesy of Nicolet Instrument Corp., Oscilloscope Div., Madison, Wis.)

dard. Some digital oscilloscopes take this a step further by offering programs on floppy disk that can be down-loaded into the oscilloscope and executed via a single push button. These programs can be extremely simple, such as three-point smoothing, or more complex, such as FFT spectrum analysis. In either case, the user needs no computer or programming knowledge; instructions are displayed automatically on the screen (see Figure 8.21).

For more sophisticated analysis, the stored signal can easily be transmitted to a computer or calculator via standard interfaces such as GPIB-488 or RS-232. The digital oscilloscope thus makes an ideal front end for a

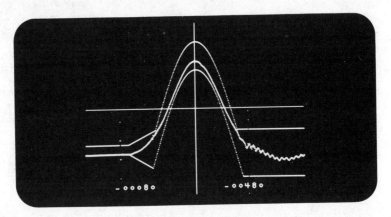

FIGURE 8.22
Comparison of a live waveform with computer-generated high–low limits. (Courtesy of Honeywell/Defense Systems Division)

complete signal analysis system. If required, the manipulated data can be returned to the oscilloscope for examination. The computer can even input calculated parameters such as high–low limits for direct comparison with acquired signals. This is a valuable asset in quality-control applications. For example, Figure 8.22 illustrates the use of a digital oscilloscope and computer combination for impact testing to military specifications. The computer "draws" the test limits directly on the screen, and the operator compares each signal with these limits. The computer then calculates relevant parameters from the waveform while the oscilloscope both provides a hard-copy output and stores the data permanently on floppy disk for future reference.

Within their bandwidth range of typically 10 MHz or below, digital oscilloscopes outperform their analog counterparts by every known criterion. They are not only easy to use but offer up to 100 times the resolution, 10 times the accuracy, and the benefits of storage, computer interfacing, and display flexibility. With these advantages, digital oscilloscopes are rapidly becoming standard instruments in such diverse fields as medical research, automotive testing, and electrical power monitoring. With the rapid advance of A/D technology, their maximum bandwidth is expected to improve dramatically, removing their only disadvantage. Already instruments are becoming available with 100-MHz digitizing rates, and researchers are demonstrating even higher speeds. Within a few years there is little doubt that the digital oscilloscope will be competing successfully in all areas of the traditional oscilloscope market.

8.6.1   Waveform Digitizer [d]

Hewlett-Packard's model 1980A/B oscilloscope measurement system (Figure 8.23) is a fully programmable, time-domain instrument based on a computer architecture. This automated HP-IB compatible instrument makes significant contributions in the viewing, measuring, and processing of time-domain waveforms. Improved measurement capability is achieved with the computer architecture design that permits internal operation to be controlled by a microprocessor. This results in an easy-to-use instrument with an extensive feature set. Internally, the instrument is divided into eight independent functional blocks that interface with each other over a bus structure. This extensive digital control permits many features not in other oscilloscopes, including programmable hardware and firmware expansion, an easy-to-use front panel, autoranging, complete programmability, and digital waveform storage.

In its basic configuration, the 1980 offers two 100-MHz analog measurement channels with continuous 2-mV to 10-V/div deflection factors, two independent and continuous main and delayed sweeps from 5 ns to 1 s/div, main or delayed trigger view, both delta time and voltage measurements, an

---

[d] The material in Section 8.6.1 is used courtesy of Hewlett-Packard Co., Palo Alto, Calif.

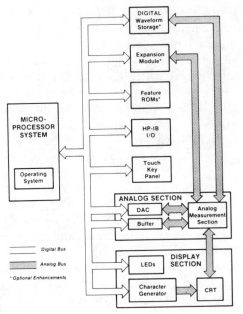

- ROM is read-only-memory
- I/O is input–output device
- DAC is digital-to-analog converter, or D/A

FIGURE 8.23

Simplified block diagram of the HP 1980A/B architecture. (Courtesy of Hewlett-Packard Co., Palo Alto, Calif.)

auto-scope feature for automatic signal scaling, and many other features. These capabilities are combined with an innovative front panel design with color-coded keys and a single rotary control. The internal microprocessor scans the front panel for any key activation and variable control changes and then sets the instrument. Because these functions are controlled digitally by the microprocessor, they can also be programmed quite easily through the HP-IB interface. This programmability allows instrument setup to be done automatically by a computer, which eliminates errors and greatly reduces setup time.

The basic instrument also contains provisions for expanded measurement and control capabilities in the form of additional hardware or firmware. Hardware expansion modules, such as the two-channel 1950A, fit into a front panel compartment and contain analog and digital interfaces for measurement integrity and programmability. Firmware expansion is through the addition of up to four preprogrammed 4K ROMs that provide either additional control capabilities or data manipulation. Digital waveform storage is another form of enhancement that provides completely automatic measurement capability.

Remote programmed operation is through the standard HP-IB port that interfaces through the internal digital bus with all functional blocks.

Delta time measurements are made using a single marker method with a direct digital readout of time interval relationships on a LED display. A delayed sweep oscillator, which is calibrated with an internal crystal oscillator, provides accurate time interval measurements and also permits measurements from the first pulse. Time interval measurements can be converted to frequency by selecting the reciprocal function $1/\Delta T$.

By using a separate delayed sweep oscillator as a time reference, time intervals relative to the first pulse leading edge are easily accomplished in the delayed sweep mode. This capability is particularly useful for high-resolution measurements of low duty cycle pulses. Because first pulse measurements can be made using the same pulse that triggers the main sweep, the sweep speed can be set for optimum resolution and accuracy with up to 100-ps resolution.

8.6.2 Waveform Analyzer

The SC61 waveform analyzer (Figure 8.24) represents the first improvement in waveform measuring techniques since the first oscilloscope was introduced over 50 years ago. Before the SC61, all oscilloscopes were strictly analog devices. Every measurement of peak-to-peak amplitude, time, or frequency required matching the waveform to CRT graticule markings, estimating parts of a CRT division or part of a cycle of the waveform, and multiplying the size of the waveform times the setting of the horizontal or vertical switches. The measurements were not very accurate (typical

FIGURE 8.24
Sencore SC waveform analyzer. (Courtesy of Sencore, Inc.)

accuracy on an analog scope is ±15% when interpretation error is considered), were time consuming, and were subject to many errors (such as forgetting to turn the vernier to the calibrated position).

The SC61 combines the waveform analyzing capability of a high-quality oscilloscope with the speed, accuracy, and freedom from errors of a computer. The SC61 has a microcomputer that monitors the vertical and horizontal circuits at all times. You just push a button when you want to measure all or part of a waveform. The microcomputer determines the reading, automatically sets the decimal, and produces a direct digital readout of the value.

The SC61 waveform analyzer starts with a high-quality oscilloscope. The CRT display section has a bandwidth of 60 MHz and is usable to 100 MHz to show the needed detail when analyzing digital signals. Specially designed sync circuits (using differential amplifiers and emitter-coupled logic stages) provide solid triggering with the fewest trigger adjustments possible. The input capabilities extend from 5 mV/div all the way to 2000 V dc or ac p-p.

Special video circuits simplify analyzing composite video waveforms. Sync separators produce stable triggering on these complex signals. Additional stages eliminate the half-line shift on interlaced signals, while others eliminate vertical sync from interfering with digital readings of signals displayed at the horizontal sweep rate. Video preset buttons allow push-button selection of vertical or horizontal scanning frequencies.

There are two types of digital measurements made through the same probe used for the CRT display. The first three functions for each channel are called the Auto-Tracking™ tests. Auto-Tracking means the microcomputer automatically tracks the CRT display at all times. The three Auto-Tracking functions are dc volts, peak-to-peak volts, and frequency. All three functions measure the entire waveform.

The second group of digital tests are the delta tests, which let you measure part of a waveform. These tests measure the peak-to-peak amplitude, time, or frequency of any part of the waveform shown on the CRT.

Both types of digital tests are unaffected by the vertical or horizontal verniers or position controls, allowing the CRT display to be any size you want without affecting the accuracy of the digital readout. All tests are automatically ranged or directly interfaced to the input attenuators for direct readings.

One final test allows the frequency of the signal applied to channel A to be compared to the frequency of the signal applied to channel B. The digital shows the ratio of the two frequencies to troubleshoot divider or multiplier stages.

8.6.2.1 SC61 Specifications.

1. **Analog CRT Section**
   a. Vertical Amplifiers

- Frequency response: ± 3 dB, dc to 60 MHz, usable to 100 MHz. AC coupled: ±3 dB, 10 Hz to 60 MHz.

- Rise time: 6 ns.

- Deflection factors: 12 calibrated ranges in 1–2–5 sequence calibrated to read direct with supplied 39G153 10× probes. Concentric vernier continuously variable between ranges with detent for calibrated position.

- Sensitivity: 50 mV/div to 200 V/div, with supplied 10× probes, 5 mV/div to 20 V/div direct.

- Calibration accuracy: ±4% from 20° to 30°C (68° to 86°F) for channel A and/or B; ±10% for A + B or B − A modes.

- Input impedance: 10 MΩ shunted by 15 pF with 39G153 10× low-capacity probes, 1 MΩ shunted by 50 pF direct.

- Maximum input voltage: 2000 V (dc + peak ac) through 39G153 probes, 500 V (dc + peak ac) direct. Derated for frequencies above 1 MHz.

- Input coupling: ac, dc, and ground.

- Timing: 70-ns delay line for both channels to show trigger point.

- Invert: available on channel A by pulling vertical position knob.

- Display modes: channel A, inverted channel A (−A), channel B, dual trace (A&B), algebraic sum (A+B) or difference (B−A), and vector (X−Y).

- Dual trace: automatically switches between dual-alternate and dual-chopped (approximately 500-kHz chopping frequency) depending on sweep rate when A&B button is pressed. Forced to dual-alternate for all sweep rates if all CRT display buttons are released to out position.

- Vector: channel A is Y axis, channel B is X axis. Bandwidth: ±3 dB from dc to 4 MHz. Phase shift: ±3° from 10 Hz to 4 MHz.

- Sensitivity: same as vertical amplifiers.

- Z-axis input: BNC input on rear panel. DC coupled: +5-V blanks trace. Frequency range: dc to 5 MHz. Protection: 35 V dc + peak ac.

b. Horizontal Sweep
  - Time base: 19 calibrated sweep rates in 1–2–5 sequence. Concentric vernier continuously variable between steps with detent for calibrated position.

- Sweep rates: 100 ms/div to 0.1 μs/div.
- Accuracy: ±4% from 20° to 30°C (68° to 86°F).
- Video presets: activated in 20th switch position. Push-button selection of two horizontal lines or two vertical fields of standard NTSC composite video signal.
- 10× expand: expands horizontal sweep 10 times. Activated by pulling horizontal position control. Accuracy: ±5% except ±8% on 0.1, 0.2, and 0.5 μs/div sweep rates.

c. Trigger circuits

- Trigger source: channel A, channel B, ac power line, or an external source.
- Trigger modes: NORM (only provides trace when trigger circuits have signal), AUTO (provides trace at all times), TV (same as auto with sync separators added to use vertical or horizontal sync pulses as trigger reference. TV mode automatically selected by video preset function, no matter where MODE switch set).
- Trigger polarity: Selectable between + and − waveform transition for nonvideo signals. Selects positive- or negative-going sync on video signals.
- Internal sensitivity: normal or auto mode. AC coupled; .5 divisions of CRT deflection from 10 Hz to 20 MHz, increasing to 1.5 divisions at 60 MHz and 3 divisions (typical) at 100 MHz. TV: 1 division of signal needed. Vertical sync selected for all "ms" sweep rates, horizontal sync selected for all "μs" sweep rates.
- External sensitivity: 100 millivolts to 40 MHz, triggerable to 100 MHz. Maximum input: 500 bolts (DC + Peak AC).

d. CRT:

- Size: 96.4 × 120 mm (approximately 5 inches) rectangular.
- Type: Post deflection tube with P31 (blue-green) phosphor.
- Model: 140CGB41. Accelerating voltage: 6 KV to match CRT requirements.
- Graticule: Internally 8 × 10 division (approx. 0.9 cm/div) etched on CRT faceplate to eliminate parallax. Special 0%, 10%, 90%, and 100% markings for rise-time measurements.
- Beam finder: Disables trigger and intensity controls, and

reduces vertical and horizontal gain to locate beam.

2. **Auto-Tracking™ Digital Tests**

a. DC Volts

- Ranges: 4, automatically selected; 0–2, 2–20, 20–200, and 200–2000, direct reading with supplied low-capacity probes.

- Accuracy: ±0.5%, ±2 digits including low-capacity probes. No more than 0.5% between channels.

- Input impedance: 15 MΩ through supplied low-capacity probes, 1.5 MΩ direct.

- Resolution: $3\frac{1}{2}$ digits (2000 counts).

- Source: Selected with channel A or B push buttons.

- Protection: 2000 V (dc + peak ac) with supplied low-capacity probes, 500 V (dc + peak ac) direct.

b. Peak-to-Peak Volts

- Ranges: 4, selected by channel A or B input attenuator (unaffected by vertical vernier): 0–8, 8–80, 80–800, 800–2000 V p-p. Direct reading with supplied 39G ± 53 10× low-capacity probes.

- Resolution: $3\frac{4}{5}$ digits (8000 counts).

- Accuracy: ±2%, ±5 counts including low-capacity probes. No more than 2% between channels. Calibrated at 1 kHz. Frequency response: ±0.5 dB from 30 Hz to 30 MHz, −3 dB at 60 MHz.

- Method (patent pending): DC coupled, microcomputer controlled successive approximation for positive and negative peak, 26 approximations per reading, worst-case.

- Source: channel A or B selected by pushbutton.

c. Frequency

- Ranges: 7 automatically selected: 1.00–9.99 Hz, 10.00–99.99 Hz, 100.0–999.9 Hz, 1.0000–99.9999 kHz, 100.000–999.999 kHz, 1.00000–9.99999 MHz, 10.0000–99.9999 MHz.

- Resolution: Up to 6 digits (microcomputer controlled), 0.01 Hz to 10 kHz depending on input frequency (see "Ranges" for details). Automatic resolution multiplier used for 1 Hz–100 kHz to reduce gate time.

- Accuracy: ±0.001%, ±1 digit from 15° to 35°C (59° to 95°F). Aging less than 0.001% per year.

- Source: selected with channel A or B push button. Micro-computer automatically selects trigger circuit output or auxiliary counter amplifier depending on trigger source selected.
- Sensitivity: same as internal trigger circuits if reading frequency of channel used as trigger source, 1.5 division of CRT deflection if reading channel other than trigger source.
- Read rate: automatically selected; <2 s for frequencies between 1 and 10 Hz, <0.5 s for higher frequencies.

d. Frequency ratio

- Method: calculates ratio of frequencies applied to channel A and B.
- Range: 1 to 999,999. A/B or B/A display annunciators indicate which input frequency is greater.
- Sensitivity: same as frequency function.
- Accuracy: 13 digits (percentage does not apply because accuracy of channel A cancels accuracy of channel B).
- Response time: 1 to 4 s depending on input frequencies.

3. **Delta Digital Tests**

a. Delta Bar

- Measurement bar: intensified area set to any portion of waveform with DELTA BEGIN and DELTA END controls; functions of controls automatically reverse if overlapped.
- Range: 1 s to 50 ns.
- Setability: ±20 nS typical.

b. Peak-to-Peak Volts

- Function: amplitude of intensified area measured.
- Range and specifications: same as peak-to-peak volts above.
- Source: selected with channel A or B push button.

c. Delta Time

- Function: actual time of intensified area measured.
- Accuracy: same as frequency function above. Accuracy unaffected by setting of horizontal or vertical controls.
- Ranging: microcomputer automatically places decimal and annunciator reading of milliseconds or micro-seconds.

d. 1/Delta Time

- Function: calculates equivalent frequency of delta time reading.

- Accuracy: same as delta time.

- Ranging: microcomputer automatically places decimal and annunciator reading of hertz, kilohertz, or megahertz.

e. Digital Display

- Type: liquid crystal with high temperature fluid for fast response time and high contrast.

- Number of digits: six (resolution controlled by microcomputer).

- Annunciators: 12 (controlled by microcomputer) Δ, A, B, MHz, kHz, Hz, μs, ms, V dc, Vp-p, A/B, and B/A.

4. **General**

- Warm-up time: Unit completely operable as soon as CRT trace appears. For maximum accuracy of frequency measurements, unit should operate for at least 30 min at room temperature.

- Construction: vinyl-clad aluminum case, fully EMI shielded, with Cycolac* bezel.

- Storage compartment: located in rear for storage of probes and other small accessories.

- Size: $9.5 \times 12 \times 17$in. (*HWD*) ($21.6 \times 30.5 \times 43.2$ cm).

- Weight: 31 lb (14.1 kg).

- Power: 105–130 V ac 50/60 Hz as delivered from factory. Field convertible to 210 to 250 V ac 50/60 Hz operation.

- Power consumption: 90 W maximum.

8.7
REVIEW QUESTIONS

1. Discuss the uses of an oscilloscope.
2. Describe how a dual-trace analog oscilloscope works.
3. Discuss the operation of a digital oscilloscope.
4. Discuss the use of a microprocessor in a waveform digitizer.
5. Draw a probe used in a waveform analyzer.

---

*Cycolac is a registered trademark of Borg-Warner.

8.8
REFERENCES

1. Clyde N. Herrick, *Oscilloscope Handbook*, Reston Publishing Co., Reston, Va., 1974.

2. Douglas Bapton, *Modern Oscilloscope Handbook*, Reston Publishing Co., Reston, Va., 1979.

3. Jim Bechtold, ed., "All about CRTs", *Hewlett-Packard Bench Brief*, vol. 21, no. 2, March–May 1981.

4. Vince Lutheran and Bernie Floersch, "Dual-beam: an Often Understood Type of Oscilloscope, *Electronic Design News*, Aug. 5, 1974.

5. Chuck DeVere, *Cathode-ray Tubes, Circuit Concepts*, 2nd ed., Tektronix, Inc., Beaverton, Ore., July 1969.

6. *Electronic Instruments and Systems*, Hewlett-Packard Co., Palo Alto, Calif., 1982 (published yearly).

7. *Tektronix 1981*, Tektronix, Inc., Beaverton, Ore. (published yearly).

8. Rick Nelson, "Storage Oscilloscopes," *EDN*, June 10, 1981, pp. 76-88.

9. Amelia Tegen, Nicolet Instrument Corp., "Trends in Digital Oscilloscope Applications," *Wescon 1981 Professional Program*, Paper 7/1, Electronic Conventions Inc., El Segundo, Calif., 1981.

10. A. W. Crooke, *et al*, Data Precision Corp., "User Interface: Complexity vs Flexibility in a Digitizing Signal Analyzer," *Wescon 1981 Professional Program*, Paper 7/3, Electronic Conventions Inc., El Segundo, Calif., 1981.

9

Analog and Digital Readout Devices

9.1
INTRODUCTION [a]

For more than the last 70 years, most laboratory and production recorders have been electrically actuated. They convert an analog electrical signal into a displacement of the writing pen. They have three general classifications—galvanometric, oscillographic, and potentiometric.

Galvanometric recorders are called direct-writing galvanometers. Their measurement system consists of rotating coils with a permanent magnetic field. When the input signal is applied, the coils rotate in response to the signal magnitude and direction. Since the writing mechanism is directly linked to the rotating coils, any change in input will be directly written or recorded on a chart.

Advantages of galvanometric recorders include economy, excellent frequency response, and sensitivity. The disadvantages include low torque, susceptibility to external noise, and nonlinearity.

---

[a] The material in Section 9.1 is excerpted in part by courtesy of *Medical Electronics*, Sept. 1981, 2994 West Liberty Ave., Pittsburgh, Pa. 15216.

Oscillographic recorders employ the galvanometer mechanisms and contain additional internal electronics to provide better torque, greater sensitivity, and good linearity. Better accuracy and extension of frequency response can be obtained. The advantage of oscillographic recorders include compatibility with a wide range of signal conditioners (device before the recorder), high speed, and multichannel capability. The disadvantage lies in the requirements for special heat- or light-sensitive paper in the recording medium.

The purpose for obtaining bioelectric analog signals is to supply physiological data in a useful form. It is important that data be recorded and displayed so as to be meaningful, perhaps at a later date. Data should be identified with name of subject, time taken, type of instrument, input and filtering conditions, paper speed, name of operator, and all other pertinent information—location of transducer, type of transducer, and so on.

Display and recording devices, including microprocessor technology used for physiological monitoring, include analog meters, oscilloscopes and graphic recorders, and computer-based devices. For most physiological work, a permanent record of the phenomena is essential. Therefore, an analog instrument is necessary.

Many factors must be considered in selecting an analog medical recorder system. They include the following:

1. Frequency response or bandwidth

2. Gain and gain settings (adjustability)

3. Accuracy

4. Noise level and common-mode rejection

5. Input impedance

6. Number and width of channels

7. Cost

Table 9.1 shows the frequency response of the five major types of analog recorders. The most economic is the permanent-magnet moving-coil (PM/MC) direct-writing recorder. It has a response to 100 Hz and a standard accuracy of 98% (2% error).

The servo potentiometer recorder is slower than the PM/MC but more accurate, with only 0.1% error. The light beam and oscilloscope types have high response frequency, but are more expensive than the PM/MC.

A useful instrument for displaying nonpermanent recordings of high-frequency phenomena is the cathode-ray oscilloscope discussed in Chapter 8. As it has no mechanical parts, a frequency response of megahertz is possible. Permanent records, however, can be obtained by taking pictures of the tracing on the scope face.

Industrial applications, however, may require the digital readout approach, especially in data acquisition and conversion applications.

In direct-writing recorders, the stylus or pen is attached to the armature of a galvanometer, which may be a rotating-coil (shown in Figure 9.1) or

TABLE 9.1
Accuracy and response of five basic recorder types

(1) Potentiometer, (2) PM|MC, Ink, (3) PM|MC, Light Beam,
(4) Oscilloscope|Camera, (5) Analog Magnetic Tape

	Servo potentiometer	PM|MC ink	PM|MC Light beam
Accuracy	0.05%	2%	2%
Response			
Full scale	DC–1 Hz	DC–30 Hz	DC–1 kHz
Reduced amplitude	DC–10 Hz	DC–100 Hz	DC–25 kHz

	Oscilloscope + Camera	Analog mag tape FM	Analog mag tape Direct
Accuracy	3%	2–5%	±3 dB
Response			
Full scale	DC–100 kHz	DC–80 kHz	300 Hz–
Reduced amplitude	DC–1.0 MHz	DC–160 kHz	1.6 MHz

Courtesy of *Medical Electronics*, vol. 1, no. 1, Jan.–Feb. 1970. *Medical Electronics*, 2994 W. Liberty Avenue, Pittsburgh, PA 15216

moving-iron type (which offers small size, low weight, and ruggedness).

Writing systems include the ink or ink jet, carbon transfer, thermal, electrodischarge, or pressure technique. Ink offers lowest medium cost; pressure systems are lowest in cost (and performance); carbon-transfer, thermal, and pressure systems are inherently rectilinear. Ink recorders can use curvilinear-coordinate conversion to achieve rectilinear presentation (see Figure 9.2).

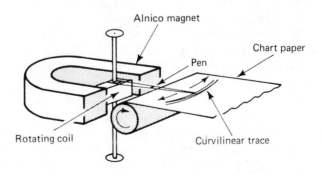

FIGURE 9.1
Galvanometric recorder. (Courtesy of Gould Inc., Instrument Systems Division, Cleveland, Ohio)

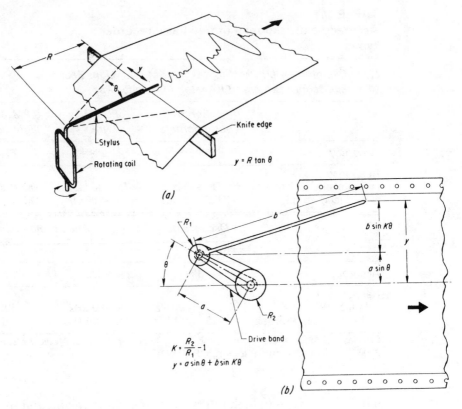

FIGURE 9.2

Two characteristic rectilinear writing systems. In (a) a heated stylus sweeps across thermal chart paper drawn over a knife edge. A different portion of the stylus end contacts the paper for each angle of displacement. In (b) a parallel-motion mechanism keeps the pen tip nearly perpendicular to the direction of chart motion and uses ordinary paper. (Courtesy of Gould Inc., Instrument Systems Division, Cleveland, Ohio. Reprinted from *Machine Design,* April 20, 1972. Copyright 1972 by Penton Publishing Co., Cleveland, Ohio)

Selection factors include frequency response, sensitivity, dynamic response, overshoot, hysteresis, linearity, stability, paper speed, medium (durability and cost), accessories, weight, size, and power requirements.

The galvanometer drive system may operate open loop, or with velocity or position feedback. In the open-loop spring-return galvanometer, pole shaping or shading may be used to increase linearity. Close control of stylus pressure is required. Position-feedback galvanometers use a feedback signal from an angle-to-voltage converter rather than a spring. Manufacturers claim improved torque, writing pressure, linearity, long-term stability, and control of overshoot.

9.2
ANALOG AND DIGITAL READOUT RECORDERS [b]

The principle of galvanometer light-beam recording is shown in Figure 9.3. As the galvanometer moves only the tiny mass of a mirror, the frequency response is increased from the 1-kHz upper response figure of direct-recording types to about 25 kHz, which is near the practical limit of light-beam recording types. The beam "writes" on photosensitive film or paper.

In light-beam recording, both rapid wet and dry processing techniques without chemicals now give almost immediate readout, eliminating the once inherent disadvantage of this technique—developing time.

The potentiometric recorder (Figures 9.4 and 9.5) operates by comparing a current or voltage input signal to a reference signal. The difference between the input and reference is termed the error signal. The error signal is converted to ac, amplified, and demodulated. The signal is further amplified and applied to a dc power stage, the output of which drives a dc balancing motor and associated slide-wire contact in a direction to reduce the error signal to zero. As the error signal is reduced to zero, the motor driving the display system moves the pointer and pen over the calibrated chart width with a distance equal to the original signal.

Present-day potentiometric recorders have automatic gain control and damping with any individual span setting. Potentiometric lab recorders today also work on the basis of linear servomotors that operate within a magnetic field produced by integral permanent magnets, instead of using the traditional slide-wire mechanisms. The advantage of potentiometric recorders include flexible measuring system, multiple chart speed possibilities, better resolution than galvanometric recorders, better sensitivity, plus compatibility with a wide range of signal conditioners available today. The disadvantage lies only in the limitation of the frequency response range; the highest frequency response is only 10 Hz.

The readout device of an electronic instrumentation system (recorder, readout device, data display device, etc.) may be one of several varieties, and the choice depends primarily upon the rate of change of the electronic instrument variable being studied. The ability of a measuring instrument to follow dynamically changing signals is most often expressed in terms of *frequency response*, that is, the fidelity with which sinusoidal input signals

[b] The material in Section 9.2 is excerpted in part from issue of September 1981 of *Medical Electronics*, 2994 West Liberty Ave., Pittsburgh, Pa. 15216, and is also based on application notes, courtesy of Beckman Instruments, Inc., Schiller Park, Ill.

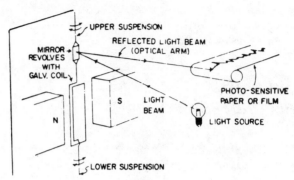

FIGURE 9.3
Principle of galvanometer light-beam recording.
(Courtesy of *Medical Electronics,* 2994 W. Liberty
Ave., Pittsburgh, Pa. 15216)

of various frequencies can be reproduced. Thus an output transducer that has a frequency response flat from 0 to 100 Hz will exactly reproduce the relative amplitudes of all input frequencies within this range. Hence it is convenient to divide output transducers into four major categories depending upon their frequency response, that is, on their ability to follow rapidly changing input signals.

The low-frequency analog recorders (0 to 2 Hz) consist of a direct-writing moving coil galvanometer or a servo system. Because the mass of the coil and writing arm is relatively great in relation to the stiffness of the restoring spring, such writers have a low-frequency response. (In general, the *natural frequency* of a mass–spring system equals K/M, where K is the spring stiffness and M is the mass.) Such devices generally consume little power (perhaps 1 to 2 W), so power amplification requirements are not great. Examples of such recorders include the Esterline-Angus recording milliameter (galvanometer) and the SID strip-chart recorders (servo).

The intermediate-frequency analog recorders (0 to 200 Hz) consist of direct-writing moving coil galvanometers. However, to increase the frequency response, much stiffer restoring springs are used, and this in turn increases the power required to drive the unit (perhaps 8 to 10 W). Such units are often called *pen-motors.*

The high-frequency analog recorders (0 to 10 kHz) employ optical galvanometers consisting of a very tiny mirror mounted on a very light moving coil. The "writing arm" itself is a beam of light that has neither mass nor friction and can be made long enough to provide considerable amplification of coil motion. Power requirements are small. Records are made on moving film, which must be processed before it is available for examination. However, a recently developed instrument (Honeywell Visicorder) uses an ultraviolet light and special paper that does not require processing and makes the records available within a few seconds.

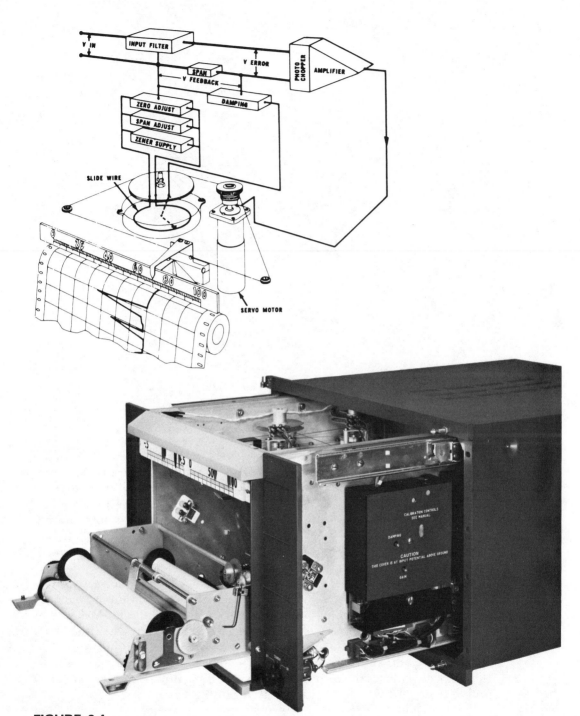

FIGURE 9.4
Conventional servo potentiometric recorder uses servomotor to control pens connected to slide-wire. (Courtesy Texas Instruments Servo / Riter II and Fisher Recordall)

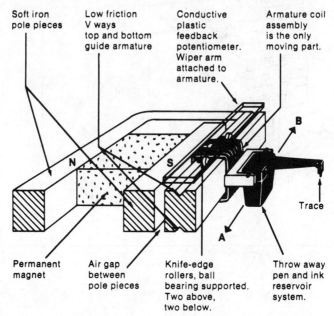

FIGURE 9.5

Speed Servo achieves 0.2-s response by use of
servomotor with moving armature. (Courtesy
Esterline-Angus Instrument Corporation)

The ultrahigh-frequency recorders (0 to 0.5 MHz or more) consist of the familiar cathode-ray oscilloscope (CRO) discussed in detail in Chapter 8. The "writing arm" consists of a beam of electrons with negligible mass or friction. Since the beam is controlled electrostatically, there is no galvanometer coil to contribute inertia. Although power amplification is not required, there is need for considerable voltage amplification since most cathode-ray tubes require an input signal of 7 to 100 V per inch of deflection. Also, displays must be photographed if a permanent record is desired.

All these readout devices are electromechanical transducers. They receive an electrical signal from the voltage or power amplifier and convert it to the mechanical motion of a galvanometer coil with attached pen or mirror or of a beam of electrons. The record displayed by all of them is a graph of writing point displacement on the vertical axis (ordinate) against time on the horizontal axis (abscissa). The time scale is provided by a chart drive, which moves the recording paper past the writing point at a predetermined rate, or by a special sweep circuit in the CRO.

The readout device as well as every other component of the recording chain must be chosen to have a frequency response adequate to follow the electronic variable under study.

If 1 V at the input of an analog recorder moves the analog recorder pen 1 in. and 10 V moves the pen 10 in., the pen should deflect 3 in. for 3 V, 7 in. for 7 V, and so on, if the recorder is linear. A good recorder has a linearity specification of $\frac{1}{2}$% for full scale of 50 millimeters (mm). If the 50-mm scale is calibrated for 0 to 5 V, then 3 V should move the pen within $\frac{1}{4}$ mm of 30 mm ($\frac{1}{2}$% of 50 mm $= \frac{1}{4}$ mm).

Looking at the same problem in a different way, if the pen moves to 30 mm, the input must be 3 ± 0.025 V. If the 50-mm scale is calibrated from 0 to 200 mm of Hg for blood pressure and the pen moves to 25 mm, the pressure must be 100 ± 1 mm of Hg.

The analog frequency response of an analog recorder is determined by connecting its input to a sine-wave generator whose frequency can be varied while the output signal amplitude from the sine-wave generator remains constant. It is desirable that the recorder output amplitude remain constant over the range of frequencies. Any type of waveform can be constructed from the addition of sine and/or cosine waves of various frequencies.

The analog recorder gain has no units but is the number of times larger or smaller that the amplifier output voltage is with respect to the input voltage to the amplifier. For example, if the input voltage is 1 mV and the gain is 1000, the output voltage is 1000 mV or 1 V. If the input voltage is 10 V and the gain is 0.1 or $\frac{1}{10}$, the output is 1 V. The maximum gain of a good preamplifier could be as high as 400. The maximum gain of a good power amplifier could be as high as 500. Therefore, the overall maximum gain is 400 times 500 or 200,000. If the recorder input is 10 μV, the output would be 2,000,000 μV or 2 V. Because 2 V is required at the galvanometer input to drive the pen on the galvanometer 1 cm, 10 μV at the preamplifier input will drive the pen 1 cm when the recorder is set at maximum gain.

The analog sensitivity is related to gain but has units, for example, volts per centimeter or centimeters per volt. Some manufacturers use the general units of volts per centimeter (V/cm). The unit of V/cm means that so many volts at the amplifier input will move the pen 1 cm. For example, if the system sensitivity is 1 mV/cm, an ECG signal at the system input with an R-wave amplitude of 1 mV will have an amplitude of 1 cm on the recording paper. If the R-wave amplitude were 5 mV, its amplitude on the paper would be 5 cm.

A drift figure for a recorder is given as the equivalent input change in volts per hour of operation at maximum gain under normal ambient conditions. The drift specification will change if the condition temperature changes, normally an increase in drift with an increase in temperature.

A typical drift specification for the analog recorder is 1 μV equivalent input per hour, or the pen tip will drift about 1 mm/h with the input shorted to ground and the gain at maximum. A good amplifier can have a maximum sensitivity of 1 μV/mm; the equivalent input to the recorder to cause the pen to move 1 mm is 1 μV.

Hysteresis is the amount of "play" in the analog recorder device that turns. For example, if a car steering wheel can be turned a little without causing the wheels to move, there is some play or hysteresis in the steering

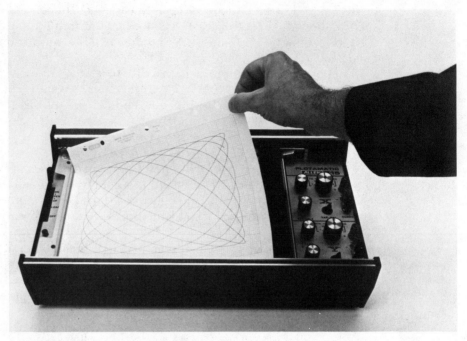

FIGURE 9.6
Allen Datagraph model 715 *XY* recorder.
(Courtesy of Allen Datagraph, Inc., Salem, N.H.)

mechanism. If a signal can be applied to a galvanometer with a pen without causing the pen to move, there is hysteresis in the galvanometer or in the linkage of a rectilinear pen. Looking at this problem from the pavement or paper end first, the wheels will wobble back and forth when the steering wheel is held still if there is hysteresis in the steering mechanism or the pen can write several different separated parallel lines with a constant input to the galvanometer if there is hysteresis in the galvanometer–pen mechanism. In either case the hysteresis is caused by the same thing, but one or the other end of the mechanism is held constant or motionless.

Conventional recorders plot one variable (the dependent variable) against time. The *XY* recorder (Figure 9.6) can plot any given variable against *any* other variable, including time. For example, the *XY* recorder can plot, in a few seconds of time, the relationship between current and voltage, lift and drag, speed and torque, stress and strain, temperature and activity, temperature and pressure, hysteresis, and so on.

Most *XY* recorders are self-balancing potentiometers, with either a flatbed or drum recording surface. The inputs are slowly varying dc voltages (from millivolts to volts). Capillary-type ink pens are most common, although high-speed plotters make extensive use of point-type recordings.

These analog signals can also be fed to a data-acquisition system for introducing these signals to a digitized system.

Linear-array recorders, as the name implies, use a linear fixed array of small recording elements under which the paper moves. This is in contrast with the conventional recorder that uses a moving pen or stylus. The stylus in the linear array recorder is a large number of fixed "styli," each one of which corresponds to one amplitude of signal to be recorded.

The obvious advantage of a linear array recorder is the absence of moving parts; nothing moves but the paper. The techniques used include thermal elements, photosensitive elements, and electrostatic elements.

The programmable light-gate array of Bell and Howell (CEC Division) is shown in schematic form in Figure 9.7. At the heart of the ECE HR-2000 Datagraph is a light-gate array, composed of modular wafers of LPZT (lanthanum-modified lead zirconate titanate) upon which electrodes defining the light gates have been deposited by photomask thin-film deposition methods. At 80 light gates per inch, each gate is 0.0125-in. wide, which provides high resolution. A 12-in.-wide array contains 960 individual light gates; the 8-in.-wide array contains 640.

The thermal linear array of Gould is shown in Figure 9.8. Thermal array writing technique puts analog traces, grid lines, trace identification and alphanumerics on plain thermal-sensitive paper. The array is composed of 512 thermal styli spaced at four per millimeter in a linear array. The array is 128 mm long by $\frac{1}{4}$ mm wide. Figure 9.9 shows the Linesis linear-array recorder Pen-less.

The environmental conditions under which a computer peripheral can operate, such as temperature and humidity, are also relevant to its choice. Additional environmental concerns that the peripheral can affect are noise (decibels), odors, cleanliness of operation, physical mobility, and power consumption.

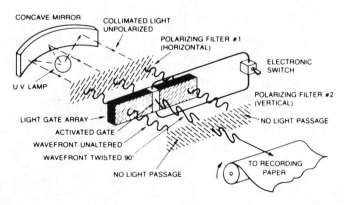

FIGURE 9.7
Principle of light-gate array. (Courtesy of Bell and Howell, CEC Division)

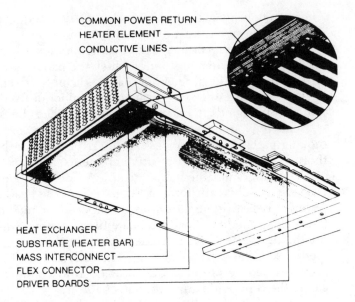

COMMON POWER RETURN
HEATER ELEMENT
CONDUCTIVE LINES

HEAT EXCHANGER
SUBSTRATE (HEATER BAR)
MASS INTERCONNECT
FLEX CONNECTOR
DRIVER BOARDS

FIGURE 9.8
Gould TA 600 linear thermal array comprises 512
thermal styli, 4/mm, for a total of 128 mm (5 in).
Each stylus corresponds to one amplitude of input signals; Gould also offers the ES1000 electrostatic linear-array recorder. (Courtesy of Gould
Inc., Instruments Division)

For plotting, terms that can have a direct effect in the cost of the device
are its accuracy and its resolution. The accuracy of a plotter is determined
by how close a line can be placed to any given point on the paper on a
repeatable basis. This figure may be expressed in thousandths of an inch
(mils) or in percent of the length and width of the plot. Typical values would
be 5 mils or 0.5%. Resolution is the number of resolvable lines that can be
plotted per inch (e.g., 100 lines per inch).

The speed of a printing or plotting device can be treated in several
ways; for devices that primarily print, the unit of measure for speed can be
characters per second or alphanumeric lines per minutes. Speed ranges are
from very slow (10 to 200 characters per second), to slow (up to 200 lines per
minute), to medium (800 to 1400 lines per minute), to fast (up to 5000 lines
per minute). Plotting speeds are normally defined in inches per second, but
in this area a confusion arises. For pen plotters, inches per second is the
pen's speed as it travels over the paper. For plotters not using movable
pens, inches per second is the paper speed as it moves across a fixed writing
device. A direct comparison may not be made between the two, since paper
speed is usually independent of plot complexity for one device, while the
time it takes a pen plotter to draw a plot varies proportionally with plot

FIGURE 9.9
Linseis linear-array recorder Pen-less. (Courtesy
of Linseis Inc., Princeton, N.J.)

complexity. Another speed specification in use with printer–plotter devices
is steps per second, where the paper is moved past a fixed writing device
in exact increments or steps. It is this specification that actually determines
the machine's final mechanical capability in terms of alphanumerical lines
per minute or plot inches per second. The correlation to alphanumeric lines
per minute is determined by the number of steps required to print one
alphanumeric line, usually between 6 and 15 steps:

$$\left(\frac{\text{steps/s}}{\text{steps/line}}\right)\left(\frac{60 \text{ s}}{\text{min}}\right) = \left(\frac{60 \text{ lines}}{\text{minute}}\right) \tag{9.1}$$

The relation of speed to data rate in a printer–plotter is of primary importance to the computer user. The final paper speed in inches per second or
the number of alphanumeric lines per minute depends, as described, on the
machine speed in steps per second. Between 600 and 1500 data bits may be
printed in one step, depending on the writing width of the printer–plotter.
(This number is a constant for any particular machine.) To reverse the speed
formula,

$$(60)\left(\frac{\text{lines}}{\text{minute}}\right)\left(\frac{1 \text{ min}}{60 \text{ s}}\right)\left(\frac{\text{steps}}{\text{line}}\right)\left(\frac{\text{bits}}{\text{step}}\right) =$$
$$\frac{\text{bits}}{\text{second}} \quad \text{data rate (or sometimes hertz)} \tag{9.2}$$

The data rate in bits per second may be converted alternately to bytes
per second by dividing by 8 for an 8-bit byte.

The term *baud* is associated with communication and has erroneously
been applied to data rates for printer–plotters. Since the term includes
control codes or start and stop bits in a serial stream, the data rate in baud
will appear greater than it actually is in terms of data transfer in bits per
second. To be sure of the exact meaning in context, the user should have
the term clarified when it is encountered. Even the communications field

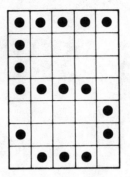

FIGURE 9.10
A dot-matrix 5 plot. (Courtesy of *Medical Electronics,* 2994 W. Liberty Ave., Pittsburgh, Pa. 15216)

is beginning to use bits or bytes per second as the preferable data rate definition.

The terms synchronous and asynchronous operation with a printing or plotting device designate the regularity of the data rate incoming to the device. *Synchronous operation* indicates constant data rates (equally spaced pulses), while *asynchronous operation* is the ability of the

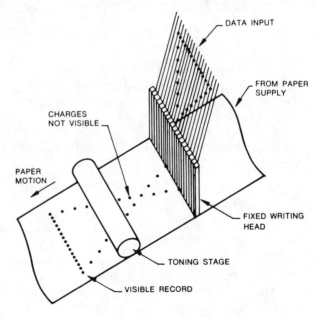

FIGURE 9.11
Electrostatic printer–plotter. (Courtesy of *Medical Electronics,* 2994 W. Liberty Ave., Pittsburgh, Pa. 15216)

printer–plotter device to assimilate data at varying and irregular intervals.

Font is a term associated with printing devices that defines the characteristic style of a set of alphanumerics. Gothic, for example, may be the style used on a particular impact line printer. Other fonts can be formed by vectors on plotting devices or by dot matrixes on nonimpact printers. A dot-matrix 5 is shown in Figure 9.10.

Electrostatic printer–plotters employ dielectric-coated paper that is electrostatically charged in the desired image areas in a dot pattern and then passed through a liquid toner suspension of charged particles. The toner particles adhere to the paper wherever a charge exists, resulting in a permanent, high-contrast image on the paper.

A fixed writing head containing multiple conducting styli is used to place electrostatic charges in the form of dots across the paper. The styli, fixed in a linear array, are individually activated by digital input data, thus producing a permanent digital record. Resolutions of 80 to 100 dots per inch are common. The paper is usually moved over the writing head in precise increments; thus the combination of scanning in the x-direction and stepping in the y-direction digitally covers every predetermined spot on the paper. A diagram of the electrostatic printer–plotter is shown in Figure 9.11. Characteristic ratings of printer–plotters are compared in Table 9.2.

Digital plotters can be described as a peripheral under the control of a digital computer or other digital data storage device. These plotters accept data in digital form and draw lines and characters on paper to provide a hard-copy graphic representation.

There are three general types of digital plotters currently available: electrostatic (shown in Figure 9.11), flatbed, and drum. Often considered as a fourth category is the computer output microfilm device (COM), which can generate plots on microfilm or microfiche.

The flatbed plotter operation is fairly straightforward. A flat sheet of

TABLE 9.2
Printer–Plotter Characteristics

	Speed	Accuracy	Reso-lution	Font	Record Quality	Relia-bility	Environ-mental	Oper-ating cost	Pur-chase cost	Figure of Merit
Line Printer (impact)	3	1	1	4	2	3	3	5	2	24
Thermal	3	2	2	2	2	3	2	2	4	22
Electrolytic	3	2	2	3	2	3	3	3	3	24
Photographic* (com)	1	5	5	3	4	3	3	2	1	27
Ink jet	2	3	3	2	3	2	2	5	3	25
Direct pen	2	5	5	1	5	3	3	2	2	28
Electrostatic	5	4	4	2	4	4	4	3	3	33

NOTE: Rating scales remain as noted in the two charts on printers and plotters.
* Computer output to microfilm.

(Courtesy of *Medical Electronics*)

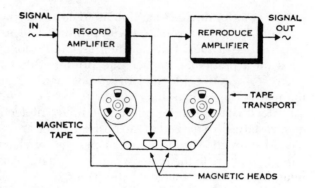

FIGURE 9.12
Tape recorder elements. (Courtesy of *Medical Electronics,* 2994 W. Liberty Ave., Pittsburgh, Pa. 15216)

paper is held in place while the writing instrument (pen) is moved in X–Y directions from point to point, generating the hard copy. The drum plotter is somewhat similar except that the writing instrument moves in one direction while the paper is moved to achieve the motion in the other direction. Drum plotters can be stepped in forward and reverse directions, permitting generation of very long drawings on a relatively compact machine. The drum plotter is the most widely used type; the flatbed is generally preferred when high accuracy is required.

Plots of electrostatic printers consist of a matrix of dots with typical resolutions of 100 and 200 dots/in. Metric versions of electrostatic plotters are also available with resolutions of 40 and 80 dots/cm. This dot-matrix approach means that read-only memories that contain 7×9 or 5×7 (or others, such as 16×8) dot-matrix images of alphanumeric characters can be used to generate printing and emulate high-speed line printers. As many as 132 characters/line, 6 or 8 lines/in., and speeds of over 1600 lines/min can be attained; plotting speeds of 3 or more in. of paper per second can be

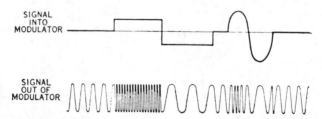

FIGURE 9.13
Frequency-modulation (FM) principle. (Courtesy of *Medical Electronics,* 2994 W. Liberty Ave., Pittsburgh, Pa. 15216)

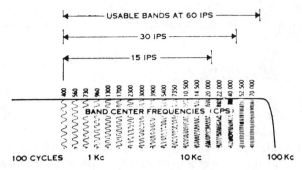

FIGURE 9.14
Frequency-division multiplexing. (Courtesy of
Medical Electronics, 2994 W. Liberty Ave., Pitts-
burgh, Pa. 15216)

obtained using electrostatic plotters, significantly faster than flatbed and
drum devices.

Magnetic tape recording is done most often by magnetizing a magnetic
material (fine iron oxide particles) coated on a nonmagnetic plastic ribbon
tape (usually cellulose acetate or polyester). Tape recorder elements are
shown in Figure 9.12.

The recording medium can be magnetized along any of its three dimensions: length (A), width (B), or thickness (C). The most common technique
is longitudinal recording, which is done along the direction of motion of the
tape.

Unfortunately, the recording characteristics of all magnetic materials
are nonlinear, that is, equal increments of magnetizing field (H) applied to
the material do not create equal increments of remanent magnetic flux (B)
when the field is removed. Thus it is necessary to use some method for
applying the signal to the tape in such a way as to operate only on the linear
portion of the tape characteristics, if linear direct recording is desired. This
can be done by applying either dc or ac bias so that the signal remains on
the linear portion of the curve. The most common method is by use of a
high-frequency ac bias (about four times the highest frequency to be
recorded). Simple unbiased recording can be used for pulses and digital
signal recording. See Figure 9.13 for the frequency-modulation (FM)
principle.

Signals recorded by either direct (biased) or saturation (pulse, FM, or
digital) recording cannot be reproduced with an amplitude accuracy of
better than ±10% because of many factors, especially head-to-tape contact;
a simple pressure pad pressing the tape against the head can cause a 20%
change in signal. In addition, tape and transport variations cause further
amplitude inaccuracy. Thus a tape recorder should not be depended on for
more than about 10% accuracy in amplitude accuracy, which is why FM,

PDM, and digital recording techniques are recommended for instrumentation work. Be sure to study Figure 9.14 for frequency-division multiplexing.

9.3
LIGHT-EMITTING DIODES[c]

Light-emitting diodes (LEDs) are now extensively used as readout devices in addition to lamps. Under forward bias conditions, certain specially compounded semiconductor materials emit light. The most common material is gallium arsenide doped with phosphorus. The wavelength of light emitted is 660 nanometers (nm) (red) for this material. By increasing the amount of diode material, more light can be produced. Green and yellow LEDs are also available. Light-emitting diodes and alphanumeric devices (letters and numbers) are used in TTL IC logic circuits and other digital applications such as counters.

Dialight bicolor LEDs are available in two basic discrete configurations, both of which consist of one red and one green chip, mounted in a common package.

The two leaded device, P/N 521-9177 (Figure 9.15), has the chips mounted in parallel with the polarities inverted with respect to each other. Reversing the voltage on terminals A and B in Figure 9.15 changes the color of the emitted light from red to green, or vice versa.

The P/N 521-9178 is a leaded device, with the chips in a common cathode configuration (Figure 9.16), which permits the user to power either chip independently.

Typical circuits using LEDs are shown in Figure 9.17.

The display in Figure 9.18 is the 745-0005 Diode-Lite[(R)] solid-state LED dot matrix alphanumeric display. This display is made of gallium arsenside phosphide and is mounted on a 14-pin dual-in-line substrate and cast within an electrically nonconductive transparent epoxy.

The alphanumeric display with typical driving circuitry is illustrated in Figure 9.19. The character displayed is a function of six input lines A_1 through A_6. The seven rows are scanned sequentially one at a time. The timing is controlled by the clock, which drives a binary counter whose outputs ABC control the row to be selected. This ensures that the read-only-memory's (ROM) five outputs C_1 to C_5 correspond to the row of dots that are enabled by the 1 out of 8 decoder. The SN74155 is a dual two to four line decoder, used here as a 1 out of 8 decoder.

For example, let us assume that the letter K is to be displayed. K in a 5 × 7 array of dots looks as depicted in Figure 9.20. The ASCII code A_1 to A_6 for selecting K is 001011. The SN7416 inverter buffer–drivers and the SN7417 buffer–driver are used to interface the inputs and outputs of the

---

[c] The material in Section 9.3 is used courtesy of Dialight, A North American Philips Co., 203 Harrison Place, Brooklyn, N.Y. 11237.

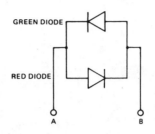

FIGURE 9.15
P/N 521-9177 Dialight bicolor LEDs. (Courtesy of
Dialight, A North American Philips Co.)

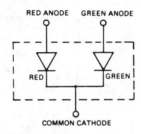

FIGURE 9.16
P/N 521-9178 Dialight LEDs. (Courtesy of Dialight,
A North American Philips Co.)

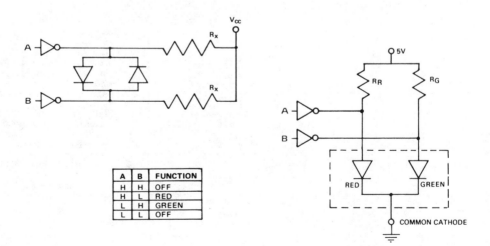

A	B	FUNCTION
H	H	OFF
H	L	RED
L	H	GREEN
L	L	OFF

FIGURE 9.17
Typical LED circuits. (Courtesy of Dialight, A
North American Philips Co.)

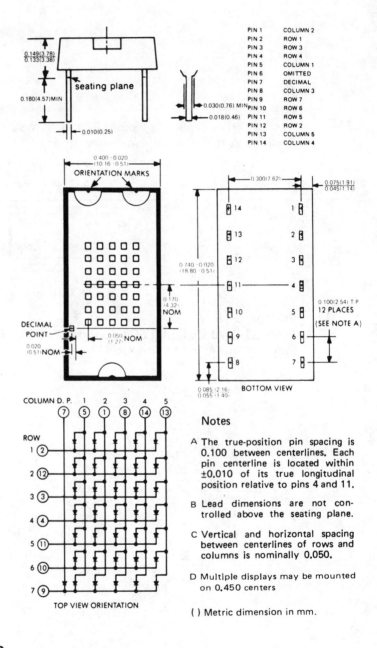

PIN 1	COLUMN 2
PIN 2	ROW 1
PIN 3	ROW 3
PIN 4	ROW 4
PIN 5	COLUMN 1
PIN 6	OMITTED
PIN 7	DECIMAL
PIN 8	COLUMN 3
PIN 9	ROW 7
PIN 10	ROW 6
PIN 11	ROW 5
PIN 12	ROW 2
PIN 13	COLUMN 5
PIN 14	COLUMN 4

Notes

A The true-position pin spacing is 0.100 between centerlines. Each pin centerline is located within ±0.010 of its true longitudinal position relative to pins 4 and 11.

B Lead dimensions are not controlled above the seating plane.

C Vertical and horizontal spacing between centerlines of rows and columns is nominally 0.050.

D Multiple displays may be mounted on 0.450 centers

( ) Metric dimension in mm.

FIGURE 9.18

The 745-0005 Diode-Lite(R) solid-state LED dot matrix alphanumeric display. (Courtesy of Dialight, A North American Philips Co.)

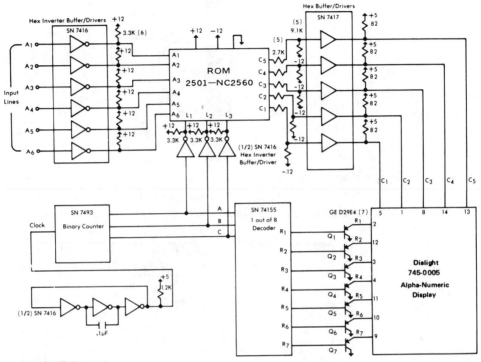

FIGURE 9.19
A typical application for the 745-0005. (Courtesy
of Dialight, A North American Philips Co.)

ROM. The binary counter's outputs ABC continuously cycle from 000 to 111 and back to 000. When ABC equals 000, then through the row inputs L_1, L_2, L_3 the ROM outputs C_1 and C_5 will be high and C_2, C_3, and C_4 will be low, as shown in Figure 9.20. Simultaneously, the 1 out of 8 decoder drives Q_1 *on* and this enables the first and fifth dots of the first row to be illuminated. On the next clock pulse, the counter advances to 001. C_1 and C_4 will become high and Q_2 will be driven *on*, thereby illuminating the first and fourth dots of the second row. This process is continued and repeated rapidly so that the final result appears to be the letter K.

The 745-0005 in conjunction with any suitable ASCII ROM will display 64 alphanumeric characters. These are shown in Figure 9.21 as a function of the 6-bit inputs A_1 to A_6.

Typical applications for the 745-0017 Diode-Lite[R] solid-state LED readout device with character height of 0.300 in. are shown in Figure 9.22.

MSI is an abbreviation for a type of electronic microminiaturization in which between 100 and 1000 transistors may be concentrated in a single-circuit chip. LSI is a type of electronic microminiaturization in which about 1000 transistors are concentrated in a single-circuit chip. SSI refers to about 100 transistors in a single-circuit chip.

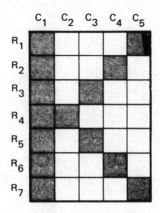

FIGURE 9.20
A 5 × 7 array of dots. (Courtesy of Dialight, A
North American Philips Co.)

A_6	A_5	A_4	A_3	A_2	A_1	Resultant Display	A_6	A_5	A_4	A_3	A_2	A_1	Resultant Display
0	0	0	0	0	0	@	1	0	0	0	0	0	
0	0	0	0	0	1	A	1	0	0	0	0	1	!
0	0	0	0	1	0	B	1	0	0	0	1	0	"
0	0	0	0	1	1	C	1	0	0	0	1	1	#
0	0	0	1	0	0	D	1	0	0	1	0	0	$
0	0	0	1	0	1	E	1	0	0	1	0	1	%
0	0	0	1	1	0	F	1	0	0	1	1	0	&
0	0	0	1	1	1	G	1	0	0	1	1	1	'
0	0	1	0	0	0	H	1	0	1	0	0	0	(
0	0	1	0	0	1	I	1	0	1	0	0	1	)
0	0	1	0	1	0	J	1	0	1	0	1	0	*
0	0	1	0	1	1	K	1	0	1	0	1	1	+
0	0	1	1	0	0	L	1	0	1	1	0	0	,
0	0	1	1	0	1	M	1	0	1	1	0	1	-
0	0	1	1	1	0	N	1	0	1	1	1	0	.
0	0	1	1	1	1	O	1	0	1	1	1	1	/
0	1	0	0	0	0	P	1	1	0	0	0	0	ϕ
0	1	0	0	0	1	Q	1	1	0	0	0	1	1
0	1	0	0	1	0	R	1	1	0	0	1	0	2
0	1	0	0	1	1	S	1	1	0	0	1	1	3
0	1	0	1	0	0	T	1	1	0	1	0	0	4
0	1	0	1	0	1	U	1	1	0	1	0	1	5
0	1	0	1	1	0	V	1	1	0	1	1	0	6
0	1	0	1	1	1	W	1	1	0	1	1	1	7
0	1	1	0	0	0	X	1	1	1	0	0	0	8
0	1	1	0	0	1	Y	1	1	1	0	0	1	9
0	1	1	0	1	0	Z	1	1	1	0	1	0	:
0	1	1	0	1	1	[	1	1	1	0	1	1	;
0	1	1	1	0	0	\	1	1	1	1	0	0	<
0	1	1	1	0	1	]	1	1	1	1	0	1	=
0	1	1	1	1	0	~	1	1	1	1	1	0	>
0	1	1	1	1	1	—	1	1	1	1	1	1	?

FIGURE 9.21
Resultant displays of alphanumeric device in conjunction with the ASCII ROM. (Courtesy of Dialight, A North American Philips Co.)

DISPLAY WITH DECODER/DRIVER

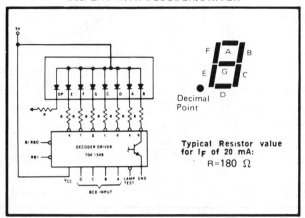

Decimal
Point

Typical Resistor value
for I_F of 20 mA:
R=180 Ω

DISPLAY WITH COUNTER-LATCH-DECODER

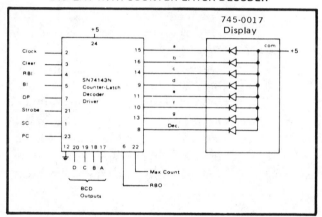

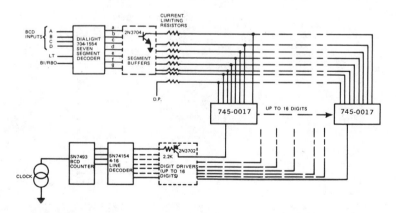

FIGURE 9.22
Typical applications for the 745-0017 Diode-Lite[R]
solid-state LED readout device. (Courtesy of Dialight, A North American Philips Co.)

The 745-0007 Diode-Lite[(R)] hexadecimal display on an integral TTL MSI circuit chip with a 0.270 in. character height is described next.

This hexadecimal display contains a 4-bit latch, decoder, driver, and 4 × 7 light-emitting-diode (LED) character with two externally driven decimal points in a 14-pin package. A description of the functions of the inputs of this device follows:

1. **Latch strobe input, pin 5:** When low, the data in the latches follow the data on the latch data inputs. When high, the data in the latches will not change. If the display is blanked and then restored while the enable input is high, the previous character will again be displayed.

2. **Blanking input, pin 8:** When high, the display is blanked regardless of the levels of the other inputs. When low, a character is displayed as determined by the data in the latches. The blanking input may be pulsed for intensity modulation.

3. **Latch data inputs (A, B, C, D), pins 3, 2, 13, 12:** Data on these inputs are entered into the latches when the enable input is low. The binary weights of these inputs are $A = 1, B = 2, C = 4, D = 8$.

4. **Decimal point cathodes, pins 4, 10:** These LEDs are not connected to the logic chip. If a decimal point is used, an external resistor or other current-limiting mechanism must be connected in series with it.

5. **LED supply, pin 1:** This connection permits the user to save on regulated V_{CC} current by using a separate LED supply, or it may be externally connected to the logic supply (V_{CC}).

6. **Logic supply (V_{CC}), pin 14:** Separate V_{CC} connection for the logic chip.

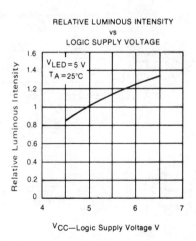

FIGURE 9.23

Relative luminous intensity versus logic supply voltage for the 745-0007. (Courtesy of Dialight, A North American Philips Co.)

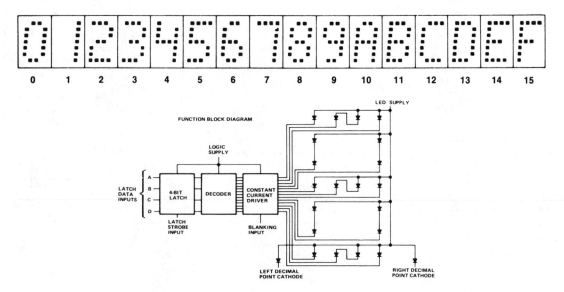

FIGURE 9.24
Resultant displays for the values of binary data in
the latches and a function block diagram for the
745-0007. (Courtesy of Dialight, A North American
Philips Co.)

7. **Common ground, pin 7:** This is the negative terminal for all logic and
   LED currents except for the decimal points.

The LED driver outputs are designed to maintain a relatively constant
on-level current of approximately 5 mA through each of the LEDs forming
the hexadecimal character. This current is virtually independent of the
LED supply voltage within the recommended operating conditions. The
drive current varies with changes in logic supply voltage, resulting in a
change in luminous intensity, as shown in Figure 9.23. The decimal-point
anodes are connected to the LED supply; the cathodes are connected to
external pins. Since there is no current limiting built into the decimal-point
circuits, this must be provided externally if the decimal points are used.

The resultant displays for the values of the binary data in the latches and
a function block diagram are shown in Figure 9.24. Operating characteristics for 745-0007 at 25°C ambient air temperature are given in Figure 9.25.

The 704-1549 (Figure 9.26) is a binary-coded decimal (BCD) to
7-segment decoder driver that drives solid-state LED and incandescent
readouts. It is designed to convert the BCD code into the appropriate
outputs, which will drive 7-segment readouts. The outputs of the 704-1549
decoder are open collector transistors with active low outputs able to drive

PARAMETER		TEST CONDITIONS	MIN	TYP	MAX	UNIT
I_V Luminous Intensity (See Note 4)	Average Per Character LED	V_{CC} = 5V, V_{LED} = 5 V See Note 5	35	100		μcd
	Each decimal	$I_{F(DP)}$ = 5 mA	35	100		μcd
λp Wavelength at Peak Emission		V_{CC} = 5 V, V_{LED} = 5 V	640	660	680	nm
B Spectral Bandwidth between Half-Power Points		$I_{F(DP)}$ = 5 mA, See Note 6		20		nm
V_{IH} High-Level Input Voltage			2			V
V_{IL} Low-Level Input Voltage					0.8	V
V_I Input Clamp Voltage		V_{CC} = 4.75 V, I_I = −12 mA			−1.5	V
I_f Input Current at Maximum Input Voltage		V_{CC} = 5.5 V, V_I = 5.5V			1	mA
I_{IH} High-Level Input Current		V_{CC} = 5.5 V, V_I = 2.4V			40	μA
I_{IL} Low-Level Input Current		V_{CC} = 5.5 V, V_I = 0.4 V			−1.6	mA
I_{CC} Logic Supply Current		V_{CC} = 5.5 V, V_{LED} = 5.5 V,		60	90	mA
I_{LED} LED Supply Current		$I_{F(DP)}$ = 5 mA, All inputs at 0V		45	90	mA

NOTES: 4. Luminous intensity is measured with a solar cell and filter combination which approximates the CIE (International Commission on Illumination) eye-response curve.

5. This parameter is measured with ⌐ displayed, then again with ⌐ displayed.

6. These parameters are measured with ⌐ .

FIGURE 9.25
Operating characteristics of the 745-0007 hex display at 25°C ambient air temperature. (Courtesy of Dialight, A North American Philips Co.)

all lamps up to a maximum of 15 V with a sinking capability of 30 mA. It features the following:

1. BCD inputs compatible with ECL and TTL devices

2. Monolithic MSI integrated circuit

3. Lamp test

4. Ripple blanking capability

5. Intensity control

9.4
SPECIAL DIGITAL READOUT APPLICATIONS

In the sections that follow we will introduce the following digital readout applications:

1. Cassette recorders

2. Waveform recorders

3. Serial alpha thermal miniprinter

4. Panel meters

5. Microprocessor-based weight monitor

Operating Characteristics (Ambient Temperature 25° C)

Supply Voltage, V_{cc}		5 ± 0.5 V
Supply Current		55 mA
Logic '1' }	BCD, LT, RBI	2.0 To 5.5 V
Logic '0' }	Inputs	0.0 To 0.8 V, Sink 1.6 mA
Logic '1' }	BI/RBO	2.4 V Min., @ 200 µA
Logic '0' }	Output	0.4 V Max., Sink 8.0 mA
OUTPUTS a-g Sink Current (Note 1)		20 mA, V_{ON}= 0.4 V Max. ('ON' State)
Voltage, V_{OFF}		15 V Max., @ 250 µA ('OFF' State)
Temperature Range		0° To 70° C

Note 1. The outputs a-g are open collector outputs which can sink 30 mA in the "ON" state and withstand +15V in the "OFF" state.

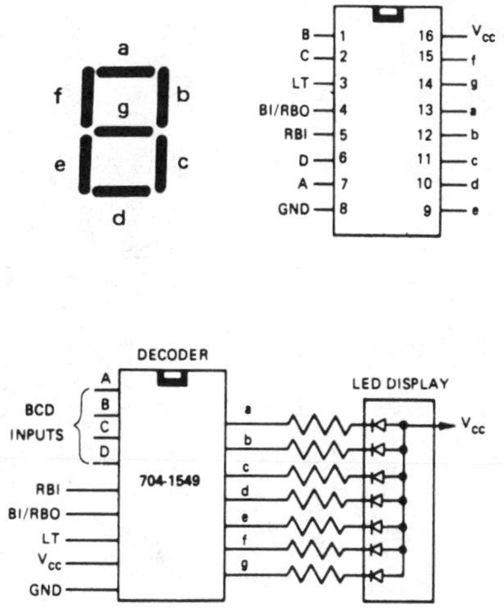

FIGURE 9.26
The 704-1549 and its operating characteristics. (Courtesy of Dialight, A North American Philips Co.)

6. Process recorder

Figure 9.27 shows the ICT-WZ CNRZ development systems of Datel-Intersil, Inc. The ICT series used in cassette recorders is discussed next.

9.4.1 Digital Cassette Recorders[d]

The ICT-series digital incremental cassette tape transport systems (Figure 9.28) has a data recorder and reader system offering a new concept to the instrument designer—miniature, removable ultra-low-power, digital data storage. Using a Philips data tape cassette, low-power incremental stepping motor, and CMOS electronics, these recorders are portable-battery-powered data-collection instruments.

The Datel-Intersil cassette systems form the nucleus of the data recorders. They complement the selection of input transducers, signal conditioners, and the computer system. Inputs to the optional A/D converter analog section have been standardized to 5- and 10-V levels commonly available from many sensors.

Reader–interface systems that are separate from the data recorder provide several output forms (full parallel, TTY/RS-232-C ASCII, $\frac{1}{2}$-in. magnetic tape) to adapt to different computers and printout devices.

Datel-Intersil also offers a new concept to the engineer or scientist

---

[d]The material in Section 9.4.1 is used courtesy of Datel-Intersil, Inc.

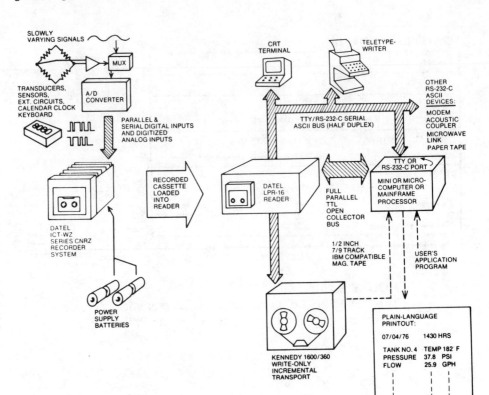

FIGURE 9.27
ICT-WZ CNRZ application development systems.
(Courtesy of Datel-Intersil, Inc.)

interested in data recording with tape systems. The ICT series offers a simplified approach with circuit card modules that are specified in terms familiar to circuit designers who know digital logic. Data entry, retrieval, and power-up, power-down considerations are simplified to the point of providing and accepting logic levels at the required times.

With very fast power turn on, the tape moves only when actually recording data or creating a gap. At all other times when not recording data, the transport and power-multiplexed electronics remain turned off to save tape and batteries.

The fast turn-on, turn-off feature means that tape is up to writing speed within 10 ms with no coasting when turned off. Data systems such as seismic recorders may be designed to record only when data are actually present, providing a very large data capacity per cassette. Hand-held stock inventory recorders move the tape only when keyboard data have been entered.

A complete cassette data-logging system from Datel-Intersil is available to record up to 16 or 64 analog channels, both as user-mounted modular systems and as stand-alone, weatherproof automatic data acquisition on logging systems (refer to Datel models LPS-16 and DL-2). Applications include oceanographic buoys and submersible probes, portable air and water quality environmental monitors, traffic and noise loggers, natural resources exploration, vehicular testing, seismic and geophysical measurements, RF field strength and transmission loggers, unmanned weather stations, and biomedical loggers.

Datel-Intersil uses a unique digital recording method. A none return-to-zero recording is abbreviated NRZ. A dual-track complementary NRZ (CNRZ) method is used for high noise immunity and self-clocking on playback (see Figure 9.29.) This high-capacity method uses both tracks of the tape simultaneously and records in only one direction. Using separate 1s and 0s tracks and very small gaps, one cassette can hold up to 2.2 million bits, or 120,000 A/D samples.

General specifications for the model ICT-WZ complementary NRZ format follow:

1. **Recording medium:** Philips magnetic tape cassette, certified for digital operation by using a bit-error check for dropouts. The transport is designed to accept cassettes complying with ANSIX3, 48, ECMA 34, and ISO/DIS3407 specs.

2. **Number of tracks:** two half-width tracks, recorded simultaneously.

3. **Recording direction:** One direction only.

FIGURE 9.28
ICT series low-power incremental cassette transport. (Courtesy of Datel-Intersil, Inc.)

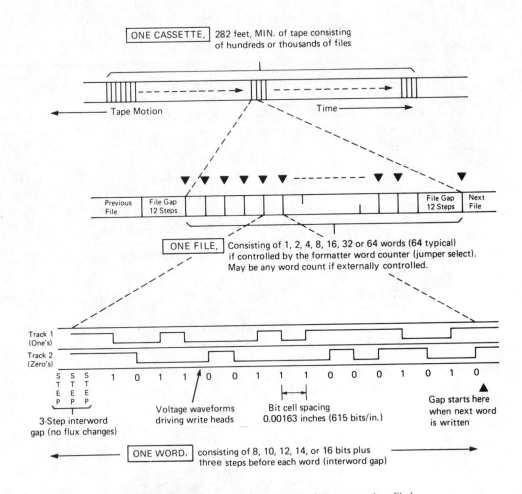

FIGURE 9.29
Complementary NRZ cassette formating.
(Courtesy of Datel-Intersil, Inc.)

4. **Motor:** four-winding, 24-pole, permanent magnet stepping motor.

5. **Equivalent tape speed range:** zero to 0.163 in./s (4 to 13 mm/s).

6. **Bit density:** 615 bits/in. (615 FRPI), 24 bits/mm.

7. **Bit cell spacing:** 0.00163 in./bit (0.0414 mm/bit).

8. **Bit writing rate:** Zero to 100 bits/s, asynchronous.

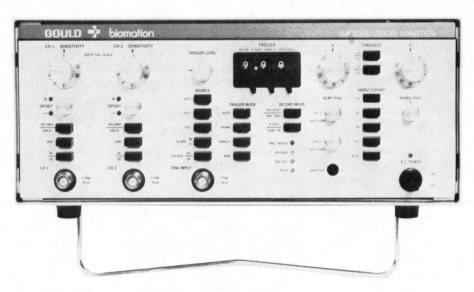

FIGURE 9.30
The Model 2805 Master Waveform Recorder manages up to three slave 2805s, providing up to eight channels of synchronous recording at 5 MHz and storing 2048 words of data per channel. (Courtesy of Gould Inc., Instruments Division, Biomation Operation, Santa Clara, Calif.)

9.4.2 Waveform Recorders[e]

The Gould model 2805 Master/Slave waveform recorder (Figure 9.30) introduces a new approach to simultaneous multichannel recording with additional channel expansion capability. Independently, the 2805 Master offers simultaneous dual-channel recording at rates up to 5 MHz (0.2 μs per sample), providing an 8-bit data word with a memory capacity of 2048 words per channel. The 2805 Slave is a Master-dependent simultaneous dual-channel recorder. The 2805 Master can manage up to three 2805 Slave units, providing up to eight channels of synchronous recording.

The pretrigger and delayed trigger recording modes enable the capture of information prior to a selected trigger event, around the trigger event, or at a selected interval downstream from that event. Samples are stored into a solid-state memory determined by the internal time-base selection. Internal sample rates are selectable from 0.2 μs to 100 ms per sample. Or an external clock may be input to the 2805 to provide the sample interval clock.

---

[e]The material in Section 9.4.2 is used courtesy of Gould Inc., Instruments Division, Biomation Operation, Santa Clara, Calif.

After the data have been stored, they may be viewed on a variety of CRT monitors or oscilloscopes, discussed in Chapter 8. The 2805 reconstructs the stored data via a digital-to-analog converter at a 1-MHz rate (repetitively for a flicker-free presentation). The information may also be output at a slow rate compatible with an *XY* or strip-chart recorder. Digital access to the memory via a bit parallel, word serial, handshake data exchange allows transfer of the data to a digital memory, calculator, or computer.

Versatile display controls allow switching off undesired channels for display, digital, or plot outputs. Up to eight channels of data may be viewed simultaneously for comparison analysis.

As shown in the block diagram of the 2805, a waveform recorder is most simply described as an analog-to-digital converter with memory (Figure 9.31). Surrounding control circuitry provides the unique transient capture flexibility of the instrument. The unit is easy to operate, more versatile than an oscilloscope, and has proved valuable in a wide variety of uses.

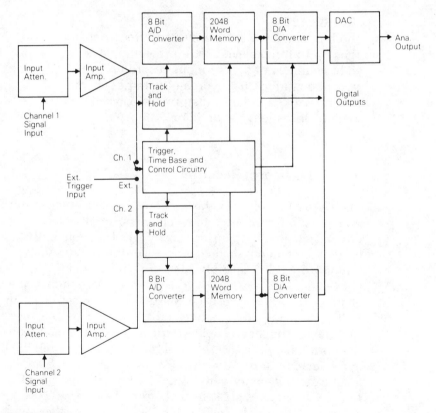

FIGURE 9.31
Block diagram of the master 2805. (Courtesy of Gould Inc., Instruments Division, Biomation Operation, Santa Clara, Calif.)

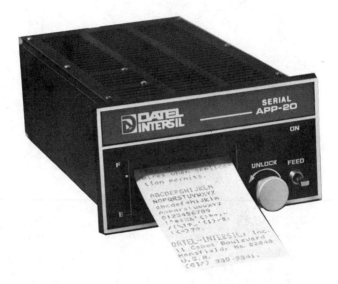

FIGURE 9.32
Serial APP-20 panel-mounted 20-column alphanu-
meric thermal printer. (Courtesy of Datel-Intersil,
Inc.)

The speed, amplitude resolution, and memory length of the Model 2805 provide performance usable over a wide range of signals. The unit is particularly well suited as a viable, and often superior, alternative to the traditional techniques, such as single trace scope camera recording, storage scope displays, analog tape recording, and oscillographic recording, and as a transient decoder and is useful in electronic instrumentation computer microprocessor applications.

9.4.3 Serial Alpha Thermal Miniprinter[f]

Models APP-20A2 and APP-20E2 are full serial input versions of the Datel-Intersil APP-20 panel-mounted 20-column alphanumeric thermal printer shown in Figure 9.32. Overall performance of the serial version is identical to the parallel input version, and all the unique programmable features are retained, including tall characters, single-character printing, inverted (text) printing, and noninverted (lister) printing.

An important feature of the APP-20 product concept is the inclusion of all electronics *inside* the miniature panel-mounted housing; therefore, the user does not need additional electronics. The serial input APP-20 extends this concept to the final step by *simplifying all external wiring* to the

---

[f]The material in Section 9.4.3 is used courtesy of Datel-Intersil, Inc.

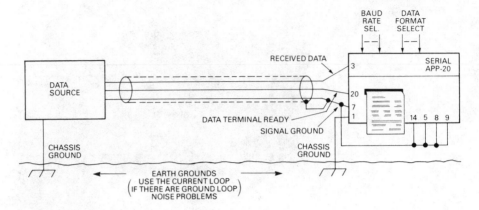

FIGURE 9.33
Finding current loops using the APP-20. (Courtesy
of Datel-Intersil, Inc.)

absolute minimum. In fact, only two wires (pins 17 and 18) are required to
operate the serial APP-20 in the 20-mA current loop mode. This mode also
adds one further advantage in that almost any type of wire may be used, and
the printer wiring can extend many hundreds of feet from the computer data
source (see Figure 9.33). The current loop input is optoisolated for large
common-mode noise rejection and elimination of ground loop problems.

9.4.4 Digital Panel Meters[g]

Digital panel meters (DDM) were discussed in Chapter 5. They are also
included here as readout devices, such as the Datel-Intersil DM-3100X,
$3\frac{1}{2}$-digit LCD with a 5 V/9 V micropowered battery (see Figure 9.34). The
DM-3100X is a $3\frac{1}{2}$-digit liquid-crystal display (LCD) digital panel meter that
uses extremely low power (+5 V at 6 mA or +9 V at 3 mA) and has a power
voltage range of +4 V to +15 V dc. The large 0.5-in. display can be seen
from many feet away under normal room lighting conditions. This DPM is
contained in a short depth case that measures only 2.15 in. (54.6 mm) deep.
Besides measuring dc voltages, components can be internally placed to
make ohm and current readings possible along with attenuators to measure
higher voltages.

This DPM accepts a dc or slowly varying input voltage between ±1.999
V and displays that input on front panel numerical indicators. It employs a
conventional dual-slope A/D converter plus 7-segment display
decoder–drivers, all in one LSI microcircuit. Since this microcircuit requires approximately 9 V to power the A/D section, an internal dc/dc
converter generates −5 V from +5-V power input. Together these two

---

[g]The material in Section 9.4.4 is used courtesy of Datel-Intersil, Inc.

FIGURE 9.34
DM-3100X digital panel meter. (Courtesy of Datel-
Intersil, Inc.)

voltage sources form a bipolar power supply to power the A/D converter. The DM-3100X may also be powered directly from a single 9-V battery at 3 mA without using the dc/dc converter.

Another feature of the DM-3100X is that it employs a balanced differential input. When used with a bridge or transducer input, it offers high noise immunity and can accurately measure very small signals in the presence of much larger common-mode noise. Another characteristic of this balanced differential input is that it will not load down sensitive input circuits owing to its high input impedance of 1000 MΩ and low 5-pA bias current.

A very noteworthy feature of this meter is that it can be operated ratiometrically. This means that it has internal circuits that can automatically compensate for reference drifts in the supplies of balanced bridge or transducer sensors and still give accurate readings.

The DM-3100X finds use in analytical instruments, industrial process controllers, portable diagnostic instruments, automatic test equipment, medical and patent monitoring instruments, airborne, marine, and ground vehicles, and data-acquisition/data-logging systems.

The digital ohmmeter circuit is shown in Figure 9.35. An external refer-

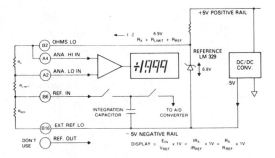

RANGE	RESOLUTION	R_LIMIT[1]	R_REF[1]	DECIMAL POINT
19.99 MΩ	10 kΩ	22 MΩ	10 MΩ	A11 to B13
1.999 MΩ	1 kΩ	3.6 MΩ	1 MΩ	A 12 to B13
199.9 kΩ	100 Ω	360 kΩ	100 kΩ	A10 to B13
19.99 kΩ	10 Ω	36 kΩ	10 kΩ	A11 to B13
1.999 kΩ	1 Ω	6.2 kΩ	1 kΩ	A12 to B13

1. ^{R}Limit and ^{R}Ref should be metal film, High Stability Resistors (AS RN60(

FIGURE 9.35
Digital ohmmeter circuit. (Courtesy of Datel-
Intersil, Inc.)

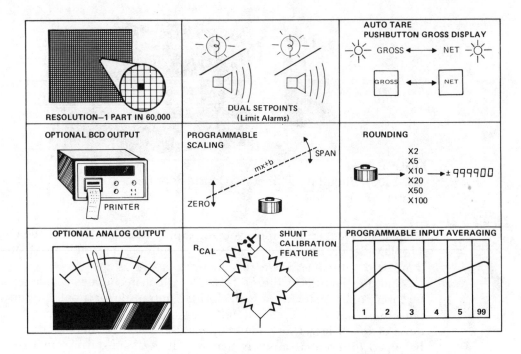

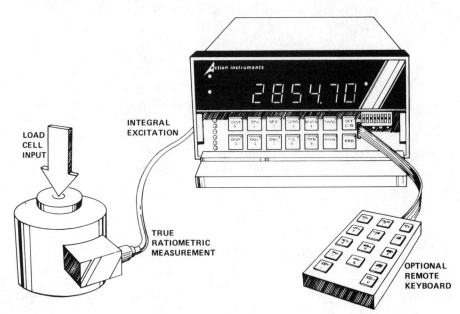

FIGURE 9.36
Visipak VIPS-524W microprocessor-based weight monitor with illustrative applications. (Courtesy of Action Instruments Co., Inc., San Diego, Calif.)

ence resistor of known resistance, accuracy, and temperature drift is connected in series with the unknown resistance. A constant, stable voltage from the DPM's internal reference diode is applied to the resistor pair to produce a constant current. This current develops two voltage drops across the resistors, which are proportional only to the ratio of the resistances since the current through them is identical.

The chart in Figure 9.31 lists recommended R_{REF} and R_{LIMIT} resistance values corresponding to different ohmmeter ranges. Values of R_{LIMIT} were selected to limit the current through R_{REF} and R_X to 1 mA maximum.

9.4.5 Microprocessor-based Digital Indicator [h]

The Visipak VIP524W is a full six-digit microprocessor-based strain gage input weight monitor, as shown in Figure 9.36. Strain gage transducer theory was discussed in Chapter 2. Although the monitor will function as a simple high-resolution digital indicator displaying the input variable, the 524W is designed primarily for the more sophisticated functions enabled by its microprocessor base. These functions include display net/gross, display with rounding, display in engineering units, display average input, alarm at set points, rate-of-change display, and peak/valley display. A programmable isolated analog output and a BCD output are also available.

The Visipak VIP524W features the following:

1. Resolution 1 part in 50,000
2. Full six-digit display
3- Dual set points–limit alarms
4. Push button: Tare, "gross" display, display scaling, input averaging, display rounding
5. Normal, peak–valley or rate-of-change display
6. Shunt calibration feature
7. Battery backup
8. Integral excitation

Other options include the following also:

1. BCD output
2. Isolated analog output
3. Remote keyboard

---

[h] The material in Section 9.4.5 is used courtesy of Action Instruments Co., San Diego, Calif.

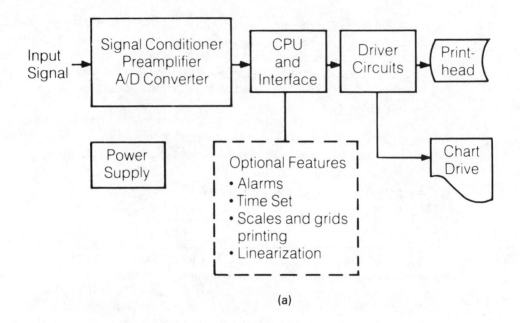

(a)

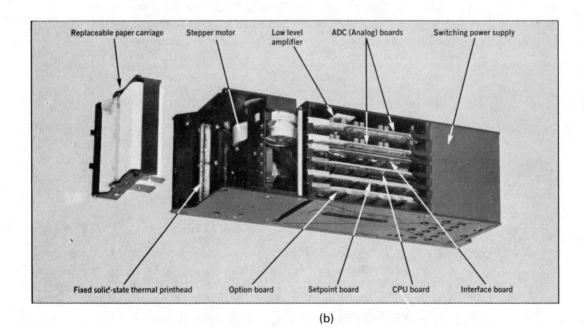

(b)

FIGURE 9.37

Tigraph 100: (a) Schematic diagram, (b) modular construction, and (c) status. (Courtesy of Texas Instruments, Inc.)

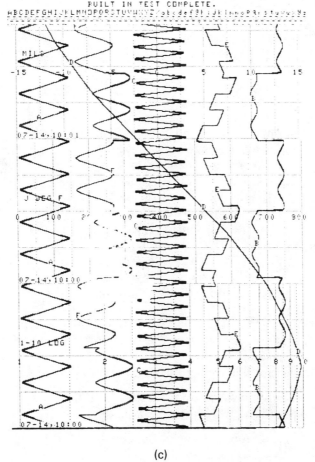

(c)

FIGURE 9.37 (Continued)

347

TABLE 9.3
Specifications of the Tigraph 100

General

Recording:
Thermal recording using stationary printhead

Printhead:
106.7 mm (4.2 in.) integrated circuit with 100 dots/inch resolution

Channels:
1, 3 or 6 channels

Channel ID:
Characters A through F

Chart Speeds:
28 speeds from 0.3 inch/hour to 3 inches/minute

Chart Width:
Maximum chart width:
114 mm (4.5 in.)
Maximum recording width:
107 mm (4.2 in.)
Calibrated recording width:
102 mm (4.0 in.)

*Grid Printing:
Latitudinal grid lines printed in 12.7 mm (0.5 in.) increments; longitudinal grid lines: 11 lines printed at 10% increments; scale 0-100

Paper Length:
25 m (82 ft.)

Time Printing:
Elapsed time every 51 mm (2 in.); time of day printing optional

Print Rate:
5 print lines/second, maximum; all channels printed simultaneously

Accuracy:
± 0.5% of full scale; all errors inclusive

Resolution:
0.25% full scale

Span:
0.5 V to 2 V standard; 1 mV to 30 V optional

Zero-Adjust:
± 100% span

Multirange:
Independent range and signal type for each channel

Input Type:
Volts, millivolts, milliamps, thermocouple, and RTD

†Input Resistance:
Greater than 10 megohms for spans up to 2 V
Greater than 2 V and less than 8 V, resistance = 40k ohms
Greater than 8 V and less than 30 V resistance = 160k ohms

Frequency Response:
0.25 Hz

Normal Mode Rejection:
60 dB at 50 or 60 Hz

Common Mode Rejection:
120 dB at dc, 50 Hz or 60 Hz

Common Mode Voltage:
30 V dc or ac peak, maximum

Power:
115 (± 15) V, 50/60 (± 4) Hz
230 (± 30) V, 50/60 (± 4) Hz optional
65 watts typical

Weight of Unit:
7.7 kg (17 pounds)

Ambient Temperature:
Rated operating conditions:
15 to 40°C
Extreme operating conditions:
−9 to 50°
Storage conditions:
−40 to 70°C

Optional

Low Levels:
Preamplifier card for millivolt spans below 500 mV, three-wire RTD with lead compensation, and thermocouples with reference junction compensators

Option Card:
Right and left event markers; 5 to 12 V ac or dc activation
Date and time-of-day printing
Accepts linearizer and chart module PROMs

Setpoint Option Card:
Independent Hi/Lo setpoints for each channel; resolution: 1% of full scale
Common alarm relay: rated 0.5 A at 24 Vdc resistive load
Alarm ID binary-coded output including Hi/Lo and read-strobe lines

*Other self-printed charts are optional
†Optional 1 megohm input resistance for spans greater than 2 V to 5 V

Courtesy of Texas Instruments, Inc.

9.4.6 Process Recorder [i]

In operation, incoming signals to the Tigraph 100 (Figure 9.37) are digitized before entering the central processing unit, where the signals are conditioned and processed for such functions as alarm set levels, linearization, and scale and grid printing. The signals then are routed through the driver circuitry to the thermal printhead and chart drive. The solid-state printhead consists of a single row of 420 individual printing elements, each less than 0.25 mm (1/100 in.) square. The microprocessor-controlled printhead, when energized, prints characters and traces on heat-sensitive paper at a resolution of 100 dots/in. (4 dots/mm).

The Tigraph 100 prints its own chart scale and grids in conjunction with recorded data. A unique chart scale and grid is possible for each of six different channels.

Upon start-up or change in operating parameters, the Tigraph 100 self-tests its print electronics, exercises the printhead, and prints a record of all operating parameters.

Important features of the Tigraph 100 Graphic Display include:

1. Multiple chart scale and grid printing.

2. Linearization for six independent inputs.

3. Chart annotation, including channel ID printing.

---

[i] The material in Section 9.4.6 is used courtesy of Texas Instruments, Inc., Houston, Texas.

4. Time-of-day printing.

5. Documentation of functions in status printout.

6. Up to six high–low set points.

Specifications for the Tigraph 100 are shown in Table 9.3.

9.5
REVIEW QUESTIONS

1. Discuss galvanometric recorders.

2. Discuss the light-gate array recorder.

3. Discuss the digital plotter.

4. Discuss the magnetic tape recorder. Draw tape recorder elements.

5. Discuss light-emitting diodes. Draw a typical circuit using light-emitting diodes.

6. Discuss an LED dot matrix alphanumeric display. Draw a diagram to illustrate the matrix alphanumeric display.

7. Discuss a digital readout recorder and complementary NRZ cassette formating.

8. Discuss how a waveform recorder works. Draw a block diagram and state where it is used.

9. Discuss the serial alpha thermal printer. Draw a block diagram of a use of this printer in current loop analysis.

10. Discuss the digital ohmmeter using a digital panel meter with a circuit.

9.6
REFERENCES

1. Peter Strong, *Biophysical Measurements*, 1st ed., Tektronix, Inc., Beaverton, Ore., June 1973.

2. "Recorders," *Medical Electronics*, 136–163, Sept. 1981.

3. *Instruments and System Handbook*, Datel-Intersil, Inc., Mansfield, Mass., 1981.

4. Clyde F. Coombs, Jr., *Basic Electronic Instrument Handbook*, McGraw-Hill Book Co., New York, 1972.

5. Nick Stadtfeld, *Information Display Concepts*, 1st ed., Tektronix, Inc., Beaverton, Ore., 1969.

6. Peter A. Howes, "Instrumentation Recorders," *Electronic Products*, 53–55, Mar. 3, 1982.

10
Input–Output Devices

10.1
INTRODUCTION

Input devices before the computer typically are either analog or digital in nature. Analog signals measure any random data while computer signals respond to an on and off state (represented by a 1 or 0). Most transducers, transferring mechanical or optical energy to electrical energy, are analog in nature and can be fed directly into an analog computer. To use a digital computer, the analog information has to be digitalized by an analog-to-digital converter (A/D), and then the information is fed to the digital computer. Many digital voltmeters have digital converters.

The digital computers are normally fed by a number of devices, such as teletype, paper-tape readers, magnetic tapes, discs, drums, data modems, and so on.

The output of digital computers feeds devices into teletype, tape punches, magnetic tape, CRT (cathode-ray tube) displays, and so on.

Computer time can be rented, as can most of the computer equipment. Various companies use this approach. Computer time-sharing people can give programs which are adaptable over local telephone lines or teletype with an acoustic coupler. This approach requires the gathering of a data-making program or selecting an existing program to fit one's needs. An example of this would be to send electrocardiogram information (the electric activity of the heart) from the home of a patient with heart disease to a hospital or to a doctor's office.

Outputs from the analog computer can be read on internal meters or external scopes (CRT displays), digital voltmeters, or XY plotters, or outputs can be digitized by analog-to-digital converters.

10.2
ELECTRONIC INSTRUMENTATION SYSTEM

In general, the sensor, which may be a transducer or electrode, is fed to a signal conditioner that amplifies the small signal and feeds the analog signal to a computer via the A/D converter or a readout device that is analog in nature. The analog readout device can consist of a voltmeter, oscilloscope, recorder, magnetic recording device, and others. Refer to Chapters 2 through 9 for further details.

10.3
COMPUTER INPUT–OUTPUT DEVICES: EXAMPLES

The following are examples of computer inout–output devices:

Input	Output
Card reader	Card punch
Tape reader	Tape punch
Magnetic ink reader	Line printer
Scanner	Typewriter
Typewriter	CRT displays
CRT displays	
Terminals to the CRT	Terminals to the CRT
Keyboard with terminals to the CRT	Keyboard with terminals to the CRT
Question displayed on CRT and choice of answers and person who answers question touches answer button by means of sensor pickup signal	Magnetic discs and tapes Magnetic cartridge Mark sensing device (IBM)
Time-sharing terminal	Time-sharing terminal
Magnetic discs and tapes	Optical character
Optical reader	

10.4
DIGITAL COMPUTER

The prime purpose of a digital computer is to serve humans. There must be a method of transmitting our desires to the computer and a means of receiving the results of the computer's calculations. The sections of the computer that carry out these functions are the input and the output units.

Input information from a number of media can be conveyed to the input section of the computer in various forms. Data may originate from a notebook, business file, telephone network, radio transmitter, teletypewriter, or perhaps a radar set. This information must first be converted into binary machine language acceptable to the input unit. One common method is to type the data on a keyboard that prepares coded punched paper tape or punched cards. For high-speed input processing, the information is encoded and then transferred to magnetic-tape handlers, which can process more than 50,000 characters per second. In addition to the data to be processed, a program of operating instructions is prepared that prescribes the manner in which operations are to be carried out. Both data and instructions are transferred to the computer input section, which selects the information in the order needed and feeds the resulting selection into the internal storage (memory) unit.

Input devices are used to supply the values needed by the computer and the instructions to tell the computer how to operate on the values. Input unit requirements vary greatly from machine to machine. A manually operated keyboard may be sufficient for a small computer. Other computers requiring faster input use punched cards for data inputs. Some systems utilize removable plugboards that can be prewired to perform certain instructions. Input may also be via punched paper tape or magnetic tape.

The output section is similar to the input, except that the processed information must not be retrieved from memory and reconverted into a form suitable for the output devices. Upon receiving the appropriate commands from the control, the output section transfers the processed data bits from storage in a logical sequence and arranged into characters, for insertion into the selected output device. As with the input, the data may first be transcribed onto magentic tape for high-speed printer or a paper or card punch. Modern high-speed printers can print out approximately 1000 lines/min, each 120 characters long. Thus more than 100,000 characters can be transferred out of the computer each minute. The printed-out final data, again, can be transmitted to a remote location by radio, telephone, or teletype.

Output devices, therefore, record the results of the computer operations. These results may be recorded in a permanent form (e.g., as a printout on the teleprinter) or they may be used to initiate a physical action (e.g., to adjust a pressure-valve setting). Many of the media used for input, such as paper tape, punched cards, and magentic tape, can also be used to receive the output.

The input and output units can be combined because in many cases the same device acts as both an input and an output unit. The teletype console, for example, can be used to input information that will be accepted by the computer, or it can accept processed information and print it as output. Thus the two units of input and output are very often joined and referred to as input–output or simply I/O.

10.5
BLOCK DIAGRAMS OF COMPUTER INPUT–OUTPUT DEVICES

In Figures 10.1 through 10.3, we see block diagrams showing the input and output devices and the intermediate blocks. The arithmetic unit of a digital computer performs the actual work of computation and calculation. The arithmetic unit counts series of pulses or by the use of logic circuits. Modern computers use components such as transistors and integrated circuits. Switches and relays were used previously. Modern computers, however, because of the speed desired, make use of smaller electronics components whenever possible, such as transistors and operational amplifiers.

The arithmetic unit of the PDP-8 (Figure 10.2) has, as its major component, a 12-bit accumulator, which is simply a register capable of storing a number of 12 binary digits. It is called the accumulator because it accumulates partial sums during the operation of the PDP-8. All arithmetic operations are performed in the accumulator.

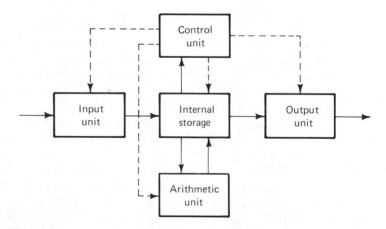

FIGURE 10.1
PDP-8 general organization. (Adapted from *Digital Introduction to Programming*, digital pdp-8 handbook series, copyright 1970 by Digital Equipment Corp., Maynard, Mass.)

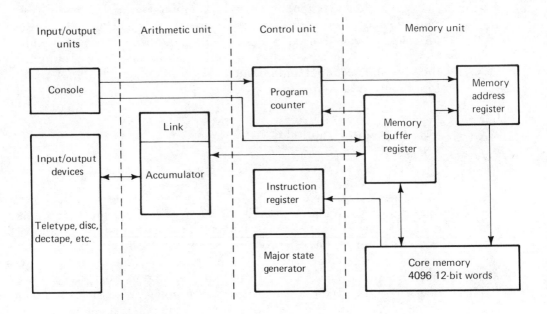

FIGURE 10.2
PDP-8 block diagram. (Adapted from *Digital Introduction to Programming,* digital pdp-8 handbook
series, copyright 1970 by Digital Equipment Corp.,
Maynard, Mass.)

The control unit of a digital computer is an administrative or switching section. It receives information entering the machine and decides how and when to perform operations. It tells the arithmetic unit what to do and where to get the necessary information. It knows when the arithmetic unit has completed a calculation, and it tells the arithmetic unit what to do with the results, and what to do next. The control unit itself knows what to tell the arithmetic unit to do by interpreting a set of instructions. This set of instructions for the control unit is called a *program* and is stored in the computer memory.

The memory unit, sometimes called the core storage unit, contains information for the control unit (instructions) and for the arithmetic unit (data). The terms core storage and memory may be used interchangeably. Some computer texts refer to external units as storage, such as magnetic tapes and disks, and to internal units as memory, such as magnetic cores. The requirements of the internal storage units may vary greatly from computer to computer. The PDP-8 memory unit is composed of magnetic cores, which are often compared to tiny doughnuts. These magnetic cores record binary information by the direction in which they are magnetized (clockwise or counterclockwise). The memory unit is arranged in such a

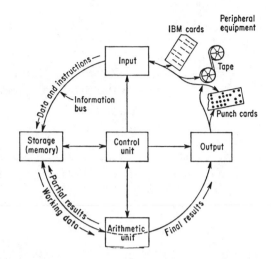

FIGURE 10.3
Functional block diagram of an automatic digital computer. (From Henry Jacobowitz, *Electronic Computers: A Made Simple Book,* courtesy of Doubleday & Co., Garden City, N.Y.)

way that it can store 4096 "words" of binary information. These words are each 12 bits in length. Each core storage location has an address, which is a unique number used by the control unit to specify that location. Storage of this type in which each location can be specified and reached as easily as any other is referred to as random-access storage. The other type of storage is sequential storage such as magnetic tape, in which case some locations (those at the beginning of the tape) are easier to reach than others (those at the end of the tape).

The PDP-11 digital computer is also manufactured by the Digital Equipment Corporation.

10.6
DIFFERENT INSTRUMENTS WITH
INPUT–OUTPUT (GENERALIZED)

Features of different instrumentation with generalized input–output devices are used in electronics, medicine, geology, and physics.

10.6.1 Electronics

You desire information on some electrical condition, on a condition of state:

1. An electric circuit can monitor voltage electrode or current. One uses a probe and attaches it to points of an electrical circuit. The output is read

on the voltmeter, such as a VTVM or a FET meter, a digital voltmeter, oscilloscope, or other output device. If you get a small shock from what you touch, a small voltage is indicated. If you get a big shock, there is a big voltage. But we know that the brain is not a good indicator, since the brain cannot respond to the stimulus.

2. As another example, suppose you are working in an electroplating plant and wish to determine the acidity of the solution. You can stick your finger in the solution, but you will not want to do this. So one uses electrodes. They are immersed in the solution, and a conductivity reading is obtained in terms of pH, which represents the negative logarithm of the hydrogen concentration.

10.6.2   Medicine

To determine the electrical activity of the heart, electrodes are externally connected to the chest and are fed to an electrocardiogram amplifier, the output of which can be digitized and fed to a computer. The output of the computer can be fed to an acoustic coupler to a telephone line. The signal can be transmitted from one point, then to another. A pacemaker that activates the heartbeat can be also monitored in the same way. In fact, there are about 100 different examples.

10.6.3   Geology

In geology, the main focus is exploration for crude oil or ores. Instruments that check the earth's magnetic field can also be used to check for ore or oil locations. The oil causes a lower magnetic flow (field is stationary, plane is moving, and breaking of the magnetic lines of forces causes a flow that will give the observer a flow chart). A seismic station is a second example; planting an explosion on the earth or the moon may be used for locating oil and water. Shock waves are caused and travel through earth and rock. Under varying conditions, there is a time delay between the point of detonation and where the transducer is located so that a plot is obtained.

10.6.4   Physics

As an example, consider nuclear physics. There are many things you may want to know, but you cannot tell by smell or feel how to use nuclear instruments. As an illustration, a high-energy chunk of uranium looks harmless and feels harmless, but it may kill you, so we cannot use human sense to make determinations of its value, harmless characteristics, and so on. Therefore, instruments are required to sense the properties present. The instruments themselves cannot tell you what is good or bad. The brain must apply human knowledge. It takes the mind of the human being to interpret test results, in spite of the fact that instruments may even talk.

10.6.5   Other Uses

Interpretation of the readout device is required to provide information for humans. And today we have computers that can tell what is good or bad. But these computers can make mistakes, too. Some computers can check themselves by sending to the input circuit some program information and comparing it to a fixed answer or solution, and the output will turn itself off if it does not match. They are getting smaller and less costly; the minicomputer is an example. In reality, the computer operates in the whole spectrum of sizes, from the computer operation of a large chemical oil checking plant down to the operation of a small computerized wristwatch.

In the chemical plant, as products are packed, they are placed on pallets for shipping. Formerly, they were hand stacked, and small companies still hand stack. Large companies have palletizing machines. As the cartons accumulate on an accumulator belt, the palletization machine calls for the producer. In turn, it spaces, orients, and places the cartons on the pallet in a certain pattern and orientation for proper interlocking to cause proper stability of the pallet load. It uses photoelectric sensors, mechanical sensors, and a simple computer system (prearranged points on selector relays).

Simple computers are used in vehicle traffic or product traffic control. Large computerized trains are becoming important, especially in the city where there is a high density of trains. In Japan computer-operated trains run at speeds in excess of 150 miles per hour. The Japanese also use computer-operated boats. Control systems (emergency braking and steering) for cars that are also coming into use.

In process control, relays and solenoids are primitive output devices. We now use much more sophisticated devices such as the CRT, printer, and numerical readout tubes. A generator can be an input or output device. As things progress, they get more complex. For example, originally to light a lamp, one used a knife switch. Then we used a rotary switch, then a toggle switch, then a tumbler switch; then a mercury switch, and then a touch (microswitch). Today we have the electronic touch switch with no moving parts.

If the output devices are solenoids and relays, input devices would be limit switch (which is a device to stop an action or motion). The microswitch is used in a similar manner. Other input devices would be electrodes and gas-filled electrodes.

The output could even be a feedback circuit (abling and disabling circuits governing a larger picture).

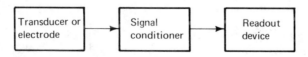

FIGURE 10.4
Block diagram showing analog input—output devices.

10.7
BLOCK DIAGRAMS FOR GENERALIZED INPUT–OUTPUT SYSTEMS, INCLUDING THE CONCEPT OF MODEM

In the Figure 10.4 through Figure 10.15, we present a complete input–output system by using block diagrams. In Figure 10.4, the input–output block diagram is represented by a sensor, which may include a transducer or electrode that is fed to a signal conditioner, which amplifies the signal. This analog signal is then fed to a readout device, such as an oscilloscope or recorder.

In Figure 10.5, the analog signal from the signal conditioner is fed to an analog computer, which can be connected to a digital voltmeter by means of an A/D converter.

In Figure 10.6, the analog signal is digitized via an A/D converter and returned to an analog signal via a D/A converter fed to an analog computer.

In Figure 10.7, the digital computer is fed to an acoustic coupler and a telephone line directly or via a telephone. The *modem* can be placed after

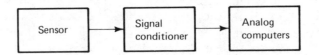

FIGURE 10.5
Block diagram showing input–output devices.

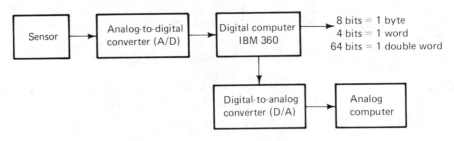

FIGURE 10.6
Block diagram showing input–output devices.

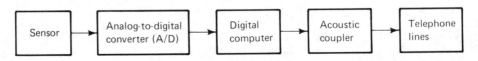

FIGURE 10.7
Block diagram showing input–output devices.

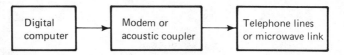

FIGURE 10.8
Block diagram showing input—output devices.

the digital computer; it is a modulator demodulator. This hookup encodes digital information data on voice frequencies for transmission and decodes the data at the receiving end. There may be more than one modem in a digital transmission system. In Figure 10.7, the acoustic coupler serves the same purpose.

In Figure 10.8, the digital information is fed to a modem or acoustic coupler and then to a telephone line or a microwave link.

In Figure 10.9, a multiplexer is used to process the data to an A/D converter and then to a PDP-8 computer. The digital information is fed to a line printer, tape deck, and magnetic disk.

In Figure 10.10, we see a complex block diagram of a data-collection and data-processing system of kinematic measurements in walking. The data are recorded via transducers and fed to a signal-conditioning device. A tape recorder is used for data playback and processing and this information can be fed to a large computer via an A/D converter. A teleprinter is also fed to the computer, and the output of the computer is fed to a line printer. A paper-tape punch is fed to a paper punch reader, and then the digital signal is sent to a Gerber plotter and is processed as graphical data.

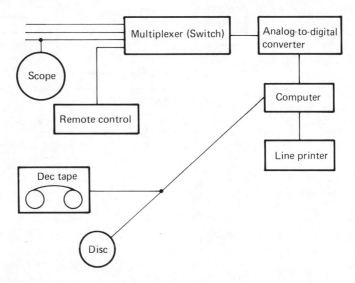

FIGURE 10.9
A computer diagram scheme showing input/output devices.

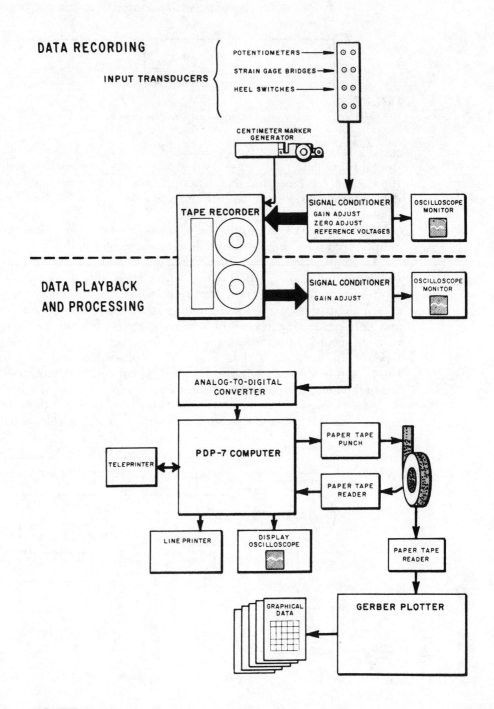

FIGURE 10.10
Block diagram of data-collection and data-processing system. (*Courtesy of Bulletin of Prosthetics Research,* BPR 10-15, p. 21, Spring 1971)

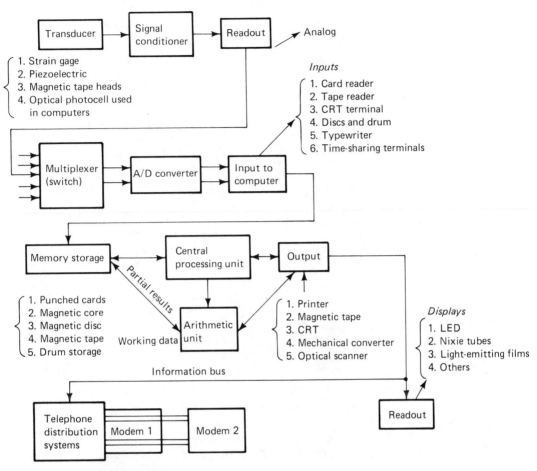

FIGURE 10.11
Generalized I–O system.

In Figure 10.11, the block diagram includes the information required to convert an analog signal to digital form and shows all the parts required. The digitized signal from the computer is fed to computer output devices as well as computer display devices.

In Figures 10.12 through 10.15, application devices are illustrated. The blocks show how the analog signals are digitized. Such processes are quite complex.

10.8
ILLUSTRATIVE PROBLEMS FOR
INPUT–OUTPUT DEVICES

Let us consider a Wheatstone bridge with a sensor in one leg, as shown in Figure 10.16. This shows a bridge amplifier with three resistors and a

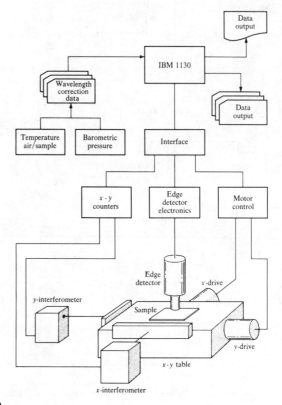

FIGURE 10.12
Block diagram of high-precision measuring
system. (From *IBM Journal of Research and
Development,* vol. 17, no. 6, Nov. 1973. Courtesy
of International Business Machines Corp. Copyright 1973 by International Business Machines
Corporation; reprinted with permission)

strain-gage sensor that has a resistance $R + \Delta R$, where ΔR is the increase in
resistance due to the pressure applied. The voltage

$$V_{a_1} = \frac{R}{R+R} V_b = \frac{R}{2R} V_b = \frac{1}{2} V_b \tag{10.1}$$

The voltage

$$V_{c_1} = \frac{R}{R+R+\Delta R} V_b = \frac{R}{2R+\Delta R} V_b \tag{10.2}$$

$$= \frac{R/R}{(2R/R)+(\Delta R/R)} V_b = \frac{V_b}{2+(\Delta R/R)} = \frac{V_b}{2+\delta}$$

where $\Delta R/R$ is denoted as δ.

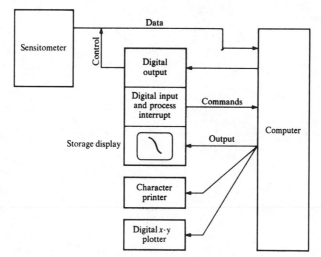

FIGURE 10.13
System diagram for computer-assisted sensi-
tometry. (From *IBM Journal of Research and
Development,* vol. 17, no. 6, Nov. 1973. Courtesy
of the International Business Machines Corp.
Copyright 1973 by International Business Machines Corporation; reprinted with permission)

In each case, the voltage-divider rule is used, since the current is not known, to find the voltage V_{ca}:

$$V_{ca} = +V_{c_1} - V_{a_1} = \frac{V_b}{2+\delta} - \frac{V_b}{2+\delta} - \frac{V_b}{2}$$

$$= V_b \left[\frac{1}{2+\delta} - \frac{1}{2} \right]$$

$$= V_b \left[\frac{2-(2+\delta)}{2(2+\delta)} \right]$$

$$= V_b \left[\frac{-\delta}{4+2\delta} \right] \tag{10.3}$$

$$= V_b \left[\frac{-\delta/4}{1-(\delta/2)} \right]$$

$$= V_b \left[\frac{-\delta}{4} \right] \left[\frac{1}{1-(\delta/2)} \right]$$

$$= -\frac{\delta}{4} V_b \left[\frac{1}{1-(\delta/2)} \right]$$

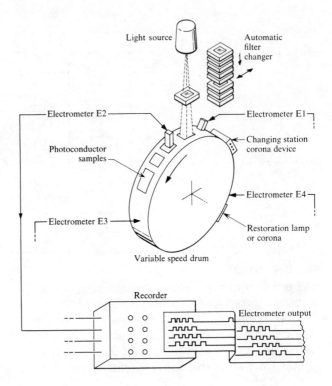

Light source

Automatic filter changer

Electrometer E2

Electrometer E1

Changing station corona device

Photoconductor samples

Electrometer E4

Electrometer E3

Restoration lamp or corona

Variable speed drum

Recorder

Electrometer output

FIGURE 10.14
Experimental apparatus for exposure sensitome-
try. (From *IBM Journal of Research and Development,* vol. 17, no. 6, Nov. 1973. Courtesy of the
International Business Machines Corp. Copyright
1973 by International Business Machines Corporation; reprinted with permission)

Since in practice, $(\delta/2) << 1$

$$V_{ca} = -\frac{\delta}{4}V_b \qquad (10.4)$$

If the differential dc amplifier has a gain of K, then the output, V_o, can be
expressed as

$$V_o = KV_{ca} = -K\frac{\delta}{4}V_b \qquad (10.5)$$

If the output, V_o, is again fed into another amplifier with a gain of K_1, the
new output V_{o1} becomes

$$V_{o1} = -KK_1\frac{\delta}{4}V_b \qquad (10.6)$$

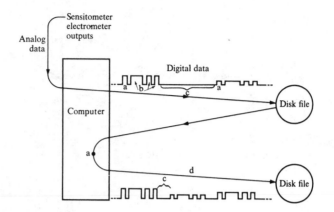

FIGURE 10.15
Preliminary data-reduction task executed automatically on completion of data acquisition. (a) Locate leading edge of first sample on each drum revolution, note file address, calculate drum speeds; (b) calculate location of other samples using sample size table; (c) discard; (d) store data and note location with respect to electrometer drum revolution and sample. (From *IBM Journal of Research and Development,* vol. 17, no. 6, Nov. 1973. Courtesy of the International Business Machines Corp. Copyright 1973 by International Business Machines Corporation; reprinted with permission)

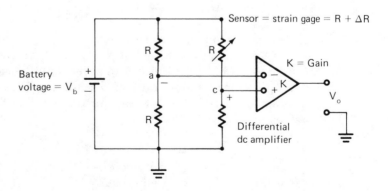

FIGURE 10.16
Bridge amplifier.

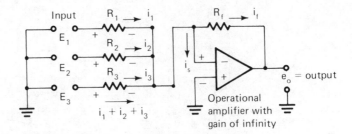

FIGURE 10.17
An analog summing amplifier.

It is interesting to note here that $R + \Delta R$ is the same as

$$R \left[1 + \frac{R}{\Delta R} \right] \quad \text{or} \quad R \left[1 + \delta \right]$$

In a practical problem in the Wheatstone bridge configuration of Figure 10.16, V_b, battery voltage, is taken as a few volts. Let us assume 5 V. The output voltage V_{ca} could be -10 mV. Using Eq. (10.4), we find that

$$V_{ca} = -V_b \frac{\delta}{4}$$

$$10 \times 10^{-3} = \frac{5}{4} \delta$$

$$= \frac{40 \times 10^{-3}}{8} = 5 \times 10^{-3} = \frac{5}{1000} = \delta$$

a quantity that is dimensionless. What does this mean? If $R = 500 \ \Omega$, then ΔR, the increase in resistance of the strain gage sensor, turns out to be

$$\Delta R = R \times \delta = 500 \times \frac{5}{1000} = 2.5 \ \Omega$$

Thus ΔR turns out to be a *very small resistance*.

Using Ohm's law and Kirchhoff's current law, one can develop an analog summing amplifier, which is another input–output device with an idealized operational amplifier (oP-aMP) with a gain of infinity.

Consider the circuit shown in Figure 10.17. Using Kirchhoff's current law, we get

$$i_1 + i_2 + i_3 + i_f = 0 \tag{10.7}$$

By replacing these current parameters by their voltage and resistance values, we find the following equation:

$$\frac{e_1}{R_1} + \frac{e_2}{R_2} + \frac{e_3}{R_3} + \frac{e_o}{R_f} = 0 \tag{10.8}$$

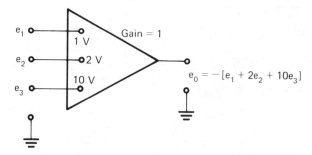

FIGURE 10.18
An analog summing inverting amplifier block.

Rearranging terms, we get

$$e_o = -\frac{R_f}{R_1}e_1 - \frac{R_f}{R_2}e_2 - \frac{R_f}{R_3}e_3 \tag{10.9}$$

Using $R_1 = 1$ MΩ, $R_2 = 500{,}000$ Ω, $R_3 = 100{,}000$ Ω, and R_f, the feedback resistor, across the operational amplifier is 1 MΩ, we arrive at

$$e_o = -\frac{1 \times 10^6}{1 \times 10^6}e_1 - \frac{1 \times 10^6}{0.5 \times 10^6}e_2 - \frac{1 \times 10^6}{0.1 \times 10^6}e_3$$

$$e_o = -[e_1 + 2e_2 + 10e_3] \tag{10.10}$$

This inverting amplifier can be redrawn as in Figure 10.18.

It is assumed that computational circuits can be designed using summing inverting amplifiers. We can also use operational amplifiers to design an analog integrator. Such a device is useful in computing, signal-processing, and also signal-generating systems.

Let us now draw an analog inverting integrator (see Figure 10.19). The current in the resistor, R, is i_R and can be calculated as follows:

$$i_R = \frac{e_{in} - e_2}{R} = i \tag{10.11}$$

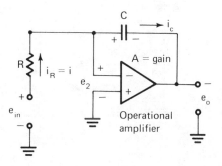

FIGURE 10.19
An analog inverting integrator.

The current in the capacitor may be found from the following relationship:

$$e_2 - e_o = \frac{1}{C} \int_0^t i\,dt = \frac{1}{RC} \int_0^t (e_{in} - e_2)\,dt \tag{10.12}$$

We can also say that

$$e_2 = \frac{e_o}{A} \tag{10.13}$$

Since $A \to$ infinity then $e_2 \to 0$ and

$$e_o = -\frac{1}{RC} \int_0^t e_{in}\,dt \tag{10.14}$$

Knowing that $e_2 = 0$ for the above case, we could also say that

$$i_c = -C\frac{de_o}{dt} \quad \text{and} \quad i_R = \frac{e_{in}}{R} \tag{10.15}$$

Since $i_c = i_R$ we see that

$$-C\frac{de_o}{dt} = \frac{e_{in}}{R} \quad \text{or} \quad de_o = \frac{e_{in}}{RC}\,dt \tag{10.16}$$

Finally, this results in

$$e_o = -\frac{1}{RC} \int_0^t e_{in}\,dt \tag{10.17}$$

which is the same as Eq. (10.14).

Equation (10.17) states that the output signal is the integral of the input signal. Interchanging R and C in Figure 10.19, we obtain the analog inverting differentiator shown in Figure 10.20.

Since the current flowing through the feedback resistor, R_f, is equal to

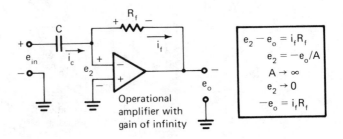

FIGURE 10.20
An analog inverting differentiator.

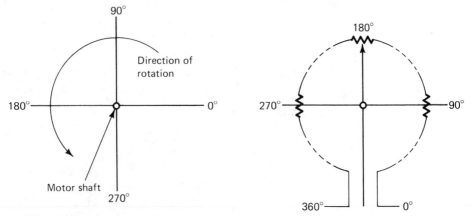

FIGURE 10.21
Potentiometer device used as a rotational displacement transducer.

the current through the capacitor, we obtain

$$i_c = C \frac{de_{\text{in}}}{dt} \tag{10.18}$$

and

$$i_f = -\frac{e_o}{R_f} \tag{10.19}$$

Since i_c and i_f are equal, we arrive at the final result:

$$-R_f C \frac{de_{\text{in}}}{dt} = e_o \tag{10.20}$$

Equation 10.20 says that the output is the differential of the input signal.

We can develop the output as the function of the input for a potentiometric device. Rotational displacement is usually connected with a motor shaft with a 360° rotation, and what we want to measure is the angle of displacement with respect to some reference point at any given point in time. Figure 10.21 shows the potentiometer used as a rotational displacement, while Figure 10.21 shows the output voltage as a function of shaft rotation. If we take a linear taper potentiometer with a rotation of 360° and apply a positive potential E at one end and ground the other end, we have created a reference voltage divider. If we then connect the pot shaft to the motor shaft at the desired middle resistance, we have a resulting output voltage proportional to the angle of the motor shaft. The calibration of the pot to the motor is quite simple. What is desired is that the pot be set at one-half its value for 180° of motor shaft rotation. It will then have a minimum resistance at a shaft angle of 0° and a maximum resistance at a shaft angle of 360°. The output of the potentiometer can then be sent directly to a recorder or other readout device.

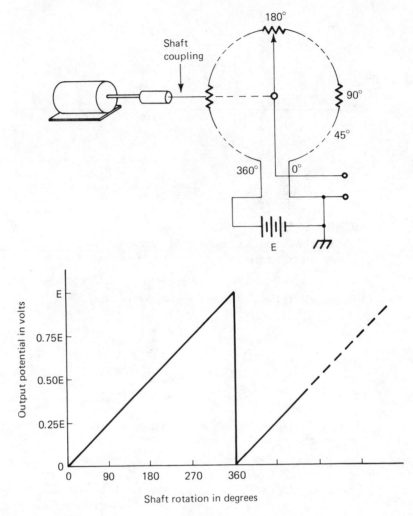

FIGURE 10.22
Voltage output as a function of shaft rotation.

We could also state that the motion of the movable slider results in resistance changes that may produce a linear, sine, cosine, logarithmic or hyperbolic wave shape.

Let us now consider a position servo as shown in Figure 10.23. The output potentiometer measures the output shaft position and converts this into a voltage. We can write the voltage equation:

$$e_o = K_o\,\Theta_s \qquad (10.21)$$

where Θ_s is the output shaft angle in radians and K_o is a constant that converts the right side of the equation to voltage and would be expressed in volts per radians. K_o can be found by dividing the total output voltage by

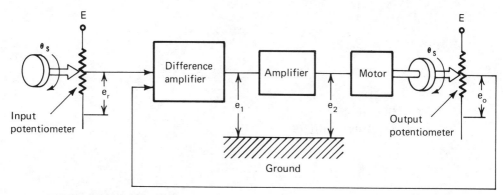

FIGURE 10.23
A position servo. The difference amplifier has a
gain of A_1 and the other amplifier a gain of A_2.

the maximum rotation of the potentiometer and can be rewritten in equation form as follows:

$$K_o = \frac{E}{\Theta_{s \text{ max}}} \quad \text{in volts/radian} \tag{10.22}$$

The input knob position can be converted to a signal voltage with an identical potentiometer as used for the output potentiometer. We can therefore find e_1, noting that the input signal to the difference amplifier has two signals, and the output of the difference amplifier is the difference between the two input signals multiplied by a gain, A_1. Mathematically, we can evaluate e_1, the output of the difference amplifier. The output, e_1, can be determined by the following equation:

$$e_1 = A_1 (e_o - e_r) = A_1 K_o (r - \Theta_s) \tag{10.23}$$

At the motor terminals the output voltage is e_2 and can be given as

$$e_2 = A_2 e_1 = A_1 A_2 K_b (r - \Theta_s) \tag{10.24}$$

Analog symbols are very useful during analog computer operation. Figures 10.24 through 10.27 illustrate some of the symbols used.

We consider a mechanical system, which we can represent by a dashpot in series with a spring, which represents a viscoelastic tissue material.

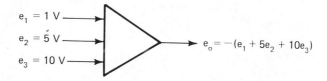

FIGURE 10.24
Analog inverting summer.

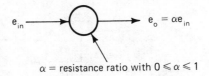

$$e_{in} \longrightarrow \bigcirc \longrightarrow e_o = \alpha e_{in}$$

α = resistance ratio with $0 \leqslant \alpha \leqslant 1$

FIGURE 10.25
Analog potentiometer.

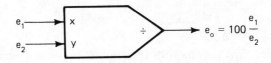

$$e_o = 100 \frac{e_1}{e_2}$$

FIGURE 10.26
Analog multiplier.

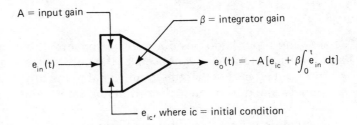

A = input gain

β = integrator gain

$$e_o(t) = -A[e_{ic} + \beta \int_0^t e_{in}\, dt]$$

e_{ic}, where ic = initial condition

FIGURE 10.27
Analog integrator.

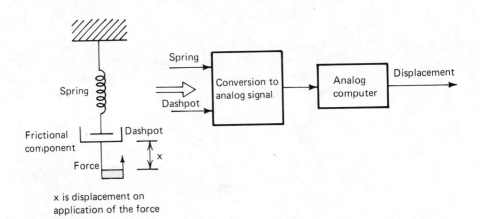

x is displacement on
application of the force

FIGURE 10.28
Representation of an analog computer.

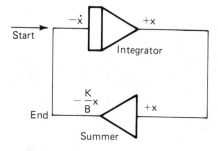

The integrator box on the right shows:

$$\dot{x} = -\frac{K}{B}x$$

$$(-x) = -\left[-\int \frac{K}{B}\,dt\right]$$

Integrator

FIGURE 10.29
An analog computer to represent a spring and a dashpot.

Assume that we rigidly tie one end down and apply a force at the end and release the elements. This can be represented as in Figure 10.28.

An equation can be written for Figure 10.28, which can be stated as follows: force exerted on the dashpot + force exerted on the spring = 0. Thus,

$$F_d + F_s = 0 \tag{10.25}$$

Mathematically, we say that

$$B\dot{x} + Kx = 0 \tag{10.26}$$

where

B = frictional element

$\dot{x}$ = derivative of the displacement with respect to time

K = force per unit of displacement

x = displacement on application of the force

We can rewrite Eq. (10.26) as

$$B\dot{x} = -Kx \tag{10.27}$$

Dividing each side of the equation by B, we obtain the following result:

$$-\dot{x} = -\left(-\frac{K}{B}x\right) \quad \text{or} \quad -\dot{x} = \frac{K}{B}x \tag{10.28}$$

An analog computer symbolic representation of this follows with initial conditions that x at time equal to 0 is 0 (see Figure 10.29). Integrator symbols may be confusing since there is no standardization for their representation.

10.9
INPUT–OUTPUT DEVICES IN HOSPITALS

Computer measurements today play a major role in the acquisition, storage, and analysis of hospital data. Large computers are installed in many hospitals and medical centers to handle monitoring of patients, processing of data, billing of patients, pharmacy control, clinical lab measurements, and so on.

The first 72 hours in the hospital environment after a heart attack are critical. Physiological monitoring and patient care are emphasized during this period. If a competent medical professional can be at the bedside within a few minutes after the onset of a cardiac catastrophe, it often is possible to resuscitate the coronary patient. As it is impossible for a physician or nurse to be with the patient continuously, computer systems are

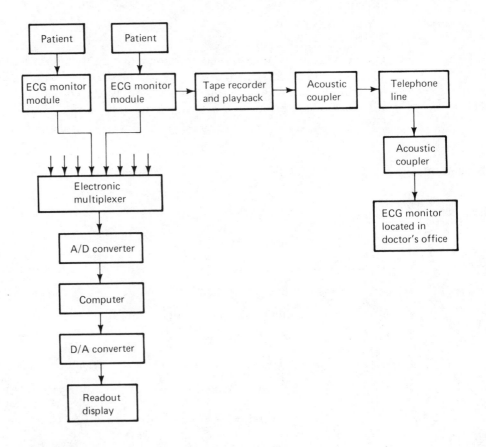

FIGURE 10.30
Block diagram of a computer system that can be linked by telephone lines.

being used in major hospitals to continuously monitor a large number of cardiac patients simultaneously. These patient-care systems are equipped with a warning system to alert medical personnel on duty of any dangerous change in the patient's condition.

In some hospitals, portable computers are also used for the intensive-care monitoring of patients. A growing array of instruments is available for signal analysis via analysis of the frequency components of the signal, called *spectrum analysis*. Still other types of computers are used to teach medical students. Computer measurements therefore are relatively important tools for the clinician and the paramedics.

One digital computer may process and analyze hundreds of different patients simultaneously by a technique known as time sharing. The results to be analyzed may be at a remote location and are linked to the computer by telephone lines. A block diagram of such a computer system is shown in Figure 10.30.

Some important definitions with reference to medical computer systems are the following:

1. **Modem:** A modem is a device that encodes serial digital information for transmission over normal communication systems, such as telephone, cable, or microwave link. The information is received and fed into a receiving modem, which decodes the information such that it can be fed into tape, printer, or the like.

2. **Acoustic coupler:** A device (modem) that converts the digital signal into a voice frequency to be transmitted over the telephone lines. (See Figure 10.30.)

3. **Multiplexer:** A device that processes the electronic signal such that it can be combined with other signals and transmitted. This device allows us to transmit hundreds of different signals over one pair of telephone lines (or cable). At the receiving end, each of these multiplexed signals can be separately decoded into useful information. (Refer to Chapter 4 for more details. Also refer to Figure 10.30.)

A microprocessor is a semiconductor device that can perform arithmetic, logic, and decision-making operations under the control of a set of instructions stored in a memory device. It can also communicate with a set of peripheral devices via some input–output structure. Large-scale integrated circuits including a central processing unit (CPU) on a single chip of a semiconductor material, memory, input–output circuits power supply, and so on, are needed to turn the microprocessor into a microcomputer (see Chapter 15 for further details).

The major significance of a microprocessor (MPU or μP) is that all the elements of a minicomputer have been reproduced on a single large-scale integrated chip of silicon, usually not much larger than a pencil eraser. Due to its small size, it is presently possible using microprocessor technology to design electronic instruments to measure and give a digital readout of physiological vital signs. Microprocessors are also in blood-pressure moni-

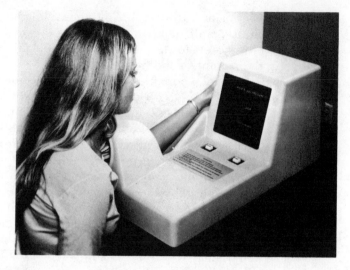

FIGURE 10.31
A commercial automated blood-pressure measuring device that uses a computer. (Courtesy of Medical Monitors, Inc.)

toring measurements. Figure 10.31 shows a commercial automated blood-pressure measuring device that uses a microprocessor.

Microprocessors can also measure the electrical impedance of the nervous system and can serve as program controllers for generating nervous system stimuli and for measuring nervous system responses. A portable microprocessor-controlled memory can also monitor the cardiopulmonary rates of high-risk infants with breathing disorders.Microprocessor control is also used in real-time sector-scan ultrasonography systems. There are many uses for microprocessors in industry and clinical medicine that have not yet been realized.

10.10
REVIEW QUESTIONS

1. List five computer input devices.
2. List five computer output devices.
3. Discuss the difference between an analog and digital signal.
4. Discuss the difference between an analog and digital computer.
5. Discuss the purpose of a modem.
6. Draw a block diagram to measure a generalized input–output system.

7. Given:

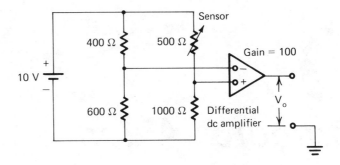

Find V_o.

8. Discuss the analog summing inverting amplifier and analog inverting integrator.

9. Discuss the analog inverting differentiator.

10. Discuss a position servo and draw a block diagram.

11. Discuss an analog computer that represents a spring and a dashpot.

12. Draw an analog multiplier.

13. Draw a mass spring–damper system.

14. Discuss input–output devices in hospitals.

15. Discuss a definition for microprocessors in the hospital.

10.11
REFERENCES

1. Maxwell G. Gilliland and analog computer staff, *Handbook of Analog Computation* (including Application of Digital Control Logic), Systron Donner Corp., Concord, Calif., June 1967.

2. Patrick Garrett, *Analog I/O Design and Acquisition: Conversion: Recovery*, Reston Publishing Co., Reston, Va., 1981.

3. Harry Thomas, *Handbook of Biomedical Instrumentation and Measurement*, Reston Publishing Co., Reston, Va., 1974.

4. Jerome E. Olefsky, and George G. Rutlowski, *Microprocessor and Digital Computer Technology*, Prentice-Hall, Inc., Englewood Cliffs, N.J., 1981.

11
Computer-Aided Instrumentation Systems

11.1
INTRODUCTION

Computer-aided or -assisted instrumentation systems require the use of simple and complex computers in electronics, medicine, research, chemistry, environmental problems, industry, and many other applications. We are now living in the computer age. For example, such systems are used in the following areas:

1. Electronic manufacturing, known as computer-aided manufacturing or CAM.

2. Drafting, known as computer-aided drafting or CAD.

3. Research in small laboratories, called MINC by the Digital Equipment Corp.

4. Computer-assisted chemistry.

5. Computer-assisted integrated-circuit design.

6. Computer-assisted management of occupational health.

7. Computer-aided medical systems.

Computer-aided systems and/or computer graphics are presently generated with the use of microprocessors and the LSI and VLSI circuit technology available and discussed in Chapter 16. The material that follows will highlight some of the most recent developments in the areas of computer-aided instrumentation systems.

Electronic instrumentation using microcomputers assists designers and production engineers in the testing of complicated automated electronic equipment. Computers are built into measuring instruments and can be interfaced with calculators and other data-processing systems through bus-connected systems.

11.2
COMPUTER-AIDED DESIGN AND COMPUTER-AIDED MANUFACTURING (CAD/CAM) IN ELECTRONICS[a]

Generally, CAD/CAM refers to the use of computers to assist people in all of the processes necessary to translate a product from concept into the final physical product. CAD/CAM plays a key role in areas such as design, analysis, detailing, documentation, NC programming, tooling, fabrication, assembly, and quality testing.

The impact of industry automation on our standard of living and quality of life has been equally dramatic. The ability to design extremely complex electromechanical systems (such as airplanes), to package hundreds of thousands of electronic circuits in a chip, and other such technological "miracles" would not have been possible without the use of CAD/CAM.

Turnkey CAD/CAM systems made their debut in the early 1970s. Readily integrated into the design and drafting rooms of printed circuit board and integrated circuit manufacturers, they rapidly became one of the industry's main design and production aids.

More recently, CAD/CAM has moved into the realm of wire and cable systems, a very natural development considering that all electronics, from the smallest IC to the largest PC board, eventually have to be connected to some type of power source and power outlet, all through wires or cables.

In the discussion of CAD/CAM that follows, we will be referring to three major areas of electronics: PC, IC, and wiring systems. At the risk of oversimplification, let us attempt to generalize about what goes into a CAD/CAM system configured for electronic applications.

---

[a] The material in Sections 11.2 and 11.3 are from *Computer Aided Design/Computer Aided Manufacture—The Decade Ahead in Electronics*, by Timothy I. Ristine, product manager, Computervision Corp., Bedford, Mass. The material appeared in Electro/81 and was given at IEEE Section 22 at the Sheraton Center Hotel, April 7–9, 1981. This material is reprinted here courtesy of Timothy Ristine, Andrew McIntosch, and the Computervision Corp.

A typical stand-alone turnkey CAD/CAM system includes a central processing facility with a minicomputer and mass memory for programs and drawing storage, as well as system software. A terminal or work station will consist of a cathode-ray tube, a graphic tablet, a function keyboard, an alphanumeric keyboard, possibly a digitizer, and a hard-copy device for printing alphanumeric or graphical output.

While the system configuration will vary depending on the application, it will probably include one or more design terminals, all sharing the computer's memory storage and processing capabilities. A number of engineers and designers can work simultaneously at various terminals, each on a different phase of development, such as design, engineering analysis, drafting, or manufacturing—for a single product or for many different products.

The heart of any CAD/CAM system is the design terminal or work station. Here the operator interacts with the system to develop a product design in detail, monitoring his or her work constantly on a TV-like graphics display. By issuing commands to the system, then responding to system prompts, the operator creates the design—manipulating, modifying, refining—all without ever having to draw a line on paper or recreate an existing design element. Once the design is to his or her liking, the designer can command the system to make a hard copy or generate a computer tape to guide computer-controlled machine tools or test equipment.

As a design is developed, the computer graphics system is accumulating and storing physically related data that identify the precise location, dimensions, descriptive text, and other properties of every design element that helps to define the new part or product. Using this design-related data, the system helps the operator to do complex engineering analyses, generate special lists and reports, and detect and flag design flaws before the part reaches manufacturing.

The operator interacts by pointing an electronic pen to a premarked menu portion of the touch-sensitive drawing tablet. This is the primary means of issuing commands to the system to add, delete, move, magnify, rotate, copy, or otherwise manipulate the entire design or any part of it.

Computer-aided design provides a powerful tool to reduce cycle time and cost in the creation of PCBs, ICs, and wiring diagrams. It greatly speeds up those work steps where repetition dominates and creativity is not required. CAD does not attempt to automate creative processes that require human intelligence. Instead, it enables the skilled designer or engineer to use the power of the computer to do the noncreative but very time consuming operations.

As a part is designed on the system, its physical dimensions are defined along with the properties of the various components or entities in the design. These data, also stored in the computer's memory, can later serve many other nongraphical needs. For example, the operator can use part-number data to help generate bills of materials for the purchasing department, as well as to generate computer tapes to guide NC machine tools, quality control, and other testing equipment.

11.3
AN ECONOMIC NECESSITY

What is there about electrical and electronics design and manufacturing that has made CAD/CAM an economic necessity for so many companies? For one thing, consider the trend toward packing more and more circuitry into less and less space. VLSI technology is rapidly approaching 1 million devices per logic chip. Current trends in PC design are toward larger boards with greater circuit density—from 1.0 to 0.40 in.^{2} per IC.

As circuit size and complexity have increased, there has been a decrease in productivity for a wide variety of layout techniques, layout being the most expensive and time-consuming portion of the development cycle. Computer-aided design and drafting are indispensable to increasing layout productivity. The improved throughput CAD makes possible enables the manufacturer to meet the severe time and cost constraints imposed by today's fierce competition. Only with CAD can engineers and designers of complex circuitry hope to achieve the high degree of accuracy—and ultimately product quality—vital to staying competitive.

Large, complex circuits require far too much preproduction checking for engineers and designers to use manual methods. For example, a typical wiring system in just one motor control system might include: 200 motors, 1500 fuses, 200 transformers, 5000 relays and contacts, 200 power supplies, and 3000 switches. It would take an impossible amount of time to manually check coil and contact cross-reference numbers among hundreds of sheets and to manually generate the voluminous report required. CAD systems can do both cross-referencing and design rules checking far faster and more accurately.

Equally important, the CAD/CAM database provides a unique standardized, centralized source of all the diverse information, graphic and nongraphic, relating to a product under development. This database is automatically updated, consistently accurate, and available, usually within minutes, for design, engineering, and manufacturing purposes.

Another major problem throughout the electronics industry is how to find and keep skilled engineers and designers. CAD/CAM enables a company to get significantly greater design and engineering throughput from the same number of people.

One way that CAD/CAM vendors are making this possible is by steadily improving the man–machine interface, making more and more powerful capabilities readily available to all users, whether in the drafting room, the engineering lab, or the manufacturing floor.

The CAD/CAM system of the future will, itself, have greater compute power: more powerful processors, much more mass memory, and larger disks. These processors will be tied together in network system configurations that will have a host database management system and multiple satellites (terminals) accessing that database.

At Computervision, they are developing an integrated approach to analysis; coupling graphics to simulation is a reality. Customers will be able to transfer compute-bound analysis tasks, not to a mainframe or time-sharing service, but directly to an *analytic processing unit* (APU). Rapid graphical feedback will be available to both designers and engineers, enabling them to more easily determine if the proposed circuit will perform properly before the actual design phase.

Terminal operators will be able to achieve faster analysis and greater throughput by working directly from the schematic, not indirectly through batch or remote-batch techniques. Large compute-bound tasks like routing and placement will be done in the analytic processing unit, as well as logic simulation and circuit analysis. Meanwhile, smaller computers networked into the system will be directly available to terminal operators for more graphically oriented tasks that tie the intelligence and creativity of people to the power of the computer.

11.3.1 Networking

A trend of the 1980s will be that companies like Computervision will produce CAD/CAM computers that will be part of an integrated network. There will be more networking software linking graphics processors to mainframes used for sophisticated manufacturing information systems.

As more and more corporations begin to introduce CAD/CAM systems into their design, drafting, and manufacturing organizations, the need arises for communicating both graphic and nongraphic information among the systems's users. Networks of interconnected systems will allow large numbers of CAD/CAM users to access a common, integrated engineering–design–manufacturing database on either a department, division, or corporate basis.

11.3.2 Data Management

A major trend in CAD/CAM software in the 1980s will be to define an integrated database management system that addresses the needs of both engineering and manufacturing. Starting from the basic design data input at the beginning of the development cycle, this database management system will keep adding, coordinating, and integrating all the information about the emerging product. Marketing requirements, engineering and geometric models, part costs and availability, available machine tools, and labor—all this, and more, must be contained in the database.

11.3.3 Interdisciplinary Applications

Electronics design is heavily influenced by mechanical packaging factors. Consider the three-dimensional mechanical design involved in the development of the chassis and cabinetry to enclose the electronics for a car engine or household appliance. In the coming decade, there will be greater merging of mechanical and electrical technology. Manufacturers will not produce just the electronics, but also the electromechanical system associated with it. Those manufacturers will be demanding CAD/CAM systems that provide multiapplication capabilities. Already many Computervision customers are running PC, mechanical, and wiring system design tasks on the same hardware and using the same common database.

Another prime example of an interdisciplinary application approach lies in the wiring diagram area. Computervision is currently developing software to expedite the routing of wire and cable through three-dimensional space—a problem of great interest to designers of airplanes, power supplies, and motor control centers.

The emergence of the gate array, a custom approach to logic design, is but one example of the growing interest among electronics manufacturers in emerging IC and PC design requirements via CAD/CAM.

So far we have concentrated on trends in CAD/CAM that will expand and refine its design and preproduction engineering capabilities. But in the decade ahead, CAD/CAM will also be moving downstream and become more and more involved with testing and manufacturing finished product. This march toward the "automated factory" will be made possible through CAD/CAM because of two factors previously discussed: more compute power in the system, and more nongraphical data in the database.

11.3.4 Computer-aided Testing

Currently, testing equipment requires large amounts of nonrecurring engineering in the form of programming. With such programming running into hundreds of thousands of dollars, it is not economically feasible to apply automatic testing except in very large runs of components.

In the coming decade, however, the programming of automatic test equipment will simply fall out of the design of the product on the CAD/CAM system. The system will directly output the programming required to drive the test equipment. No skilled programmers will be required, so the cost will be nominal.

11.4
CAD/CAM IN PCB MANUFACTURING [b]

The printed circuit board (PCB) is the workhorse component for electronics manufacturers. A sandwich of plastic, glue, and delicate wiring, the PCB

---

[b] The following section is from Happy Holden, Engineering Manager, "CAD/CAM in PCB Manufacturing." Courtesy of Hewlett-Packard Co., Palo Alto, Calif. Reprinted from *ASSEMBLY ENGINEERING*, July 1981. By permission of the Publisher, © 1981 Hitchcock Publishing Co. All rights reserved. Also appears in *Circuit Manufacturing*, May 1980, © 1980 by the Morgan Publishing Co. Reprinted here with permission.

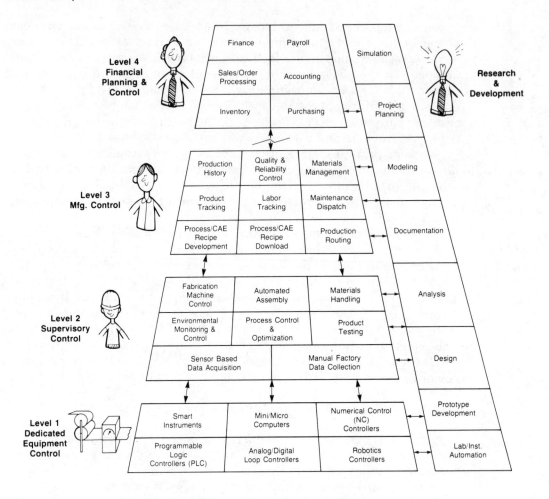

Computers in Manufacturing

FIGURE 11.1
The hierarchy of computers dedicated to specific tasks. (Courtesy of Hewlett-Packard Co., Palo Alto, Calif.)

makes it possible to interconnect thousands of circuits in a relatively small space. While the technology of PCBs is well understood, today's increasingly complex integrated circuits have required the boards themselves to increase in complexity. Their fabrication has become a costly, high-technology process that requires professional engineers, capable management, trained personnel, and expensive machinery. The ratio of the cost of a PCB to the cost of a completed unit—the I/O (input–output) coefficient—has grown to more than 5%.

To support its broad line of computers and electronic instrumentation, Hewlett-Packard Company requires more than a thousand different types of PCBs. Many of them have six or eight layers, while the commercial state-of-the-art is four. To ensure the continuous high quality and performance that their complex systems require, they make 90% of their own PCBs. To keep costs down while staying on the leading edge of technology, HP has built a new PCB plant in Sunnyvale, California, that employs computer-aided manufacturing (CAM) techniques, based on the HP 1000 and HP 3000 families, throughout the fabrication cycle. The new plant represents one of the most extensive uses of CAM in PCB manufacture. The CAM strategy is based on a hierarchy of computers (see Figure 11.1) dedicated to specific areas.

Hewlett-Packard combines computer-aided design and manufacture to improve the quality and performance of complex printed circuit boards. Figure 11.2 shows a technician monitoring a chemical process line on a computer-generated diagram of board progression.

FIGURE 11.2
Technician monitors chemical process line on computer-generated diagram of PC board progression. (Courtesy of Hewlett-Packard Co., Palo Alto, Calif. Reprinted from *ASSEMBLY ENGINEERING,* July 1981. By permission of the Publisher, © 1981 Hitchcock Publishing Co. All rights reserved.)

Automation of PCB fabrication began at Hewlett-Packard in 1971 when they built automatic equipment to increase production at their Palo Alto, California, plant. The new Sunnyvale plant is their response to the need for a completely new facility to support the next generation of digital products. The design goals were high: high technology, high quality, and high flexibility to meet future requirements.

From the beginning Hewlett-Packard felt that the secret lay in the use of CAM: the linking of information retrieval with process control. The plan was to preserve information by letting computers talk to computers and to improve yield and quality by refining the process through subtle manufacturing changes. Cost control depends as much on quality as on productivity, and with a CAM system, Hewlett-Packard expects to approach their goal of zero defects. The goal is defined to be a full computer-integrated-manufacturing (CIM) facility such as outlined by Figure 11.3.

Although Hewlett-Packard uses both computer-aided design (CAD) and computer-aided manufacturing (CAM) in the fabrication of printed circuit boards, the two processes are carried out through different databases. Only the largest systems houses have the computing power, people, and systems approach to solutions that are necessary to produce a truly

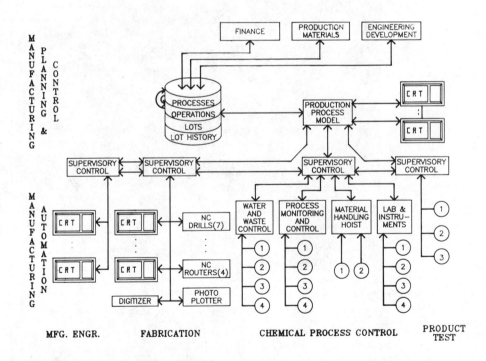

FIGURE 11.3
Computer-integrated-manufacturing (CIM) strategy evolves from computer-aided manufacturing.
(Courtesy of Hewlett-Packard Co., Palo Alto, Calif.)

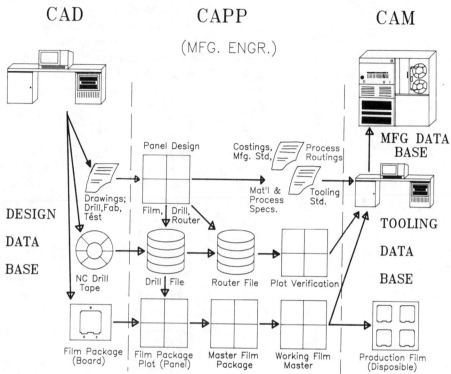

FIGURE 11.4
CAD/CAM interface via computer-aided process planning (CAPP). (Courtesy of Hewlett-Packard Co., Palo Alto, Calif.)

integrated CAD/CAM system. And even these systems have been limited to those that make relatively simple machine parts.

PCBs are not simple. There are 32 steps in the equipment matrix for the fabrication of multilayered boards. The CAD input that defines what the boards will look like and how they will perform is not complete enough to define how they should be made. Usually the input concerns artwork and drilling and does not address the plating process, which is most critical to PCB performance, or even the total plating area.

Before CAD and CAM can be linked, they need to share a common database. A CAD/CAM interface system is given in Figure 11.4. But even that will not be enough unless the CAM equipment can read the CAD output. CAD systems differ widely; there is no standard format for output. PCB hole sizes, for instance, are standardized industrywide. There are EIA standards for databases. But individual systems houses may format programs differently so that, as an example, a tape punch or magnetic tape deck can read a database, but a terminal in a CAM system cannot. A manufacturer could standardize on one CAD system, but only at the expense of losing access to all the others.

The fault is not necessarily only that of the CAD producers. There are many numerically controlled routing and drilling machines, and they all use different formats. The CAD engineer cannot be expected to design for all of them. In addition, PCB manufacturers have never told their CAD counterparts just what inputs they need for their database. Naturally, everyone's databases differ, too, so that some kind of agreement is needed on a CAM database standard before CAD people can design for it. That would be the first step toward a truly integrated CAD/CAM system—a standard CAM database.

As Figure 11.4 shows, the boundary between the CAD and CAM portions of Hewlett-Packard's system is presently being crossed by humans. The output of the CAD system is film, drawings, and a tape drill program. These must be manually translated into the correct format and entered into the CAM system. To make this interface transparent to the machines, the photoplotter, *XY* plotter, and tape unit that produce the CAD output have to be taken out of the CAD system across the boundary into the CAM system.

Before that can happen, there must be some kind of agreement on what the CAM data should look like. The CAM engineers presently control many

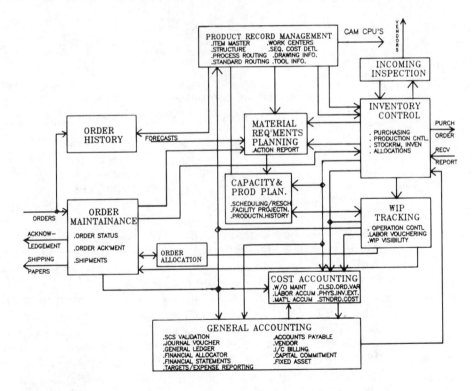

FIGURE 11.5
Printed circuit management system. (Courtesy of Hewlett-Packard Co., Palo Alto, Calif.)

parameters that are totally outside the realm of the CAD programs. For instance, CAD delivers a film of the network but has no information on how to step-and-repeat it for CAM purposes. Instead of the CAD designer saying, "This is what I designed—I don't care how you make it," everyone in the design process has to know and care about what everyone else is doing. Then the boundary between CAD and CAM will dissolve.

11.4.1 Building a Database

The systems required the construction of a large database. The engineering database is actually a subset of the master database maintained by a printed circuit management system (see Figure 11.5). This master database tracks

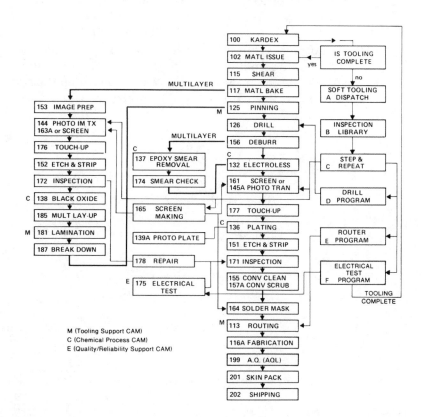

FIGURE 11.6
Flow chart shows progress of PC boards CAM systems. (Courtesy of Hewlett-Packard Co., Palo Alto, Calif. Reprinted from *ASSEMBLY ENGINEERING,* July 1981. By permission of the Publisher, © Hitchcock Publishing Co. All rights reserved.)

orders, production, bills of material, costing, and material management for approximately 1400 part numbers.

The engineering data base, on 95 megabytes of disc storage, supports 27 machines—computers, controllers and manufacturing equipment—that produce finished PCBs. Data on tooling, chemical processing, and testing are stored for each part. Tooling data, for instance, includes information on artwork, drill and routing drawings, hole and drill sizes, and production history. Tool use also is tracked so that worn drills can be replaced before they slip out of tolerance. The chemical process database supplies information such as the size of plated areas, trace width, plating thickness, chemical concentrations, current, and voltage.

11.4.2 Equipment Matrix and Stream Matrix

The CAM process consists of a series of operations interspersed with directions for getting from one step to the next. The operations, which are the actual steps in the manufacture of a PCB, make up the *equipment matrix*. The directions that guide the product through every aspect of manufacture

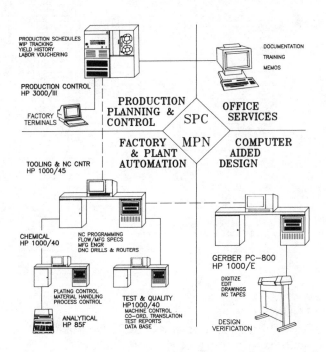

FIGURE 11.7
Printed circuit manufacturing matrix. (Courtesy of Hewlett-Packard Co., Palo Alto, Calif.)

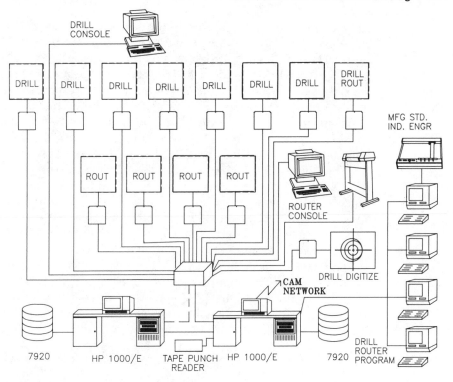

FIGURE 11.8
Tooling and NC programming with direct numeric control. (Courtesy of Hewlett-Packard Co., Palo Alto, Calif.)

from order processing to shipping compose the *stream matrix*. The flow chart Figure 11.6 shows the equipment matrix, supported by the engineering database, as boxes, while the arrows represent the stream matrix, supported by the master database. Their relationship is shown in Figure 11.7. The business section controls the stream through direct interaction with the HP 3000 series III, and the technical section controls the manufacturing with a network of HP 1000s via DS/3000 hardware and software.

The manufacturing intelligence has been distributed into three main areas: chemical processing, tooling, and quality control and reliability. The central system is an HP 1000 Model 45 with an F-series processor and 55 megabytes of disc store. This system's manufacturing function is to support the numerical control (NC) machines that perform drilling and routing. It has vector instructions set firmware to perform the matrix manipulations required by the NC software. The Model 45 also supports chemical processing and electrical testing through DS/1000 links to two HP 1000 Model 40 computers, each with 20 megabytes of disc.

Although Hewlett-Packard has computerized the design of printed circuit boards, the CAD data are entered into the manufacturing database manually. The CAD output consists of a film, engineering drawing, and cassette tape of the drill program. The film is mounted on a step-and-repeat camera, which turns out multiimage, multicopy production films. Data from the drawings are entered through a terminal, and the Model 45 then generates a routine program. This program can be displayed on a plotter for visual inspection and verification. The cassette data are also entered on a terminal to generate the CAM drill program. Both the routing and drill program make use of information from the step-and-repeat process.

The Model 45 is accessed through an array of peripherals, including an HP 7920 disc drive, plotters, a tape punch and reader, an HP 2645 system console, a step-and-repeat terminal, an HP 3497 scanner to monitor multilayer lamination, and an OPIC III programmer that can derive *XY* data directly from a film. The system controls drill and router banks through HP 2113 E-series computers (see Figure 11.8).

The most critical part of PCB manufacture is chemical processing. The chemical processing center controls five main subsections: two plating lines, their power supplies, an electroless line system similar to the electrical power supply in that it monitors chemical levels and replenishes them

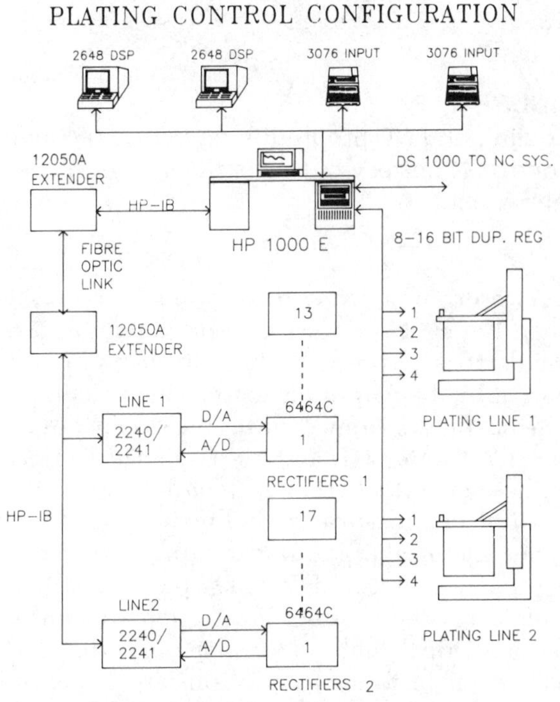

PLATING CONTROL CONFIGURATION

FIGURE 11.9
Computer-controlled chemical processing center.
(Courtesy of Hewlett-Packard Co., Palo Alto, Calif.)

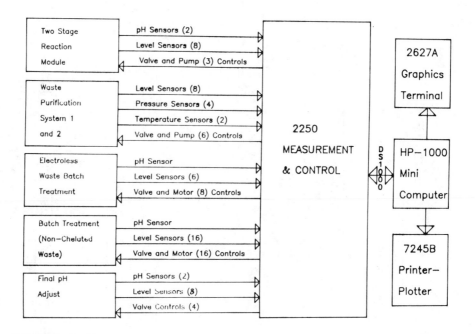

FIGURE 11.10
Computer-controlled environmental and water effluent system. (Courtesy of Hewlett-Packard Co., Palo Alto, Calif.)

when necessary, waste treatment, and a chemical control laboratory. The Model 40 computer that controls this center has its own console, printers, plotters, backup computer, and monitoring terminals. (See Figure 11.9 for a computer-controlled chemical processing center.)

The main function of the plating line is to step the boards through chemical baths according to the sequence determined by the database for a given part. One HP 2240 measurement and control processor determines temperature, bath concentration, and valve openings, and also runs an alarm system. Another HP 2240 provides the appropriate current and voltage settings for 23 ten-kW dc power supplies.

The electroless line, which metalizes plastic boards, is so critical and sensitive that it must be continuously monitored by a spectrophotometer. Information reported by this instrument to the main processor determines the flow of chemicals into the vats.

Another HP 1000 computer, connected over a distributed link to the Model 40, controls waste treatment by monitoring conductivity, pH, level and flow, and valve and pump activity. Forty percent of the water supply, after treatment, is recycled (see Figure 11.10).

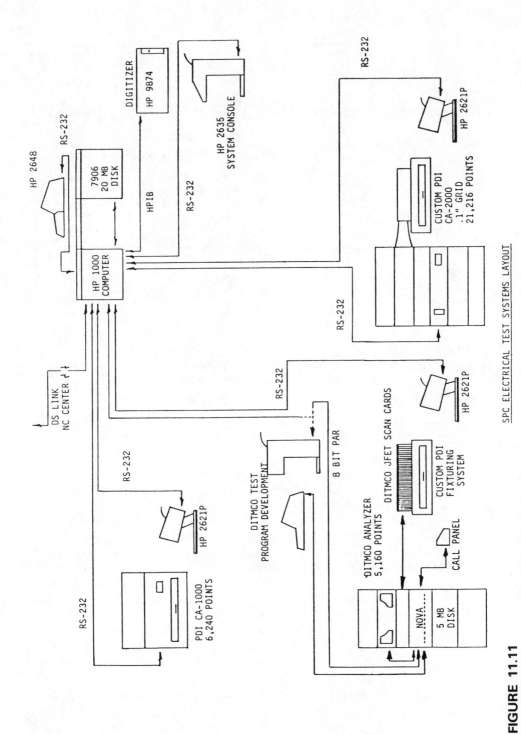

FIGURE 11.11

Electrical test and quality-control system.
(Courtesy of Hewlett-Packard Co., Palo Alto, Calif.)

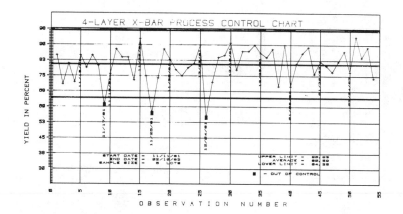

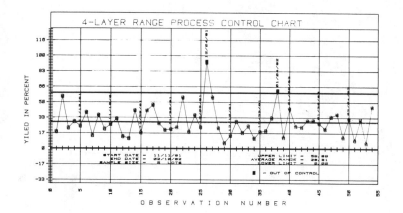

FIGURE 11.12

Statistical quality-control charts. (Courtesy of
Hewlett-Packard Co., Palo Alto, Calif.)

A third HP 1000 system supports electrical testing in the inspection and
quality control center (see Figure 11.11). It, too, has an array of peripherals,
including a printer and interactive display terminal. Graphic information
can be entered via an HP 9874 digitizer. Graphic output is available such
as the statistical quality charts shown in Figure 11.12.

The fabrication of printed circuits at Hewlett-Packard is now solely a
function of the CAD/CAM databases. But the CAM model in Figure 11.11
can be generalized to typical manufacturing, as shown in the automation
diagram in Figure 11.13.

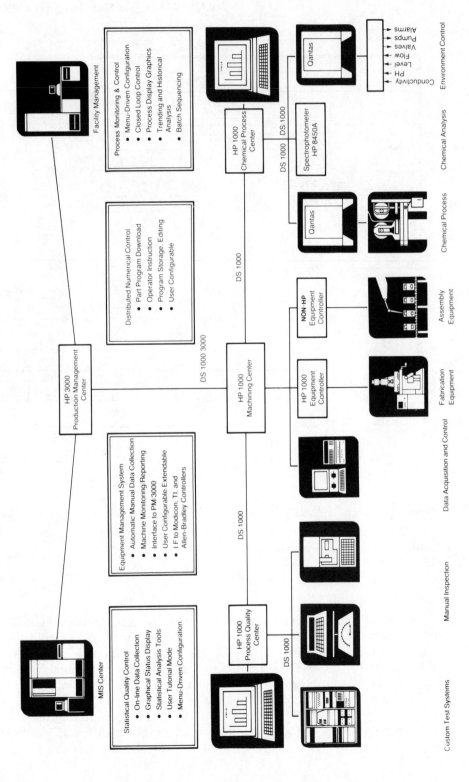

FIGURE 11.13
Typical printed circuit automation system.
(Courtesy of Hewlett-Packard Co., Palo Alto, Calif.)

11.5
APPLICATIONS OF SMALL COMPUTERS IN LABORATORY RESEARCH[c]

The computer is now at work in laboratories involved in virtually all industrial fields, including electronic instrumentation. Applications of Digital's MINC(TM) computer systems include the following:

1. Sprinklers in flaming fires

2. Weather applications

3. Developing diodes and monolithics integrated circuitry

4. Testing analog and hybrid circuit boards

5. Real-time study of electrochemical interaction between brain cells

6. Many other electronic, mechanical, medical, and environmental applications

Engineering scientists at Raytheon Company Research Division are using four MINCs in their research and development related to semiconductors, signal processing and the development of a ring laser gyroscope. A giant in the field of radar research and development, Raytheon has applied many radar principles to the design and installation of systems for airport air traffic control and the large-screen displays used in flight control rooms.

The Semiconductor Laboratory of the Research Division is involved in the development of state-of-the-art microwave semiconductor devices such as IMPATT diodes, field-effect transistors (FETs) and monolithic integrated circuits. Much of the group's work involves high technology and requires sophisticated design and measurement processes. The semiconductor group uses MINC to control network analyzers and microwave test setups that enable engineers to measure the complex impedances of circuits and devices as a function of frequency and power level. Using an interactive program, written in FORTRAN, MINC permits control of the measurement frequency (which covers from 2 to 100 GHz in several bands), of the signal level, as well as of the different instruments involved in the measurement. Repeatable errors are calibrated out of the system, and the measurement results displayed are corrected automatically.

The two MINC systems acquire, reduce, and store data from the network analyzers 8 hours a day. The engineers can analyze these data in real time by using the foreground–background monitor under RT-11. Once properly formated, the data are sent via telephone line to a central CDC

---

[c] The material in Section 11.5 is from "The Indispensable Applications of Small Computers in Laboratory Research," basic introduction to new small computers, with fundamental and practical laboratory applications of interest to scientists and engineers. Compiled and published as a service by the information staff of the manufacturers of Digital's MINC(TM) computer systems. Reprinted here courtesy of the Digital Equipment Corporation.

(Control Data Corporation) Cyber computer to become part of a comprehensive computer-aided design (CAD) facility. Dan Masse, the scientist responsible for the microwave semiconductor measurements, has said, "MINC has become an integral part of our research facility. Without it, most of our work would be made much more difficult if not impossible at times."

Another research group, the Microwave Ultrasonics Group, uses a MINC-11 system to design and build analog electronic devices such as oscillators. They use MINC to make short-term frequency stability measurements and to control frequency versus temperature and frequency versus phase measurements with the goal of improving the performance of the oscillators.

Figure 11.14 shows an automated network analyzer (2 to 18 GHz) using the MINC-11. Figure 11.15 shows the 30- to 50-GHz automatic network analyzer for the IMPATT device circuit characterization using Digital MINC-11 computer.

The electro optics lab uses the MINC-23 in designing and developing a high-accuracy, low-cost, ring laser gyroscope. Commonly used for directional reference on board ships and airplanes, gyroscopes are also used as a basis for guiding all space shots. These researchers hope to surpass the accuracy of mechanical gyroscopes by using a laser with a circular beam path; thus the term ring laser gyroscope. Traditional gyroscopes are delicate

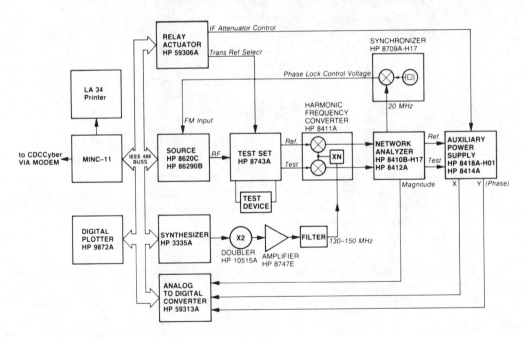

FIGURE 11.14
Automated network analyzer (2 to 18 GHz) using the MINC 11. (Courtesy of Digital Equipment Corp.)

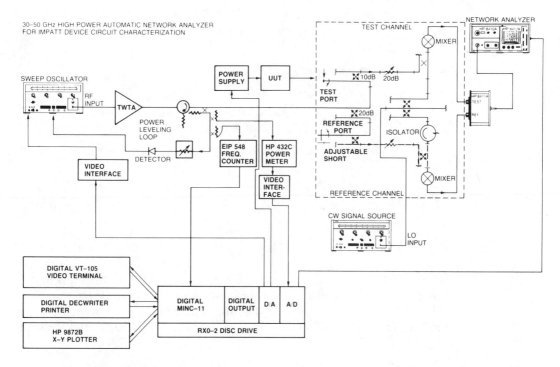

FIGURE 11.15

The 30- to 50-GHz high-power automatic network analyzer for IMPATT device circuit characterization using a digital MINC-11 computer. (Courtesy of Digital Equipment Corp.)

instruments consisting of many mechanical parts. Experimenters believe a laser gyroscope is inherently more reliable and more accurate than a traditional gyroscope since it uses a beam of light rather than a mechanical device as reference. Thus the ring laser gyroscope has the mechanical ruggedness typical of devices with no moving parts.

Hughes Aircraft Company, known worldwide for their contributions to the scientific, commercial, and military communities, has seven MINCs installed in the Production Test Engineering Department, located in Fullerton, California. This department uses MINCs as part of a test system through which 90% of all analog and hybrid analog–digital circuit boards pass. These circuit boards are used in radar and sonar systems, tactical data display systems, and computers manufactured by Hughes Aircraft.

The MINC systems accomplish three goals:

1. Increase the reliability of the test system.

2. Decrease test time.

3. Provide for expansion of the system.

The engineering group uses the MINCs for a variety of purposes. Four of the MINCs are part of general analog test (GAT) stations, which include the following instruments:

1. Two electronic pulse generators.

2. Function generator, which generates various wave forms.

3. "Smart" timer counter, which measures frequency–time delays, rise times–fall times. (The timer counter has a built-in microprocessor.)

4. "Smart" digital multimeter. (The multimeter also has its own microprocessor.)

5. Twelve power supplies, two of which are controlled by the IEEE bus.

6. Hughes-built custom interfacing box that controls signal routing and the unit under test.

As shown in Figure 11.16, MINC, as the controller, sends and receives data over the IEEE bus. Two pulse generators and a function generator provide the stimulus, and a custom-built interfacer sends the signals to the unit under test. The multimeter and timer counter measure the responses.

The MINC controls the system through a test station executive program written in MINC BASIC. This program allows the user interactive control

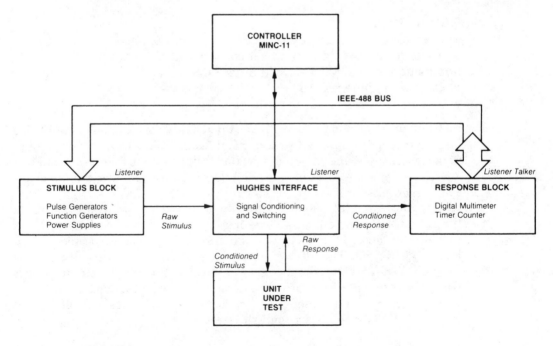

FIGURE 11.16
Functional block diagram for general analog tester. (Courtesy of Digital Equipment Corp.)

of the system through English commands, eliminating the necessity of the user having any programming experience.

Test Engineering uses one of the MINC-based GAT systems to develop and debug the adapter, which holds the circuit board, and to refine the final version of the test procedure.

Another group, Design Engineering, uses a MINC-based GAT to design and debug engineering prototypes of newly designed circuit boards.

Another MINC is used for off-line programming that includes general engineering programs as well as programs used to generate the initial test data files that will become test procedures on the GAT.

Systems Test Engineering uses MINC in a dedicated test position that tests part of a radar system.

The MINCs used in the GAT stations replace an 8080-based microcomputer. Upgrading the MINC provides the following benefits:

1. Speed.

2. Enhanced capabilities of the test stations by allowing for larger programs; for example, programs that exceed the work space can be written through use of the CHAIN command.

3. Insert and delete steps in a test procedure capability.

4. Ability to add to HELP facility.

5. Access to more functions of the instruments used in the test station; for example, with MINC, all the capability of the multimeter can be used. Before MINC, they ran out of memory and could not access the options.

6. Better editing of the test procedure through virtual array files.

The MINC system has significantly reduced the test time, increased the reliability of the test system, and provided for expansion of the test system.

Digital has installed more computers in laboratories than any other manufacturer, thus far. Now with Enhancement II, MINC provides color graphics capability, greater memory capacities, and higher speed.

11.6
COMPUTER-ASSISTED MANAGEMENT
OF OCCUPATIONAL HEALTH
PROGRAMS [d]

Modern comprehensive occupational health programs involve several integrated activities, all involving detailed data collection and analysis using computer-assisted systems. They include the following:

---

[d] The material in Section 11.6 is from Ronald E. Costin, "Computer Assisted Management of Occupational Health Programs," *National Safety News*, August 1981. It is reprinted here courtesy of *National Safety News* and Ronald E. Costin, M.D.

1. Surveys of the workplace to determine or estimate the level of agents such as noise and chemicals that may be hazardous.

2. Estimation of exposure times and levels for each individual worker.

3. Development of measures to eliminate or reduce the hazard or to provide personal protective equipment if possible.

4. Application of OSHA, TOSCA, and government standards, NIOSH criteria documents, and sound medical research data to develop a medical monitoring program for individuals unavoidably exposed or potentially exposed.

5. Correlation of medical data and exposure data, along with personal medical history information, to detect any adverse effects. This correlation serves as a basis for decisions to intervene by changes either in the workplace or in placement of the worker.

6. Record maintenance for 30 years as required under current OSHA regulations.

7. Each step of such a program is dependent upon the storage, correlation, and update of massive amounts of ever-changing criteria and data.

11.7
COMPUTER-ASSISTED CARDIAC LABORATORY[e]

Advances in integrated circuits and microcomputer technology make the approach of a computer-assisted cardiac laboratory [9] a reality.

Reduced costs, processors, memory chips, and equipment miniaturization have made the concept of special-purpose microcomputers possible. Each program performs a given function of the particular analytical system, which may be coupled by one central microcomputer for final processing, collecting, and report generation. Essentially, this frees a central minicomputer to accomplish the data, correlations, and interpretive functions, while the peripheral dedicated instrumental microcomputers prepare the raw data and provide instant quality control. This system is modular and flexible in hardware components and function. Such a computer will release the laboratory technicians and/or physicians from the more routine but time-consuming operations, calculations, and evaluations that the computer can do more efficiently at higher speeds and often more accurately.

Another consideration is the difficulty in evaluating and storing physiological measurements in large amounts of quantified data. Beginnings have only been made in correlating and evaluating this quantitative physiolog-

[e] The material in Section 11.7 is excerpted from F. W. Schoonmaker, N. K. Vhay and James Utzerath, Facility Report, St. Luke's Hospital, Denver, Colo., *CVP*, vol. 7, no. 3 April–May 1979, published by the Brentwood Publishing Corp., Los Angeles, Calif.

ical information, and with this type of electronic assistance the physician's decision capability is greatly improved.

Data conversion is performed by a microprocessing system rather than by a central computer. If the data conversion is complex or is a high-speed process, the expense might not be justified; however, in most cases the use of solid-state electronic modules with conversion of raw segments to test results is quite simple and economic. Presently, large-scale integrated circuits use solid-state memory. Microprocessors are best applied in the control of calculations involving specialized functions in the bioelectronic instrumentation system.

Midwest Analog and Digital, Inc., have used a microprocessor system at St. Luke's Hospital in Denver, Colorado, to serve in the computer cardiac assist laboratory. Two catheterization laboratories at St. Luke's Hospital have been a computer-assisted system in conjunction with a cardiac operating room. The sharing of a single microprocessor developed by Midwest Analog and Digital, Inc., among the three areas has kept reduplication of equipment to a minimum and physician and/or technician utilization time to maximum efficiency. By staggering caseloads, the proficiency and output of the areas operating from this single control unit have increased. In the three-way system, data obtained from a microprocessor with readouts may be copied from a disk or tape playback for instantaneous report generation. The video portion is done by color coding and may be used by tape drive or floppy disk from the actual arteriography.

The light-pen method is used for ventricular-function determinations. Accuracy depends on the displayed image size. The smaller the image, the more accurate are the calibrations. All pressure contours and left ventricular functions are directly seen or measured and may be copied during playback.

In any computer malfunction, conventional laboratory techniques using transducers and a recorder of cinefilming in any of the laboratories or in the adjacent surgical suite may be substituted. Time sharing, not reduplication, of the contour pressure recorder, tape, and disk is the rule, and one microprocessor assists in computerization of the two laboratories and the operating room.

11.8
COMPUTER-ASSISTED PREVENTIVE
MAINTENANCE PROGRAM

With the growing number of hospitals developing preventive maintenance capabilities, and the increasing number of pieces of equipment and instruments needing preventive maintenance [6], automatic data processing has emerged as a tool to aid the clinical engineer and hospital in planning preventive maintenance programs. With the use of a computer-assisted preventive maintenance system, many functions can be performed, such as:

1. Accurate records of preventive maintenance.

2. Safety testing and the storage of the testing information required.

3. Year-to-year maintenance costs.

4. Information on availability of equipment for services.

5. Priorities.

6. Printout of the procedural forms for use in preventive maintenance.

The computer-assisted system is useful in hospital safety and preventive maintenance programs. Such a system can provide services to many user sites throughout the hospital with additional remote stations.

11.9
AUTOMATIC TEST SYSTEM
FOR THE BIOELECTRONIC PACEMAKER [f]

Automatic testing is used by many medical manufacturers to test and evaluate their bioelectronic instruments. Let us examine, as an example, the heart pacemaker [10]. Medtronix, Inc., at their Rice Creek facility, operates an automatic test system that uses data processing, resource sharing, and a standardized instrument bus system.

Their system, which is called the minicomputer electronic data network (MED-NET), performs long-term, high-volume receiving inspection testing. Performance tests include life test, quality tests, and incoming tests and are used in conjunction with the minicomputer system of Medtronix, Inc.

Life testing, which takes on a special meaning in the pacemaker context, is handled directly by the central computer. The units are power-on in manufacture and permanently sealed. Life tests require sensing and measurement of the signal produced.

For the tests, pacemaker samples are placed in tanks containing a saline solution kept at temperatures and salinities that simulate the environment and impedance of human tissue. These battery-operated pacemakers are typically monitored indefinitely; some have run for 10 years. The computer monitors several parameters (pulse width, amplitude, sensitivity, etc.) for each unit.

Each device under test has its own tank address and test parameters. The computer keeps track of these tank maps and automatically invokes the proper test sequence for each unit when called by the test program.

Besides running the tests, the role of the central computer is to collect, process, and store the data on disc, and produce results on hard copy. Later the data are transferred to magnetic tape for archival storage or for input to the company's time-sharing system for further analysis.

---

[f] The material in Section 11.9 is based on "Distributed Computers Put Pacemakers Through their Paces," reprinted with permission from *Quality Magazine*, February 1979.

The central unit is also used for test program development, checkout of test hardware, and modification of existing programs to accommodate specification changes with minimal interruption. In addition, it acts as the central file for programs that are called up and down-loaded for use by the distributed satellites.

Qualification testing of hybrid circuits and batteries offered by potential vendors is performed by the first satellite computer. The procedure entails initial electrical tests, and then mechanical, electrical, and environmental stress of the part before it is retested. This process repeats several times during the evaluation. The initial, interim, and final test results are compared to determine if there has been deterioriation of the part. Qualification tests are run on lots of 100.

The satellite that handles the new hybrid qualification testing also performs qualification tests on new battery configurations, as well as on process changes proposed for current components.

A second satellite computer, located in the inspection testing area, monitors the high volume of parts that the company receives from its suppliers and its own hybrid circuit subsidiary. Two test stations are operated by the computer. The tests entail measurement of up to 50 parameters, which can take from 30 to 120 seconds per device.

The slow test throughput is dictated by two constraints. First, low-level parametric measurements require instrumentation and device settling time. Second, the parts are designed to operate at the same rate as the human heart. Test time increases with the number of programmable features for a device type.

The units for test are preheated, and heaters on the test sockets maintain the 37°C (body temperature) level during the tests.

Most tests are of the closed loop controlled, stimulus–response type. Typical parameters include sensitivity, current drain, refractory, leakage, and output pulse rate, width, and amplitude. Hard-copy data are available on request, a feature that is particularly useful when returning parts to a vendor with test data.

One operator controls both stations from the CRT terminal. Eight pre-programmed soft keys (four for each station) are set up for normal testing. This minimizes the need for discretionary operator interference with the computer operating system. The system is configured such that dissimilar devices can be tested simultaneously.

11.10
REVIEW QUESTIONS

1. Discuss CAD/CAM in electronic instrumentation.

2. Discuss CAD/CAM in printed circuit manufacturing.

3. Discuss computer-aided systems in laboratory research.

4. Discuss computer-assisted occupational health programs.

5. Discuss briefly a computer-assisted cardiac laboratory.

6. List five uses of a computer-assisted preventive maintenance system.

7. Discuss a computer-assisted test system for pacemakers.

8. Draw a simplified block diagram of a small computer system in a laboratory research laboratory for analog testing.

11.11
REFERENCES

1. Carl Machover and Robert E. Blauth, *The CAD/CAM Handbook*, Computervision Corp., Bedford, Mass., 1980.

2. Happy Holden, "CAD/CAM in PCB Manufacturing," *ASSEMBLY ENGINEERING*, July 1981.

3. John K. Krouse, "Computer Time-Sharing for CAD/CAM," *Machine Design*, April 23, 1981.

4. Computer Graphics: What is it? What Can it Do for You? Session 22, Electro 1981 Professional Program Papers, Electronic Convention, Inc., 1981.

5. *The Indispensable Applications of Small Computers in Laboratory Research*, Digital Equipment Corp., Maynard, Mass., 1981.

6. Charles W. Beardsley, "Computer Aids for IC Design, Artwork and Mash Generation," *IEEE Spectrum*, Sept. 1971.

7. Ronald E. Costin, "Computer Assisted Management of Occupational Health Programs," *National Safety News*, Aug. 1981.

8. Larry Schwartz, "A Computer Assisted Preventive Maintenance Program," *Journal of Clinical Engineering*, vol. 4, no. 1, Jan.–Mar. 1979.

9. F. W. Schoomaker, N. K. Vhay and James Utzerath, Facility Report, St. Luke's Hospital, Denver, Colo., *CVP*, vol. 7, no. 3, April–May 1979, Brentwood Publishing Corp., Los Angeles, Calif.

10. "Distributed Computers Put Pacemakers Through Their Paces," *Quality Magazine*, 26–28, Feb. 1979.

11. Mitchell Waite, *Computer Graphics*, 3rd printing, Howard W. Sams & Co., Indianapolis, 1980.

12. Philip H. Abelson and Mary Dorfman, "Computers and Electronics," *Science*, vol. 215, no. 4534, ISSN 0036-8075, Feb. 12, 1982.

13. Happy Holden, "Computerized PCB Fabrication," *Circuit Manufacturing*, © May 1980.

12

Computer-Based Systems, Including Transducers

12.1
INTRODUCTION

In this chapter, data-acquisition systems are analyzed. With electronic instrumentation required to interface transducers under the control of a computer-based system, implementation of automatic testing, measurements, and control problems is feasible. Examples include channel weather monitoring, digitizing transient waveforms from transducers, production tests, and energy management. Transducers of all the types discussed in Chapter 2 are applicable for such data acquisition on conversion systems. Measurement functions can include the following:

1. Temperature transducers
   a. Thermocouples
   b. Resistance temperature detectors
   c. Thermistors
   d. Integrated-circuit sensors
2. DC voltage

3. Resistance, including the strain gage

4. Frequency, period, and pulse width

5. Digital inputs and outputs

6. Power supply voltages

12.2
DATA-ACQUISITION INSTRUMENTS FOR
TRANSDUCER MEASUREMENTS [a]

When specifying data-acquisition instrumentation, performance can be measured by system integrity, precision, versatility, and throughput. Seldom are two applications identical. The HP Model 3497A Data Acquisition/Control Unit (Figure 12.1) provides the user with a versatile HP-IB system that has a wide range of precision measurement and control plug-in assemblies. This instrument serves as the eyes, ears, and hands for a computer-controlled system that acquires data from transducers and controls equipment and processes.

The versatile HP-IB system is the HP implementation of the IEEE 488 (1978) interface bus connections. The versatility and some of the capability of the 3497A can be shown by looking at three of the many ways it can be used. First, the instrument may accept commands and supply data as requested one at a time by the computer in control. Second, a scan sequence consisting of a start and stop channel, voltmeter setup, time between scan sequences, and internal buffer storage for up to 100 readings may be specified. When each scan sequence is complete and the 100 readings are ready to be output, the 3497A will indicate a service request to the HP-IB. This allows optimum use of processing time in the computer since it can process previous data while the 3497A is gathering the next sequence. Third, a minimum system configuration can be established using a listen-only printer or cartridge tape unit and the 3497A in a talk-only mode. A scan sequence, as described, can be set up with the data being logged on the external device rather than in the internal memory.

The 3497A has the capability of taking up to 300 readings per second in the $3\frac{1}{2}$-digit mode and storing these in its internal memory. In addition, a dedicated scan-only mode was devised to provide the high-speed scanning that solid-state switches are capable of achieving. In this mode the 3497A can scan up to 5000 channels per second. Synchronization lines are provided on the back panel for interfacing to other instruments such as the HP 3437A System Voltmeter and 3456A Digital Voltmeter.

---

[a] The material in Section 12.2 is from James S. Epstein and Thomas J. Heger, "Versatile Instrument Makes High-Performance Transducer-Based Measurements," *Hewlett-Packard Journal*, 9–15, July 1981. It is reprinted with the courtesy of Hewlett-Packard Co., Palo Alto, Calif.

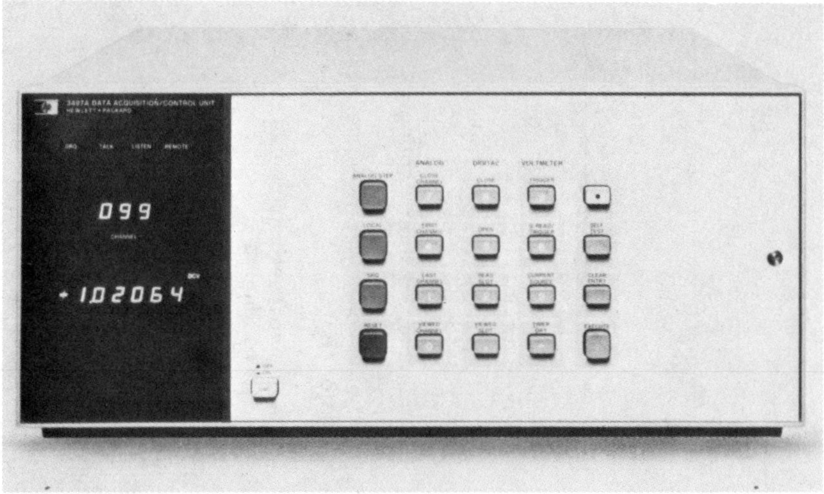

FIGURE 12.1
The 3497A Data Acquisition/Control Unit is an
easy-to-use system for precision measurement of
transducers and thermocouples and for control of
equipment. (Courtesy of Hewlett-Packard Co.,
Palo Alto, Calif.)

The precision of the 3497A begins with its fully guarded five-digit digital voltmeter (DVM) with a 0.003% basic accuracy specification. Analog scanning adds less than 1 μV of offset error. The precision of the system is also extended to the accompanying family of digital input, digital output, counter, and analog-to-digital (A/D) converter plug-in assemblies. All plug-ins have individual channel isolation that guarantees system measurement integrity.

Transducer-based measurements can detect the presence of typical laboratory and factory noise environments. As an example, a type K thermocouple, as shown in Figure 12.2, has a sensitivity on the order of 40 μV/°C. Mounting the thermocouple to a transformer induces common-mode and normal-mode noise levels on the order of volts. Thus, to discern a temperature difference of 0.1°C, it is necessary to detect a microvolt signal that is less than 0.0001% of the noise level. The HP 3497A and its plug-in assemblies incorporate multiple noise-reduction techniques to yield high-quality measurements in severe noise environments. These are guarding, free switching, and signal integration periods equal to one power line cycle. Figure 12.2 gives the diagram of the thermocouple connections.

Additional capability to condition transducer measurements before the A/D conversion is often needed. An isothermal block whose temperature is known to 0.1°C is required to handle thermocouples. A precision current source is required for resistance temperature detectors (RTD), thermistors,

TABLE 12.1
Two Application Examples for a 3497A Based System

3497A and Plug-In Assemblies	Measurement Function	Engine Design Analysis	Facility Monitoring
DVM, 44411A	Temperature: Thermocouples	Coolant and oil temperature	Heating and air conditioning
DVM, 44421A	RTDs	Hot spot analysis	Fire detection
DVM, 44421A	Thermistors	Exhaust temperature	Motor overloads
DVM, 44421A	IC Sensors	Cooling system analysis	Power or facilities use
DVM, 44421A	DC Voltage	Battery voltage	Backup lighting
DVM, 44421A	Resistance	Continuity or isolation	Continuity, isolation, lighting level
44426A	Frequency, period, pulse width, totalize	Fuel/air flow, r/min, event counter	Air or water flow, power use, traffic monitoring
44425A	Digital input	Status lines, limit switches	Status lines, security checks
44425A	Digital interrupt	Interrupts	Alarm conditions
44428A	Digital output	On and off valve control	On and off valve and vent control, data display
44428A	Actuator output 0 to ±10 V programmable voltage	Engine load control	Pumps, exhaust fans, alarms
44429A	4- to 20-mA programmable current	Proportional throttle control	Lighting control
44430A	High-speed digitization and scanning	Current control loop	Heating or cooling control
44421A	Time	Continuous real-time monitoring	Continuous real-time monitoring
Real-time clock		Data logging	Logging events, data

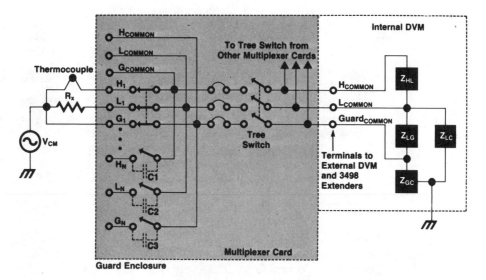

FIGURE 12.2
Thermocouple connections to a 44421A
20-channel multiplexer plug-in assembly in the
3497A, showing the internal circuits for the tree
switch and the guard. (Courtesy of Hewlett-
Packard Co., Palo Alto, Calif.)

and other resistance measurements. Frequency counting is needed for flow measurements and the determination of digital on–off states is essential. Strain gages require a bridge-completion network, and control, both proportional and on–off, is needed to provide stimulus and to complete measurement feedback loops.

Because of this assortment of requirements, a system must be able to be tailored to a customer's application. The 3497A Data Acquisition/Control Unit and its optional plug-in assemblies provide user flexibility in all the areas mentioned while maintaining performance accuracy and isolation for data integrity. Figure 12.3 shows a block diagram of the 3497A system and Table 12.1 outlines two possible 3497A system applications using the plug-in assemblies.

A highly stable fully floating current source in the voltmeter assembly provides excitation for RTD, thermistors, and other resistance measurements. User programmable output levels of 10 μA, 100 μA, and 1 mA combined with the DVM provide resistance measurements from 1 mΩ to 1 MΩ. An important consideration when making highly accurate resistances is how the four-wire technique is performed. The current source is derived by chopping the +5-V low-referenced power supply at 25 kHz. This signal is put through two pulse transformers coupled by a single turn. The single turn reduces the feed-through capacitance to 1 pF and thus minimizes

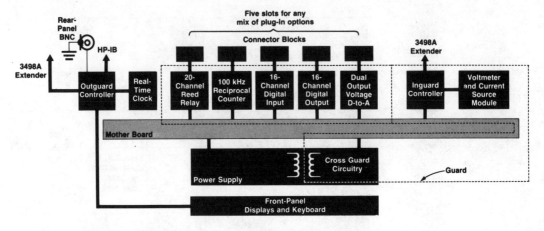

FIGURE 12.3
Block diagram of the 3497A Data
Acquisition/Control Unit. (Courtesy of Hewlett-
Packard Co., Palo Alto, Calif.)

injected current. The current source is completed by using a stable voltage
reference and some custom fine-line resistors.

12.3
THERMISTOR DATA LOGGER
INSTRUMENTATION[b]

By placing a precision scanning voltmeter together with a completely self-
contained scientific computer, a data logger is obtained. Figure 12.4 shows
a data logger with the ultimate in flexibility.

The HP3054DL performs difficult transducer measurements with confidence by taking advantage of the $5\frac{1}{2}$-digit A/D resolution and its excellent
noise rejection. You can linearize the data and display it in a form you can
use—plots, graphs, strip charts.

The computer is a valuable assistant to help with scientific problem-
solving and statistical analysis, even when you are through logging data.
The 3054DL is a data-acquisition system useful for thermistor manufacturing applications.

The HP3054 data logger is used by Thermometrics, Inc., consisting of
the HP 3497A data acquisition/control unit and an HP85 computer. This
system is available from Hewlett-Packard with a variety of subprograms for
making measurements. Thermometrics use the one called "ohm" for mak-

---

[b] The material in Section 12.3 is from "3054 DL Data Logger, Technical Data," October
1980, and is reprinted courtesy of Hewlett-Packard Co., Palo Alto, Calif.

ing four-wire resistance measurements at a measurement current of 10 μA. In each of the measurements, consideration of at least three measurements that have to fall within $\pm0.05\%$ of the average value are made. Sampling of a standard precision resistor and a standard thermistor on each measurement seam is performed. Most of the measurements involve 20 channels (two relay cards) of four-wire resistance measurements taking into account two different temperature baths. When excessive fluctuations in both temperatures (i.e., greater than $\pm0.025°C$) occur, seven measurements are used to obtain a median value. The three readings representing the greatest deviations from the median are eliminated. The remaining four readings are then averaged and accepted if they fall within $\pm0.05\%$ of the average.

In the remaining sections, we shall discuss the following:

1. Schaevitz Engineering computer-based systems for linear variable differential transformers.

2. Dynamic Measurement Corporation's analog-to-digital techniques for high-performance data-acquisition systems using transducers.

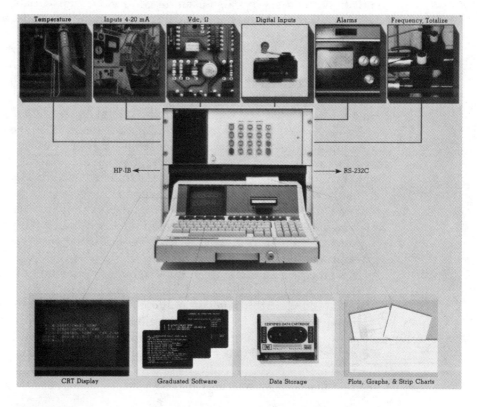

FIGURE 12.4
HP 3054DL data logger. (Courtesy of Hewlett-Packard Co., Palo Alto, Calif.)

3. Validyne Engineering Corporation's practical high-speed data-acquisition system.

4. Datel-Intersil SineTrac concept with use of computer-based systems for transducers.

12.4
DATA ACQUISITION AND CONVERSION SYSTEMS FOR LINEAR VARIABLE DIFFERENTIAL TRANSFORMERS [c]

Computer-based systems provide much greater versatility and flexibility than simple analog systems. In some instances they are mandatory to achieve certain objectives. An important factor that has made computer-based systems economically feasible is the proliferation of powerful mini computers.

In some installations, the computer-based system is used only for data acquisition; in others it is also used for control. Of course, data acquisition is basic to all applications, because data must be acquired before any control can be initiated.

Figure 12.5 is the block diagram of a typical data-acquisition system. The transducer inputs are scanned sequentially under program control. Under certain circumstances, the sequential scan may be interrupted and certain variables may be addressed selectively. As each input is sampled, its reading is converted to a digital value by the analog-to-digital (A/D) converter. The speed with which the inputs are scanned and converted is a function of process dynamics. In most instances, a scan and conversion rate of 30 to 40 points per second is more than adequate. If there are a large number of inputs, or if each input, or certain ones, must be sampled more frequently, a higher sampling and conversion rate is used.

When the A/D conversion of an input is completed, the A/D converter signals the computer that a reading is available. This causes the computer to branch to a program that accesses and processes the A/D converter reading.

The processing program can also compensate for nonlinearity of the LVDT. The accuracy of the LVDT can be improved by passing the reading through a second- or third-degree polynomial whose coefficients have been determined previously by an on-line calibration program. For this calibration process to be meaningful, the transducer must have inherently repeatable characteristics. This is an attribute of the LVDT not shared by many other commonly encountered transducers.

The illustrated system will also accept inputs from thermocouples and

---

[c] The material in Section 12.4 is from Edward E. Herceg, *Handbook of Measurement and Control*, Schaevitz Engineering, Pennsauken, N.J., 1976.

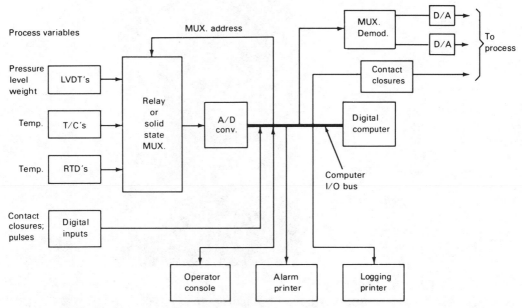

FIGURE 12.5
Block diagram of typical data-acquisition system.
(Courtesy of Schaevitz Engineering)

other dc-output transducers that may be needed in a particular system application.

After each input is processed, it is compared with stored alarm limits for each variable. An appropriate message is printed out if the variable goes into alarm status or returns to normal. The inputs may be averaged over variable intervals. The averages are then printed out periodically to provide a time history. These averages of the various variables can be entered into programmed equations for periodic computation of plant performance.

The operator may communicate with the system by means of an operator's console. This console usually comes in one of three forms:

1. Teletypewriter

2. Projection displays with specialized keyboard

3. CRT display and associated keyboard

In a typical system (as shown in Figure 12.6) the operator may be permitted to change alarm limits, request the continuous display of critical variables, or request a trend log of certain variables.

The data-acquisition system becomes a control system by adding digital-to-analog (D/A) converters and/or contact closures, which are controlled by the computer program after it has processed the inputs. The D/A converter is the link that closes the process loop in a continuous system.

FIGURE 12.6
A typical computer-based measurement system.
(Courtesy of Schaevitz Engineering)

The contact closures delivery start–stop or open–close commands in a batching system. In some cases the system may be hybrid and require both D/A converters and contact closures.

Either of two modes of control are possible: direct *digital control* (DDC) or *supervisory control* (SC). In the DDC mode, the computer output manipulates the process control elements directly. In the SC mode, the computer output is the set point for a dedicated analog controller. In either case, the D/A converter or contact outputs can be used to generate the signals required by the process interface.

Various control methods can be implemented by the software. This permits the user to change the control method as she or he gains experience or as the process changes without being constrained by a hard-wired control system.

12.4.1 Application of Computer-based Systems

The following illustrate how a computer-based system can solve difficult measurement and control problems. The activities of a materials testing laboratory are a cross between batch processing and continuous processing. They are batch in that various specimens are tested over a finite and relatively short time interval. However, they are continuous in the sense that

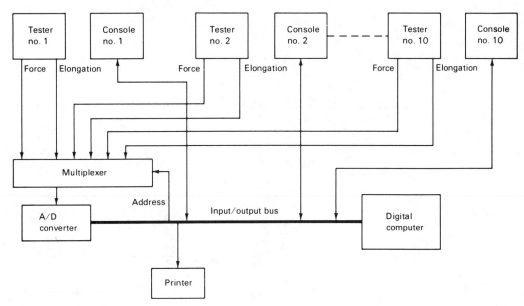

FIGURE 12.7
Block diagram of a computer-based materials
testing system. (Courtesy of Schaevitz
Engineering)

many readings are taken on each sample during this interval. Figure 12.7 is
a block diagram of a laboratory that contains a number of tensile test stations. Each station may perform tension, compression, or flexure tests on
various specimens. A CRT display and a keyboard at each station enable
the operator to enter fixed data about the specimen before starting the test.
The computer accepts this information and notifies the operator by means
of the CRT and an audible signal when the test may be started.

The purpose of this feedback from the computer is to assure that a test
is not started when the computer is already engaged in a large number of
tasks. This prevents an overload situation. (This is analogous to a telephone
switching system designed for a reasonably large peak traffic load, which
does not return a dial tone to a caller when the system is at peak traffic.)
Thus a computerized laboratory test facility with 10 stations might provide
computer capacity in terms of speed and storage for only seven concurrent
tests, because the probability of eight or more stations running simultaneously is remote.

An LVDT load cell measures the applied force, and the resulting deformation is determined by an LVDT extensometer. During the test process,
the load cell and the extensometer are sampled at a relatively high rate. As
prescribed changes in applied force occur, the program stores the corresponding deformation measurement. In this way, only the data required to

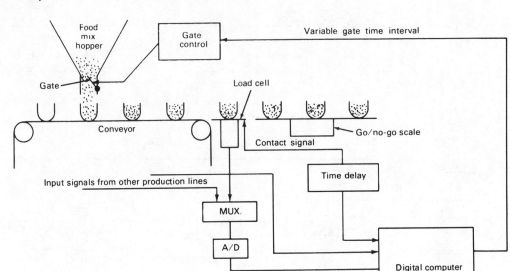

FIGURE 12.8
Block diagram of a computer-based food weighing system. (Courtesy of Schaevitz Engineering)

reconstitute the stress–strain curve after the test is completed are stored,
saving a great deal of computer storage.

At the conclusion of each run, the program computes the important
physical properties such as tensile strength, yield point, modulus of elastic-
ity, and proportional limit. These quantities can be displayed on the CRT
for visual observation and/or printed out on a central teleprinter at an
operator request.

In some applications a number of runs may be made on the same
material, and then an average for each physical property may be computed
from the data sets. The standard deviation and other statistical information
may also be computed using the sets of data obtained from the various runs
on the same material.

Figure 12.8 is an example of a computer-based system controlling food
packaging. The object is to minimize overweight. A food mix is automatically fed from a hopper into individual packages as they pass under the
hopper. The packages are then sealed and weighed on a go–no go scale at
the end of the line. To minimize rejections, the gate in the hopper is kept
open for an interval longer than that required when the mix is at its lowest
density. Consequently, whenever the material density is high, there is a
considerable amount of loss in overweight.

The remedy is to install an LVDT load cell at the end of each production
line and feed the output of the load cells into the computer-based control

system. It is not practical to control the hopper gate as a function of each weighing. Instead, the program averages a significant number of measurements. It then feeds back a signal to the hopper that modulates the gate time as a function of the averaging computation. The process of averaging is preceded by a statistical weighting process that rejects any obviously extraneous values so that they do not contribute to the average.

In this kind of application, the computer would have much free time, even if it were handling 10 production lines. Therefore, the computer can be used simultaneously for other tasks. For instance, in the process of computing averages, the number of packages going through each line must be counted. The counter for each line can be a location in the core memory of the computer, providing an automatic inventory record for each line that can be printed out periodically, or as an operator requires.

Computers can also greatly facilitate the testing of automotive or aircraft engines. The computer is used in two ways:

1. It sequences the engine test in accordance with scheduled speeds and loads.

2. It records all the variables of interest for each of the sequences (e.g., fuel and oil consumption at various temperatures).

A block diagram for a typical test configuration is shown in Figure 12.9. An operator at the console adjacent to the test stand types in the pertinent header information (engine type and serial number), which is transmitted

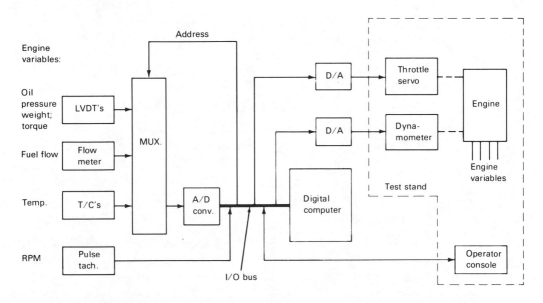

FIGURE 12.9
Block diagram of a computer-based engine test stand. (Courtesy of Schaevitz Engineering)

to the computer. The computer sends the information back to the teletype-writer or CRT display so that the operator can verify the accuracy of the header information the computer has received. The operator places the engine in an idling condition and signals the computer that the engine is ready for test. The computer has various test sequences for the various engine types stored in it. The program selects the proper sequence and transmits set points for the speed and dynamometer controllers. The computer then samples the engine speed and dynamometer torque. When these variables reach their steady-state value, the computer starts to integrate fuel flow and averages the values for critical pressures and temperatures.

Before starting the test, an LVDT load cell weighs the oil in the engine very accurately. The weight is stored in the computer memory. At the conclusion of the test, the oil is drained into the weighing tank and measured again. The computer prints out oil consumption along with all the other test variables. In the engine test stand, thermocouples measure temperatures, while LVDTs are used for weighing, pressure measurements, and dynamometer torque measurements.

12.5
A/D TECHNIQUES FOR HIGH-PERFORMANCE DATA-ACQUISITION SYSTEMS [d]

An analog-to-digital converter (A/D or ADC) for a high-performance data-acquisition system (DAS) must simultaneously cope with fast-changing analog input signals while interfacing with external, often computer-based, logic.

To properly select an ADC for this application, the designer should understand the following:

1. The operating environment a DAS presents to an ADC.

2. The conversion process, the constraints it imposes upon maximum input frequency, and the ability of the sample and hold amplifier to ease these constraints.

3. The significance of key hardware specifications and their possible cost–performance trade-offs.

4. Desirable ADC interface capabilities for "intelligent" applications.

The intent of this section is to simplify the designer's and student's task of selecting an ADC for fast-signal, computer-oriented DAS applications. To better achieve this, we will confine the discussion to ADC and DAS

---

[d] The material in Section 12.5 is from "Application Techniques AT-801, Analog-to-Digital Techniques for High Performance Data Acquisition Systems," and is used courtesy of Dynamic Measurements Corp., Winchester, Mass.

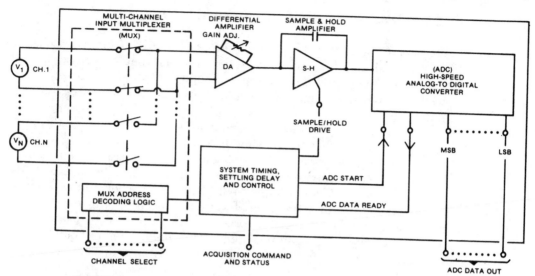

FIGURE 12.10

Simplified diagram of differential-input data acquisition system. (Courtesy of Dynamic Measurement Corp., Winchester, Mass.)

characteristics that pertain to speed of response and computer interfacing.

Basically, a DAS is a signal-conditioned, switched ADC that digitizes multiplexed analog inputs. Moderate- to high-speed designs (Figure 12.10) are implemented in the following manner:

1. A single-ended or differential-input analog multiplexer (MUX) under digital command connects one of many external signals to a differential amplifier (DA).

2. The DA rejects common-mode input signals, impedance-buffers the MUX to minimize input source loading, and gain-scales the analog signal inputs. The DA output provides the analog input signal to a sample and hold amplifier (S-H).

3. Under digital command, the S-H produces an output that either tracks a time-varying analog input or retains the signal level at, or very near, the time a HOLD command is given. This output provides an ADC with an essentially constant input during conversion, permitting precise sampling of fast signals.

4. The ADC, which digitizes the S-H output, also generates a bilevel DATA READY signal that indicates whether conversion is in process or complete. Conversion can be initiated by an external digital START command, which is a useful feature for DAS applications.

5. Finally, system timing and control logic coordinate channel selection and conversion in response to external acquisition request.

12.5.1 Factors That Affect System Speed

The speed of a DAS, its throughput rate, is generally expressed in channels per second. It should be pointed out that system throughput rate defines the number of sequential channels per second a DAS can digitize, not the maximum allowable signal frequency; the ADC/S-H combination determines that.

Each DAS component requires time to perform its task, and the sum of these times establishes the maximum system throughput rate. For a sense of proportion, consider a DAS utilizing an ADC that can convert 12 bits in $2\mu s$, with a corresponding conversion throughput rate of 0.500 MHz (DMC's Model 2809, for example). While the MUX can channel-switch very quickly, the DA settling time, which varies with gain setting, can far exceed the conversion time of the ADC; to a far lesser degree, the S-H settling time introduces some small additional delay. Thus the throughput rate of a high-speed DAS is generally limited by the analog settling times rather than the ADC.

12.5.2 The Need for a High-speed A/D

Analog-to-digital conversion is a sampling process that digitizes the level of an analog input signal at different points in time. The duration and frequency of sampling (the conversion time and throughput rate of the ADC)

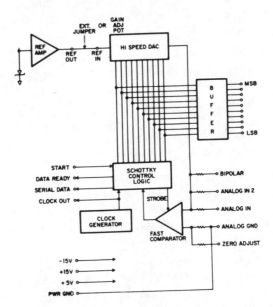

FIGURE 12.11
Analog-to-digital converter. (Courtesy of Dynamic Measurement Corp., Winchester, Mass.)

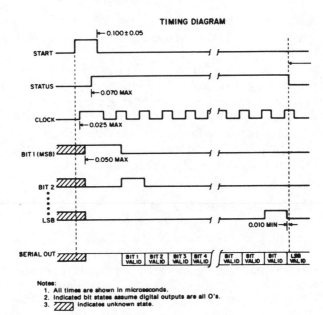

Notes:
1. All times are shown in microseconds.
2. Indicated bit states assume digital outputs are all 0's.
3. ▨ indicates unknown state.

FIGURE 12.12
A/D conversion cycle. (Courtesy of Dynamic
Measurement Corp., Winchester, Mass.)

assume crucial importance for increasing rates of change of the input signal. The reasons for this are twofold:

1. The Nyquist criterion states that to avoid *aliasing* or *foldover* errors the sampling rate must be no less than twice the highest frequency component of the input signal. Unless the input waveform is known, two points will give insufficient data, and higher sampling rates will be required. The short conversion time and the corresponding higher throughput rate of a high-speed ADC offer a practical and direct solution. The Model 2809, for example, can sample at a 500-kHz rate.

2. An equally important consideration relates the signal and ADC dynamics. The ADC under discussion utilizes the successive approximation technique (Figure 12.11).

This approach compares the analog input signal to the output of a high-speed, precision digital-to-analog converter (DAC). An internal clock first turns on the most significant bit (MSB) of the DAC, and a high-speed comparator determines whether the analog signal is less than, equal to, or greater than the MSB contribution. If the analog input is less than the MSB, the comparator assumes a state that causes the MSB data output line to be at a logical 0 level; the MSB data output will be a 1 when the analog input equals or exceeds the DAC input.

On succeeding clock pulses, each lower-order DAC bit is compared in sequence and in the same manner until, finally, the least significant bit

(LSB) is weighed. Following this, the STATUS or DATA READY output line, which changed state at the start of conversion, resumes its normal, nonconverting state.

If the analog input signal changes while conversion is in process, it changes by as little as $\frac{1}{2}$ LSB, the lowest-order bit reading of the ADC becomes meaningless. The timing diagram of Figure 12.12 illustrates the method of successive approximation and should help to reinforce this single, fundamental rule: *For an N-bit ADC to deliver N-bit performance, the input signal variation during conversion must be less than $\pm\frac{1}{2}LSB$.* This brings us to an interesting question.

12.5.3 How Fast Is Fast?

A 12-bit, 2-μs ADC capable of digitizing a $\pm$10-V analog signal is considered a "fast" (if not state-of-the-art) successive approximation device. It can take 500,000 signal samples per second and that, too, is "fast." But without further input conditioning, this candidate for high-speed DAS service would be hard pressed to perform a 12-bit accurate conversion on even a 20-Hz signal. This can be demonstrated by developing a quantitative expression for the fundamental rule presented in the discussion of Figure 12.12.

Consider the ADC response to a 10-V peak sinusoidal input signal:

$$e = A \sin \omega t \tag{12.1}$$

where A = the peak input voltage = 10 V. The rate of change of input will be

$$\frac{de}{dt} = A\omega \cos \omega t = 10\omega \cos \omega t \tag{12.2}$$

The maximum rate of change occurs at the signal zero crossover, where $\cos \omega t = 1$, or

$$\frac{de}{dt}_{\max} = 10\omega = 10(2\pi F) \tag{12.3}$$

where

$$f = \text{maximum frequency for } \tfrac{1}{2} \text{ LSB error}$$

$$de \rightarrow \Delta e \leq 2.44 \text{ mV for a } \tfrac{1}{2} \text{ LSB error}$$

$$dt \rightarrow \Delta t = 2 \ \mu\text{s, the ADC conversion time.}$$

Then the maximum frequency for maintaining an error voltage $\leq \frac{1}{2}$ LSB is

$$f \leq \frac{\Delta e}{20\pi \ \Delta t} \leq \frac{2.44 \times 10^{-3}}{20 \times 3.14 \times 2 \times 10^{-6}} \approx 19 \text{ Hz} \tag{12.4}$$

This response is totally inadequate for *any* application involving analog signals of even moderate frequency. As will be shown, the S-H will provide orders of magnitude of improvement.

12.5.4 Speed up with a Sample and Hold

The S-H (Figure 12.13) is a digitally commanded, two-mode, high-speed amplifier. For example, a logic 1 applied to the control input of a DMC 1410 series sample and hold amplifier causes the output to *sample* or follow the analog input signal. For a logic 0 control input, the S-H output will *hold* or remain fixed at a level corresponding to the input very shortly after the HOLD command occurs. Though the analog signal continues to vary, the S-H presents the ADC with an essentially constant input. While the S-H does not respond instantaneously to a SAMPLE-TO-HOLD command, the delay (the window of time uncertainty) is far less than that of the ADC. The delay consists of an essentially fixed component called *aperture delay time* (ADT) and a tolerance component known as *aperture uncertainty time* (AUT). Both components are specifications that describe the time uncertainty for an S-H to achieve a sample-to-hold transition. These input parameters only define the time interval during which the S-H will reconfigure so as to hold an input valve following assertion of a HOLD command; the S-H output signal will still require time to settle. The sample and hold amplifier abbreviation, S-H, is the same as S/H used in Chapter 4.

Because of its nearly constant nature, ADT can be compensated for by

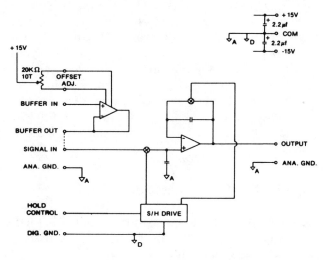

FIGURE 12.13
Simplified diagram of 1410 series sample and holds. (Courtesy of Dynamic Measurement Corp., Winchester, Mass.)

introducing an external system delay. Even if uncompensated, ADT would merely delay each repetitive sample by the same interval of time. For many applications, this presents no problem.

The reason for and the power of the S-H is AUT. Typically 200 ps in the 1410 series, an S-H so endowed can quickly capture the value of a fast-moving input signal and, shortly after, present the ADC with an essentially constant input. This increases the maximum permissible input frequency for a successive approximation ADC by the ratio of conversion time (t_{con}) to AUT. This *speed-up factor* can be expressed as $t_{con}/$AUT.

The impact of this result is best shown by an example. Connecting a 1410 series S-H to the input of the 12-bit, 2-μs ADC of the previous example, we have

$$\text{ADC maximum input frequency} = f_{adc} \approx 19 \text{ Hz}$$

$$\text{ADC conversion time} = 2 \ \mu\text{s}$$

$$\text{S-H AUT} = 200 \text{ ps}$$

$$\text{Speed factor} = \frac{t_{con}}{\text{AUT}} = \frac{2 \ \mu\text{s}}{200 \text{ ps}} = 1 \times 10^4$$

The maximum input frequency, F_{max}, for which a S-H/ADC combination will maintain errors to within $\frac{1}{2}$ LSB is

$$f_{max} = \text{speed factor} \times f_{adc}$$

$$= 1 \times 10^4 \times 19 \text{ Hz} = 190 \text{ kHz} \tag{12.5}$$

This performance is based upon a 10-V p-p sinusoidal input. With a 1-MHz full power bandwidth, a 20 V/μs skew rate, and a 7.5-MHz small signal bandwidth, a 1410 series S-H is amply equipped to deliver this level of improvement. However, power supply bypassing, proper grounding practices, and avoidance of excessive load capacitance on the S-H output are crucial to success.

The S-H/ADC system described can accommodate even higher input frequencies provided that the input amplitude is reduced. But care must be taken not to exceed the large and small signal capabilities of the amplifier internal to the S-H.

12.5.5 Final Considerations

The S-H/ADC data-conversion system has been discussed as a DAS system component. The system of Figure 12.14 utilizes two external tristate data output buffers for microcomputer bus interfacing. The buffers permit 2-byte (or nibble) data transfers and, for bus unloading, assume a high-impedance state when disabled. The external Schmitt triggers provide required sequence delays.

When considering a DAS application, the designer should realize that

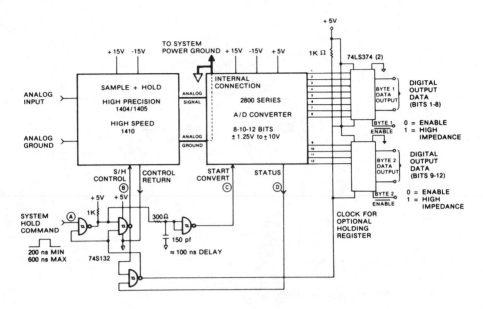

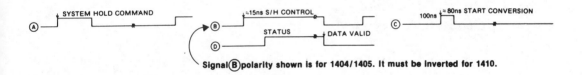

Signal Ⓑ polarity shown is for 1404/1405. It must be inverted for 1410.

FIGURE 12.14
A high-speed sample and hold, analog-to-digital
converter system with CMOS-compatible tristate
data outputs. (Courtesy of Dynamic Measurement
Corp., Winchester, Mass.)

the multiplexed analog inputs can present the conversion system with full-
scale step inputs. The *acquisition time* delay resulting from these input
level variations is specified and must be allowed for. In a high-speed S-H,
the voltage across the storage capacitor changes with temperature, and
quite rapidly with time; this is specified as *droop rate*. The designer must
limit time in the HOLD mode to avoid problems with this error.

Although we have focused on the dynamics of high-speed A/D conversion, static and environmental performance criteria must not be overlooked.
Only the proper combination of both will provide the required system
response.

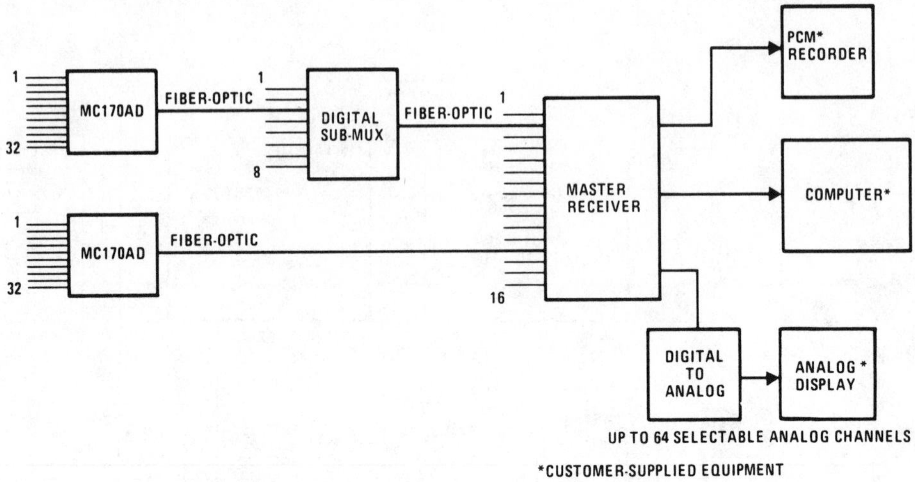

FIGURE 12.15
HD310 High-Speed Data Acquisition System.
(Courtesy of Validyne Engineering Corp.)

12.6
A PRACTICAL HIGH-SPEED DATA-ACQUISITION SYSTEM[e]

The Validyne Engineering Corporation's HD310 High-Speed Data Acquisition System is shown in Figure 12.15. This high-speed, low-cost modular system is specifically designed to provide remote data-gathering and signal-conditioning capability and to transmit the multiplexed data to a central location for access by a computer, as well as any other data-handling or bulk storage peripherals. This system could also be treated as a computer-based system for transducers.

The HD310 High Speed Data Acquisition System can process up to 4096 channels of data, multiplex the data, and transmit it over fiber-optic cables to a master receiver. This receiver acts as a central data bank, which may, under manual or computer control, output any or all data to any type of data-handling equipment, including most computers.

The MC170AD Remote Multiplexer Unit shown in Figure 12.15 provides the signal-conditioning, multiplexing and A/D conversion capabilities to process and transmit up to 32 channels of data. Each channel of data may be an analog signal or four digital (on–off) signals (up to 100 per MC170AD unit). The MC170AD unit transmits the data directly, or through a MX211 Digital Sub-Multiplexer, to the master receiver, which can receive

---

[e] The material in Section 12.6 is used courtesy of Validyne Engineering Corp. and is reprinted with their permission.

data from up to 16 MC170AD or MX311 units. As each digital submultiplexer can handle up to 8 MC170AD inputs, a maximum configuration of 128 MC170AD units can be accommodated, for a total of 4096 input channels.

The master receiver's random access memory in Figure 12.15 stores one scan of each input channel and continuously updates the data for each channel a minimum of 260 times every second. The output from the master receiver may be digital [to computer, pulse code modulation (PCM) recorder, disk, etc.] or analog through digital-to-analog (D/A) converters. Input to a computer is accomplished through a 16-bit parallel transmission link on a continuous or single-scan basis. The data are output to the computer's direct access memory according to a computer-programmable channel list stored in the master receiver, and transmission is initiated by direct computer command. This allows the computer to receive data for any or all channels as slow or as fast as required, limited only by the computer's capabilities. The HD310 system may also interface to the computer via a serial data link (RS232 type) if desired.

The master receiver may also output data to a PCM recorder or disc through an output port, independent of the computer interface. This port consists of 16 parallel data channels and one time channel. Each data channel passes randomized serial data directly to the PCM recorder or disc as it comes into the master receiver. To retrieve the data, the PCM recorder or disc data are simply played back through the master receiver and accessed just as if they were live data.

12.6.1 System Operation

The HD310 system is designed to have a constant word transmission rate of 71,500 words per second for each of 16 possible input links into the master receiver. Each input link may handle from 1 to 256 separate data channels such that each channel is scanned from 262 to 23,800 times each second, depending on the number of channels in the link. A block diagram of the HD310 High-Speed Acquisition System is shown in Figure 12.16.

Although the scanning rate for the HD310 system is fixed for a given system configuration, all channels are held in a random-access memory for recall by the computer. The data in the memory are continually updated by new scans of the system; therefore, even at a very low sample rate, the computer receives the data from the most recent scan of the HD310. By controlling the rate the computer accesses the HD310 memory, the computer has complete control over the effective scan rate.

The HD310 system offers two methods for performing a limit checking on the data channels being monitored. One method consists of using digital-to-analog converters (DAC) to monitor the signals that require limit checking. Each DAC can store high and low limits and can perform a limit check on one data channel automatically. When a limit is exceeded, the DAC will automatically activate an alarm bus, which may signal the computer, turn on

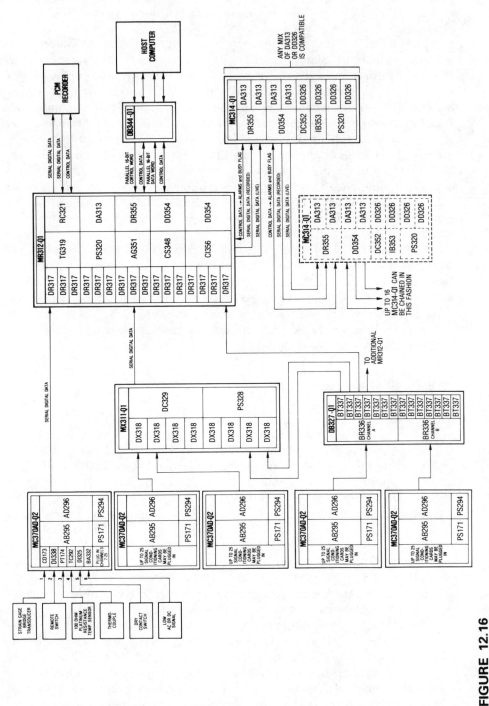

FIGURE 12.16

Block diagram of the HD310 High-Speed Data Acquisition System. (Courtesy of Validyne Engineering Corp.)

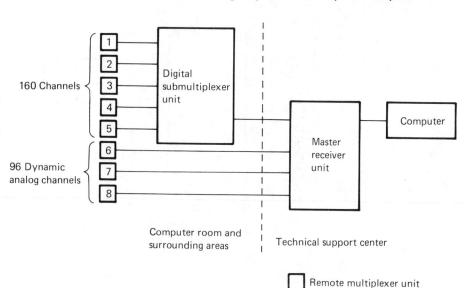

FIGURE 12.17
A 256-channel system, plan A. (Courtesy of Validyne Engineering Corp.)

a PCM recorder, and so on. A total of 128 signals may be automatically monitored in this manner.

Another method currently under development uses a separate limit check module, which will store separate high and low limits for each of the 4096 possible data channels. This module will check each data scan against its limits and send an interrupt signal to the computer whenever an out-of-limit condition is encountered.

12.6.2 Expandability

Since the HD310 system is completely modular (all modules can be mounted in standard 19-in. racks), the system may be expanded at any time without any hardware modification. Furthermore, the MC170AD remote multiplexer unit is seismically and environmentally qualified to IEEE 323-1974, and when used with appropriate signal-conditioning modules may be connected directly to class 1E circuits without additional isolation.

Figures 12.17 through 12.20 show sample system configurations for 256- and 1500-channel systems. Individual system configurations should take into account the type and location of signal taps, future expansion possibilities, and maximum scan rates desired for individual signals. The HD310 system may be expanded to handle a total of 4096 channels of data

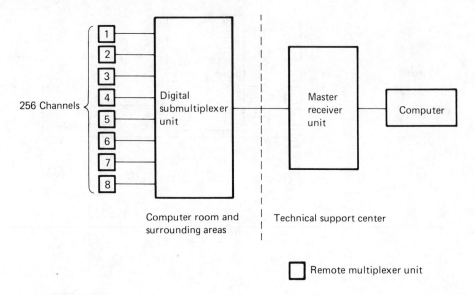

256 Channels

Digital submultiplexer unit

Master receiver unit

Computer

Computer room and surrounding areas

Technical support center

☐ Remote multiplexer unit

FIGURE 12.18
A 256-channel system, plan B. (Courtesy of Validyne Engineering Corp.)

for a maximum integrated throughput of 1,073,000 data points per second (see Figure 12.10).

Another feature of the HD310 system is its ability to provide independent, redundant inputs to several computers, if required. This is accomplished by using DB327 digital buffer units at either the output of the MC170AD remote multiplexer or the output of the MX311 digital multiplexer. Each input to the DB327 is isolated, split, and retransmitted over from two to seven parallel lines to the required number of master receivers. Each master receiver receives identical inputs but interfaces with separate computer systems independent from each other, as shown in Figure 12.21.

12.7
THE SINETRAC CONCEPT[f]

There is a simple method of transferring real-world physical data (temperature, pressure, etc.) into and out of a computer. Since a computer cannot directly accept analog signal voltages from temperature and pressure transducers and other sensors, an analog-to-digital and/or digital-to-analog conversion system such as Datel's SineTrac System is needed for

---

[f]The material in Section 12.7 is used courtesy of Datel-Intersil, Inc.

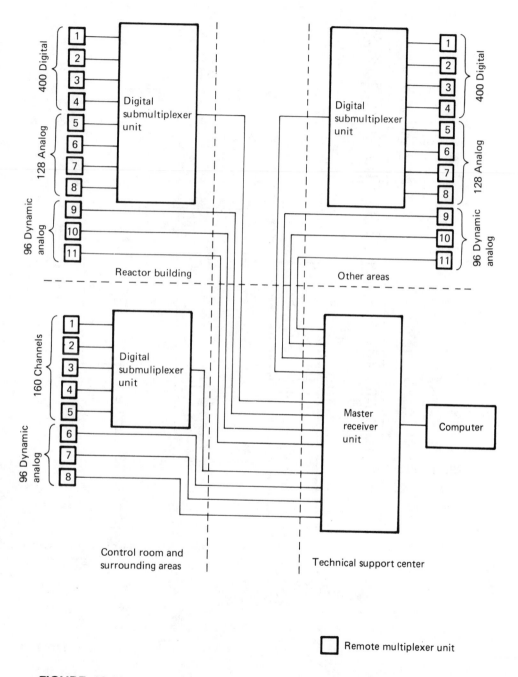

FIGURE 12.19
Expanded plan A. (Courtesy of Validyne Engineering Corp.)

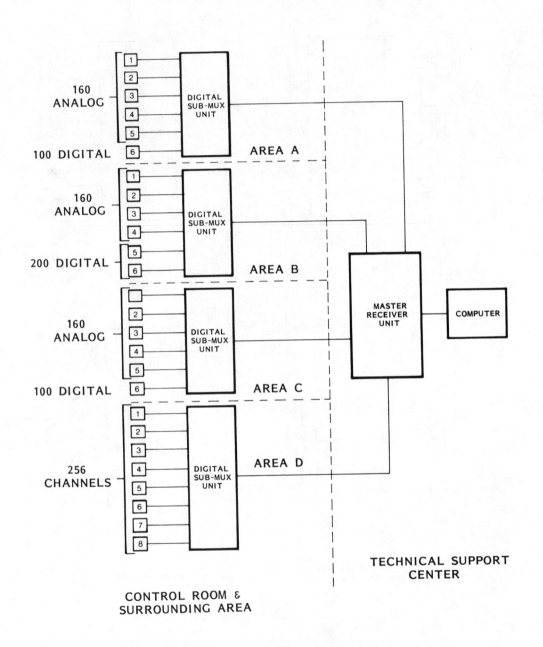

FIGURE 12.20
Expanded plan B. (Courtesy of Validyne Engineering Corp.)

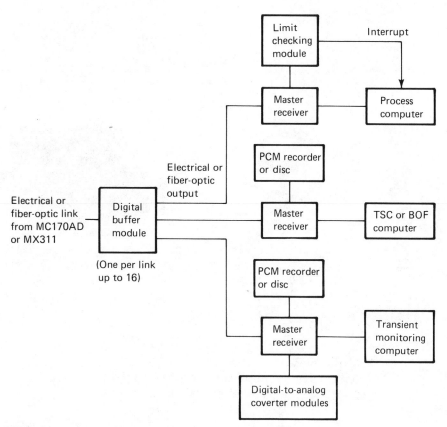

FIGURE 12.21
Expanded data output scheme. (Courtesy of Validyne Engineering Corp.)

interfacing to external analog signals. The Datel SineTrac concept, shown in Figure 12.22, places analog-to-digital and digital-to-analog converters *inside* a computer housing on slide-in circuit cards containing their own controller interface logic. These cards are dedicated to a specific computer family and are mechanically compatible to the computer's card guides, bus compatible to the backplane electrical pinout, and software compatible to that computer's assembly language and higher-level programs. The A/D–D/A board appears to the computer as either a register-addressable peripheral I/O device or as a reserved portion of core memory (a memory-mapped peripheral).

The idea of placing the A/D and D/A components inside the computer was not possible before the advent of Datel's miniature high-performance microcircuit converters. The SineTrac system concept immediately solves two application problems that have hampered low-cost data-acquisition

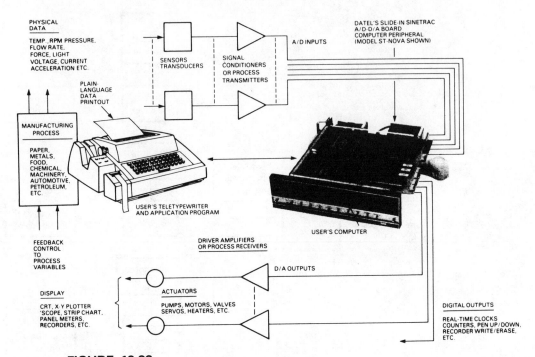

FIGURE 12.22

A typical system application of the SineTrac concept. (Courtesy of Datel-Intersil, Inc.)

designs for years. SineTrac relieves the user of risky, tedious interface circuit design with a proven, production-volume system. This also eliminates expensive digital cabling normally required for connecting to external A/D–D/A components. Only analog signal cabling is required.

The SineTrac hardware includes diagnostic software in assembly language on paper tape for convenient teletypewriter entry into the computer. SineTrac's diagnostic software gives a strong head start in the other problem areas of application programming. The paper tapes allow analog input signals to be printed out on the TTY just as soon as the signals are connected and the diagnostic program is loaded. Complete program listings in the supplied manuals allow the user to extract useful portions into the final program.

A dramatic new application area, particularly for SineTrac-equipped microcomputers, is distributed processing, whereby on-line factory personnel using a local SineTrac control computer can closely monitor and control the manufacturing process they are responsible for. Yet the existing central computer control room can retain observation of all plant processes by communicating through remote SineTrac controllers at the process site.

Most SineTrac peripherals incorporate many hardware functions to

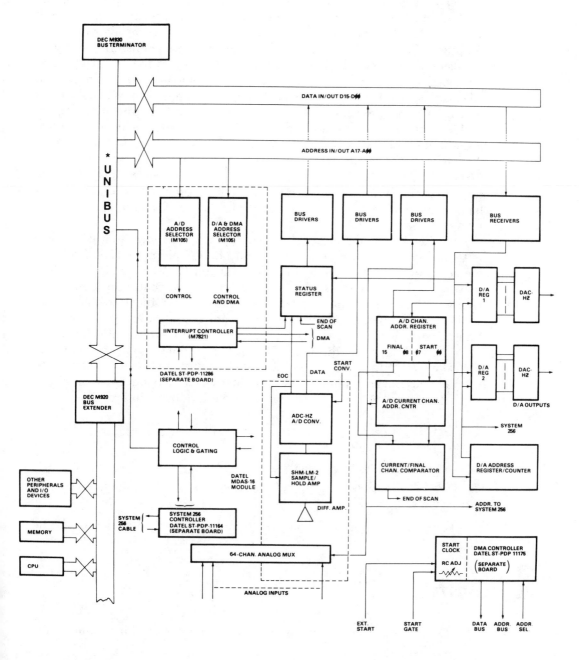

FIGURE 12.23
Typical block diagram of the Datel ST-PDP Sine-Trac concept. (Courtesy of Datel-Intersil, Inc.)

relieve software requirements for highest speeds and easiest programming. Such features as autostart, autosequencing of channel addressing, on-board start, and DMA clocks and end of scan–end of conversion interrupt logic simplify programming while working in many applications. This architecture is used because high-performance A/D–D/A peripherals (many channels, high speeds) require short, fast routines for maximum throughput. Off-loading the hardware functions onto software routines would slow down the throughput rate. This situation is similar to hardware multiply–divide. Although a subroutine can multiply or divide, hardware is much faster and eliminates a separate routine.

These high speeds are especially useful if other fast peripherals are involved (such as CRT terminals, D/As, etc.) or devices with lengthy protocol (e.g., communication with a remote host processor through a wide bandwidth modem) or complex signal processing (fast Fourier transform for spectral analysis).

The block diagram for Datel's ST-PDP SineTrac shown in Figure 12.23 is typical of the internal organization of other SineTrac series. Common elements to all these systems include A/D and D/A conversion sections, power supply, bus address decoders, and interface controller logic. Such computer-based systems are applicable for use for the transducers described in Chapter 2. The temperature transducers, strain gages, voltage transmitters, any ac signal, and other transducers are usually fed to an interface and signal-conditioning device. The output is fed to an integrating dual slope A/D converter providing high immunity to line frequency noise components and an automatic auto zero in each conversion. Once the signal is digital using logic programming, it can be fed to the following:

1. Display
    a. Decimal point selection
2. Digital status and control signals
3. Parallel binary-code decimal display, printer or automatic control systems.

The power supply can convert the output signal to analog or digital signals.

12.8
REVIEW QUESTIONS

1. Describe a data acquisition or conversion system for thermocouple application.
2. Describe the use of a data logger.
3. Describe a computer-based LVDT measuring system for automobile engine testing.

4. Describe briefly the concepts of an analog-to-digital high-performance data-acquisition system.

5. Discuss the use of a receiver in a high-speed data-acquisition system.

6. Draw a simplified block diagram of a high-speed data-acquisition system.

7. Describe a system to measure noise in a temperature transducer system using thermistors.

8. Describe the use of fiber-optics in a high-speed data-acquisition system.

9. Discuss a method of transferring real-world physical parameters, such as pressure, in and out of a computer.

10. List 10 measurement functions usable in a data acquisition or conversion system.

12.9
REFERENCES

1. Daniel H. Sheingold, *Transducer Interfacing Handbook,* Analog Devices, Norwood, Mass. 1980, 1981.

2. *Hewlett Packard Journal,* vol. 32, no. 7, July 1981.

3. Edward E. Herceg, *Handbook of Measurement and Control,* 2nd printing, Schaevitz Engineering, Pennsauken, N.J., April 1980.

4. "SineTrac A/D–D/A Peripherals," Datel-Intersil, Inc., Mansfield, Mass., July 1978.

5. Kepco Power Supply Catalog and Handbook, Flushing, N.Y. 1982.

6. *Digital Panel Instruments, Designer's Handbook and Catalog 1980/81,* 1st ed., Analogic Corp., Wakefield, Mass., 1979.

7. *1982 Design Engineers Handbook and Selection Guide A/D and D/A Signal Conditioning and Control Modules,* Analogic Corp., Wakefield, Mass., 1982.

8. Daniel H. Sheingood, "Analog-Digital Conversion Notes," Analog Devices, Inc., Norwood, Mass., 1980.

9. *The Analogic Data-Conversion Systems Digest,* 4th ed., Analogic Corp., Wakefield, Mass., 1981.

10. Eugene L. Zuck, *Data Acquisition and Conversion Handbook,* 4th Printing, Datel-Intersil, Inc., Mansfield, Mass., Sept. 1981.

11. *Bell and Howell Pressure Transducer Handbook,* 3rd Printing, Bell and Howell CEC Division, Pasadena, Calif., 1977.

13
Electronic
Counters[a]

13.1
INTRODUCTION

In this chapter, we will analyze electronic counters. It is an important electronic instrumentation standard where counting is required. In 1976, Global Specialties Corporation responded to the increasing demand for a low-cost frequency counter with the first battery-operated portable frequency counter, the MAX-100.

Taking advantage of new large-scale-integrated (LSI) circuit technology, the Model 5001 universal counter-timer was introduced. It is the intent of the authors to acquaint our audience with these counters.

It is also important to realize that many manufacturers have produced electronic counters and they vary from manufacturer to manufacturer.

All frequency counters, with their inherent need for precision, challenge the present engineering advances. As a result, an innovative temperature-controlled crystal oscillator that combines precision with the

---

[a] The material in Chapter 13 is reprinted courtesy of the Global Specialties Corp., New Haven, Conn.

frequency counter has been developed by the Global Specialties Corporation, the Model 6001 benchtop 650-MHz frequency counter.

Recognizing that a stable, cost-effective time base had extensive applications, the Model 4401 precision frequency standard has been designed for the calibration of time and frequency counters, oscilloscopes, or as a precision clock source for microprocessors, and more. It provides a unique, inexpensive source of discrete, selectable precision frequencies, which are factory calibrated to NBS standards. These electronic counters will further offer to electronic engineers an insight into the theory and problems of present counter principles.

In the sections that follow we will examine a 100-MHz and 650-MHz frequency counters, universal counter-timers, and a frequency standard.

13.2
THE 100-MHz FREQUENCY COUNTER

The latest LSI technology and advanced engineering technology have made possible the MAX-100 frequency counter (Figure 13.1), which provides precision, accurate frequency measurements, and simple operation. The following shows how the 100-MHz frequency counter works. Figure 13.2 shows a block diagram of the MAX-100. It consists of four sections:

1. Front end (input amplifier)
2. Main counter
3. Display
4. Time base

The front end processes the input signal and conditions it for the high-speed decade counter that follows it. The counter output is then fed into the main counter, where the major portion of signal processing is done. Measurements are sorted into digits and updated periodically by precise timing commands from the time-base section. Once a measurement is ready for

FIGURE 13.1
MAX-100 frequency counter. (Courtesy of Global Specialties Corp.)

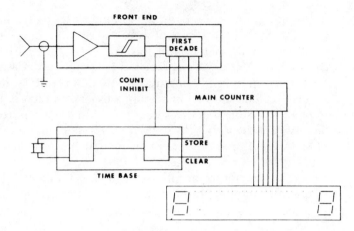

FIGURE 13.2
Block diagram of the MAX-100 frequency counter.
(Courtesy of Global Specialties Corp.)

display, it is converted from a binary code to a display code, and finally fed to the eight 7-segment LED readouts.

The time-base section generates a precision 1-s gate pulse for the main counter and the required signals to process the input signal. Each section will be dealt with in more detail, but first consider the basic *timing cycle* of the MAX-100.

Figure 13.3(a) shows a timing diagram, with the series of pulses required to process the input signal. The top waveform shows a 1-s downtime between points A and B, followed by a $\frac{1}{6}$-s pulse width. This is the *count inhibit waveform* that controls the first decade counter in the front end. Since all signals coming in must first pass through the first decade divider, control of the main counter can be achieved by switching the first decade divider on and off.

Assume that the main counter has been cleared and all digits are zero. At point A in the Figure 13.3(a), the first decade is switched on and starts to accumulate (count) incoming pulses. At point B in the Figure 13.3(a), the first decade counter is switched off and all pulses that have entered it during this time interval have been sorted and stored in the main counter.

At this point the update cycle begins. Figure 13.3(b) shows the *updated waveform* in detail. Shortly after the start of the update pulse, a *store pulse* commands the main counter (which has held all the pulses in temporary registers) to feed data into the storage latches. The latches (which have been storing data from a previous measurement) now accept the new measurement. The latches will store the new measurement until the next store pulse occurs.

Due to the manner in which the input signals are scanned, the display is always one measurement period (1 s) behind. This is not a major concern since the input frequency should be fairly constant and not change widely from measurement to measurement.

Next the *reset pulse* instructs the main counter to reset all temporary registers to zero and prepare for the next measurement cycle.

In asyncronous systems such as MAX-100, there is an inherent plus or minus one-count error. This occurs due to the random starting point of the timing cycle with respect to the input signal; that is, the timing cycle can start just before or after one cycle of the input signal.

The front end of the counter connects to the signal source under test and must therefore be able to withstand the wide range of conditions the owner or operator may encounter and still function properly. The MAX-100 is equipped with a diode-protected wideband FET preamp that can withstand vast overloads and still continue to function, while the high input impedance of the FET allows it to couple to any circuit with a minimum of loading.

Once a test signal has passed the FET preamp it is presented to a series of low-noise preamps that amplify it to a suitable level to operate a Schmitt trigger. To improve its stability and reduce noise, the signal is converted into a series of fast rise-time pulses, after which it passes on to the first decade divider.

The first decade divider is a specially designed high-speed decade counter whose output data are in a coded form that the main counter cannot decipher. Therefore, a decoder is inserted between the decade divider and the main counter to recode the data into a format that the main counter can process.

This switching of codes back and forth occurs quite often in digital circuitry and is used several times in the MAX-100.

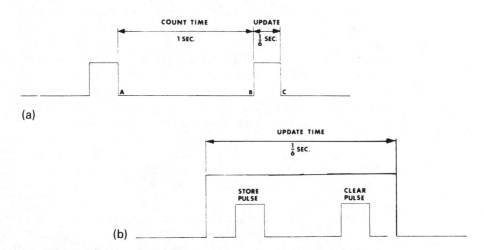

(a)

(b)

FIGURE 13.3

Timing cycle of the 100-MHz frequency counter.
(Courtesy of Global Specialties Corp.)

The main counter is a 40-pin large-scale integrated-circuit chip containing all the latches and registers required to display an 8-digit binary coded digital (BDC) output. It also contains the logic to produce an 8-bit strobe system, scan oscillator, lead zero blanking system, overflow latch, and decimal point blanking.

Here the data are processed as described previously. Held in registers during the measurement period, the new measurement is finally transferred to the display latches. Each latch services one digit of the display and has four output lines. The display scanning logic sequentially samples the four lines. The displays are made up of seven individual LED segments per digit, which in combination can produce any number from 0 to 9. To produce a number on the display, the BDC outputs from the master counter must be converted into a code that is compatible with the 7-segment LED displays.

In most designs each display latch would require its own 7-segment decoder driver, with each decoder driver requiring a limiting resistor for each segment driven, and thus 7 segments times 8 digits, or 56 resistors and 8 decoder drivers. To reduce the circuitry cost, a multiplexing system has been incorporated that requires a single 7-segment decoder driver, seven limiting resistors, and eight driver transistors.

The system sequentially switches each 4-bit latch to the decoder and simultaneously turns on the associated display digit. Therefore, only one digit is active at any one time. The switching from digit to digit is extremely rapid and beyond the ability of the human eye to discern, thus the eye "sees" a continuous 8-digit display.

There is a twofold benefit in multiplexing the LED displays: first, circuit simplification and, second, an increase in the *efficiency* of the LED display.

Since each digit is on only one-eighth of the time, to achieve the same brightness as a continuous display, the LED current must be increased eight times. Interesting results are produced via this process. LED efficiency increases with current, a pulsed LED giving off more light than a steady-state LED for the same current. The human eye does not average pulsed light linearly. It tends to see pulsed light as actually *brighter* than it normally is, thus allowing the reduction in display current (which is a major portion of the power drain) and still maintaining a bright discernible display.

The time-base section is the master "clock" for the counter, and, as such, it must produce highly accurate stable pulses. It also must produce all the store, clear, reset, and update pulses needed to route the digital traffic correctly.

A specially selected 3.58-MHz crystal provides the precision frequency source for the time base. When used in conjunction with a special oscillator divider IC, it produces an ultrastable 60-Hz output, which is divided down to 1 Hz. The general specifications for the 100-MHz 8-digit MAX-100 frequency counter include the following:

1. **General specifications:**
   - Frequency: 5 Hz to 100 MHz
   - 8-digit display, 7-segment LED

2. **Input characteristics:**
   - Impedance: 1.5-MΩ resistance shunted with 10 pF
   - Connector: Phono jack
   - Coupling: AC
   - Sine-wave sensitivity: 40 mV, 5 Hz to 1 kHz
                              30 mV, 1 kHz to 100 kHZ
                              10 mV, 100 kHz to 10 MHz
                              40 mV, 10 MHz to 60 MHz
                              120 mV, 60 MHz to 100 MHz
   - Count accuracy: Time-base accuracy $\pm$ 1 count
   - Overload: Peak ac or dc

        5 Hz to 30 kHz, 200 V p-p
        30 kHz to 70 kHz, 150 V p-p
        70 kHz to 100 kHz, 125 V p-p
        100 kHz to 1.0 MHz, 42 V p-p
        1 MHz to 10 MHz, 13.8 V p-p
        10 MHz to 60 MHz, 6.4 V p-p
        60 MHz to 100 MHz, 5.4 V p-p

   - Time base:

        Crystal oscillator: 3.579545 MHz
        Temperature stability: 5° to 45°/°C
                               0.1 ppm/°C
        Aging rate: 10 ppm/yr
        Accuracy: $\pm$3 ppm at 25°C
        Trimmer adjustable: $\pm$40 ppm
        Gating time: 1.0 s
        Update time: $\frac{1}{6}$ s

   - Lead zero blanking for all but the least significant digit.
   - Automatic MHz decimal point; when the display shows greater than 1 MHz, decimal point appears between the sixth and seventh digit.
   - Low-battery indicator: when the battery voltage drops to 7.5 V, display will pulse on and off at a 1-s rate.
   - Batteries: 6 NiCad (fast charge) AA-type
               6 Alkaline AA-type

There are certain rules to follow for reliable readings. These are based on knowing the following:

1. Signal source characteristics

2. Counter input characteristics

3. Coupling between source and counter

The MAX-100 has an input sensitivity of less than 40 mV from 5 Hz to 50 MHz. However, if the signal source is basically noisy, the counter will read a meaningless jumble of numbers. The first rule is to know the signal-to-noise ratio. *Forty decibels or greater* (100:1) is needed to get a meaningful reading. Knowing the signal source also means knowing its frequency stability.

When testing CB transmitters, a clean crisp reading is always achieved because the transmitter is crystal controlled. However, many signal or function generators are *not* very frequency stable, being approximately in the 1% to 5% range. They are prone to frequency modulation, which appears to the counter as noise and causes jumbling of the lower-order digits.

The MAX-100 input circuit of a 330-Ω resistor in series with a 1-MΩ resistor shunted by a 10-pF capacitor (see Figure 13.4). At audio frequencies, the capacitive reactance of the 10-pF capacitor is so high that, effectively, the input network looks like several hundred kilohms to any circuit it is testing.

At RF frequencies such as CB, the 10-pF capacitor is now approaching 560-Ω impedance. Although the input impedance can never go below the 100-Ω input resistor, it can approach it. Therefore, some care should be taken in hooking up to RF systems.

When measuring the frequency of an RF oscillator coil, an effective impedance of 600 Ω is seen by the coil. In some cases the 600-Ω load may inhibit oscillation. There are, however, no loading problems when connecting to a 50-Ω transmission line termination, and the hookup does *not* affect the power match. The coupling device between the signal source and the counter can be a critical point in acquiring accurate readings.

Most shielded cable has approximately 25 pF of capacity per foot. A 4-ft cable will appear as a 100-pF capacitor. When it is connected to a source generator with a 100-kΩ output impedance that is generating a 150-kHz signal, the cable will load down the signal output to one-tenth its open-

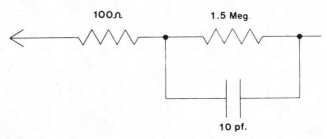

FIGURE 13.4
MAX-100 frequency counter input characteristics.
(Courtesy of Global Specialties Corp.)

circuit voltage. Therefore, it is important to (1) keep all lines as short as possible, and (2) understand that cables can load high-impedance signal sources and *detune* resonated signal sources.

Audio signals can have their special problems. One of the most common is high-frequency noise spikes on the audio signal. A small series of "glitches" or pulses at 30 MHz may get lost in the total 1-s count time. However, at 1 kHz the same series of glitches may represent 30% of the count. Therefore, it is most important to keep audio signals free of spikes. One remedy for this is to place a small bypass capacitor across the counter input leads (50 to 1000 pF).

Another problem encountered more often in audio than in RF testing is outside interference being sensed by the test setup. Audio circuitry in general uses higher impedance sources than RF and is, therefore, more sensitive to random power line spikes and outside interference. A good rule to follow is *always to use a shielded cable* such as the one provided with the MAX-100 when testing audio (100-IPC). *Keep unshielded connections as short as possible.*

The frequency counter can also be connected to a spectrum analyzer to make frequency measurements. When the input signal frequency exceeds 100 MHz, the lead zero blanking is removed and the most significant digit will blink. The counter will still read correctly, in most cases up to 110 MHz. However, input signals of *greater* than 300 mV may be needed in order to obtain a reading.

When the input frequency exceeds 1 MHz, a decimal point automatically appears between the sixth and seventh digits to indicate a signal in the megahertz range.

When the battery voltage drops below 8.5 V, the display will blink at a 1-s rate. The display will still read correctly and the counter will be usable for some time. The blinking display not only indicates a low battery voltage, but also reduces the display current by 50%, thereby extending the remaining battery life. The remaining usable life in a set of batteries depends on several factors: intermittent or continuous use, and the type of batteries used.

13.2.1 The 650-MHz Frequency Counter

The Global Specialties Model 6001 benchtop frequency counter is a high precision, 650-MHz frequency counter (Figure 13.5) designed for professional use where an accurate measurement of frequency is required. Its versatility and state-of-the-art design make it an ideal addition to and extremely useful in engineering laboratories, quality-control departments, calibration and service departments, scientific applications, transmitting and receiving stations, recording studios, as well as ham, CB radio, telephony, RC control, and in audio sales and service centers.

The 6001 measures frequency from 5 Hz up to 650 MHz in two over-

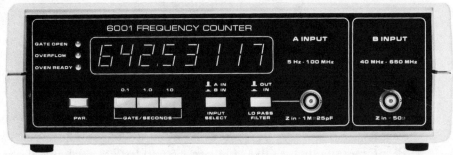

(a) Front panel

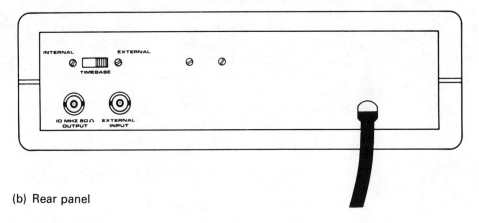

(b) Rear panel

FIGURE 13.5
The 6001 650-MHz frequency counter. (Courtesy
of Global Specialties Corp.)

lapping ranges with superior accuracy. The A input covers frequencies from
5 Hz through 100 MHz with an input impedance of 1 MΩ at 10 pF. In
addition, the A input has a push-button-activated low-pass filter that provides a 3 dB/octave roll-off starting at 50 kHz to limit high-frequency noise
from affecting audio and ultrasonic measurements. The B input is a constant
50-Ω input impedance with fuse protection providing a 50- to 650-MHz
frequency range. Both inputs are TTL compatible. The 6001 also features
a unit count mode. In this mode the 6001 counts and totals input pulses up
to 1 billion.

The 6001 features a bright, 0.43-in.-high, 8-digit LED display with a
built-in contrast enhancement filter to ensure legibility in high-ambient-
light environments. Display features include lead-zero blanking and will
present frequency at resolutions of 1 Hz on the low range and 10 Hz on the
high range up to 650 MHz.

For ease of use, the front panel controls have been kept to a minimum, which include three switch-selectable gate times providing 0.1-, 1.0-, and 10-s intervals, an input selection control (A input or B input), and a push-button-activated low-pass filter. Three panel LED indicators are included indicating gate, overflow, and oven-ready functions.

The internal time base for the 6001 is a uniquely designed, precision 10-MHz crystal oven oscillator with an accuracy of ±0.5 ppm from 0° to 40°C. Alternately, an external reference may be selected via a rear panel switch and connected through a BNC connector also mounted on the rear panel. A second rear panel BNC provides a buffered TTL-level oven oscillator output.

Typical applications include measuring transmitter frequency, testing and repairing audio equipment, measuring rpm, testing and repairing broadcast equipment, checking FCC compliance to transmission variance, calibrating signal generators and test equipment, telephony, and measuring microprocessor time bases.

Figure 13.6 shows a block diagram of the 6001. It consists of eleven sections:

1. Low-frequency front end (5 Hz to 100 MHz)
2. High-frequency front end (40 to 650 MHz)
3. Prescaler
4. Input selection logic
5. First decade counter
6. Second decade counter
7. Main counter
8. Display
9. 10-MHz "crystal oven" time base
10. Time-base divider chain
11. Gate selection logic

The dual-path front ends (high- and low-frequency inputs) process the input signals along with the divide-by-10 prescaler. The input selection logic decides which front end it will allow the signals to pass through. These signals are passed through the decade dividers into the main counter where the measurements are sorted into digits and are updated periodically (0.1, 1, and 10 s) by the internal or external time base. Once a measurement is ready for display, it is converted from binary code to display code and fed to the eight 7-segment LED readouts.

The time-base section generates a precision gate pulse for the main counter and the required signals to process the input signal. The time of the gate pulse is switch-selectable from 0.1, 1, and 10 s.

An external time base may be substituted via a switch-selectable input for more accuracy or scaling.

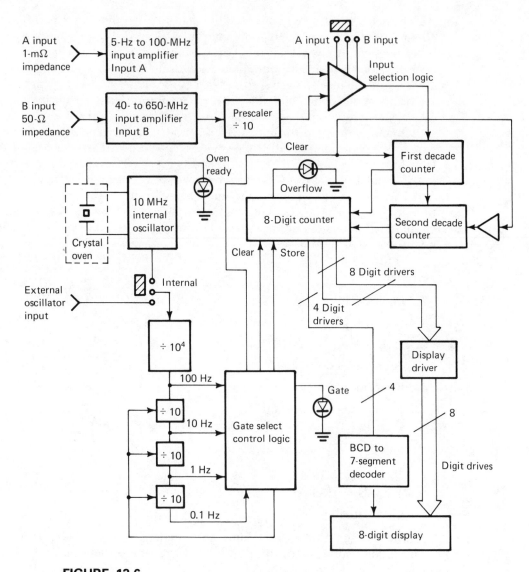

FIGURE 13.6
Block diagram of the 6001 650-MHz frequency
counter. (Courtesy of Global Specialties Corp.)

The 6001 specifications include the following:

1. Inputs:
   a. Two inputs, A and B, ac coupled, BNC connectors
   b. Impedance: 1 MΩ at 25 pF at A, 50 Ω at B
   c. Response: 5 Hz to 100 MHz at A, 40 to 650 MHz at B

d. Sensitivity:

- A input: 40 mV, 5 Hz to 1 kHz; 40 mV, 1 to 100 kHz; 20 mV, 100 kHz to 10 MHz; 40 mV, 10 to 60 MHz; 120 mV, 60 to 90 MHz; 200 mV, 90 to 100 MHz

- Maximum input voltage:

$$V \text{ (max) rms} \pm 0.707 \sqrt{11 + \frac{1.8 \times 109}{104 + \text{freq.}}}$$

Switchable low-pass filter for audio–ultrasonic range provides 3-dB/octave roll-off at approximately 50 kHz.

- B input: 100 mV, 40 to 450 MHz; 200 mV, 450 to 600 MHz; 250 mV, 600 to 650 MHz; fuse protected.

- Maximum input voltage: 5 V p-p

e. Auxiliary time-base input. Switch selectable, BNC connector.

f. Impedance: 50 Ω

g. Response: 1 to 25 MHz

h. Sensitivity: Standard TTL levels or sine greater than 2.5 V_{rms}, 10 V p-p max.

2. Output: Buffered time base (10-MHz crystal oven oscillator), dc-coupled through BNC connector on rear panel, standard TTL-level square wave.

- Drive: Buffered to drive 10 TTL loads.

3. Reference: Time base: 10-MHz crystal oven oscillator, ±0.5 ppm from 0° to 40°C; oven temperature, 55°C.

4. Modes:

a. Frequency mode: Indicates input frequency in megahertz; use internal or external 10-MHz time-base reference.

b. Scaling mode: Multiplies input frequency by factor of 0.1 to 2.5 to indicate in units other than megahertz; use 1- to 25-MHz external time base.

5. Controls: Power, gate time select (0.1, 1.0, 10 s); A or B input select, low-pass filter in–out, internal–external time base (rear panel).

6. Display: 8-digit, 7-segment, 0.43-in. LED display; decimal point indicates frequency in megahertz; lead-zero blanking; discrete overflow (counter overflow), gate (gate open) and oven-ready LEDs.

7. Power: 105 to 135 V ac, 57 to 63 Hz, 10 VA maximum (215 to 250 V ac, 50 to 60-Hz version available)

8. Operating temperature: 0° to 40°C

The Global Specialties 6001 frequency counter contains one of the most

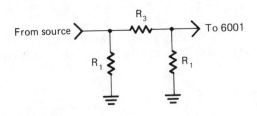

Attenuation in dB	Shunt R_1	Series R_3
3	292	18
6	151	37
10	96	71
20	61	248

FIGURE 13.7

Table of various values for some common 50-Ω attenuators. (Courtesy of Global Specialties Corp.)

advanced front ends available in a frequency counter today. It utilizes ECL logic to obtain high amplification with noise suppression circuitry to process the input signal source into a signal the 6001's counting logic can handle. And as a user convenience, it has on its low-frequency front end a low-pass filter to eliminate high-frequency noise that may be superimposed on low-frequency signals.

This noise can come from many sources: static pickup, unterminated cables, RF fields, ground loops, or the signal source itself. To eliminate this, you can shield the cable (the 6001 has special internal shielding surrounding its front end to stop RFI noise pickup), terminate the cable to stop ringing, break grounds to eliminate loops, and attenuate and filter the incoming signal.

An external attenuator can easily be added to the 6001, which will cut down both the signal and noise. The counter will now read just the signal and ignore the noise. Figure 13.7 shows a table of various values for some common 50-Ω attenuators.

The 6001 incorporates a low-pass filter for the low-frequency front end with a 50-kHz roll-off. If a low-pass filter is needed with a cut-off frequency other than 50 kHz, a simple low-pass filter can easily be constructed. Use the following formula and the diagram shown in Figure 13.8 to design your filter. Assume that you need a filter to eliminate all signals above 100 kHz; the following values apply. Assume that $R_1 = 1$ kΩ:

$$C = \frac{1}{(2)(3.14159)(1000)(100,000)}$$

$$= 1.59 \times 10^{-9}$$

$$= 1.590 \text{ pF}$$

By inserting this into the input of the 6001, all frequencies above 100 kHz will be rejected, while all frequencies below 100 kHz can be read. The 6001 is also a useful tool in the service and repair of DB, business radio, ham radio, RCC, and others. An optional antenna is available for quick testing of transmitters. It adapts quickly to the counter inputs via a BNC connector.

A carrier frequency can be measured simply and accurately by using

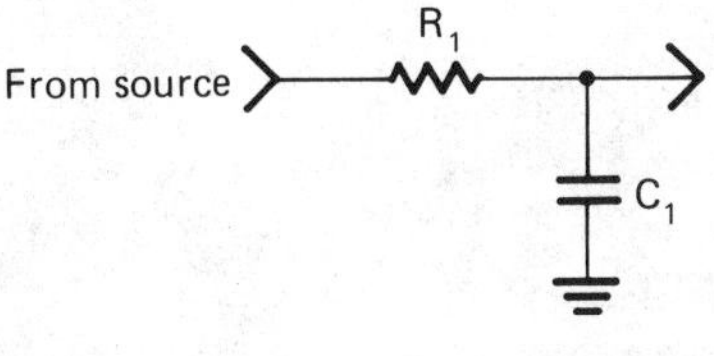

From source >———R_1———•———> To 6001

C_1

$$C_1 = \frac{1}{2\pi R_1 F}$$

where F = frequency of rolloff

FIGURE 13.8
Formula and diagram to design a filter. (Courtesy
of Global Specialties Corp.)

a sample-and-hold function available to the 6001 by use of the
external–internal time base. Set the counter to the resolution desired, attach
the antenna, key the transmitter until a stable reading is obtained, and
switch to the external time-base mode. The counter will now display the
reading until you return the counter to the internal time-base mode. This
eliminates the problems encountered of tying up a channel while testing
and calibrating a transmitter.

The 6001 is also useful in measuring audio as well as RF signals. It can
read down to 5 Hz (guaranteed) with 1-Hz resolution and its built-in low-
pass filter eliminates high-frequency noise. Its switchable gate times provide even greater accuracy.

The 6001 can be used to test and calibrate function generators, tone
burst oscillators, audio amplifiers, and the like. By varying the external
clock input, the 6001 can directly read in units other than cycles per second,
such as feet per minute, gallons per hour, and revolutions per minute.

13.3
THE UNIVERSAL COUNTER-TIMER

The Model 5001 universal counter-timer, shown in Figure 13.9, is designed
for the electronic measurement and display of frequency, period, interval,
and counted events. It features full signal conditioning on both input channels, including attenuators, slope selection, and variable trigger level. A
unique variable delay hold circuit causes a 75-ms to 7.5-s delay between
measurement cycles, during which time the results of the last measurement
cycle remain displayed; a detent position on this control provides infinite
delay, maintaining the results of one measurement cycle on the display
indefinitely.

Measurement capabilities of the Model 5001 suggest broad applications
in industry, laboratories, education, process control and production:

1. **Frequency.** Measurements to 10 MHz in four ranges. Selectable gate
   times of 0.01, 0.1, 1.0, or 10 s. Display indicates frequency at A input in
   kilohertz.

FIGURE 13.9
The 5001 universal counter-timer. (Courtesy of
Global Specialties Corp.)

2. **Period.** Measures period of signal at A input, 400 ns to 10 s. Measures single cycle or averages over 10, 100, or 1000 cycles. Display indicates time in microseconds.

3. **Frequency ratio.** Counts number of cycles appearing at A input (to 10 MHz) during one cycle at B input (to 2 MHz), or averages this ratio over 10, 100, or 1000 cycles at B. Useful for scaling measurements, the display indicates the ratio F_A / F_B.

4. **Time interval.** Measures time between selected signal edge appearing at A input (which starts the measurement) and selected signal edge appearing at B input (which completes the measurement) in a range from 200 ns to 10 s. In addition to single interval measurements, an average over 10, 100, or 1000 intervals may be selected.

5. **Event count.** Counts up to 99,999,999 events (each event is marked by the appearance of the selected edge of a signal at the A input) appearing at the A input at up to 10 MHz. The RUN push button allows counting to begin. When the HOLD push button is pressed, the display freezes while counting continues; pressing RUN returns the display to following the accumulated total counts. The RESET push button returns the count to zero.

The Model 5001 features a bright 8-digit 0.43-in. (11-mm) LED display. Decimal points are positioned to give frequency measurements in kilohertz and time measurement in microseconds. Discrete LEDs show overflow (count exceeds 99,999,999) and gate-open conditions.

Applications for the Model 5001 include tachometry, process control, production line counting, flow metering, stroboscopy, chronometry, signal analysis, and more. The heart of the Model 5001 universal counter-timer is a fully integrated LSI universal counter and LED display driver circuit, the Intersil ICM7226. The block diagram is given in Figure 13.10 and the schematic diagram is given in Figure 13.11.

The ICM 7266 has been designed to function as a frequency counter, a period counter, a frequency (F_A/F_B) ratio gating counter, an interval timing counter, and an accumulating event counter. With a 10-MHz crystal oscillator, resolution in the nanosecond and fractional hertz range is achievable. Gating, reset, hold, and storage signals are available to increase operational flexibility.

The 5001 provides identical input signal processing on both inputs to fully condition signals for optimum operation of the ICM 7226. Selectable attenuation, slope selection, overload protection, trigger level dc offset adjustment, and buffering are incorporated in both channels. Both input preamplifiers are designed to handle signal frequencies up to 35 MHz; limitations of the ICM 7226, however, restrict operation to 10 MHz in the A channel and 2 MHz in B.

The mode, function, and range control switch matrix, in addition to formating the operation of the ICM 7226, provides for automatic clearing and resetting of the counters and the *display hold/measurement cycle delay* timer whenever the selected mode changes. This also occurs when power is first turned on. The *display hold/measurement cycle delay* timer

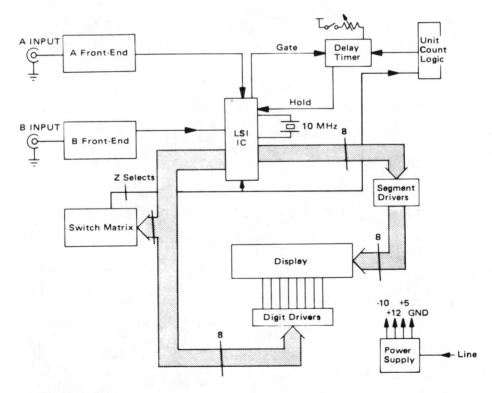

FIGURE 13.10
Block diagram for the Model 5001. (Courtesy of Global Specialties Corp.)

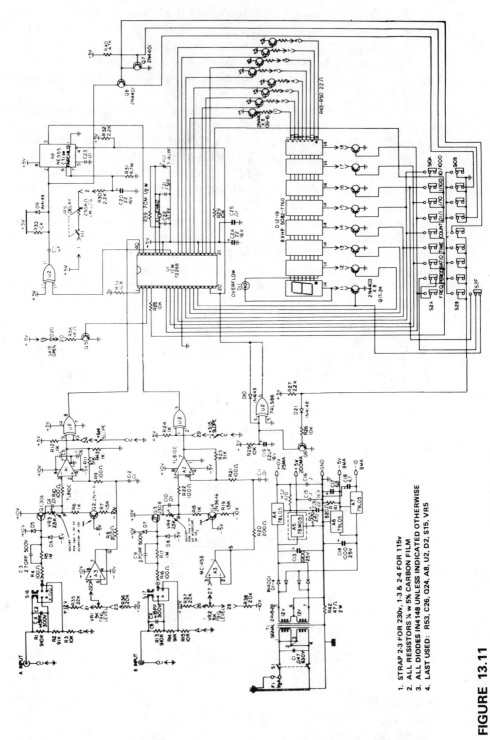

FIGURE 13.11

Schematic of the 5001. (Courtesy of Global Specialties Corp.)

456

provides a user-variable delay in the ICM 7226 RESET output before it reaches the IC's HOLD input.

Global Specialties has incorporated discrete transistor drivers on every segment and digit display line to permit higher drive levels, and thereby a higher brightness display. Eight 0.43-in. digits are included in the 5001 display, in addition to discrete LED indicators for counter overflow and gate activity. Four separate regulators are used in the 5001 power supply to assure optimum isolation between digital circuitry and input channel amplifier stages.

Refer to the block diagram (Figure 13.10) and schematic diagram (Figure 13.11) of the Model 5001 during this discussion of circuit operation.

Input signals at the A and B inputs are connected directly to three-position $\times 1 / \times 10 / \times 100$ attenuators ($R1$, $R2$, $R3$, and $S16$ at A; $R13$, $R14$, $R15$, and $S17$ at B). Input protection is provided by 100-Ω series resistors ($R4$, $R16$) and diodes $D5$ to $D8$.

Signals are then buffered by JFET source follower preamplifiers $Q1$ and $Q3$; optimum biasing of these transistors is assured through the use of constant current sources $Q2$ and $Q4$.

Signals then reach Schmitt triggers $A1$ and $A2$, where they are compared to levels set by *trigger level* controls $VR1$ and $VR2$ through voltage followers in dual-amplifier $A3$. The Schmitt trigger incorporates approximately 20 mV of hysteresis to assure clean transitions even with some noise on the signal. Exclusive-OR gates (on $U2$) in conjunction with *slope* switches $S14$ and $S15$ determine whether rising or falling edges will be counted.

Note that the entire preamplifier circuit for both preamplifiers is designed to work with signal frequencies up to 35 MHz; internal limitations on the LSI counter IC, however, limit A-channel response to 10 MHz and B-channel response to 2 MHz.

LSI IC $U1$ is an Intersil ICM7726B, which executes all counting, gating housekeeping, and display drive signal chores for each universal counter-timer basic function. While no detailed accounting of its operation will be attempted here, highlights follow.

A *measurement in progress* output at pin 3 turns on gate open LED $D20$ through transistor $Q5$. Function, range, and controinputs are time multiplexed onto the display digit drive lines. Selector push-button switchbank $S2$ through $S10$ provides appropriate codes to $U1$ through switching in the A bank; the B bank performs other special functions. For FREQUENCY, PERIOD, INTERVAL, RATIO, and COUNT mode selectors $S2$ to $S6$, a signal is sent to transistor $Q6$ any time a new mode is selected. This is combined with a power-on signal at a third $U2$ exclusive-OR gate to generate a reset for $U1$ and for DELAY timer $A8$ (see Figure 13.12).

This same function is accomplished slightly differently in the event counter mode. Here the base of $Q6$ is not grounded directly through the B switchbank, as for other functions, but instead via the base–emitter junction of $Q8$; the emitter of $Q8$ is connected to ground in every case except when the RESET push button is selected in the event counter mode. In this case a

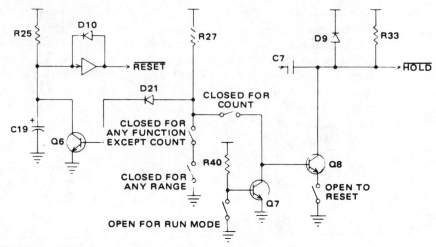

FIGURE 13.12
Detail from schematic shows coordination of reset modes. (Courtesy of Global Specialties Corp.)

reset will occur and $Q8$ will turn off, allowing the input of $A8$ to swing high. When the RUN push button is selected, ground will be disconnected from the base of $Q7$, which will then be biased on by $R40$; this will turn $Q8$ off, releasing $A8$.

Delay timer $A8$ is a 555 timer configured as a monostable multivibrator. It drives the HOLD input of $U1$, stopping any measurement in progress, resetting the main counter, and making the chip ready to initiate a new measurement. Also, the main counter data latches are not updated, so the display continues to show the last complete measurement. But because the input to $A8$ is the RESET output of $U1$, $A8$ permits one complete measurement cycle (only) before imposing the HOLD condition.

Delay control $VR5$ (logarithmic) varies the monostable pulse width at the output of $A8$ from 75 ms to 7.5 s; switch $S11$ extends it indefinitely.

Transistors $Q9$ to $Q24$ buffer digit and segment display drive lines, permitting higher driving currents for optimum display brightness. Overflow LED $D11$ is multiplexed in place of the most-significant-digit decimal point.

The power supply makes liberal use of integrated three-terminal regulators. Note that power to the input channel preamplifiers is fully isolated from power to the digital circuit sections in order to minimize spurious coupling. $A6$ and $A7$ are cascaded to provide -5 V dc and -10 V dc using two 5-V negative regulators. $A3$ supplies $+5$ V dc to the logic circuitry. $A4$ is floated to provide a clean $+10$-V dc supply using a 5-V dc regulator; this feeds the front end circuitry.

$R42$ prevents ground loops possible through connection of the 5001 to other equipment. The shell of one BNC input connector should always be electrically connected to the ground of the circuit supplying the input signal.

The 5001 specifications include the following:

1. Inputs:
   a. Two inputs A and B, dc coupled, BNC connector
   b. Impedance: $M\Omega$ + 25 pF
   c. Response: 10 MHz maximum at A, 2 MHz maximum at B
   d. Sensitivity: 20 $mV_{rms}R$
   e. Controls: $\times 1/\times 10/\times 100$ attenuators, slope select, variable trigger level

2. Reference: 10-MHz crystal oscillator, ± 4 ppm from 5° to 35°C
   a. Modes
   b. Frequency: 10 MHz maximum, four ranges with gate times of .01/0.1/1.0/10 s, display in kilohertz, A input only
      - Period: 400 ns to 10 s, four ranges with 1/10/100/1000 cycle average, display in microseconds, A input only
      - Frequency ratio: 10 MHz maximum at A, 2 MHz maximum at B, four ranges, counts cycles at A during 1/10/100/1000 cycles at B
   c. Time interval: 200 ns to 10 s, four ranges, measurement cycle begins with first selected edge at A, ends with next selected edge at B for single interval measurement; cycle is repeated 10, 100 or 1000 times and average for all cycles displayed for multiple time interval average measurements.
   d. Unit count: Maximum count 10^8, maximum frequency 10 MHz, A input only, one range, RUN button starts and displays running count or returns display to running count. HOLD button freezes display while running count continues. RESET button resets count to zero.

3. Controls: Power, five-mode selector switches, four range selector switches, run, hold, reset, display delay, plus trigger level, slope select, and attenuator in both A and B input channels.

4. Display: 8-digit 7-segment 0.43-in. LED display; decimal point indicates time in microseconds, frequency in kilohertz; discrete LEDs indicate overflow (counter overflow) and gate (gate open); delay feature varies period between measurement cycles from 75 ms to 7.5 s with delay control; detent position holds next measurement reading indefinitely.

5. Power: 105 to 135 V ac, 57 to 63 Hz, 10 VA maximum, 215 to 250 V ac, 50- to 60-Hz version available)

6. Operating temperature: 0° to 50°C, calibrated at 25°C $\pm$ 5%.

13.3.1 Applications of the Universal Counter-timer

Always measure oscillator frequencies after at least one stage of buffering; measurements taken at the oscillator components themselves can pull the oscillator frequency. When this is unavoidable, using a ×10 oscilloscope probe connected to the 5001 A input will minimize circuit loading, reducing this frequency shift.

Certain critical applications may require that the Model 5001 input impedance be identical to circuit impedance; for example, a 50-, 75-, or 300-Ω termination may be required to properly load the circuit under test. Since both 5001 inputs are high impedance, this termination may be applied at the 5001 input directly.

When measuring transmitter and other radiated field frequencies, a short piece of wire or telescopic antenna connected directly to the 5001 input may provide adequate coupling for a reliable reading. Do not attempt to connect the 5001 input in place of the transmitter antenna or anywhere directly in the transmission line.

To measure a frequency that is only available for a short time, set the delay control to its hold (detent, fully counterclockwise) position, press in the FREQUENCY push button, and select a gate time of 0.1, 1.0, or 10 (but not 0.01) s. Disregard the display at this point. As soon as the signal is available, press in the 0.01-gate time push button. The signal then only need remain valid for 10 ms; the 5001 will complete one measurement cycle in that time and will display the results of that measurement indefinitely.

For fast, accurate measurements of low frequencies, use the period measurement mode and calculate the frequency as the reciprocal of period:

$$\text{Frequency (megahertz)} = \frac{1}{\text{period (microseconds)}} \qquad (13.1)$$

In measuring microprocessor clock frequency, locate the processor's buffered clock output. Assuming the signal is at TTL levels, set the 5001 A input attenuator to ×10, set the trigger level control to +2 V (approximately the 1 o'clock position), and connect clock and ground directly to the 5001 A input connector. Follow the operating instruction manual.

The accuracy of the period measurement is very dependent on the amount of noise present on the input signal, even when this noise is below the 5001's 3-mV maximum noise threshold. To substantially reduce this dependency, you can average over a larger number of cycles, increase the input amplitude of the waveform, and/or select the steeper edge of the waveform as the significant edge.

Period or multiple period average measurements yield much quicker and more accurate results than frequency measurements for low phenomena. For example, it would take 10 s to provide 0.1-Hz resolution in determining the frequency of a 110.4-Hz (0.1104-kHz) signal in the frequency counter mode; in the period measurement mode, a 100-cycle average of 9057.971 μs is available in less than 1 s. If the period were instead to be measured as 9057.946 μs, the frequency could be calculated as 100.4003 Hz, yielding an additional three accurate digits.

The speed of a motion picture projector may be measured indirectly using the 5001 more simply, more conveniently, and more accurately than through direct measurements of mechanical motion. In the case of a motion picture projector, the variable of interest is frame rate in frames per second. Nominally, most projectors run at either 16 or 24 frames / s; in television film chains, every fifth frame is projected twice, converting nominal 24-frame / s motion pictures to nominal (standard) 30-frame / s video rates. Accurate framing of video picture information within sync requires that projector speeds be especially accurate. The 5001 permits accurate measurement of these low frequencies through reciprocal calculation of period, as measured by leading edges of photosensitive transducer outputs in response to each projected frame. A photovoltaic cell may be connected directly to the 5001 input, but the response of these is usually slower than that of other photodetectors. A photodiode or phototransistor can be used with the generalized input sensor conditioning circuit shown. With no film in the projector and both motor and light turned on, place the sensor directly in front of the lens. Some degree of optical filtering, limiting, or shadowing may be desired, depending on specifics of the detector.

Unlike measurements of multiple period averages, multiple time interval average measurements accumulate the total errors occurring in all measured intervals; this implies both that some jitter is inevitable when averaging multiple intervals, increasing with the averaged number of intervals, and that variance between single interval measurements is to be expected.

The time interval measurement mode provides a highly useful facility for determining pulse parameters. This is done by externally commoning the A and B inputs. Pulse width can then be determined by selecting the rising edge on the A-input slope selector and the falling edge on the B-input slope selector. Pulse spacing can be determined by selecting the falling edge on the A-input slope selector and the rising edge on the B-input slope selector. Pulse period can be determined by taking the sum of pulse width and pulse spacing or by selecting the period measurement mode of the 5001. Pulse repetition rate can be determined by taking the reciprocal of pulse period or by selecting the frequency counter mode of the 5001. Duty cycle can be determined mathematically. The following formulas apply:

$$\text{Pulse period} = \text{pulse width} + \text{pulse spacing} \qquad (13.2)$$

$$\text{Duty cycle} = \frac{\text{pulse width}}{\text{pulse period}} = \frac{\text{pulse width}}{\text{pulse width} + \text{pulse spacing}} \times 100\%$$

$$(13.3)$$

$$\text{Duty cycle} = \frac{\text{pulse width}}{\text{pulse period}} = \text{pulse width} \times \text{pulse frequency} \times 100\%$$

$$\tag{13.4}$$

$$\text{Repetition rate} = \text{pulse frequency} = \frac{1}{\text{pulse period}} \tag{13.5}$$

$$\text{Repetition rate} = \frac{1}{\text{pulse width} + \text{pulse spacing}} \tag{13.6}$$

Time of flight and time of fall for ballistic projectiles or falling objects may be directly measured by the 5001. The display hold feature is especially useful because it always permits one and only one measurement cycle, which cannot begin until the appearance of the selected edge at the A input in the time-interval mode; once the measurement cycle is completed with the appearance of the selected edge at the B input, the display is held indefinitely. Arranging electrical contacts directly in the path of the object is the simplest means of providing appropriate input signals to the 5001, and is acceptable if the force associated with the mass and velocity of the object is large compared to the contact breaking (or making) force required. Note that by using contact opening as the significant event, errors due to contact bounce problems are avoided. Alternatively, photoelectric beam-break detectors may be used or other methods more convenient to the specific test. Once clean input signals are available, the measurement can be taken 10, 100, or 1000 times using multiple interval averaging and the average for all tests displayed; however, beware of contact bounce when switches, for example, are restored between measurements.

In many cases, the 5001 will be used in conjunction with a signal derived from the metering of a physical phenomenon, such as rotational speed, flow, linear speed, or weight. The 5001 can be configured to display measurements in terms of the actual units of interest (e.g., rpm, gallons per minute, or kilometers per hour) by taking advantage of the scaling (also called rescaling) capabilities of the 5001 in the frequency ratio mode. In this mode, the 5001 indicates how many selected signal edges appear at the A input between any two subsequent selected signal edges at the B input. In other words, the B-input signal gates the A-input clock. So instead of displaying frequency measurements in terms of cycles per second (hertz), they are in cycles per B signal period. For example, if an encoding disk on a motor shaft gives 1024 pulses/revolution and the motor is rotating at 1 revolution/s, the 5001 will see a 1024-Hz signal at the A input. If a 1/1024-s period (1024-Hz) signal appears at the B input, the 5001 display will read 1. To convert to rpm, which means sampling the A input during 60 such periods of 1.1024 s, divide the B-input frequency by 60 (1024/60 = 17.07 Hz). Generally, you can determine the frequency to apply to the B input as follows:

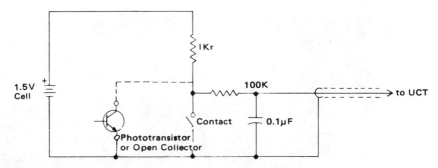

FIGURE 13.13
Circuit conditions input events. (Courtesy of
Global Specialties Corp.)

1. Determine how many pulses occur per event of interest, converting into
    the units you wish the event expressed in on the 5001 display. For
    example, if you know how many pulses are delivered per pint, multiply
    by 16 to get pulses per gallon.

2. Express 1 s in the time units you wish to use. For example, $\frac{1}{60}$ min, 1000
    ms, or 1/3600 h.

3. Determine the frequency needed at the B input as

    Frequency at B = (pulses per event) × (time units in seconds) (13.7)

4. Using the Model 4001 pulse generator or Model 2001 function genera-
    tor discussed in Chapter 7, or equivalent, provide the appropriate signal
    frequency to the B input. Note that the B input cannot accept a fre-
    quency greater than 2 MHz.

    As in other modes of operation, the flexibility of the 5001 input circuitry
permits a great many devices to be used as sensors, requiring only very
simple interface circuitry (the inputs respond only to voltages, so simple
contact closures and impedance changes require some conditioning).

    Additional conditioning of input signals (Figure 13.13) may be required
to get accurate counts if the signals include glitches, contact bounce, or a
great deal of noise. This additional circuitry might be as simple as low-pass
filtering or monostable contact conditioners. Additionally, purely electronic events could be counted, such as the number of error signals coming
from a microprocessor.

13.4
THE FREQUENCY STANDARD

The 4401 frequency standard (Figure 13.14) provides a unique source of
discrete selectable precision frequencies that can be used as either a time
or frequency standard or as a highly accurate signal source. Use it as an

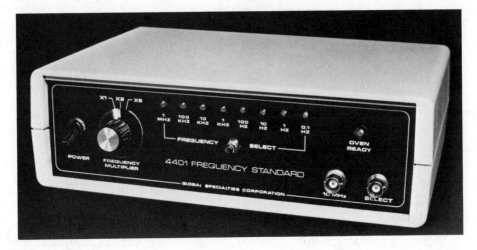

(a) Front panel

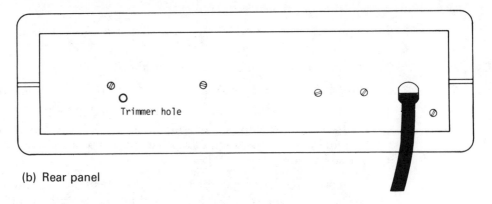

Trimmer hole

(b) Rear panel

FIGURE 13.14

The 4401 frequency standard. (Courtesy of Global Specialties Corp.)

oscilloscope time-base calibrator, as a precision clock source for microprocessors, as a precision reference for time keeping, frequency counter calibrator, or any other application where a precision frequency standard is required.

The heart of the 4401 is its unique 10-MHz precision crystal oven oscillator. It boasts an accuracy of ±0.5 ppm (±0.00005%) from 0° to 40°C. This reference is calibrated at the factory to the National Bureau of Standards through WWVB. An oven-ready LED on the front panel indicates when the unit has come up to operating temperature and is locked on to frequency (3 to 5 min).

Two BNCs provide the square-wave outputs from 10 MHz to 0.1 Hz in

9-decade steps. A frequency multiplier control at the front panel enables fractional frequencies of decades in three steps: 1x, 2x, and 5x. For example, if you select the 1-MHz range, frequencies of 1, 2, and 5 MHz are available at the SELECT BNC output.

Both outputs are 50-Ω TTL-compatible square waves and are short circuit protected. A 10-MHz square wave is always present at the 10-MHz

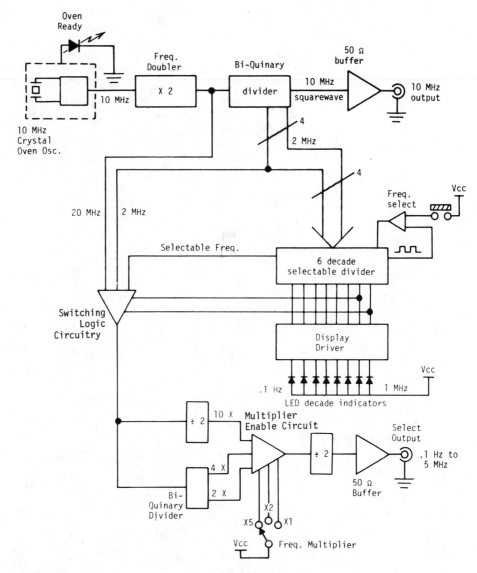

FIGURE 13.15
Block diagram of the 4401 frequency standard.
(Courtesy of Global Specialties Corp.)

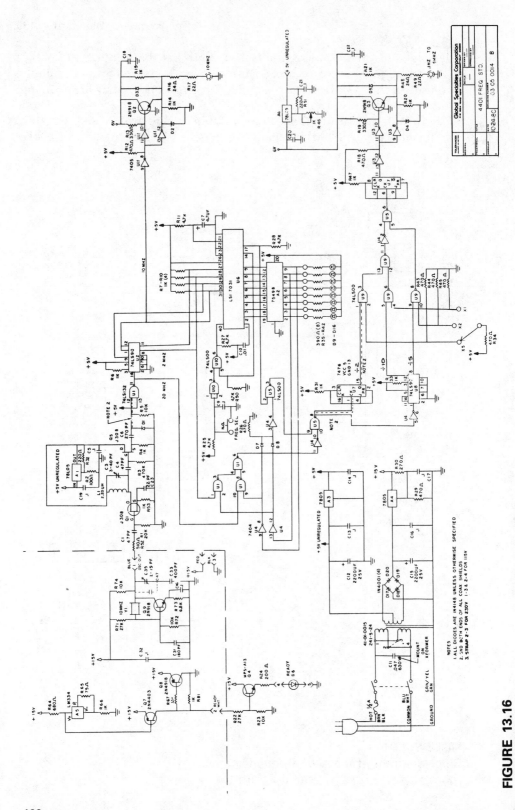

FIGURE 13.16

General schematic of the 4401 frequency standard. (Courtesy of Global Specialties Corp.)

output. At the SELECT output, the available frequency is the selected decade times the frequency multiplier (from 0.1 Hz to 5 MHz).

The 4401 is as easy to operate as it is accurate. The only controls other than the POWER-ON switch are the FREQUENCY SELECT push button and a FREQUENCY MULTIPLIER switch.

The 4001 is recommended for applications such as the calibration of time and frequency counters, oscilloscopes, and is outstanding as a precision clock source. It is suggested for use in laboratory, test bench, field service, classroom, data acquisition, information processing and communications environments, just to name a few.

Figure 13.15 is a block diagram of the 4401, which consists of 11 sections:

1. Ten-megahertz "crystal" oven time base
2. Frequency doubler
3. Biquinary divider
4. Ten megahertz, 50-Ω output buffer
5. Switching circuitry
6. Six-decade divider
7. Frequency select circuitry
8. LED drivers
9. Two flip-flops and a biquinary divider
10. Multiplier enable circuitry
11. 0.1-Hz to 5-MHz 50-Ω buffer

The heart of the 4401 is the 10-MHz crystal oven. This accurate signal source is fed into the frequency doubler where the frequency is increased to 20 MHz. The first biquinary divider divides the signal down to the proper processing frequencies, a 10-MHz square wave for the 10-MHz output buffer. Signals of 20 and 2 MHz are gated through the switching logic to the remaining dividers. The 6-decade selectable divider produces the frequencies from 0.1 Hz to 100 kHz, as well as providing for the LED displays and the strobe for the FREQUENCY SELECT push button. The remaining circuitry divides down the 20-MHz and the 2-MHz signal to obtain frequencies from 100 kHz to 5 MHz. The MULTIPLIER ENABLE circuitry chooses the proper division to obtain the ×1, ×2, ×5 multiples, and, finally, the 0.1-Hz to 5-MHz buffer processes the signals for the output. The entire schematic diagram is shown in Figure 13.16.

The time base of the frequency standard (Figure 13.15) controls the operation of the 4401 and, as such, its accuracy affects the entire frequency standard and its performance. The 4401 uses a unique design to control its time base by incorporating a specially designed crystal oven to precisely control the temperature surrounding its crystal oscillator. The oscillator is kept at a constant 55°C to ensure optimum stability across the entire operating range (0° to 40°C). Its main temperature sensor (LM-334) is an adjustable current source and, as such, will not induce spurious noise when turned on and off.

The signal from the time base, a precision 10-MHz sine wave, is fed into the frequency doubler comprised of $Q1$ and LC tank circuit, $L1$ and $C3$, and a buffer $Q5$. This 20-MHz signal is now squared up by $U1$ and fed into the first biquinary divider $U2$. $U2$ divides the 20 MHz by 2 and squares it further, subsequently passing the 10-MHz square wave to $U11$ *and* $Q2$ to provide the 50-Ω TTL output that is fed to the 10-MHz output BNC.

$U2$ also divides the 20-MHz signal from the doubler by 10 to get a 2-MHz signal, which is fed both to the 6-decade divider $U6$ and to the $U1$ switching circuitry. $U6$ handles most of the decade division, the strobe functions, and display decoding.

By depressing the FREQUENCY SELECT switch, $U10$ is enabled, allowing a 2-Hz signal to be clocked into the multiplex input of $U6$. This slowly scans the display driver $A2$, lighting the appropriate LED, and enables the internal divider outputs, which are a multiple of the desired frequency, to be fed into the switching logic $U5$.

The switching logic is a combination of $U1$, $U4$, and $U5$. When the 0.1-Hz to 10-kHz decades are selected, the signal is routed from $U6$ through $U5$ to $U8$ and $U7$, and then on to $U9$, where the multiple is selected and passed on to the 50-Ω TTL buffer. When the 100-kHz decade is selected, $A2$ and $U4$ enable $U1$ to pass the 2-MHz square wave through to either $U7$ or $U8$, where it is divided and further passed on to the multiplier select circuitry and finally to the 50-Ω TTL buffer.

The same signal processing is utilized with a 20-MHz signal to provide the 1-, 2-, and 5-MHz outputs.

With an accuracy of 0.5 ppm, the 4401 can easily be used to calibrate a frequency counter. To start, allow all instruments that will be used an ample warm-up period (approximately 30 min). Then simply tie the 10-MHz output to the frequency counter's input and adjust the frequency counter's time base according to the manufacturer's recommendations.

Many frequency counters and signal generators have available either as an option or standard an external time-base input. This is used when more accuracy is needed than is available with the standard time base. With this feature, many orders of magnitude in accuracy can be obtained with an accurate frequency generator such as the 4401.

To use the 4401 as such a time base, select the appropriate output frequency (usually 1 or 10 MHz), tie into the external time-base input on the unit, and follow the manufacturer's instructions. Remember to keep all cables as short as possible.

For calibrating an oscilloscope, most manufacturers recommend that an accurate square-wave generator be used in the test. Since most scopes have an accuracy of about 2%, the 4401 is 400 times more accurate and, therefore, ideal for this use. Make sure you follow the manufacturer's suggested calibration procedure.

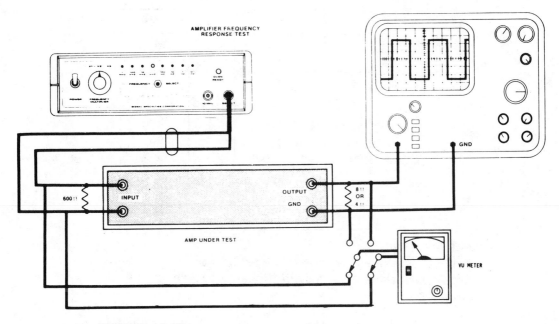

FIGURE 13.17
Test setup for amplifier. (Courtesy of Global
Specialties Corp.)

Applications of the 4401 frequency standard include the following:

1. Microprocessor clock
2. Audio testing
3. Testing transmission lines
4. Calibration of frequency counters
5. Calibration of oscilloscopes
6. Precision signal generator
7. Frequency counter time base
8. Signal generator time base

The 4401 is a unique, versatile piece of laboratory equipment, which
can be used in a multitude of various applications that require a precision
square-wave generator or frequency source. The preceding list is by no
means comprehensive; its intent is merely to assist the user with some of
its potential applications.

By substituting your 4401 for the microprocessor system clock, you can
give your microprocessor the capability of stepping through its microprogram either a step at a time (in 0.1-s mode) or at much reduced speed, by
using long timing periods such as 10 Hz, 100 Hz, 1kHz or 10 kHz. Note,
however, that some microprocessors have a minimum clock speed below

POWER RATINGS (RMS)			OUTPUT (RMS)	
8 Ω	4 Ω	E²	FULL OUTPUT	10 dB BELOW MAX. OUTPUT
1 W	.5 W	8 V	2.8 V	2.3 V
5 W	2.5 W	40 V	6.3 V	5.2 V
10 W	5.0 W	80 V	8.9 V	7.3 V
20 W	10 W	160 V	12.6 V	10.3 V
50 W	25 W	400 V	20 V	16.3 V
100 W	50 W	800 V	28 V	23.1 V

FIGURE 13.18
Table of power and voltage relationships for 4-
and 8-Ω speakers. (Courtesy of Global Specialties
Corp.)

which correct operation is not assured. If in doubt, check the data sheet for the microprocessor that you are using.

The 4401 again shows its versatility in testing audio amplifiers. Square waves are used in audio testing to display a wide range of frequencies simultaneously. Square waves consist of a fundamental frequency and a series of odd harmonics to square off the wave shape.

For an amplifier to reproduce a square wave it must have a flat frequency response from $0.1F$ to $10F$, where F is the fundamental frequency of the square wave. Connect the 4401 to the amplifier under test as shown in Figure 13.17 and observe the output voltage level for 10 dB below the maximum output of amplifiers with different power ratings. Choose the rating closest to the amplifier under test. Testing at 10 dB below maximum output ensures that the amplifier will not be in saturation. This level is approximately two-thirds of the full power output.

If a transmission line is not terminated at the far end by its characteristic impedance, reflections will occur. This phenomenon can be used to find faults on transmission lines. Using the 4401, you can find out if the cable under test is open or short circuited, and with some simple calculations you can find the length of the cable. This technique is similar to the plan proposed for the pulse generator discussed in Chapter 7. The hookup is also the same.

The specifications for the 4401 include the following:

1. Outputs: Two outputs, dc coupled, BNC connectors
   - 10-MHz drive: 50-Ω TTL-compatible square wave, buffered to drive up to 10 TTL loads, short circuit protected, 20 ns rise and fall into 50 Ω.

- Select drive: 50-Ω TTL-compatible square wave, buffered to drive up to 10 TTL loads, short circuit protected, 20 ns rise and fall into 50 Ω.

2. Reference: Time base: 10-MHz crystal oven oscillator, ± 0.5 ppm from 0° to 40°C, oven temperature 55°C, aging less than 1 ppm/yr; internal calibration user-accessible; factory calibrated to National Bureau of Standards via WWVB.

3. Controls: POWER; FREQUENCY SELECT push button scans output through 1 MHz, 100 kHz, 10 kHz, 100 Hz, 10 Hz, 1 Hz, 0.1 Hz; MULTIPLIER SELECT multiplies selected frequency ×1, ×2, or ×5.

4. Displays: Eight discrete LEDs indicate selected FREQUENCY decade, selected FREQUENCY LED also serves as POWER pilot; additional LED indicates OVEN READY.

5. Power: 105 to 135 V ac, 57 to 63 Hz, 5 VA maximum (215 to 250 V ac, 50- to 60-Hz version available).

6. Operating temperature: 0° to 40°C

13.5
REVIEW QUESTIONS

1. Discuss the purpose of a 100-MHz frequency counter.
2. Draw a block diagram of a 100-MHz frequency counter.
3. Discuss how to make reliable measurement with a frequency counter.
4. Discuss the purpose of a 650-MHz frequency counter.
5. Draw a block diagram of a 650-MHz frequency counter.
6. Discuss the applications of a 650-MHz frequency counter.
7. Discuss the purpose of the universal counter-timer.
8. Draw a block diagram of the universal counter-timer.
9. Discuss the applications of the universal counter-timer.
10. Discuss the frequency standard used as a precision clock source for microprocessors.

13.6
REFERENCES

1. "Model 5435A Electronic Counter," Hewlett-Packard Co., Palo Alto, Calif., April 1980.
2. "7260A/7261A Universal Counter/Timers," John Fluke Manufacturing Co., Mountlake Terrace, Wash., 1979.

3. "Instruments for Testing and Design," Global Specialties Corp., New Haven, Conn., 1981.

4. "Model 5335A Universal Counter, New Universal Measurement Power," Hewlett-Packard Co., Palo Alto, Calif., April 1980.

5. Clyde F. Combs, Jr., ed., *Basic Electronic Instrument Book*, McGraw-Hill Book Co., New York, 1972.

14

Data-Processing Systems[a]

14.1
INTRODUCTION

Technological advance in data processing [1, 2] is both dynamic and extensive in industry and medicine. The methods in which data-processing systems can be used seem almost boundless. Each new application demonstrates how such systems can be used to help us enlarge our capabilities.

Data-processing systems ordinarily consist of a combination of programs and physical equipment designed to handle business or scientific data at electronic speeds with self-checking accuracy. The physical equipment (Figure 14.1) consists of various units, including input, storage, processing, and output devices. Figure 14.2 pictures a *teleprocessing* (telecom-

---

[a] The material in this chapter is excerpted and reprinted from *Introduction to IBM Data Processing Systems, Student Text*, 5th ed., GC20-1684, June 1977, reprinted July 1978. Throughout this chapter you will be reading about data processing concepts and devices supported by IBM. Because of the dynamic nature of data processing, where changes and improvements are being made at a very rapid pace, the reader is advised to refer to the IBM Systems Reference Library and other IBM publications for the most current information. IBM Data Processing Division, 1133 Westchester Ave., White Plains, N.Y. 10604, (914) 696-1900.

FIGURE 14.1
IBM System/370 Model 168 data-processing
system. (Courtesy of International Business Machines Corporation)

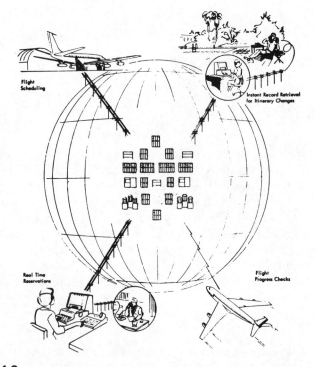

FIGURE 14.2
Data-processing system application. (Courtesy of
International Business Machines Corporation)

munications plus data processing) system applied to airline reservation activities.

Machines are devised for a purpose. In the case of data-processing machines, the purpose can be expressed simply: they offer a means to increase productivity. They do this in two ways:

1. They enable an increase in the output per hour and the quality of the output (this is true whether it be in research, production, problem solving, or the distribution of goods and services).

2. They increase productivity by encouraging careful and intelligent planning.

During the last quarter-century, further changes have taken place. Science has moved into the forefront of human activity. Research has grown to a multibillion-dollar-a-year undertaking. New technology has provided a new impetus for corporate growth. Service industries have multiplied. Patterns of consumer spending have changed.

As these changes gained force, they manifested themselves in many ways. Informational needs greatly increased. Data assumed new importance. Clerical tasks multiplied. It seemed that paper handling alone would overwhelm all productive activities, for clerical mechanization had not kept pace with production line developments in the factory. Great opportunities and challenges lie ahead.

14.1.1 Present-day Data Processing

A significant advance in computer input–output technique is the development of the various types of graphic displays. These are similar in appearance to television sets, but the "picture" is computer output in the form of printed characters or graphic designs specified by the program and data. In some cases, the user can change the output display by using a *light pen* to erase a character (or part of a line) from the display screen. Then, by using the light pen or the associated keyboards, or both, the user can alter the displayed information. More and more frequently, display and other devices are used to enter data into the computer from remote locations by means of communications lines (usually telephone lines) connecting the devices known as *terminals* with the computer. When multiple terminals are connected to the computer, concurrent usage of the computer for problem solving and program development on an *interactive* basis often increases substantially.

Along with other advances in computer technology, computer data storage capacity has increased immensely. This has led to the development of large on-line interrelated collections of data referred to as *data bases*. Many large centralized databases utilizing remote terminals have evolved in recent years. Such systems of large networks have integrated widely scattered business operations. This has occurred since, in general, information can be communicated and processed more accurately and with less cost by network-integrated processing systems.

More recently, data communications systems with processing and disk storage capabilities have been developed. These systems located at sites remote from the central (host) computer are capable of processing most of the input data locally. Data that have been deemed necessary or desirable for processing or maintenance at the central site are transmitted to the host computer. This is often referred to as *distributed function.*

Computers of the future, as well as programs, will probably be quite different from those of today. Storage and processing units will be drastically reduced in physical size, yet speed and capacity will be greatly increased.

Research scientists have already advanced to still further stages in design. Some are studying magnetic bubbles on a chip for use as secondary storage in the computer. Others are considering electron-beam techniques for main storage.

As always, the objective is to develop a better, more versatile, more useful computer—one that will work faster, store more information, require less power, occupy less space, and cost less. Computers of the future will inevitably introduce changes in the way we work, in the way we learn, and even in the way we provide for our armed defense.

Data processing is a series of planned actions and operations upon information to achieve a desired result. The procedures and devices used

FIGURE 14.3

Data-processing system. (Courtesy of International Business Machines Corporation)

FIGURE 14.4
Sources of data. (Courtesy of International Business Machines Corporation)

constitute a data-processing system (Figure 14.3). The devices may vary: all operations may be done by machine or the devices may be only pencil and paper. The procedures, however, remain basically the same.

There are many types of IBM data-processing systems. These vary in size, complexity, speed, cost, levels of programming systems, and application. But regardless of the information to be processed or the equipment used, all data processing involves at least three basic considerations:

1. The source data or *input* entering the system

2. The orderly, planned processing within the system

3. The end result or output from the system

Input may consist of any type of data: commercial, scientific, statistical, engineering, and so on (Figure 14.4).

Processing is carried out in a preestablished sequence of instructions that are followed automatically by the computer. The plan of processing is always of human origin. By calculation, sorting, analysis, or other operations, the computer arrives at results that may be used for further processing or recorded as reports or sets of data.

14.1.2 Stored Programs

Each data-processing system is designed to perform a specific number and type of operations. It is directed to perform each operation by an instruction. The instruction defines a basic operation to be performed and identifies the data, device, or mechanism needed to carry out the operation. The entire series of instructions required to complete a given procedure is known as a *program.*

The possible variations of a stored program provide the data-processing system with almost unlimited flexibility. A computer can be applied to a great number of different procedures simply by reading in or loading the proper program into storage. Any of the standard input devices can be used for this purpose, because instructions can be coded into machine language just as data can.

The stored program is accessible to the machine, providing the computer with the ability to alter the program in response to conditions encountered during an operation. Consequently, the program selects alternatives within the framework of the anticipated conditions.

The concept of maintaining optimum computer usage by interleaving and interspersing processing programs under the direction of control programs gives rise to the use of two terms: time sharing and multiprogramming.

14.1.3 Multiprogramming and Time Sharing

Briefly, *time sharing* may be thought of as the cooperative use of a central computer by more than one user (company, division or branch of a company, institution, or government agency). Each user receives a share of the time available, with the result that many jobs are being performed within a congruent time (either simultaneously or seemingly simultaneously). Two computers are joined to permit the sharing of each other's facilities (multiprocessing).

Multiprogramming is usually thought of as a system of control programs and computer equipment that permits many processing or operating programs of one or more users to go on concurrently. This is accomplished by interleaving the programs with each other in their use of the central processing unit, storage, and input–output devices. To do this, the control programs and equipment must be able to identify the point at which a problem program that is being executed must "wait" for the completion of some event. At that point, the control program begins another processing task that is ready to be executed. When that is done, the control program must be able to go on to something else or go back to the former (unfinished) program, if it is ready to continue. Since many programs may be in stages of partial completion, successful multiprogramming usually requires scheduling levels of priority for the different tasks.

Time sharing, multiprogramming, and multiprocessing are closely linked, and may be combined in many ways. While one user has the computer on a time-sharing basis, his problem may involve several different tasks that can be interleaved by a computer and programming system that provides for multiprogramming. It is also perfectly possible for teleprocessing messages to be coming in and going out of certain types of computers at the same time that process (problem) programs are being run. These are

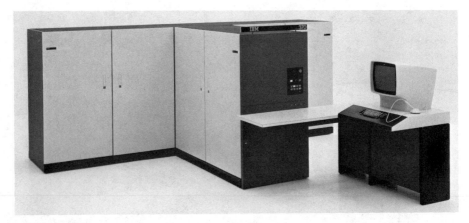

FIGURE 14.5
Central processing unit and console. (Courtesy of
International Business Machines Corporation)

but two examples of possible combinations of time sharing and multiprogramming.

Data-processing systems can be divided into four types of functional units: central processing unit, storage, input devices, and output devices.

14.1.4 Central Processing Unit

The central processing unit (Figure 14.5) is the controlling center of the entire data-processing system. It can be divided into two parts, the arithmetic–logical unit and the control section, which are discussed in Section 14.4.

14.1.5 Storage

Storage is somewhat like an electronic filing cabinet, completely indexed and instantaneously accessible to the computer. Storage is designed in such a way that information can be put there in many forms—as complete records, portions of records, digits, symbols, characters, code patterns, signals, and so on. However, capacity is usually stated in characters, meaning letters of the alphabet, digits, and special symbols of accounting, scientific notation, and report writing. In System/370, the word *byte* is used instead of "character." It is possible to *pack* two numeric digits into the same storage space that is required for letters of the alphabet, special characters, and the other symbols usually referred to as characters.

(a) IBM 3203 printer

(b) IBM 3277 display station

(c) IBM 2549 card read punch

(d) IBM 3420 magnetic tape unit

FIGURE 14.6
Input–output devices. (Courtesy of International
Business Machines Corporation)

(e) IBM 3330 disk storage drive

(f) IBM 3653 point-of-sale terminal

(g) IBM 3793 display printer

(h) IBM 3286 printer

FIGURE 14.6 (Continued)

FIGURE 14.7
IBM System/370 model 168 with IBM 3066
system console. (Courtesy of International Business Machines Corporation)

14.1.6 Data-processing Input and Output Devices

The data-processing system requires, as a necessary part of its information-handling ability, features that can enter data into the system and record data from the system. These functions are performed by input–output devices (Figure 14.6) linked directly to the system. Computer input–output has been discussed in detail in Chapter 10.

Input devices *read* or sense coded data that are recorded on a prescribed medium and make this information available to the computer. Data for input are recorded in cards and paper tape as punched holes, on magnetic tape as magnetized spots along the length of the tape, on paper documents as characters or line drawings created with the light pen and associated keyboards, and so on.

Output devices record or *write* information from the computer on cards, paper tape, and magnetic tape; they print information on paper; generate signals from transmission over teleprocessing networks; produce graphic displays, microfilm images; and take other specialized forms. The number and type of input–output devices connected directly to the computer depend on the design of the system and its application.

FIGURE 14.8
IBM System/3 Model 15. (Courtesy of International Business Machines Corporation)

14.1.7 Console

The console (Figure 14.7) is an input–output device that provides external control of the data-processing system. Keys turn power on or off, start or stop operation, and control various devices in the system. Data may be entered directly by manually depressing keys. Lights are provided so that data on the system may be displayed visually.

On some systems, a console printer (and/or display screen) and keyboard provide limited output or input. The input–output device may print or display messages, signaling the end of processing or an error condition. It may also print or display totals or other information that enables the operator to monitor and supervise operation, or it may give instructions to the operator. On the other hand, it may be used to key in meaningful information (such as altering instructions) to a data-processing system that is programmed to respond to such messages.

14.1.8 IBM System/3

IBM System/3 (Figure 14.8) is a low-cost, general-purpose system for commercial data processing and interactive problem solving. The system offers integrated card processing, stored-program capability, calculating and logic capabilities, and the flexibility of disk storage. In addition, local and binary synchronous communications adapters are available for attaching a wide variety of teleprocessing devices. System/3, which is primarily for small business, extends the use of stored-program data processing to the small data-processing user.

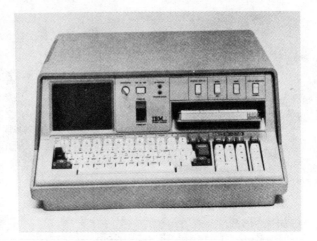

FIGURE 14.9
IBM 5100 portable computer. (Courtesy of International Business Machines Corporation)

Processor (main) storage ranges in capacity from 8192 to 262,144 bytes. All models of System/3 can communicate with System/370 over appropriate communications lines.

14.1.9   IBM 5100 Portable Computer

The IBM 5100 portable computer (Figure 14.9) provides the scientist or engineer with personal, local computing using an *interactive* high-level language, either APL or BASIC. With maximum memory in the machine, the user has an active workspace of approximately 60K bytes; and with one or two tape cassettes, an additional 200K or 400K bytes are at his disposal.

In addition to the standard keyboard, character screen, and tape cassette unit, the 5100 has available as optional equipment a second tape cassette unit, a printer, a communications adapter, and a serial input–output (SIO) adapter. The communications adapter enables the 5100 to be used as a terminal in conjunction with large, interactive systems. The SIO adapter provides for the attachment and local control of devices with a communications feature.

14.2
DATA REPRESENTATION

Symbols convey information; the symbol itself is not the information but merely represents it. The printed characters in Figure 14.10 are symbols that convey one meaning to some persons, a different meaning to others, and no meaning to those who do not know their significance.

Presenting data to the computer system is similar in many ways to communicating with another person by letter. The intelligence to be conveyed must be reduced to a set of symbols. In the English language, these are the familiar letters of the alphabet, numbers, and punctuation. The symbols are recorded on paper in a prescribed sequence and transported to another person who reads and interprets them.

Similarly, communication with the computer system requires that data be reduced to a set of symbols that can be read and interpreted by data-processing machines. The symbols differ from those commonly used by people, because the information to be represented must conform to the design and operation of the machine. The choice of these symbols (and their meaning) is a matter of convention on the part of the designers. The important fact is that information can be represented by symbols, which become a language for communication between people and machines.

Information to be used with the computer systems can be in the form of punched cards, paper tape, magnetic tape, direct-access storage devices (DASD), magnetic ink characters, optically recognizable characters, microfilm and display screen images, communication network signals, and so on. The list is growing larger each year. Some are pictured in Figure 14.11.

Data are represented on the punched card by the presence of small holes in specific locations of the card. In a similar manner, small circular holes along a paper tape represent data. On magnetic tape or on DASD, the symbols are small magnetized areas, called spots or bits, arranged in spe-

FIGURE 14.10
Symbols for communication. (Courtesy of International Business Machines Corporation)

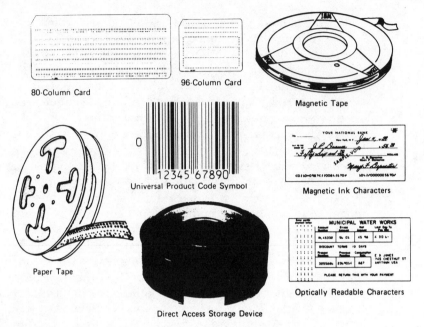

80-Column Card

96-Column Card

Magnetic Tape

Universal Product Code Symbol

Magnetic Ink Characters

Paper Tape

Direct Access Storage Device

Optically Readable Characters

FIGURE 14.11
Data-recording media. (Courtesy of International
Business Machines Corporation)

cific patterns. Magnetic ink characters are printed on paper. The shape of the characters and the magnetic properties of the ink permit the printed data to be read by both people and machines. The shape of the optical characters, together with the contrast with the background paper, permits optical characters to be read by the machine and by people.

An input device of the computer system is a machine designed to sense or read information from one of the recording media. In the reading process, recorded data are converted to, or symbolized in, electronic form; the data can then be used by the machine for data-processing operations.

An output device is a machine that receives information from the computer system and records it on the designated output medium. All input–output devices cannot be used directly with all computer systems. However, data recorded on one medium can be transcribed to another medium for use with a different system. For example, data on cards or paper tape can be transcribed onto magentic tape. Conversely, data on magnetic tape can be converted to cards, paper tape, printed reports, or plotted graphs.

As there is communication between people and machines, there is also communication from one machine to another (Figure 14.12). This intercommunication may be the direct exchange of data (in electronic form) over wires, cables, or radio waves, or recorded output of one machine or system to be used as input to another machine or system.

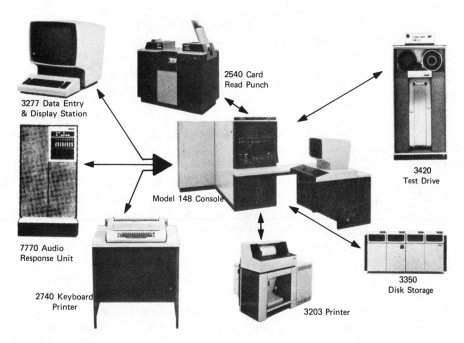

FIGURE 14.12
Machine-to-machine communication. (Courtesy of
International Business Machines Corporation)

14.2.1 Computer Data Representation

Not only must there be a method of representing data on cards, on paper
tape, on magnetic tape, and in magnetic ink characters, but there must also
be a method of representing data within a machine.

In the computer, data are represented by many electronic components:
transistors, magnetic cores, wires, and so on. The storage and flow of data
through these devices are represented as electronic signals or indications.
The presence or absence of these signals in specific circuitry is the method
of representing data, much as the presence of holes in a card represents
data.

14.2.2 Binary States

Computers function in *binary* states; this means that the computer components can indicate only two possible states or conditions. For example,
the ordinary light bulb operates in a binary mode; that is, it is either on or
off. Likewise, within the computer, transistors are maintained either conducting or nonconducting; magnetic materials are magnetized in one direction or in an opposite direction; and specific voltage potentials are present
or absent (Figure 14.13). The binary states of operation of the components

are signals to the computer, as the presence or absence of light from an electric light bulb can be a signal to a person.

Representing data within the computer is accomplished by assigning or associating a specific value to a binary indication or group of binary indications. For example, a device to represent decimal values could be designed with four- electric light bulbs and switches to turn each bulb on or off (Figure 14.14).

The bulbs are assigned decimal values of 1, 2, 4, and 8. When a light is on, it represents the decimal value associated with it. When a light is off, the decimal value is not considered. With such an arrangement, the single decimal value represented by the four bulbs will be the numeric sum indicated by the lighted bulbs.

Decimal values 0 through 15 can be represented. The numeric value 0 is represented by all lights off; the value 15, by all lights on; 9, by having the 8 and 1 lights on and the 4 and 2 lights off; 5, by the 1 and 4 lights on and the 8 and 2 lights off; and so on.

The value assigned to each bulb or indicator in the example could have been something other than the values used. This change would involve assigning new values and determining a new scheme of operation. In a computer, the values assigned to a specific number of binary indications become the code or language for representing data.

Because binary indications represent data within a computer, a binary method of notation is used to illustrate these indications. The binary system of notation uses only two symbols, zero (0) and one (1), to represent specific values. In any one position of binary notation, the 0 represents the absence of a related or assigned value, and the 1 represents the presence of a related

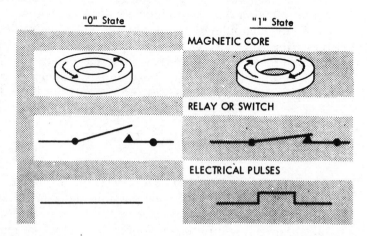

FIGURE 14.13
Binary components. (Courtesy of International
Business Machines Corporation)

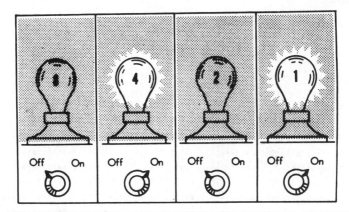

FIGURE 14.14
Representing decimal data with binary components. (Courtesy of International Business Machines Corporation)

or assigned value. For example, to illustrate the indications of the light bulb in Figure 14.14, the following binary notation would be used: 0101.

The binary notations 0 and 1 are commonly called *bits*. Properly, they are called 0 bit and 1 bit. Occasionally, however, they are loosely spoken of as no bit (0 bit) and bit (1 bit). For example, the binary notation 0101 of Figure 14.14 would be described as having a 1 bit in the 1- and 4-bit positions and a 0 bit in the 2 and 8 positions.

In some computers, the values associated with the binary notation are related directly to a binary system. This system is not used in all computers, but the method of representing values using this numbering system is useful in becoming familiar with the general concept of data representation.

14.2.3 Data-recording Media Cards

The 80-column punched card is one of the most successful media for communication with machines. Information is recorded as small rectangular holes punched in specific locations in a standard-sized card (Figure 14.15). Information, represented (coded) by the presence or absence of holes in specific locations, can be read or sensed as the card is moved through a card-reading machine.

Reading or sensing the card is basically a process of automatically converting data, recorded as holes, to an electronic impulse and thereby entering the data into the machine. Cards are used both for entering the data into the machine and for recording or punching information from a machine. Thus the card is not only a means of transferring data from some original source to a machine, but also is a common medium for the exchange of information between machines.

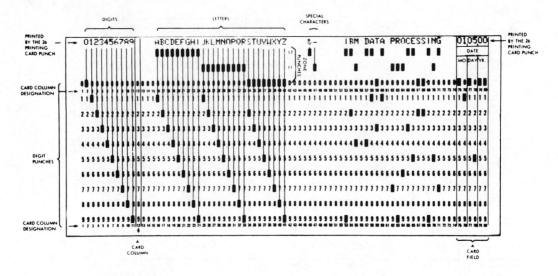

FIGURE 14.15
IBM 80-column punched card, standard hole pattern. (Courtesy of International Business Machines Corporation)

Eighty-column cards provide 12 punching positions in each column. The 12 punching positions form 12 horizontal rows across the card. One or more punches in a single column represents a character. The number of columns used depends on the amount of data to be represented.

The card is often called a *unit record* because the data are restricted to the 80 columns, and the card is read or punched as a unit of information. The actual data on the card, however, may consist of part of a record, one record, or more than one record. If more than 80 columns are needed to contain the data of a record, two or more cards may be used. Continuity between the cards of one record may be established by punching identifying information in designated columns of each card.

Information punched in cards is read or interpreted by a machine called a *card reader* and is recorded (punched) in a card by a machine called a *card punch.* Data are transcribed from source documents to punched cards by manually operated card-punch machines.

The standard *card code* uses the 12 possible punching positions of a vertical column on a card to represent a numeric, alphabetic, or special character (Figure 14.15). The 12 hole positions are divided into two areas, numeric and zone. The first 9 hole positions from the bottom edge of the card are the numeric hole positions and have an assigned value of 9, 8, 7, 6, 5, 4, 3, 2, and 1, respectively. The remaining three positions, 0, 11, and 12 are the zone positions. (The 0 position is considered to be both a numeric and a zone position.)

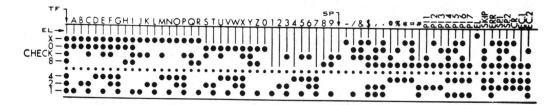

FIGURE 14.16
Paper tape, eight-channel code. (Courtesy of International Business Machines Corporation)

The numeric characters 0 through 9 are represented by a single hole in a vertical column. For example, 0 is represented by a single hole in the 0 zone position of the column.

The alphabetic characters are represented by two holes in a single vertical column, one numeric hole and one zone hole. The alphabetic characters A through I use the 12 zone hole and a numeric hole 1 through 9, respectively. The alphabetic characters J through R use the 11 hole and a numeric hole 1 through 9, respectively. The alphabetic characters S through Z use the 0 zone hole and a numeric hole 2 through 9, respectively.

The standard special characters $, *, %, and so on, are represented by one, two, or three holes in a column of the card and consist of hole patterns not used to represent numeric or alphabetic characters.

14.2.4 Paper Tape

Punched paper tape serves much the same purpose as punched cards. Developed for transmitting telegraph messages over wires, paper tape is used for data-processing communication as well. For long-distance transmission, machines convert data from cards and keyboard strokes to paper tape, send the information over telephone or telegraph wires to produce a duplicate paper tape at the other end of the wire, and reconvert the information to punched cards for later processing.

Data are recorded as a special arrangement of punched holes, precisely arranged along the length of a paper tape. Figure 14.16 shows an eight-channel-code paper tape. Paper tape is a continuous recording medium, as compared to cards, which are fixed in length. Thus, paper tape can be used to record data in records of any length, limited only by the capacity of the storage medium into which the data are to be placed or from which the data are received.

Data punched in paper tape are read or interpreted by a paper-tape reader and recorded by a paper-tape punch.

14.2.5 Magnetic Tape

Magnetic tape is one of the principal input–output recording media for computer systems. It is also used extensively for compact storage of large files of data.

Magnetic tape units offer high-speed entry of data into the computer system, as well as efficient, extremely fast recording of processed data from the system. Highly reliable input–output data rates of up to 2.5 million numeric characters per second are possible.

The magnetic tape unit functions as both an input unit and an output unit for the computer system. It moves the magnetic tape across a read–write head and accomplishes the actual reading and writing of information on the tape.

Information is recorded on magnetic tape as magnetized spots called *bits*. The recording can be retained indefinitely, or the recorded information can be automatically erased and the tape reused many times with continued high reliability.

So that tape can be easily handled and processed, it is wound on individual reels or in dust-resistant cartridges. Tape on the individual reels is $\frac{1}{2}$ in. wide and is supplied in lengths of up to 2400 ft per reel.

Several features built into the IBM 3420 magnetic tape unit ensure reliability and ease of operation. An automatic reel latch mechanically seats the file reel in position and pneumatically locks it on the hub for tape movement. With automatic threading and cartridge loading, tape mounting and demounting times are significantly reduced. Optical tachometers, built into the drive, sense small variations in the speed of the capstan and the tape and generate corrective signals. This precise control is one of the keys to the 3420's fast read access and rewind times. (Rewinding of a full 2400-ft reel takes only 45 s.)

14.3
STORAGE DEVICES

Several types of IBM storage are presently available: semiconductor, magnetic disk, and mass storage (Figures 14.17 and 14.18). Sometimes magnetic tape is thought of as storage rather than as input–output medium.

Information can be placed into, held in, or removed from computer storage as needed. The information can be:

1. Instructions to direct the central processing unit

2. Data (input, in-process, or output)

3. Reference data associated with processing (tables, codes, charts, constant factors, and so on)

Storage is classified as main or auxiliary, as in System/370 (Figure

IBM 3340 Disk Storage

IBM 2305 Fixed Head Storage

IBM 3330 Disk Storage

FIGURE 14.17
IBM storage devices. (Courtesy of International
Business Machines Corporation)

14.18). Main storage is sometimes called memory. It can consist of core storage, but in today's larger systems such as System/370, it is usually made up of semiconductor integrated circuits.

Auxiliary refers to all other storage and is of two types:

1. **Direct access:** Disk and mass storage devices in which records can be accessed without having to read from the beginning of a file to find them.

2. **Sequential:** Tape units where reels must be read from the beginning in order to read or write a desired record.

Main storage accepts data from an input unit, exchanges data with and supplies instructions to the central-processing unit, and can furnish data to an output unit. All data to be processed by any system must pass through main storage. This unit must therefore have capacity to retain a usable amount of data and the necessary instructions for processing.

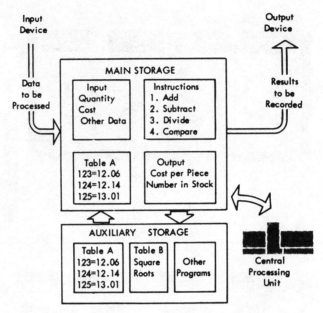

FIGURE 14.18
Schematic of main and auxiliary storage.
(Courtesy of International Business Machines
Corporation)

Disk storage offers the advantages of lower direct cost to offset slower speed. Disk devices also offer the advantage of capacity in billions of digits. The largest storage system is the IBM 3850 Mass Storage System, which has a capacity of 472 billion 8-bit characters.

14.3.1 Main Storage

The main storage of most computers, such as IBM System/370, consists of microminiature integrated circuits that are designed to store data in addressable locations called *bit cells*. Electrical devices, such as transistors, in each bit cell are capable of being in one of two states: on or off. The two states are used, therefore, to represent 0 or 1, plus or minus, yes or no, and so on. This is the basis of the computer's binary system of storing information.

Since any specified location of storage must be instantly accessible (called *random access*), the bits are arranged so that any combination of ones (1s) and zeros (0s) representing a character can be electronically written into or read back from main memory when needed. Other electronic circuits are used to locate the desired bit cell as an xy intersection on an imaginary grid covering all the main storage. Thus any memory bit can be addressed (or accessed) in a few nanoseconds.

In a complicated manufacturing process, the electronic memory circuits are built into a single chip of semiconductor material. They are termed *monolithic* (derived from the Greek words denoting a "single stone") because the electrical components are formed within a single chip of silicon. Transistors, diodes, resistors, and capacitors can all be produced within such a monolithic structure.

Large numbers of electrical components on a single silicon chip are connected to create integrated electronic circuits by applying a thin metal film to the top of the chip to provide contacts to the components and interconnections between them. A typical IBM semiconductor chip measures about $\frac{1}{6}$ in. per side and contains 2048 bit cells formed by almost two meters of aluminum "wire" connecting over 14,000 electrical devices.

Due principally to the miniaturization of the bit cells made possible by semiconductor technology, main storage is usually found in the same box as the central processing unit. In contrast, auxiliary storage is located in separate boxes that are cabled to the central processor. Since electronic signals travel at the constant speed of light, shrinking the size of the main memory devices and locating them very close to the processor allows the computer to operate at extremely fast speeds.

In all semiconductor main storage the basic principle is to use tiny integrated electronic circuits to store zeros (0s) and ones (1s), resulting in binary codes that can represent any data required by the computer.

14.3.2 Magnetic Disk Storage

Magnetic disk storage provides IBM data-processing systems with the ability to record and retrieve stored data sequentially or randomly (directly). It permits immediate access to specific areas of information without the need to examine sequentially all recorded data. Magnetic tape operations do not have this ability; tape searching must start at the beginning of the tape reel and continue sequentially through all records until the desired information area is found.

The high-speed access to data-storage locations provided by direct-access data processing permits the user to maintain up-to-date files and to make frequent direct reference to the stored data.

The magnetic disk is a thin metal disk coated on both sides with magnetic recording material. Disks are mounted on a vertical shaft; they are slightly separated from one another to provide space for the movement of read–write assemblies.

Data are stored as magnetized spots in concentric tracks on each surface of the disk. Some units have 808 tracks on each surface. The tracks are accessible for reading and writing by positioning the read–write heads between the spinning disks.

The magnetic disk data surface can be used repetitively. Each time new information is recorded and stored on a track the old information is erased.

The recorded data may be read as often as desired; data remain until written over.

14.3.3 Storage and Data-processing Methods

IBM data-processing systems use two methods of data handling: sequential or batch processing and in-line or direct-access processing (see Figure 14.19). The application requirements determine which method is needed. In either case, all data pertaining to a single application are maintained in files (often called *data sets*).

In sequential processing, these files are stored outside the computer (usually on magnetic tape) and they can be arranged in a predetermined sequence. The data may concern inventory, accounts receivable, accounts payable, payroll, and the like. Each file (data set) is made up of records, each containing information required to describe completely a single item. The sequence may be by item number, name, or person number, but all files pertaining to a single application must be in the same sequence.

In many cases, processing involves not only performing calculation on some parts of each record to arrive at balances, amounts, or earnings, but also involves adding, changing, or deleting records as new transactions occur. However, before transactions can be applied against the main or

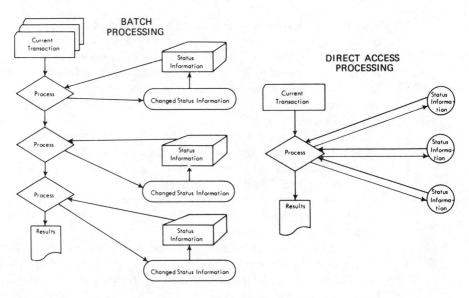

FIGURE 14.19
Batch and direct-access processing. (Courtesy of International Business Machines Corporation)

master file, they must also be arranged in the same sequence as the master file. For this reason, they are accumulated in convenient groups or batches.

The two files (data sets), master and transaction, now become input to the data-processing system. One record or a small group of records (also called a block) is read into storage at a time. These are processed, and the result is written as output. When magnetic tape files are used, the output records with the updated results of current processing must be recorded on a separate tape, producing a new master that will be used as input the next time the job is to be done. The next group of records is read in, and the process is repeated. The series of repetitive operations continues under the direction of program instructions, record by record, until the input files are exhausted. The results form a revised master file, updated according to the current transactions. The new master file is in the same sequence as the original files.

14.4
CENTRAL PROCESSING UNIT (CPU)

The central processing unit controls and supervises the entire computer system and performs the actual arithmetic and logical operations on data. From a functional viewpoint, the CPU consists of two sections: control and arithmetic–logical (Figure 14.20).

The *control section* directs and coordinates all operations called for by instructions. This involves control of input–output devices, entry or removal of information from storage, and routing of information between storage and the arithmetic–logical section. Through the action of the control section, automatic, integrated operation of the entire computer system is achieved.

In many ways, the control section can be compared to a telephone

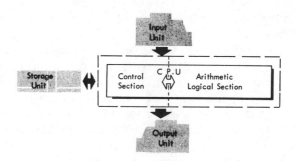

FIGURE 14.20
Central processing unit in the data-processing system. (Courtesy of International Business Machines Corporation)

FIGURE 14.21
Telephone exchange system. (Courtesy of International Business Machines Corporation)

exchange. All possible data transfer paths already exist, just as there are connecting lines between all telephones serviced by a central exchange (Figure 14.21).

The telephone exchange has a means of controlling instruments that carry sound pulses from one phone to another, ring the phones, connect and disconnect circuits, and so on. The path of conversation between one telephone and another is set up by appropriate controls in the exchange itself. In the computer, execution of an instruction involves opening and closing many paths or gates for a given operation. Some functions of the control section are to start or stop an input–output unit, to turn a signal device on or off, to rewind a tape reel, or to direct a process of calculation. In some System/370 models, a part of this section consists of a control device called *read-only storage* that contains circuits for performing operations designated by the operation codes. It also houses the *emulator* circuits that the user may select to make the System/370 perform programming instructions written for other computers.

The *arithmetic–logical section* contains the circuitry to perform *arithmetic* and *logical* operations. The former portion calculates, shifts numbers, sets the algebraic sign of results, rounds, compares, and so on. The latter portion carries out the decision-making operations to change the sequence of instruction execution.

14.4.1 Functional Units

A *register* is a device capable of receiving information, holding it, and transferring it as directed by control circuits. The electronic components used may be magnetic cores or transistors.

Registers may be named according to their function: an accumulator accumulates results; a multiplier-quotient holds either multiplier or quotient; a storage register contains information taken from or being sent to storage; an address register holds the address of a storage location or device; and an instruction register contains the instruction being executed (Figure 14.22). System/370 has general-purpose registers, which are used for several functions, including storage addresses, index addresses, and data that are to be processed logically or arithmetically.

Registers differ in size, capacity, and use. In some cases, extra positions detect possible *overflow* conditions during an arithmetic operation. For example, if two 11-digit numbers are added, it is possible that the result is a 12-digit answer (Figure 14.23).

In Figure 14.23, register A holds one factor, and register B holds the

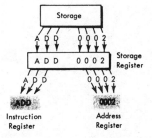

FIGURE 14.22
Register nomenclature and function. (Courtesy of International Business Machines Corporation)

Register A 5 7 3 2 8 0 3 1 2 2 1

Register B 4 3 1 0 5 7 2 6 4 2 0

Register C 1 0 0 4 3 3 7 5 7 6 4 1

↑
Overflow Position

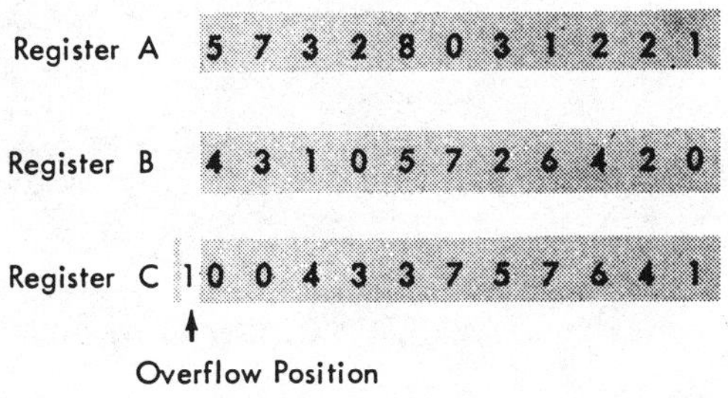

FIGURE 14.23
Overflow condition resulting from addition.
(Courtesy of International Business Machines
Corporation)

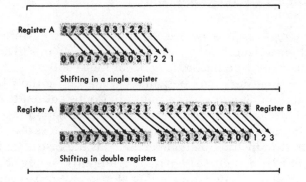

FIGURE 14.24
Types of computer register shifting. (Courtesy of
International Business Machines Corporation)

FIGURE 14.25
Typical system console (System/370 Model 148).
(Courtesy of International Business Machines
Corporation)

other factor. The two factors are combined, and the result is placed in register C, where an overflow condition is indicated by the presence of data in the overflow position. The contents of other registers can be *shifted* right or left within the register and, in some cases, even between registers. Figure 14.24 shows shifting of register contents three positions to the right. Positions vacated are filled with zeros, and numbers shifted beyond register capacity are lost.

In other instances, a register holds data while associated circuits analyze the data. For example, an instruction can be placed in a register, and associated circuits can determine the operation to be performed and locate the data to be used. Data within specific registers may also be checked for validity.

The more important registers of a system, particularly those involved in normal data flow and storage addressing, may have small lights associated with them. These lights are located on certain machine consoles (Figure 14.25) for visual indication of register contents and various program conditions.

The *counter* is closely related to a register and may perform some of the same functions. Its contents can be increased or decreased. The action of a counter is related to its design and use within the computer system. Like the register, it may also have visual indicators on the system console.

The *adder* receives data from two or more sources, performs addition, and sends the result to a receiving register or accumulator. Figure 14.26 shows two positions of an adder circuit, with input from registers A and B. The sum is developed in the adder. A carry from any position is sent to the next-higher-order position. The final sum goes to the corresponding positions of the receiving register.

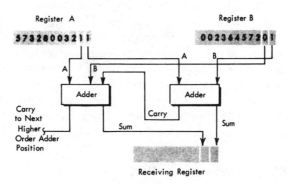

FIGURE 14.26
Adders in a computer system. (Courtesy of International Business Machines Corporation)

14.4.2 Machine Cycles

All computer operations take place in fixed intervals of time. These intervals are measured by regular pulses emitted from an electronic clock at frequencies as high as 17.2 million/s. A fixed number of pulses determines the time of each basic machine cycle.

In computer usage, time references are stated in such terms as *milliseconds* (ms), *microseconds* (μs), and *nanoseconds* (ns). These terms may convey no meaning unless it is realized just how short an interval a millisecond is. For example, the blink of an eye takes about one-tenth of a second or 100 milliseconds!

The following table establishes some additional terms and abbreviations:

$$0.1 \quad = \tfrac{1}{10} \text{ s}$$
$$= 100 \text{ ms}$$
$$0.001 \quad = 1/1000 \text{ s}$$
$$= 1 \text{ ms}$$
$$0.000001 \quad = 1/1,000,000 \text{ s}$$
$$= 1 \ \mu\text{s}$$
$$0.000000001 = 1/1,000,000,000 \text{ s}$$
$$= 1 \text{ ns}$$

Within a machine cycle, the computer can perform a specific machine operation. The number of operations required to execute a single instruction depends on the instruction. Various machine operations are thus combined to execute each instruction.

Instructions usually consist of at least two parts, an operation and an operand. The *operation* tells the machine which function to perform: read, write, add, subtract, and so on. The *operand* can be the address of data or an instruction in main storage, the address of data or programs in secondary storage, or the address of an input–output unit. It can also specify a control function, such as shifting a quality in a register, or backspacing and rewinding a reel of tape. In System/370, most instructions contain two operands.

To receive, interpret, and execute instructions, the central processing unit must operate in a prescribed sequence, which is determined by the specific instruction and is carried out during a fixed interval of timed pulses.

14.4.3 Instruction Cycle

The first machine cycle required to execute an instruction is called an *instruction cycle*. The time for this cycle is instruction or I-time. During I-time,

1. The instruction is taken from a main storage location and brought to the central processing unit.

2. The operation part is decoded in an instruction register. This tells the machine what operation is to be performed.

3. The operand is placed in an address register. This tells the machine what factors are to be used in the operation.

4. The location of the next instruction to be executed is determined.

At the beginning of a program, an instruction counter is set to the address of the first program instruction. This instruction is brought from storage, and, while it is being executed, the instruction counter automatically advances (steps) to the location corresponding to the space occupied by the next stored instruction. If each instruction occupies one storage position, the counter steps one; if an instruction occupies five positions, the counter steps five. By the time one instruction is executed, the counter has located the next instruction in program sequence. The stepping action of the counter is automatic. In other words, when the computer is directed to a series of instructions, it executes them one after another until instructed to do otherwise.

Assume that an instruction is given to add the contents of storage location 2 to the contents of a general-purpose register that will be used as an accumulator register. Figure 14.27 shows the main registers involved and the information flow lines.

I-time begins when the instruction counter transfers the location of the instruction to the address register. This instruction is selected from storage and placed in a storage register. From the storage register, the operation part is routed to the instruction register and the operand to the address register. Operation decoders then condition proper circuit paths to perform the instruction.

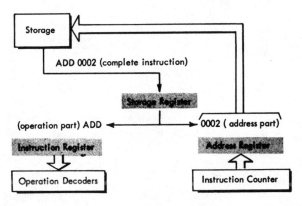

FIGURE 14.27
Computer I-cycle flow lines. (Courtesy of International Business Machines Corporation)

14.5
DATA-PROCESSING INPUT–OUTPUT DEVICES

An input–output unit is a device for putting in or getting out data from storage (Figure 14.28). Usually, device operation is initiated by a program instruction that generates a *command* to an input–output *channel*. A *control unit* decodes the command and effects operation of the device.

Input devices sense or read data from cards, magnetic tape, paper, magnetic ink characters inscribed on paper documents, images on 33mm microfilm, or remote terminals via communication lines. The data are made available to the main storage of the system for processing. Output devices record or write information from main storage on cards, magnetic tape, and paper tape, prepare printed copy, produce microfilm images, make graphic displays, or transmit information over a teleprocessing network.

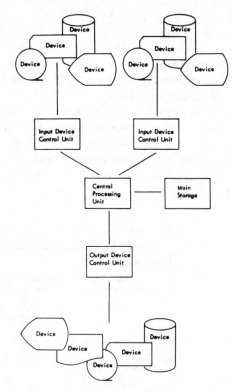

FIGURE 14.28
Input–output units in the data-processing system.
(Courtesy of International Business Machines Corporation)

Most input–output devices are automatic; once started, they continue to operate as directed by the stored program until the entire file is processed. Instructions in the program select the required device, direct it to read or to write, and indicate the storage location into which data will be entered or from which data will be taken.

Some I/O devices are used for manual entry, and no medium for recording data is involved. Instead, data are entered directly into storage using a keyboard or switches. Locally, these devices may be a console keyboard, local terminals (such as the IBM 2740s), or display terminals. Remotely, many types of teleprocessing terminals may be used. Instead of a recording medium, these terminals may require some amount of internal storage for holding (and perhaps analyzing) signals until a short message is completed, or until the terminals are polled (requested to transmit) and selected for data transfer.

14.5.1 Control Units

The type of information buffering required to coordinate the operations of the input–output device with the central processing unit (sometimes through transmission hookups) is one of the functions of the control unit. Other common functions are checking, coding, and decoding. If several similar devices are operating through one control unit, two principal functions are (1) determining priority of servicing, and (2) signaling device identification when requesting service for the input devices. Conversely, on the way out, the control unit directs the data to the address output unit.

Whereas the control unit is either included under the cover of an input–output device or located very close to a group of such devices, the channel (or channels) is contained within the central processing unit or is a separate piece of equipment near the CPU. The channel relieves the CPU of the burden of communicating directly with I/O devices and permits data processing to proceed concurrently with I/O operations. It might be thought of as the computer's control unit for one or more input–output control units. It is almost a separate, small CPU devoted exclusively to managing the input–output control units and devices assigned to it. After the channel has once been activated by an initializing instruction from a program being executed in the CPU, it carries out one or more commands that are similar to a section (subroutine) of a program, but the important difference is that of overlapping operations. The program in the CPU can be continuing with other jobs while the channel is carrying out its own program of bringing data into or out of the main storage. Sometimes it is interleaving input and output in a seemingly simultaneous fashion, working with several input–output control units at once, and maintaining the proper destinations for the messages, whether they be storage allocation (for input) or control unit and device (for output).

The steps in a program in the CPU are called instructions; the steps in a program for a channel are called *commands.* Each command has an operation code that tells the channel what to do (for instance, read, write, control, sense, etc.); if it is a command that involves a data transfer, the command also has an address telling where to get or where to put the data in the storage system of the computer; if it is a control command that does not involve a data transfer, either it contains the order to be passed on to the control units, or (in some computer systems) it contains the address of a location in storage where the order is located.

Just as the CPU is free to continue with its program once it has given an instruction to start a channel on its independent program of commands, so a channel is free to step through other commands (probably starting or terminating some other input–output transfer of data) as soon as it has commanded the control unit what to do and given it an order specifying the particular device. Thus a channel is an intermediary input–output device that is constantly juggling the various input–output operations to make the most efficient use of time, not only by overlapping different input and output but by doing so without tying up the CPU.

As soon as a particular input–output transaction is completed, the device control unit signals the channel, which, in turn, signals the CPU with an *interrupt,* meaning: "My particular job is done. As soon as convenient, use the data I have given you [if it was an input operation] and give me another command."

This idea of automatic interrupts (built into the design of the data-processing system components), combined with carefully preprogrammed commands to the channels and orders to the control units, leads to a far greater total amount of data handling per unit of time (sometimes described as *throughput*) than used to be possible.

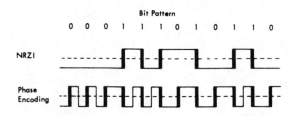

In the NRZI method of recording, a change in magnetic flux is interpreted as a 1 bit; lack of a flux change is interpreted as a 0 bit. The phase-encoded method of recording results in a continuous wave pattern, even when a record contains all zeros.

FIGURE 14.29
Comparison of NRZI and phase-encoded bit patterns. (Courtesy of International Business Machines Corporation)

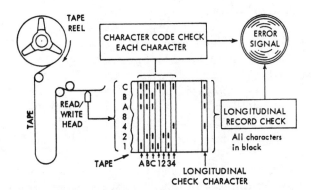

Information read from tape is checked two ways. A character code check (vertical check) is made on each column of information to ensure that an even number of bits exists for each character read. If an odd number of bits is detected for any character or column of bits, an error is indicated, unless the computer operates in odd parity. A longitudinal record check is made by developing an odd or even indication of the number of bits read in each of the seven bit tracks of the record, including the bits of the check character. If any bit track of the record block indicates an odd number of bits after it is read, an error is indicated, unless odd parity is required by system design.

FIGURE 14.30
Seven-track validity checks, BCD mode, even parity. (Courtesy of International Business Machines Corporation)

14.5.2 Checking Magnetic Tape Deck

Data recorded on magnetic tape must be accurate so that errors are not sent through the system. Data are therefore checked to ensure that valid characters are recorded and to verify that the recorded bits are of effective magnetic strength.

Two methods of recording are used on IBM magnetic tape. The phase-encoding method is used on newer IBM tape units; other magnetic tape units use the non-return-to-zero-IBM (NRZI) method (see Figure 14.29).

The NRZI method of data recording is very reliable, but it has given way to the phase-encoding method because of the increased densities of recording possible on the newer units. The tape-error-detection system used on NRZI tapes uses the principle of simple parity checking. With this system, it is possible to detect virtually all tape reading and writing errors (Figure 14.30).

Simple parity checking indicates the error, but not the kind of error. Similar double-bit errors in two characters of a record could conceivably cancel each other and indicate correct parity. However, this coincidence is extremely rare.

14.5.3 Printers

IBM printing devices provide a permanent visual record of data from the computer system. Speeds of printing vary from 15.5 to 68,136 characters per second.

As an output unit, the printer receives data, symbolized in electronic form, from the computer system. The electronic symbols enter appropriate circuitry and cause printing elements to be actuated. All printing devices have a paper transport that automatically moves the paper as printing progresses.

The major printing devices consist of the electrophotographic printer, chain–train–belt printer, serial wire matrix printer, and the typewriter.

The fastest and most versatile IBM printer is the IBM 3800 Printing Subsystem (Figure 14.31). The IBM 3800 is a high-speed, nonimpact printer that produces characters on paper through electrophotographic and laser technology.

Features of the 3800 shown in Figure 14.31 include the following:

1. Continuous forms input, transport, and stacking mechanism.

2. Fifty-two Kbyte storage for page buffering and control of printer operations.

3. Eighteen different character sets, including four special underscored sets, and 10-, 12-, and 15-pitch (characters per inch) sets, all of which may be printed separately, or any combination of up to four may be mixed on a line. (Print line maximums are 136 positions at 10 characters per inch (cpi), 163 positions at 12 cpi, and 204 positions at 15 cpi.)

4. Writable character generation storage is organized into two 64-character writable generation modules to hole 128 characters. (An additional increment of 127 writable character generation storage positions is op-

FIGURE 14.31
IBM 3800 printing subsystems. (Courtesy of International Business Machines Corporation)

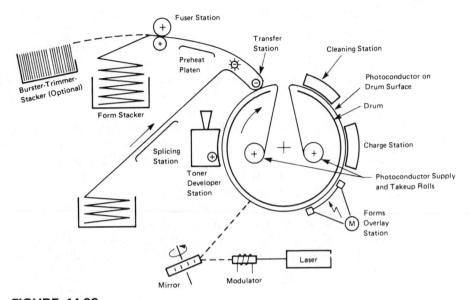

FIGURE 14.32
Path of paper forms through the IBM 3800 printing subsystem. (Courtesy of International Business Machines Corporation)

tional. This represents up to 255 graphics on line with no throughput loss.)

5. The electrophotographic process, which includes the following:

   a. A continually revolving drum on which a charged photoconductive surface is selectively discharged by a low-power laser to produce images of the printed data.

   b. A developer station where black toner is attracted to the images.

   c. A transfer station where the toner is transferred to the paper forms.

   d. A fuser station to fuse the toner into the paper.

   e. A cleaning station to remove any residual toner from the drum after the page has been printed.

   f. A charge station to prepare the photoconductor.

6. A forms overlay station to expose the drum with form images or other fixed data.

In operation, data to be printed are moved from the CPU to the 3800 a line at a time, are translated into graphic code using a set of translate tables, and are stored in the page buffer. When the page buffer contains a full page, the code is used, through interaction with character generation storage, to modulate the laser in exposing the revolving drum. Exposure is by horizontal line scanning, similar to the way a cathode-ray gun scans a TV screen to produce a picture. The image is developed with toner, transferred to paper, and fused. The photoconductor surface of the drum is cleaned and reconditioned after each exposure. Finished copies are refolded and stacked in the continuous forms stacker, complete with job separation marking (optional). Figure 14.32 is a schematic diagram of the 3800 printing mechanism.

Graphic character modification allows user- or IBM-designed characters to take the place of an equal number of standard characters in character-generation storage. Line spacing is 6, 8, or 12 lines per inch and can be intermixed within a page.

The IBM 3800 uses a single-ply, edge-punched, perforated, and stacked continuous forms in any combination of five lengths and ten widths (common-use sizes). Preprinted forms may be used, or the form image can be printed simultaneously with text by the use of a forms overlay negative, by character formating, or by any combination of these to suit the application.

14.5.4 Data Buffering

All data-processing procedures involve input, processing, and output. Each phase takes a specific amount of time. The usefulness of a computer is often directly related to the speed at which it can complete a given procedure. Any operation that does not use the central processing unit to full capacity prevents the entire system from operating at maximum efficiency. Ideally, the configuration and speed of the various input–output devices should be so arranged that the CPU is always kept busy with useful work.

The efficiency of any system can be increased to the degree in which input, output, and internal data-handling operations can be overlapped or allowed to occur simultaneously. Input is divided into specific units or logical associations of data that enter storage under control of the program. A number of output results may be developed from a single input, or conversely, several inputs can be combined to form one output result. Figure 14.33(a) shows the basic time relationship between input, processing, and output with no overlap of operations. In this type of data flow, processing is suspended during reading or writing operations. Inefficiency is obvious, because much of the available time of the central processing unit is wasted.

Figure 14.33(b) shows a possible time relation between input–output and computing when a buffered system is used. Data are first collected in an external unit called a *buffer*. When summoned by the program, the contents of the buffer are transferred to the main storage unit. The transfer takes only a fraction of the time that would be required to read the data directly from an input device. Also, while data are being assembled in the buffer, internal manipulation or computing can occur in the computer. Likewise, processed data from main storage can be placed in the buffer at high speed. The output device is then directed to write out the contents of the buffer. While writing occurs, the central processing unit is free to continue with other work. If several buffered devices are connected to the system, reading, writing, and computing can occur simultaneously [Figure 14.33(c)].

Further development of the buffering concept has led to the use of main storage as the primary buffer. Data are collected from, or sent to, the input–output devices in words or in fixed groups of characters. Transmission of words is interspersed automatically with computation, but the time required for the transmission of single words is relatively insignificant. The effect is that of overlapping internal processing with both reading and writing. The principal advantage here is that the size or length of the data handled is restricted only by the practical limits of main storage. When external buffers are used, the amount of data handled at any time is limited to the capacity of the buffer. Overlapping operations up to this point have demonstrated a principle of synchronous operation; that is, the action of the

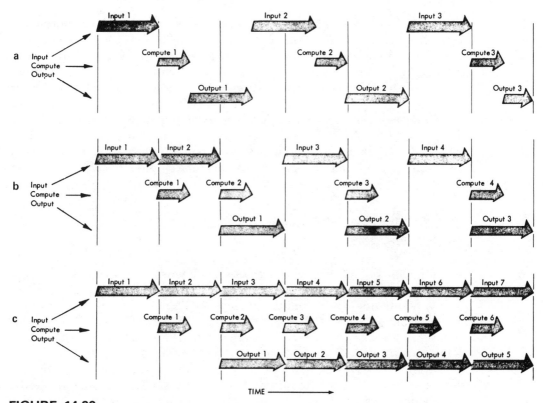

FIGURE 14.33
Data buffering. (Courtesy of International Business Machines Corporation)

input–output devices is made to occur at fixed points in the program and in a sequence established by the programmer.

In some computers, design features allow for automatic interruption of processing by the input–output devices; synchronous operation is not required. The input or output device signals the central processing unit when it is ready to read or to write. The central processing unit responds to these signals and either accepts the data as input or transmits the required information as output. In real-time teleprocessing systems, this type of input–output is likely to be nonsequential and unpredictable.

The problem arises of how to fill in the gaps in central processing unit time. The answer is to somehow queue the various tasks and programs to step in and, without interfering with one another, to use the otherwise idle time. This is the basis for multiprogramming, a subject described in more detail in Section 14.1.3

14.6
TELEPROCESSING

Teleprocessing is the process of receiving or sending data to remote locations by way of communications facilities. A teleprocessing *network* consists of a number of communications lines (communications facilities) connecting a central data processing system with remote teleprocessing devices (Figure 14.34). Such devices can be terminals, control units, or other data-processing systems. In this overview, any machine or group of machines capable of generating and/or receiving signals transmitted over communications lines will be referred to as a terminal or terminals. Thus terminals may be data-processing systems, communications systems such as the IBM 3270 Information Display System, or a single unit such as the IBM 2740 Communication Terminal.

14.6.1 Elements of a Teleprocessing Network

The elements of a teleprocessing network (Figure 14.34) consist of a host processor (central data processing system), communications control devices, modulation–demodulation devices (modems), communications lines, other terminals, and programming systems. Three of these elements, the communications control devices, modems, and communications lines, constitute a *data link* (Figure 14.35).

System/370 is designed so that it can serve as the host processor in a teleprocessing network. Requirements for the host processor include multiprogramming capability, adequate storage capacity, storage protection, adequate speed for the applications required, and, for planning purposes, the potential for expanding storage capacity and speed.

The host processor for a teleprocessing network must be able to handle

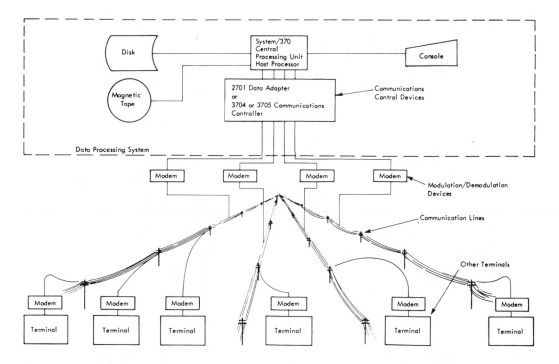

FIGURE 14.34
A teleprocessing network. (Courtesy of International Business Machines Corporation)

random and unscheduled input, as well as serialized and scheduled input.

Communications control devices are hardware components that link the communications lines to the host processor. These devices can be external to the processor, such as the IBM 3704 or 3705 Communications Controller, or they can be a part of the processor of a System / 370, such as the integrated communications adapter feature. When control devices are external units, they can be classified as data-transmission multiplexers.

The transfer of data requires noninformation transmissions for setting up, controlling, checking, and terminating information exchange. These noninformation exchanges constitute *data link control.* Communications control devices handle data link control; thus functions of these devices include:

1. Synchronization (getting the receiver in step with the transmitter)

2. Identifying the sender and receiver

3. Delimiting the beginning and ending of information (code translation)

4. Error detection and recovery

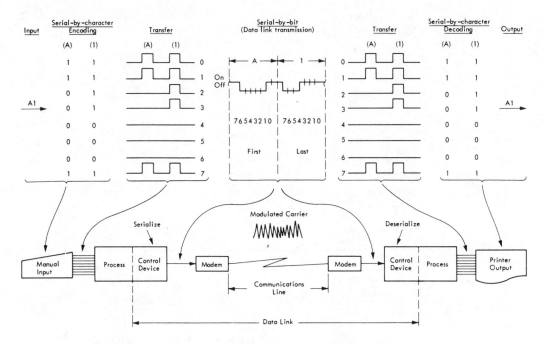

FIGURE 14.35
Data conversion for data transmission. (Courtesy of International Business Machines Corporation)

For a host processor to send data over communications lines, the data must be converted (serialized) to a stream of binary digits. Likewise, when the host processor receives data from a remote terminal, these data must be reconverted (deserialized) into machine language for processing (Figure 14.35). Control devices perform this function.

14.6.2 Modulation–Demodulation Devices

After data to be transmitted are serialized by the control device, the binary signals must be converted to audiofrequency signals (modulated) for transmission over communications lines and reconverted (demodulated) at the other end. A modulation–demodulation device, or modem, performs this function. One modem is required at each end of a data link (Figure 14.35). Data sets and line adapters have the same function as a modem.

Depending upon the type of communications lines and modem equipment, transmission of data can be voice grade (permits transmission of both data and human voice) or subvoice-grade (transmission of data only). A modem can be an integral part of a control device or terminal, or it can be an external unit.

Communications lines are classified according to configuration, transmission direction, type, and transmission mode.

14.6.3 Configuration

Two basic communications line configurations are (1) point-to-point (connects two terminals), and (2) multipoint (connects multiple terminals). In a multipoint configuration, one terminal must always be designated as the primary (control) terminal, and all others are secondary (tributary) terminals.

A communications line that transmits data in either direction, but not simultaneously, is called *half-duplex*. A line that transmits in both directions at the same time is called duplex or *full duplex* (Figure 14.36).

14.6.4 Teleprocessing Applications

The types of applications that are provided by a teleprocessing network are many and varied. Some of the most widely used applications include:

1. **Data Entry:** Entry of data from a remote terminal into a host processor via a communications link by a remote terminal.

2. **Record update:** Alteration, deletion, or addition of data contained on

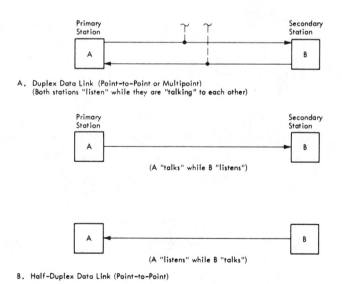

FIGURE 14.36

Communications lines configurations. (Courtesy of International Business Machines Corporation)

existing data files stored at the host processor site via a communications link from a remote terminal.

3. **Remote job entry:** Entry of logic functions from a remote terminal to be executed at the host processor location via a communications link.

4. **Message switching:** The ability to relay a message from one remote terminal to one or more remote terminals via a host processor and a series of communications links.

5. **Time sharing:** The allocation of host processor resources so that many remote terminals may execute programs concurrently and may interact with the programs during execution.

6. **Data acquisition and process control:** A high-speed data-acquisition system is designed to maintain constant communication with a process for such purposes as:

   a. Determining whether the process is operating within acceptable limits.

   b. Providing records for accounting or management decisions.

   c. Providing a record of data obtained during a research experiment.

A process control system usually incorporates data-acquisition facilities and has the additional capability of using the acquired data as a basis for supervising and controlling the process.

14.7
REVIEW QUESTIONS

1. Describe in your own words the meaning of a present-day data-processing system.

2. Discuss the meaning of stored programs and multiprogramming.

3. Discuss the meaning of a central processing unit.

4. Discuss the function of an arithmetical–logical unit.

5. Discuss the meaning of data-processing input and output devices.

6. Discuss the uses of a portable computer.

7. Discuss the meaning of data representation in data processing.

8. Discuss the meaning of binary states and uses in computers.

9. Discuss the use of punched cards, paper tape, and magnetic tape in data-processing systems.

10. Discuss the meaning of storage devices.

11. Discuss the meaning of main storage and magnetic disk storage.

12. Discuss in your own words the meaning of the central processing unit.

13. Discuss in your own words the uses of a data-processing input–output device.

14. Discuss the meaning of data buffering.

15. Discuss the meaning of teleprocessing.

14.8
REFERENCES

1. *Introduction to IBM Data Processing Systems, Student Text*, IBM No. GC20-1684-4, 5th ed., IBM publication, White Plains, N.Y., July 1978.

2. *Vocabulary for Data Processing, Telecommunications and Office Systems*, IBM No. GC20-1690-6, 7th ed. IBM, White Plains, N.Y., July 1981.

3. Don Cassell and Martin Jackson, *Introduction to Computers and Information Processing*, Reston Publishing Co., Reston, Va., 1980.

4. Frank Clark, *Mathematics for Data Processing*, Reston Publishing Co., Reston,Va., 1974.

5. Robert Condon, *Data Processing with Applications*, 2nd ed., Reston Publishing Co., Reston, Va., 1981.

15
Microprocessors and Microcomputers

15.1
INTRODUCTION

The microprocessor (μP) is an integrated circuit that accepts coded instructions for execution; the instructions may be entered, integrated, or stored internally. The microcomputer (μC) is a computer system whose processor is a μP. A basic μC includes a μP, storage, and an input–output (I/O) facility, which may or may not be on one chip.

The μP and μC are truly important electronic instrumentation devices of the computer age, which will grow and change greatly in the next 25 years. In other applications besides the computer, the μP can act as a memory and/or controller device in one chip. The μP chip appears at the beginning or the end of a circuit, such as a data acquisition and conversion system. As an integrated-circuit device, the μP can also serve many purposes. It can be used in industry, communication, and medicine but it is also found in the home in microwave ovens, television games and sets, and safety devices; it can also be processed in many other applications. The microprocessor was introduced in previous chapters. In this chapter we will introduce in detail the operation of the μP and μC by providing sections on the following:

1. Integrated circuit terminology
2. Binary numbers
3. Arithmetic and logic
4. Mnemonics and memory
5. Eight, sixteen, and thirty-two bit μPs
6. Microprocessor applications
7. Emulation for the μP
8. Dynamic digital board testing
9. The IEEE-488 bus interface

The most important large-scale-integrated circuits (LSI) in μP technology include the following:

1. *P*-channel metal oxide silicon chip, abbreviated PMOS.
2. *N*-channel metal oxide silicon chip, abbreviated NMOS.
3. Silicon-on-sapphire MOS, called the SOS / MOS.
4. Complementary MOS, called CMOS chips with the lowest power dissipations.
5. HMOS is the trademark of the Intel Corp. It is a High Density *N*-Channel Silicon Metal Oxide Chip.

Standard LSI products handle data in 4-, 8-, 16-, and 32-bit word lengths and multichip μPs. Many μPs are available, and there are more than 50 manufacturers. We shall consider the Intel 8086 (Figure 15.1), which is a representative 16-bit μP. Differences in μPs are found in their operating speed, input–output interfaces, and mnemonics and machine language used, but the basic principles are the same.

The latest available μP design is the 32-bit device, among which are the following new designs:

1. Intel Corporation's iAPX-32 micromainframe, a three-chip set in HMOS for high level-language (Ada) programmability discussed in Section 15.6.3.
2. Hewlett-Packard's single-chip NMOS with 450,000 transistors with a speed of 18 MHz and dissipating nearly 7 W.
3. Bell Laboratories' Bellmae-32, a single-chip μP fabricated in the twin-tub CMOS technology dissipating less than 1 W of power. This μP domino circuit operates at twice the speed of previously designed CMOS circuits.
4. Nippon Telegraph and Telephone's CMOS μP with 2-μm line widths

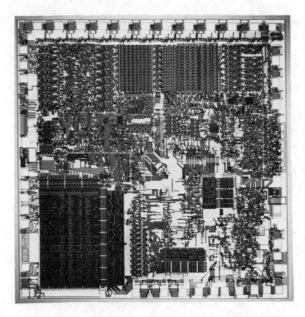

FIGURE 15.1
Intel HMOS 16-bit, 8086 μP designed to deliver 10
times the processing power of the 8080, yet it is
an evolution of the 8080/8085 and software convertible. Chip size is 165 square mils. (Courtesy of
Intel Corporation)

consisting of 20,000 gates on one chip 12 mm^2 and dissipating of 750 mW of power.

Use of CMOS and NMOS for increased μP performance appears to be accelerating. Logic designs use P-channel and N-channel devices.

High-speed very large integrated circuits (VLSI) with NMOS designs are used in electronic μP and μC applications. VLSI are also used in manufacturing today.

15.1.1   Very Large Scale Integrated Circuits[a]

The electronic instrumentation technology required for manufacturing control is becoming ever more sophisticated. The trends toward very large scale integration (VLSI) with small device geometries, larger chip areas, greater wafer diameters and more complex and expensive processing equipment increase the demands and distributions, silicon defects, wafer-

---

[a] The material in Section 15.1.1 is from the *EDN* silver anniversary issue of *Electronic Technology ... The Next 25 Years*, Oct. 14, 1981, and is reprinted courtesy of *EDN*, a Cahners Publication.

flatness and distortion, oxide defects, and critical dimensions. VLSI techniques also depend on control of oxide and metal-film thickness, design tolerances (rules), level-to-level registration, submicrometer particulate contamination, and process-induced defects. In addition, the increasing cost and complexity of manufacturing equipment make more thorough evaluation before procurement both necessary and cost effective.

The trend toward higher levels of integration is expected to continue in the near future. Circuits with 100 million or even 1 billion elements are forecast for the year 2000. Electronic systems will therefore consist of many fewer individual parts, each of enormous complexity and embodying design considerations traditionally regarded as being within the province of the system-design engineer. Thus circuit and system designers will need to become more aware of the materials properties and manufacturing aspects that will govern chip performance. In this context they will encounter metrology, which has traditionally been of concern only to device manufacturers and to suppliers of materials and manufacturing equipment.

Silicon technology is the general-purpose VLSI technology of the 1980s. Every year, U.S. suppliers increase silicon-IC production, and the sale of integration rises. Current systems employ chips containing the equivalent of 500,000 transistors. But systems in the planning and development stages will require higher clock speeds than silicon technology is expected to achieve. Thus gallium-arsenide (GaAs) technology should have broad applications in the next generation of VLSI and LSI systems. This includes the field of integrated optoelectronics.

Most GaAs-logic LSI circuits fabricated to date use Schottky-diode FET logic (SDFL). But buffered-FET logic (BFL) provides higher speed, and enhancement depletion-mode logic (EDL) furnishes the lowest power per gate available. Gate lengths of GaAs transistors now range from 0.2 to 1.0 μm and propagation delays of less than 50 ps have been achieved.

As monolithic techniques mature over the next 25 years, the cost of GaAs chips will decrease and the technology will expand. New applications will also appear in μC. The wrist radio, for example, will use monolithic microwave ICs and provide personal communications via satellite.

15.1.2 The Coprocessor Concept[b]

Suppose a master chip architect lays out a CPU so large and grand and with so many instructions that the VLSI chip designer finds that, even with today's tightest geometric layout rules, this grand edifice just cannot fit on a reasonable plot of chip real estate. The chip designer can then split the edifice in two, put it on two lots, and build a bridge between the two halves.

---

[b] The material in Section 15.1.2 is from Robert H. Cushman, "Arithmetic Chips Assume Greater Importance as μC Users Demand Faster Response," *EDN*, April 14, 1982, and is reprinted courtesy of *EDN*, a Cahners Publication.

The resulting two-chip system still executes like the single-CPU system the master architect envisioned. But physically the CPU is in two packages. The main CPU fetches the instructions from memory as usual. Instruction registers in both chips simultaneously decode each instruction. Then, if it is a regular instruction, the main CPU executes it; if it is a special instruction, the coprocessor executes it.

This arrangement works well in practice. The main CPU remains physically realizable given the state of the art, so its price also remains reasonable. And only customers who want the additional features of the coprocessor chip must pay for it.

The value of the coprocessor concept is especially evident in the case of a floating-point math. Floating-point arithmetic requires extremely wide (64-bit) registers and a correspondingly wide arithmetic logic unit (ALU) for high-speed parallel execution. As a result, the 8087 coprocessor, for example, has almost double the area of the 8086 and is inherently much more expensive.

The main drawback to coprocessing is that the coprocessor must duplicate some of the CPU functions and cannot always handle interrupts. Most designers, perhaps optimistically, except that VLSI techniques will eventually put today's coprocessors back on tomorrow's CPUs.

15.2
TERMS USED FOR UNDERSTANDING MICROPROCESSORS AND INTEGRATED CIRCUITS

Terminology includes the following:

1. **Integrated circuit** is an electronic configuration containing many interconnected elements in a single piece or chip of semiconductor material. Thousands of transistors and other components are produced and wired on printed circuit board.

2. **Bit** is a binary (two-stage). The digit, the smallest unit of digital information, is usually represented as binary digits 0 and 1.

3. **Byte** is a sequence of bits used as the "operating word" of a computer; usually the number of bits in a single memory location. An 8085AH microprocessor-based computer is an 8-bit computer; its byte is 8 bits. The 8085AH is the latest Intel version of the 8080.

4. **Nibble** is a sequence of bits less than a byte, such as 4 bits.

5. **Computer** is a high-speed data processor comprising memory, control, arithmetic, logical, and input–output capabilities.

6. **Microcomputer** includes an 8-, 16-, and 32-bit using a microprocessor.

7. **Bubble memory** is the newest type of the semiconductor memory family. As a magnetic memory bubble, it is considered a large-capacity non-violative device. The stored information is retained in a magnetic material by very small cyclindrical magnetic domains called bubbles. Magnetic bubble devices could find uses in future μP data-storage systems.

8. **CAM** is an abbreviation for content addressable memory. CAMs are ideal for quick data searches, correlation checks, sorting by value or attribute, and large virtual memory systems.

9. **CCD** is an abbreviation for a charge coupled device memories that are shift registers moving cluster of electrons along a channel formed in a silicon chip. CCDs is a high-density semiconductor memory component.

10. **CMOS** is the complementary metal oxide semiconductor using both P- and N-channel devices on the same silicon substrate. It can be used as a μP.

11. **DIP** is an abbreviation for dual in-line package, the most popular IC packaging technique in use today.

12. **EAROM** is an abbreviation for the electrically alterable ROM. EAROM is similar to EPROM except that the electrical current erases the content.

13. **E²PROM** is an abbreviation for an electrically erased read-only memory. E²PROMS use 5-V operation.

14. **EPROM** is an abbreviation for the electrically programmable read-only memory. EPROM is the erasable PROM that can also be erased with ultraviolet light and reprogrammed.

15. **MOS** is an abbreviation for the metal oxide semiconductor.

16. **PROM** is an abbreviation for programmable read-only memory and refers generically to all ROMs that can be programmed by the user and, in some cases, reprogrammed. Used in μP work.

17. **RAM** is a μP term and is an abbreviation for the random-access memory in μPs and is a collection of semiconductor cells that can be either a 1 or 0 logic state. The stored information is referred to as a bit, and the series of sequential bits create a word. The memory whose contents can be altered is called a read–write (R/W) memory because it can both be read from (accessed) or written into (i.e., data stored). Since this type of memory also features random access, it is also called a random-access memory, or RAM. If information in a program is ever to be changed, it must be stored in the RAM, not the ROM.

18. **ROM** is a μP term and is an abbreviation for read-only memory. In a μP, ROM is used to store program steps and certain types of data that are never changed.

15.3
BINARY DIGIT CONCEPTS[c]

There are various ways in which binary digits (1, 0) can represent decimal numbers and/or alphabetic characters. Codes for *numbers* alone are related to the μP. All μPs use binary-coded signals for both operation and data control.

Alphabetic characters are related to input–output devices. Codes that are now in use contain numeric and alphanumeric types.

Numeric codes include the following:

1. Binary: base 2
2. Binary-coded decimal: BCD
3. Octal: base 8
4. Hexadecimal (hex): base 16
5. Gray (cyclic): one-bit-change
6. 1, 2, 4, 7: Uses only two 1's
7. Excess 3: binary plus 3

Alphanumeric codes include the following:

1. ASCII: American Standard Code for Information Interchange (7 bits)
2. BCDIC: Binary-Coded-Decimal Interchange Code (obsolete)
3. EBCDIC: Extended BCDIC (IBM, 8 bits)
4. BAUDOT: Teletype (5-bit) code

We shall limit our discussion to binary octal and hexadecimal numbers.

Straight binary uses the two symbols 1 and 0 in a conventional place-value system wherein the place of any symbol in a sequence of symbols indicates its value as a power of 2. In the conventional decimal notation, which uses 10 symbols (0 to 9), the place of the symbol in the number indicates its value as a power of 10. For example, decimal number 9031 means $9(10^3) + 0(10^2) + 3(10^1) + 1(10^0) = 9000 + 000 + 30 + 1 = 9031$.

Similarly, the binary number 1101 means $1(2^3) + 1(2^2) + 0(2^1) + 1(2^0) = 8 + 4 + 0 + 1 = 13$.

The straight binary representation of decimal numbers 0 to 15 is shown at the left in Table 15.1 Table 15.2 shows powers of 2.

The binary symbol 0 or 1 is called a *bit* (binary digit). Note that 2 bits can represent decimal numbers 0, 1, 2, 3, but no others. Two bits can represent only 4 numbers (counting 0 as a number).

---

[c] The material in Section 15.3 in part is from "Microprocessor Course No. 1," *Medical Electronics*, December 1978, 2994 W. Liberty Ave., Pittsburgh, Pa. 15216. Courtesy of the Publisher, Mr. Milt Aronson.

It requires 3 bits to represent the decimal numbers 4, 5, 6, 7. Three bits can represent any of the *eight* numbers from 0 to 7.

It requires 4 bits to represent the decimal numbers 8, 9, 10, 11, 12, 13, 14, 15. Four bits can represent any of the 16 numbers from 0 to 15.

Note that the maximum value of any binary number (11, 111, 1111, etc.) is always an odd number because of the 1 in the least-significant place. Since 4-, 8-, and 16-bit numbers are widely used in μPs, the largest value of each is of interest:

$$1111 = 15 = 2^4 - 1$$

$$1111\ 1111 = 255 = 2^8 - 1$$

$$1111\ 1111\ 1111\ 1111 = 65,535 = 2^{16} - 1$$

The largest value of an n-bit number is obviously $2^n - 1$. A 32-bit μP has $(2^{32} - 1)$ bit number.

A 16-bit μP is a lot more "powerful" than an 8-bit μP because it can handle a much larger number of individual numbers.

The decimal system uses 10 symbols (0 to 9). Binary-coded decimal expresses each decimal digit in straight binary form, using 4 bits because numbers 7, 8, and 9 require 4 bits. From 0 to 9, the BCD number is identical with straight binary.

TABLE 15.1
Binary, octal, and hexadecimal representation of decimal numbers

Binary	Decimal	Octal	Hexadecimal
000	0	:0	:00
001	1	:1	:01
010	2	:2	:02
011	3	:3	:03
100	4	:4	:04
101	5	:5	:05
110	6	:6	:06
111	7	:7	:07
1000	8	1:0	:08
1001	9	1:1	:09
1010	10	1:2	:10
1011	11	1:3	:11
1100	12	1:4	:12
1101	13	1:5	:13
1110	14	1:6	:14
1111	15	1:7	:15
10000	16	2:0	01:00
10001	17	2:1	01:01
10010	18	2:2	01:02

TABLE 15.2
Powers of 2, 8, 16

16	8	2	decimal
16^{-4}		2^{-16}	.0000152587890625
	8^{-5}		.000030517578125
			.00006103515625
			.0001220703125
16^{-3}	8^{-4}		.000244140625
			.00048828125
		2^{-10}	.0009765625
	8^{-3}		.001953125
16^{-2}			.00390625
			.0078125
	8^{-2}		.015625
		2^{-5}	.03125
16^{-1}			.0625
			.125
			.25
		2^{-1}	.5
			0
		2^{0}	1
			2
		2^{2}	4
		2^{3}	8
			16
		2^{5}	32
	8^{2}		64
			128
16^{2}			256
	8^{3}		512
		2^{10}	1024
			2048
16^{3}	8^{4}		4096
			8192
			16384
	8^{5}		32768
16^{4}		2^{16}	65536

But decimal 22 in BCD is 0010:0010.

Decimal 998 in BCD is 1001:1001:1000.

The largest 4-bit binary number to appear in a BCD number is 1001, which represents decimal 9; the binary numbers 1010, 1011, 1100, 1101, 1110, and 1111 are never used. This makes BCD less efficient than binary. The largest 16-bit number in BCD represents decimal 9999; the largest 16-bit binary number represents 65,535.

Arithmetic calculations (addition, multiplication, etc.) are possible in BCD, and early computers operate with BCD code, as do some microprocessors. The code is still widely used in digital voltmeters, frequency meters, and other instrumentation, but microprocessors today usually are programmed in octal or hex.

The octal code is a base-8 system, using 3-bit groups of binary numbers. A binary number, say 100001, is first written as 100:001, and each 3-bit group is converted into its decimal equivalent separately (4:1). The colon indicates that this is some sort of code—in this case, octal code.

The place value of octal code is a power of 8 (which is why it is called octal). Thus the octal-coded number 4:1 means $4(8^1) + 1 (8^0) = 33$, which is the same as straight binary $100001 = 1(2^5) + 1(2^0) = 32 + 1 = 33$.

Let us convert a few binary numbers into their octal equivalents:

101	:101	:5
11101	11:101	3:5
1101	1:101	1:5

Note that 1:1 in octal is decimal 9. In binary it is 001001 (which is the same as 1001) and also can be written as the octal equivalent 01:01 (or 1:1). It is essential to know what code is being used, because decimal number 10 might be written as 1010 (straight binary), or 01:10 (octal code), or 1:2 (octal *representation*), not to mention 0001:0000 (BCD), or any other code.

The advantage of using octal code is that only 3-bit numbers are used, the largest single number used being 111. Simple 3-bit circuitry can be designed to handle this small byte.

The term *hexadecimal* refers to decimal number 16, and the hexadecimal code, commonly called hex, is a base-16 code. Like octal, it is widely used in microprocessors and computers.

As shown in Table 15.1, a 4-bit number can represent any decimal number from 0 to 15, a total of 16 numbers (the reason the system is called hexadecimal).

Note (Table 15.1) that the decimal number 18 could be represented by:

Binary code	10010
Hexadecimal equivalent	1:2
Octal equivalent	2:2
BCD	0001:1000

It is obvious that the code being used must be known.

A word about the colon is pertinent at this point. It indicates that some code is being used. A familiar code used by everybody is the 60-s minute, 60-min hour code used for time. Time given as 02:10:15 means 2 hours, 10 minutes, 15 seconds.

To convert this to an equivalent number of seconds we simply convert hours into minutes (by multiplying by 60) and add up all the minutes:

$$2 \text{ hours, } 10 \text{ minutes} = (2 \times 60) + 10 = 130 \text{ minutes,}$$

which gives us

$$02:10:15 = 130:15$$

We then convert 130 min into seconds, and add all the seconds:

$$130:15 = 130 \times 60 + 15 = 7815 \text{ s}$$

We use the same procedure for any code. To convert octal 02:10:15 into decimal, we use the fact that every number to the left of a colon represents 8 units. Octal 02:10:15 is

$$(02 \times 8) + 10{:}15 = 26{:}15$$

$$(26 \times 8) + 15 = 223$$

$$\text{Octal } 02{:}15{:}15 = \text{decimal } 223$$

To convert hexadecimal 02:10:15, using the same technique:

$$02{:}10{:}15 = (02 \times 16) + 10{:}15$$

$$= 42{:}15$$

$$= (42 \times 16) + 15$$

$$= 687$$

Note that the value of any sequence of numbers depends on the base (place value) of the system:

$$\text{Decimal } 3461 = 3(10^3) + 4(10^2) + 6(10^1) + 1(10^0) = 3461$$

$$\text{Octal } 3461 = 3(8^3) + 4(8^2) + 6(8^1) + 1(8^0)$$

$$= 1841$$

$$\text{Hex } 3461 = 3(16^3) + 4(16^2) + 6(16^1) + 1(16^0)$$

$$= 13{,}409$$

The "words" used with a μC often contain 8- or 16-bit numbers, such as 10000001 (an 8-bit number) or 1000000000000001 (a 16-bit number). Since it is obviously cumbersome to handle such numbers, we simplify them by grouping them into 3-bit numbers (if octal code is to be used) or into 4-bit numbers (if hex code is to be used), and then express each nibble in decimal form. From Table 15.2, we see that the binary 8-bit number 10000001 is $2^7 + 2^0 = 128 + 1 = 129$; the 16-bit number is $2^{15} + 2^0 = 32{,}769$.

$$10000001 = 010{:}000{:}001 = 2{:}0{:}1 \quad \text{(octal)}$$

$$1000{:}0001 = 8{:}1 \quad \text{(hex)}$$

We see that each indeed represents the same binary number because

$$\text{Octal } 2{:}0{:}1 = 2(64) + 1 = 129$$

$$\text{Hex } 8{:}1 = 8(16) + 1 = 129$$

$$\text{Binary } 10000001 = 2^7 + 2^0 = 129$$

For a 16-bit number (1, 14 zeros, 1),

$$1{:}000{:}000{:}000{:}000{:}001 = 1{:}0{:}0{:}0{:}0{:}1 \quad \text{(octal)}$$

$$1000{:}0000{:}0000{:}000\pm \ = 8{:}0{:}0{:}1 \quad \text{(hex)}$$

TABLE 15.3
Decimal, octal, and hex values for 2^4 to 2^{12}

Binary	Decimal	Octal	Hex
10000	16	2:0	1:0
100000	32	4:0	2:0
1 (six zeros)	64	1:0:0	4:0
1 (seven zeros)	128	2:0:0	8:0
1 (eight zeros)	256	4:0:0	1:0:0
1 (nine zeros)	512	1:0:0:0	2:0:0
1 (ten zeros)	1024	2:0:0:0	4:0:0
1 (eleven zeros)	2048	4:0:0:0	8:0:0
1 (twelve zeros)	4096	1:0:0:0:0	1:0:0:0

Double check:

$$\text{Octal } 1{:}0{:}0{:}0{:}0{:}1 = 8^5 + 8^0 = 2^{15} + 1 = 32{,}769$$

$$\text{Hex } 8{:}0{:}0{:}1 = (8 \times 16^3) + 16^0 = 2^{15} + 1 = 32{,}769$$

The simplification resulting from the use of octal or hex is obvious from Table 15.3.

It is important to remember that the octal or hex value is only a short-hand way of expressing the longer binary number, which is what the μP actually uses. Thus the decimal number 1024, which is binary 1 followed by 10 zeros, is immediately obtained from

$$\text{Octal } 2{:}0{:}0{:}0 = 010{:}000{:}000{:}000 = 1 \quad (10 \text{ zeros})$$
$$\text{Hex } 4{:}0{:}0 = 0100{:}0000{:}0000 = 2 \quad (10 \text{ zeros})$$

The conversion of a long binary number to shorter octal or hex code is seen to be a simple process: first write the binary number in either 3- or 4-bit segments (starting at right); then express each segment in decimal form (0 to 7 for octal; 0 to 15 for hex).

15.4
ARITHMETIC AND LOGICAL
FUNCTIONS[d]

The elements of the first 8-bit μP, the 8080, are shown in Figure 15.2. A binary digit is called a bit, each 4-bit segment is called a nibble, and each 8-bit number is called a byte. A 16-bit unit is sometimes called a word. Figure 15.3 shows the elements of bytes and words.

---

[d] The material in Section 15.4 is from "Microprocessor Course No. 2." © 1979 *Medical Electronics*, 2994 W. Liberty Ave., Pittsburgh, Pa. 15216. Courtesy of the publisher, Mr. Milt Aronson.

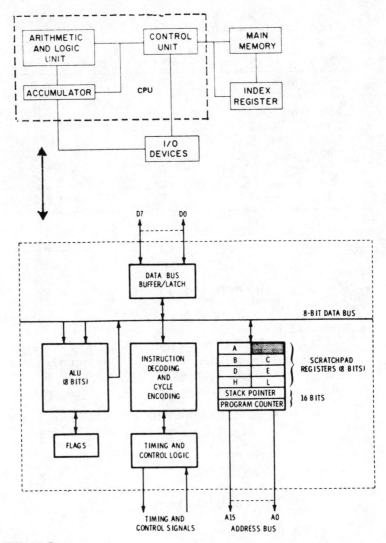

FIGURE 15.2
(a) Elements of a computer, showing CPU (dotted
box); (b) 8080 CPU elements. ALU is arithmetic
and logic unit. (Courtesy of Intel Corporation)

The functions of the μC are of two types: *arithmetic* and *logical*. Although a μC is a calculating machine that performs many arithmetic calculations at a high rate, the computer spends most of its time doing logical operations, that is, making a decision about what to do next; it spends only a small percentage of its time actually doing the arithmetic.

Let us look at these two basic functions of all computers—arithmetic and logical.

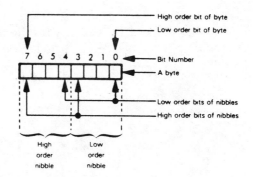

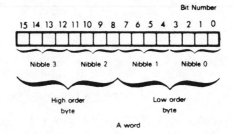

FIGURE 15.3
(a) Terminology of an 8-bit byte, and (b) a 16-bit "word." (Courtesy of *Medical Electronics,* 2994 W. Liberty Ave., Pittsburgh, Pa. 15216)

15.4.1 Binary Arithmetic

Binary is perhaps the easiest set of numbers in which to perform arithmetic; in fact, this is the reason why the binary system is used in computers and logical networks. Addition is particularly simple.

In decimal addition, we are forced to remember that 9 plus 8 is 17; that 5 plus 6 is 11; and so on and on. In binary, we need remember only the following two simple rules:

1. 0 plus 1 is 1

2. 1 plus 1 is 0 and carry a 1 to next column left.

With these two rules we can perform addition. Thus, if we are to add 10 and 11,

$$
\begin{array}{ll}
10 & \\
\underline{11} & \\
1 & \\
0 & \text{carry 1 left} \\
\underline{1} & \text{the carry number} \\
101 &
\end{array}
$$

Let us see if this is consistent with decimal addition; 10 is binary for the decimal number 2, and 11 is binary for the decimal number 3. The sum of these should give decimal 5. The sum has been determined as 101 and thus

$$101 = 1\,(2^2) + 0\,(2^1) + 1\,(2^0) = 5$$

The two results check.

As another example, let us add 27 (or binary 11011) and 13 (or binary 1101):

$$
\begin{array}{ll}
11011 & 27 \\
+1101 & 13 \\
\hline
0 & \text{carry } 1 \\
0 & \text{carry } 1 \\
0 & \text{carry } 1 \\
1 & \text{carry } 1 \\
0 & \text{carry } 1 \\
1 & \text{the carry number} \\
\hline
101000 & \text{is the resultant number}
\end{array}
$$

Converting this to decimal, $101000 = 1\,(2^5) + 1\,(2^3) = 32 + 8$. The sum is indeed 40.

15.4.2 Binary Subtraction

Whereas each borrow in the decimal system changes the value of the specific column by 10, 100, 1000, and so on, each borrow in a binary number changes the value of the column by 2, 4, 8, 16, and so on. This follows from the fact that binary

$$10 = 2$$

$$100 = 4$$

$$1000 = 8$$

Each shift of a binary number means a doubling of the number (i.e., a multiplication by "10").

Let us subtract 1 from binary 100 (which we know must result in 11, since $4 - 1 = 3$).

$$
\begin{array}{r}
100 \\
-1 \\
\hline
?
\end{array}
$$

The rules for binary subtraction are essentially the same as those for addition:

$$
\begin{array}{cccc}
0 & 1 & 1 & {}^{1}0 \\
-0 & -0 & -1 & -1 \\
\hline
0 & 1 & 0 & 1
\end{array}
$$

Since 1 cannot be subtracted from 0, a borrow from somewhere must be provided to convert the 0 into a (10), permitting a subtraction:

$$\begin{array}{cc} 10 & 0^1 0 \\ \underline{1} & \underline{1} \\ 1 & 1 \end{array}$$

Let us subtract 1 from 100:

$$\begin{array}{c} 100 \\ \underline{-1} \\ ? \end{array}$$

Since we cannot subtract a 1 from a 0, we find the nearest 1, in the third (2^2) column, and borrow it for transfer to the second (2^1) column:

2's column   2's column

4's column———→100 = 0 (10) 0

Since we have 2 in the 2's column, the number (decimal 4) remains unchanged.

We now can borrow a 1 from the 10 in the 2's column, leaving a 1 in that column $(10 - 1 = 1)$, and simultaneously transfer the borrowed 1 to the units column:

$$100 = 01^1\,0$$

————units column

$$= 01\,(10)$$

Since we have 1 units in the units column, the number is unchanged:

$$010 + 10 = 100$$

We are at last in position to borrow 1 (the original problem) from 100:

$$\begin{array}{cc} 100 = & 01\,(10) \\ \underline{-1} & \underline{-1} \\ & 11 \end{array}$$

Each borrow in binary changes a 1 into a 0, and makes a 0 in the adjacent column into a "10." Let us subtract 1 from 1000:

$$\begin{array}{c} 1000 \\ \underline{-1} \\ ? \end{array}$$

The first borrow takes a 1 from the 8's column, leaving it zero, and places a "10" in the 4's column:

4's column

$$1000 = 0^1000 = 0\,(10)\,00$$

The second borrow takes a 1 from the (10) in 4's column, leaving a 1 there, and places (10) in the 2's column:

$$2\text{'s column}$$

$$1000 = 01^100 = 01\ (10)\ 0$$

The third borrow takes a 1 from the (10) in the 2's column, leaving a 1 there, and places a (10) in the units column:

$$2\text{ in the units column}$$

$$1000 = 011^10 = 011\ (10)$$

We now can subtract the 1 (the original problem) from the (10) in the units column:

$$
\begin{array}{cc}
1000 & 011\ (10) \\
-1 & -1 \\
\hline
 & 0111
\end{array}
$$

The borrowing process is seen to be simply a technique for expressing a number in different ways. In the decimal example (subtract 1 from 2000), the consecutive borrowing effectively altered the number as follows:

$$
\begin{aligned}
2000 &= 1000 + 1000 \\
&= 1000 + 900 + 100 \\
&= 1000 + 900 + 900 + 10 \\
&= 1000 + 900 + 90 + 9 + 1
\end{aligned}
$$

at which point we subtracted the 1 to leave 1999.

Similarly, in the binary example, the number 1000 was changed by each borrow as follows:

$$
\begin{aligned}
1000 &= 0100 + 0100 \\
&= 0100 + 0010 + 0010 \\
&= 0100 + 0010 + 0001 + 0001
\end{aligned}
$$

at which point we subtracted 0001 to leave 0111.

Let us subtract 11001100 from 110011000:

$$
\begin{array}{ccccccccc}
{}^01 & {}^{10}1 & {}^10 & 0 & 1 & {}^{10}1 & {}^10 & 0 & 0 \\
 & 1 & 1 & 0 & 0 & 1 & 1 & 0 & 0 \\
\hline
 & 1 & 1 & 0 & 0 & 1 & 1 & 0 & 0
\end{array}
$$

We may wish to subtract 1 from 0. We are dealing with 4-bit numbers = 3 bits + sign. Thus +1 = 0001. We wish to subtract it from zero:

$$
\begin{array}{r}
0000 \\
-0001 \\
\hline
?
\end{array}
$$

The fifth place is not involved in the calculations, it is dropped in all

TABLE 15.4
Positive and negative numbers
in 4-bit signed notation

Decimal	Binary	Decimal Numbers		Binary Numbers	
		Positive	Negative	Positive	Negative
-7	1001	0	0	0000	0000
-6	1010	1	1	0001	1111
-5	1011	2	2	0010	1110
-4	1100	3	3	0011	1101
-3	1101	4	4	0100	1100
-2	1110	5	5	0101	1011
-1	1111	6	6	0110	1010
0	000	7	7	0111	1001
1	0001				
2	0010				Number bits
3	0011				Sign bits
4	0100				
5	0101				
6	0110				
7	0111				

calculations. So we can let 0 = 10000; then the subtraction becomes

$$\frac{\begin{array}{r}(1)0000 \\ -1\end{array}}{1111} = -1$$

The values of positive and negative numbers from 0 through 7, as found in this manner, are listed in Table 15.4. These values are now logically and numerically consistent with the basic rules of binary arithmetic.

Referring to Table 15.4, we see that the largest negative number listed (−7) is 1001. However, a computer can interpret 1000 as −8* (rather than 0) because 10000 is never actually used as zero (we used the number 10000 for 0 in the subtraction only for illustrative purposes). Thus −7 (0111) to −8 (1000) can be represented by 4-bit signed numbers that can be used in arithmetic calculations.

15.4.3 Decimals and Fractions

The decimal point in decimal arithmetic is a shorthand technique for indicating negative powers: 10^{-1}, 10^{-2}, 10^{-3}, and so on, which are the fractions $\frac{1}{10}, \frac{1}{100}, \frac{1}{1000}$, and so on. Digits to the left of a decimal point represent positive exponents; digits to the right indicate negative exponents:

$$11.11_{10} = 10^1 + 10^0 + 10^{-1} + 10^{-2}$$

---

*8 = 1000. Subtracting from zero to get −8,

$$\frac{\begin{array}{r}(1)0000 \\ -0000\end{array}}{-8 = 1000}$$

The decimal point in binary arithmetic similarly divides positive and negative powers of 2:

$$11.11_2 = 2^1 + 2^0 + 2^{-1} + 2^{-2}$$
$$= 3\tfrac{3}{4}$$
$$= 3.75$$

The decimal point in octal code divides positive and negative powers of 8:

$$11.11_8 = 8^1 + 8^0 + 8^{-1} + 8^{-2}$$
$$= 9 + \tfrac{1}{8} + \tfrac{1}{64}$$
$$= 9.1406 \ldots$$

The decimal point in a hexadecimal number divides positive and negative powers of 16:

$$11.1_{16} = 16^1 + 16^0 + 10^{-1} + 16^{-2}$$
$$= 17.066 \ldots$$

15.4.4 Multiplication and Division

We have seen that a μC subtracts by adding the negative of the number (i.e., the two's complement) to implement a subtraction. It was also mentioned that several arithmetic flags are used to indicate certain characteristics of the result that might affect further calculations: carry flag, zero flag, sign flag, and others.

Multiplication by two is implemented simply by shifting each bit of the binary number one place to the left (i.e., adding a 0 bit at the right):

$$9_{10} = 1001$$
$$18_{10} = 10010$$
$$5_{10} = 101$$
$$10_{10} = 1010$$

This is exactly analogous to multiplication by decimal 10; to multiply a decimal number by 10, we *shift* each digit one position to the left.

$$762 \times 10 = 7620$$

Note that use of a 0 (zero) indicates place value in all numbering systems, so that multiplication by 10 is a shift in place for all systems.

Division of a binary number by two (10) is accomplished by shifting each digit one position to the right.

$$111_2 \div 2 = 11.1_2$$

Multiplication and division of binary numbers by any number other than 2 (10) can be performed in several different ways, based on the principles discussed. One technique is *successive addition* (i.e., if a binary number is to be multiplied by 4, it is simply added to itself 4 times). This "brute-force" method of multiplication is obviously useful only for relatively small multipliers.

A second technique is *successive shifting* (i.e., converting the multiplication into successive multiplications by 2). To multiply binary number X by 10, one would multiply by 2 (one shift to left) and save this result $(2X)$. Then multiply X by 8 (three shifts to left) to get $8X$; then add $8X$ to $2X$ to achieve $10X$.

A third method, a general technique that works with any binary number, is identical with the technique taught for multiplying decimal numbers (i.e., individual multiplication by each bit of the multiplier, shifting as required to account for the power of the multipliers). To multiply 111 by 101,

$$
\begin{array}{r}
111 \\
\underline{101} \\
111 \\
\underline{111} \\
100011
\end{array}
$$

15.4.5 Logical Operations

The English mathematician George Boole, in 1854, published a book *Laws of Thought* that showed the simple yet powerful logical decisions that can be based on AND/OR relations. Modern computers make extensive use of basic Boolean logic concepts.

George Boole showed how complex decisions can be made by systems that use only binary (YES/NO; ON/OFF) elements and *three* logic elements: AND, OR, and NOT. As we shall see, these three logical relations are the only ones possible with binary signals, which is why complex decisions can be made by computers using binary arithmetic elements and Boolean logic elements.

Two possible conditions of any one binary element exist. These conditions can be referred to as ON or OFF, YES or NO, UP or DOWN, and so on. The most common way of expressing the condition of the element is by use of the symbols 0 and 1.

If we have two binary elements, we can combine them in only two ways; we may have (1) one OR the other or (2) one AND the other. An OR gate, symbolized by a +, recognizes the first condition; an AND gate, symbolized by a dot, recognizes the second condition.

The third of the three basic logic elements is the NOT gate or element. This simply inverts the state of the binary element. It is symbolized by a line over the bit.

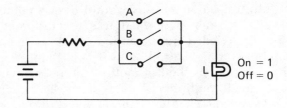

FIGURE 15.4
The OR circuit (L, lamp).

$$\text{NOT } 1 = 0 = \bar{1}$$

$$\text{NOT } 0 = 1 = \bar{0}$$

Today, every technical worker should know the fundamentals of Boolean algebra, which is based on these three logical concepts: AND, OR, and NOT.

Logic fundamentals are shown in the nine circuits of Figures 15.4 through 15.13.

The basic functions for AND, OR, and NOT circuits are as follows:

1. AND: $Y = A \cdot B$ or AB; means $Y = A$ and B.

2. OR: $Y = A + B$; means $Y = A$ or B.

3. NOT: NOT A is written as $\bar{A}$.

Truth Table for OR Gate

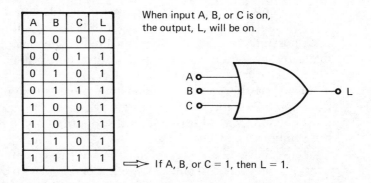

A	B	C	L
0	0	0	0
0	0	1	1
0	1	0	1
0	1	1	1
1	0	0	1
1	0	1	1
1	1	0	1
1	1	1	1

When input A, B, or C is on, the output, L, will be on.

⟹ If A, B, or C = 1, then L = 1.

FIGURE 15.5
The OR gate symbol and OR gate truth table. The truth table gives input and output signals in 0s and 1s.

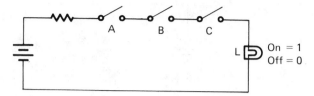

FIGURE 15.6
The AND circuit.

Truth Table for AND Gate

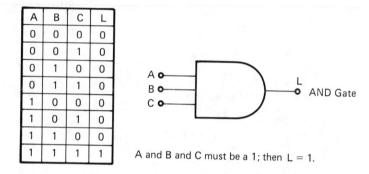

A	B	C	L
0	0	0	0
0	0	1	0
0	1	0	0
0	1	1	0
1	0	0	0
1	0	1	0
1	1	0	0
1	1	1	1

A and B and C must be a 1; then L = 1.

FIGURE 15.7
The AND gate symbol and AND gate truth table.

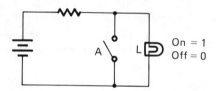

FIGURE 15.8
The NOT circuit.

Truth Table for NOT Gate

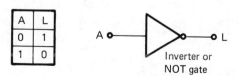

A	L
0	1
1	0

Inverter or
NOT gate

FIGURE 15.9
The inverter or NOT gate symbol and NOT gate
truth table.

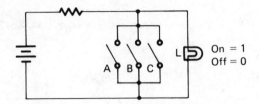

FIGURE 15.10
The NOR circuit.

Postulates and theorems include the following properties:

1. Commutative property:

   a. $AB = BA$

   b. $A + B = B + A$

2. Associative property:

   a. $A(BC) = (AB)C$

   b. $A + (B + C) = (A + B) + C$

3. Distributive property:

   a. $A(B + C) = AB + AC$

There are many combinations of gates that give countless numbers of possibilities. An example of combining gates is shown in Figure 15.14.

Boolean algebra is governed by a set of rules that are not always equal to those valid for ordinary algebra. Students who wish to further explore μC logic theory should refer to a more advanced text.

Truth Table for NOR Gate

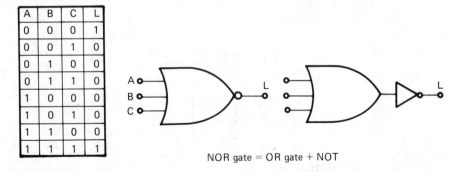

A	B	C	L
0	0	0	1
0	0	1	0
0	1	0	0
0	1	1	0
1	0	0	0
1	0	1	0
1	1	0	0
1	1	1	1

NOR gate = OR gate + NOT

FIGURE 15.11
The NOR gate symbols and NOR gate truth table.

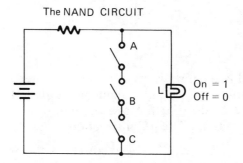

The NAND CIRCUIT

On = 1
Off = 0

FIGURE 15.12
The NAND circuit.

Truth Table for NAND Gate

A	B	C	L
0	0	0	1
0	0	1	1
0	1	0	1
0	1	1	1
1	0	0	1
1	0	1	1
1	1	0	1
1	1	1	0

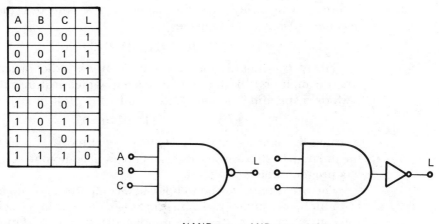

NAND gate = AND gate + NOT

FIGURE 15.13
The NAND gate symbols and NAND gate truth
table.

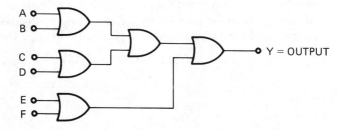

Y = OUTPUT

FIGURE 15.14
Gate combination example.

15.5
MNEMONICS[e]

In this section, we shall introduce mnemonics used in assembly-language instructions. The mnemonic is the part of the instruction that tells the μC *what* to do (but not always where or with what, the other part of the instruction). The μC instruction usually consists of two parts:

[DO SOMETHING]   [WITH WHAT]

All computers, regardless of size or type can "read" or use only straight binary numbers. All other techniques for expressing computer data or instructions are but shorthand techniques for straight binary numbers. No matter how an instruction is given, as words, mnemonics, octal code, or hex code, that instruction must be converted into a straight binary number before it can be of any use in the computer. The preceding instruction might become in "machine language"

[011011011]   [100100100]

The *instruction list* for a given computer is a dictionary of a bit or (for larger computers) 16-bit numbers, each of which instructs the computer to perform a specific function. For example, the 8-bit byte

0111   0110

instructs the 8080 μP or 8085 AH μ^P to stop or HALT. This instruction causes the program to stop as soon as it is received; it commonly appears at the end of a program being executed.

The *program* required to operate a computer ends up as a series of 8- or 16-bit binary numbers that instruct the computer to, say, obtain data from some given source, add certain numbers together, or perform certain logical operations. A six-instruction program for the first generation Intel 8080 might look as shown in Table 15.5.

The computer may understand every word in this list of machine-language instructions, but a human being, taught to think with words, has difficulty in interpreting machine-language numbers; there simply are too many 1s and 0s in each word of the language. This is why the straight-binary numbers are converted into simpler octal or hexadecimal notation (a technique we learned previously). For example, the HALT instruction can be expressed in octal (base 8) or hexadecimal (base 16) as

$$01110110 = 01\text{-}110\text{-}110 = 166_8$$

$$= 01111\text{-}0110 = 76_{16}$$

[e] The material in Section 15.5 is from "Microprocessor Course No. 3," issue of April 1979 of *Medical Electronics*, 2994 W. Liberty Ave., Pittsburgh, Pa. 15216. Courtesy of the publisher, Mr. Milt Aronson.

TABLE 15.5
Six-instruction program in machine language

1	10010111
2	11010011
3	11000110
4	11000011
5	00000000
6	00000001

The machine-language program shown in Table 15.5 can be written in any of the three ways shown in Table 15.6.

The octal and hex numbers are obviously easier to remember or interpret than the longer straight-binary strings of 1s and 0s, but they are still not as recognizable as an English word. A human recognizes and remembers the word HALT much more easily than the code 01110110 (although the machine can use the number, not the word). Even if we express the code for HALT in its octal form (166) or hexadecimal form (76), the word HALT still has more meaning to a human.

For this reason, computer instructions are identified in yet another way, using 3-letter or 4-letter word-related *mnemonics* (memory aids) that identify the operation. Mnemonics of typical instructions are:

$$HLT = halt$$
$$ADD = add$$
$$SUB = subtract$$
$$MOV = move$$
$$LDA = load\ the\ accumulator$$
$$JMP = jump$$

Note that the mnemonic, like the octal or hex representation of a straight binary number, is simply a shorthand technique for expressing the 8-bit

TABLE 15.6
Machine-language program can be written in octal or hexadecimal

Instruction	Binary	Octal	Hex
1	10010111	227	97
2	11010011	323	D3
3	11000110	306	C6
4	11000011	303	C3
5	00000000	000	00
6	00000001	001	01

straight-binary number, the only "word" the machine understands. At this point, we now have *four* ways of expressing the operation HALT:

HLT

01110110

166_8

76_{16}

These four ways are really two ways, (1) by mnemonic or (2) by binary numbers, because the three numerical codes are simply different ways of expressing the same binary number. A computer can be programmed in either of these two basic ways, using either mnemonics or binary numbers for the desired computer operations. Programming with the numbers is called, appropriately, *machine-language programming* because we are using the only language the machine understands directly.

Programming with mnemonics requires some method for converting each mnemonic into its machine-language equivalent. This relatively simple job can be done manually or by a computer program called an *assembler,* that is, a program that assembles the list of mnemonics into its machine-language equivalents (Figure 15.15). Programming with mnemonics is called *assembly-language programming*.

Since μPs usually are programmed with machine or assembly language, we shall examine these "languages" in detail, but mention might be made at this time of the fact that there are more rapid ways to program computers using shorthand higher-order languages such as FORTRAN or BASIC. For example, one FORTRAN instruction (such as GO TO) will be translated into

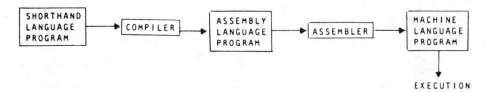

EXECUTION

FIGURE 15.15
A shorthand high-level programming language such as FORTRAN can first be compiled into assembly language and then assembled into machine language, the only language a computer understands. Machine language consists of straight binary numbers; assembly language consists of mnemonics; a high-level shorthand language consists of (symbolic) words. (Courtesy of *Medical Electronics,* 2994 W. Liberty Ave., Pittsburgh, Pa. 15216)

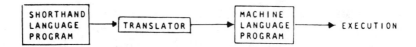

FIGURE 15.16
A translator performs the function of both compiler and assembler. (Courtesy of *Medical Electronics*, 2994 W. Liberty Ave., Pittsburgh Pa. 15216)

several assembly-language instructions, which, in turn, are assembled into the machine-language instructions that actually operate the computer. Figure 15.15 shows graphically the terminology relating the three types of computer-programming languages: shorthand (symbolic) language, assembly (mnemonic) language, and machine (binary) language.

It is possible to combine the functions of the compiler as well as the assembler into a single processor called a *translator*, as shown in Figure 15.16. The compiler can create a machine-language program from a short-

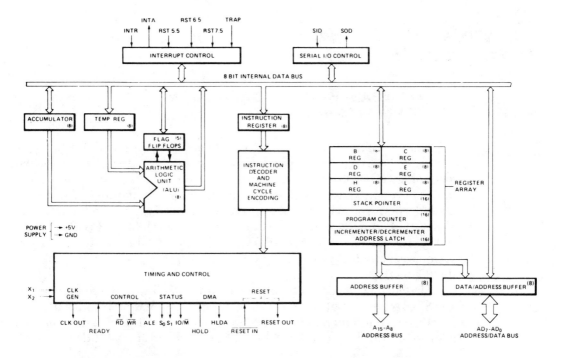

FIGURE 15.17
CPU functional block diagram for the 8085AH.
(Courtesy of Intel Corporation)

hand language without the explicit creation of an assembly-language program. However, we shall discuss machine-language and assembly-language programming of the μP at this time. To review a few important definitions:

- **Machine language:** straight binary representation of computer instructions.

- **Assembly language:** Mnemonic representation of computer instructions

- **Assembler:** Computer program that assembles mnemonics into machine language.

- **Shorthand language:** High-level language using English words as symbols that permit rapid programming; typical shorthand languages include FORTRAN, BASIC, ALGOL, and COBOL.

Mnemonic instruction depends on the 8-, 16-, and 32-bit μP used. Specifications for all instruction are available from the manufacturer. An 8-bit μP such as the Intel 8080 requires 244 octal instructions. The instructions cause data to be transferred between various elements in the μP, and a knowledge of these elements is essential for understanding the instructions. Refer to reference 10.

15.6
THE INTEL 8-, 16-, AND 32-BIT MICROPROCESSOR

15.6.1   8085AH / 8085AH-2 / 8085AH-1 8-Bit HMOS Microprocessors[f]

The Intel 8085AH (Figure 15.17) is a complete 8-bit parallel central processing unit (CPU) implemented in N-channel, depletion load, silicon gate technology (HMOS). Its instruction set is 100% software compatible with the 8080A microprocessor, and it is designed to improve the present 8080A's performance by higher system speed. Its high level of system integration allows a minimum system of three ICs [8085AH (CPU), 8156H (RAM/IO) and 8355/8755A (ROM/PROM/IO)] while maintaining total system expandability. The 8085AH-2 and 8085AH-1 are faster versions of the 8085AH.

The 8085AH incorporates all the features that the 8224 (clock generator) and 8228 (system controller) provided for the 8080A, thereby offering a high level of system integration. The 8085AH uses a multiplexed data bus. The address is split between the 8-bit address bus and the 8-bit data bus. The

---

[f]The material in Section 15.6.1 is from Intel Application Notes APN-01242C, April 1981, and is used courtesy of Intel Corp.

on-chip address latches of 8155H/8156H/8355/8755A memory products allow a direct interface with the 8085AH.

The features of these microprocessors include the following:

1. Single +5-V power supply with 10% voltage margins.

2. 3-, 5-, and 6-MHz selections available.

3. 20% lower power consumption than 8085A for 3 and 5 MHz.

4. 1.3-μs instruction cycle (8085AH); 0.8 μs (8085AH-2); 0.67 μs (8085AH-1).

5. 100% compatible with 8085A.

6. 100% software compatible with 8080A.

7. On-chip clock generator (with external crystal, *LC* or *RC* network).

8. On-chip system controller; advanced cycle status information available for large system control.

9. Four vectored interrupt inputs (one is nonmaskable), plus an 8080A-compatible interrupt.

10. Serial in–serial out port.

11. Decimal, binary, and double precision arithmetic.

12. Direct addressing capability to 64K bytes of memory.

Figure 15.18 gives the 8085AH pin configuration. Table 15.7 provides the pin description, while Table 15.8 gives the interrupt priority, restart address, and sensitivity.

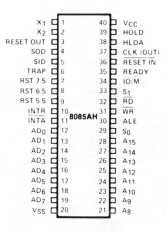

FIGURE 15.18

Pin configuration for the 8085AH. (Courtesy of Intel Corporation)

TABLE 15.7
Pin description

Symbol	Type	Name and Function
A_8–A_{15}	O	**Address Bus:** The most significant 8 bits of the memory address or the 8 bits of the I/O address, 3-stated during Hold and Halt modes and during RESET.
AD_0—$_7$	I/O	**Multiplexed Address/Data Bus:** Lower 8 bits of the memory address (or I/O address) appear on the bus during the first clock cycle (T state) of a machine cycle. It then becomes the data bus during the second and third clock cycles.
ALE	O	**Address Latch Enable:** It occurs during the first clock state of a machine cycle and enables the address to get latched into the on-chip latch of peripherals. The falling edge of ALE is set to guarantee setup and hold times for the address information. The falling edge of ALE can also be used to strobe the status information. ALE is never 3-stated.
S_0, S_1, and IO/$\overline{M}$	O	**Machine Cycle Status:**

IO/$\overline{M}$	S_1	S_0	Status
0	0	1	Memory write
0	1	0	Memory read
1	0	1	I/O write
1	1	0	I/O read
0	1	1	Opcode fetch
1	1	1	Opcode fetch
1	1	1	Interrupt Acknowledge
*	0	0	Halt
*	X	X	Hold
*	X	X	Reset

* = 3-state (high impedance)
X = unspecified

S_1 can be used as an advanced R/$\overline{W}$ status. IO/$\overline{M}$, S_0 and S_1 become valid at the beginning of a machine cycle and remain stable throughout the cycle. The falling edge of ALE may be used to latch the state of these lines.

Symbol	Type	Name and Function
$\overline{RD}$	O	**Read Control:** A low level on $\overline{RD}$ indicates the selected memory or I/O device is to be read and that the Data Bus is available for the data transfer, 3-stated during Hold and Halt modes and during RESET.
$\overline{WR}$	O	**Write Control:** A low level on $\overline{WR}$ indicates the data on the Data Bus is to be written into the selected memory or I/O location. Data is set up at the trailing edge of $\overline{WR}$. 3-stated during Hold and Halt modes and during RESET.
READY	I	**Ready:** If READY is high during a read or write cycle, it indicates that the memory or peripheral is ready to send or receive data. If READY is low, the cpu will wait an integral number of clock cycles for READY to go high before completing the read or write cycle. READY must conform to specified setup and hold times.
HOLD	I	**Hold:** Indicates that another master is requesting the use of the address and data buses. The cpu, upon receiving the hold request, will relinquish the use of the bus as soon as the completion of the current bus transfer. Internal processing can continue. The processor can regain the bus only after the HOLD is removed. When the HOLD is acknowledged, the Address, Data $\overline{RD}$, $\overline{WR}$, and IO/$\overline{M}$ lines are 3-stated.
HLDA	O	**Hold Acknowledge:** Indicates that the cpu has received the HOLD request and that it will relinquish the bus in the next clock cycle. HLDA goes low after the Hold request is removed. The cpu takes the bus one half clock cycle after HLDA goes low.
INTR	I	**Interrupt Request:** Is used as a general purpose interrupt. It is sampled only during the next to the last clock cycle of an instruction and during Hold and Halt states. If it is active, the Program Counter (PC) will be inhibited from incrementing and an $\overline{INTA}$ will be issued. During this cycle a RESTART or CALL instruction can be inserted to jump to the interrupt service routine. The INTR is enabled and disabled by software. It is disabled by Reset and immediately after an interrupt is accepted.
$\overline{INTA}$	O	**Interrupt Acknowledge:** Is used instead of (and has the same timing as) $\overline{RD}$ during the Instruction cycle after an INTR is accepted. It can be used to activate an 8259A Interrupt chip or some other interrupt port.
RST 5.5 RST 6.5 RST 7.5	I	**Restart Interrupts:** These three inputs have the same timing as INTR except they cause an internal RESTART to be automatically inserted. The priority of these interrupts is ordered as shown in Table 2. These interrupts have a higher priority than INTR. In addition, they may be individually masked out using the SIM instruction.

TABLE 15.7 (Continued)

Symbol	Type	Name and Function
TRAP	I	**Trap:** Trap interrupt is a non-maskable RESTART interrupt. It is recognized at the same time as INTR or RST 5.5-7.5. It is unaffected by any mask or Interrupt Enable. It has the highest priority of any interrupt. (See Table 2.)
RESET IN	I	**Reset In:** Sets the Program Counter to zero and resets the Interrupt Enable and HLDA flip-flops. The data and address buses and the control lines are 3-stated during RESET and because of the asynchronous nature of RESET, the processor's internal registers and flags may be altered by RESET with unpredictable results. RESET IN is a Schmitt-triggered input, allowing connection to an R-C network for power-on RESET delay (see Figure 3). Upon power-up, RESET IN must remain low for at least 10 ms after minimum V_{CC} has been reached. For proper reset operation after the power-up duration, RESET IN should be kept low a minimum of three clock periods. The CPU is held in the reset condition as long as RESET IN is applied.

Symbol	Type	Name and Function
RESET OUT	O	**Reset Out:** Reset Out indicates cpu is being reset. Can be used as a system reset. The signal is synchronized to the processor clock and lasts an integral number of clock periods.
X_1, X_2	I	**X_1 and X_2:** Are connected to a crystal, LC, or RC network to drive the internal clock generator. X_1 can also be an external clock input from a logic gate. The input frequency is divided by 2 to give the processor's internal operating frequency.
CLK	O	**Clock:** Clock output for use as a system clock. The period of CLK is twice the X_1, X_2 input period.
SID	I	**Serial Input Data Line:** The data on this line is loaded into accumulator bit 7 whenever a RIM instruction is executed.
SOD	O	**Serial Output Data Line:** The output SOD is set or reset as specified by the SIM instruction.
V_{CC}		**Power:** +5 volt supply.
V_{SS}		**Ground:** Reference.

Courtesy of Intel Corporation

TABLE 15.8
Interrupt priority, restart address, and sensitivity

Name	Priority	Address Branched To (1) When Interrupt Occurs	Type Trigger
TRAP	1	24H	Rising edge AND high level until sampled.
RST 7.5	2	3CH	Rising edge latched.
RST 6.5	3	34H	High level until sampled.
RST 5.5	4	2CH	High level until sampled.
INTR	5	See Note 2	High level until sampled.

Courtesy of Intel Corporation

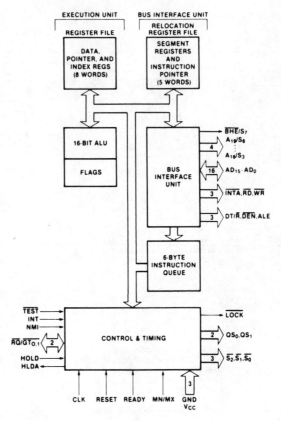

FIGURE 15.19
The 8086 CPU functional block diagram.
(Courtesy of Intel Corporation)

15.6.2 iAPX 86/10 16-Bit HMOS Microprocessor[g]

The Intel 8086-1 (Figure 15.19) is the highest performance version of Intel's standard iAPX 86/10 microprocessor implemented in N-channel, depletion load, silicon gate technology (HMOS), and packaged in a 40-pin CerDIP package. The processor has attributes of both 8- and 16-bit microprocessors. It addresses memory as a sequence of 8-bit bytes, but has a 16-bit wide physical path to memory for high performance.

The features of the iAPX 86/10 include the following:

---

[g] The material in Section 15.6.2 is from Intel Application Notes AFNO 1497A, May 1980, and is used courtesy of Intel Corp.

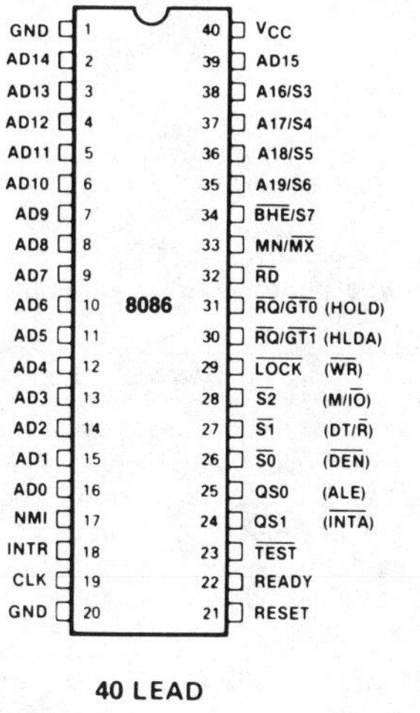

```
        GND ▢ 1         40 ▯ Vcc
       AD14 ▢ 2         39 ▯ AD15
       AD13 ▢ 3         38 ▯ A16/S3
       AD12 ▢ 4         37 ▯ A17/S4
       AD11 ▢ 5         36 ▯ A18/S5
       AD10 ▢ 6         35 ▯ A19/S6
        AD9 ▢ 7         34 ▯ BHE/S7
        AD8 ▢ 8         33 ▯ MN/MX
        AD7 ▢ 9         32 ▯ RD
        AD6 ▢ 10  8086  31 ▯ RQ/GT0 (HOLD)
        AD5 ▢ 11        30 ▯ RQ/GT1 (HLDA)
        AD4 ▢ 12        29 ▯ LOCK  (WR)
        AD3 ▢ 13        28 ▯ S2    (M/IO)
        AD2 ▢ 14        27 ▯ S1    (DT/R)
        AD1 ▢ 15        26 ▯ S0    (DEN)
        AD0 ▢ 16        25 ▯ QS0   (ALE)
        NMI ▢ 17        24 ▯ QS1   (INTA)
       INTR ▢ 18        23 ▯ TEST
        CLK ▢ 19        22 ▯ READY
        GND ▢ 20        21 ▯ RESET

              40 LEAD
```

FIGURE 15.20
Pin diagram for the 8086. (Courtesy of Intel
Corporation)

1. 10-MHz clock rate: 8086-1.

2. Direct addressing capability to 1 Mbyte of memory.

3. Assembly language compatible with 8080-8085 via a translator.

4. 14-word by 16-bit general register set.

5. 24 operand addressing modes.

6. Bit, byte, word, and block operations.

7. 8- and 16-bit signed and unsigned arithmetic in binary or decimal including multiply and divide.

8. Multibus system compatible interface.

Figure 15.20 contains the 8086 pin diagram. Table 15.9 provides a pin description of the 8086. Table 15.10 gives the absolute maximum ratings of the 8086-1. The dc and ac characteristics of the 8086-1 are given in Table 15.11 and 15.12.

15.6.3 iAPX 43201 and iAPX 43202 VLSI General Data Processor[h]

The Intel iAPX 432 general data processor (GDP) is a 32-bit microprocessor that consists of two VLSI devices, the 43201 and the 43202. These

---

[h] The material in Section 15.6.3 is from Intel Application Notes 171873-001 Rev. A, 1981, and is used courtesy of Intel Corp.

TABLE 15.9
Pin description: 8086-1

The following pin function descriptions are for iAPX 86 systems in either minimum or maximum mode. The "Local Bus" in these descriptions is the direct multiplexed bus interface connection to the 8086 (without regard to additional bus buffers).

Symbol	Pin No.	Type	Name and Function
AD_{15}-AD_0	2-16, 39	I/O	**Address Data Bus:** These lines constitute the time multiplexed memory/IO address (T_1) and data (T_2, T_3, T_W, T_4) bus. A_0 is analogous to $\overline{BHE}$ for the lower byte of the data bus, pins D_7-D_0. It is LOW during T_1 when a byte is to be transferred on the lower portion of the bus in memory or I/O operations. Eight-bit oriented devices tied to the lower half would normally use A_0 to condition chip select functions. (See $\overline{BHE}$.) These lines are active HIGH and float to 3-state OFF during interrupt acknowledge and local bus "hold acknowledge."
A_{19}/S_6, A_{18}/S_5, A_{17}/S_4, A_{16}/S_3	35-38	O	**Address/Status:** During T_1 these are the four most significant address lines for memory operations. During I/O operations these lines are LOW. During memory and I/O operations, status information is available on these lines during T_2, T_3, T_W, and T_4. The status of the interrupt enable FLAG bit (S_5) is updated at the beginning of each CLK cycle. A_{17}/S_4 and A_{16}/S_3 are encoded as shown. This information indicates which relocation register is presently being used for data accessing. These lines float to 3-state OFF during local bus "hold acknowledge." | A_{17}/S_4	A_{16}/S_3	Characteristics
---	---	---
0 (LOW)	0	Alternate Data
0	1	Stack
1 (HIGH)	0	Code or None
1	1	Data
S_6 is 0 (LOW)			
$\overline{BHE}/S_7$	34	O	**Bus High Enable/Status:** During T_1 the bus high enable signal ($\overline{BHE}$) should be used to enable data onto the most significant half of the data bus, pins D_{15}-D_8. Eight-bit oriented devices tied to the upper half of the bus would normally use $\overline{BHE}$ to condition chip select functions. $\overline{BHE}$ is LOW during T_1 for read, write, and interrupt acknowledge cycles when a byte is to be transferred on the high portion of the bus. The S_7 status information is available during T_2, T_3, and T_4. The signal is active LOW, and floats to 3-state OFF in "hold." It is LOW during T_1 for the first interrupt acknowledge cycle. | $\overline{BHE}$	A_0	Characteristics
---	---	---
0	0	Whole word
0	1	Upper byte from/to odd address
1	0	Lower byte from/to even address
1	1	None	
$\overline{RD}$	32	O	**Read:** Read strobe indicates that the processor is performing a memory of I/O read cycle, depending on the state of the S_2 pin. This signal is used to read devices which reside on the 8086 local bus. $\overline{RD}$ is active LOW during T_2, T_3 and T_W of any read cycle, and is guaranteed to remain HIGH in T_2 until the 8086 local bus has floated. This signal floats to 3-state OFF in "hold acknowledge."
READY	22	I	**READY:** is the acknowledgement from the addressed memory or I/O device that it will complete the data transfer. The RDY signal from memory/IO is synchronized by the 8284A Clock Generator to form READY. This signal is active HIGH. The 8086 READY input is not synchronized. Correct operation is not guaranteed if the setup and hold times are not met.
INTR	18	I	**Interrupt Request:** is a level triggered input which is sampled during the last clock cycle of each instruction to determine if the processor should enter into an interrupt acknowledge operation. A subroutine is vectored to via an interrupt vector lookup table located in system memory. It can be internally masked by software resetting the interrupt enable bit. INTR is internally synchronized. This signal is active HIGH.
$\overline{TEST}$	23	I	**TEST:** input is examined by the "Wait" instruction. If the $\overline{TEST}$ input is LOW execution continues, otherwise the processor waits in an "Idle" state. This input is synchronized internally during each clock cycle on the leading edge of CLK.

TABLE 15.9 (Continued)

Symbol	Pin No.	Type	Name and Function
NMI	17	I	**Non-maskable interrupt:** an edge triggered input which causes a type 2 interrupt. A subroutine is vectored to via an interrupt vector lookup table located in system memory. NMI is not maskable internally by software. A transition from a LOW to HIGH initiates the interrupt at the end of the current instruction. This input is internally synchronized.
RESET	21	I	**Reset:** causes the processor to immediately terminate its present activity. The signal must be active HIGH for at least four clock cycles. It restarts execution, as described in the Instruction Set description, when RESET returns LOW. RESET is internally synchronized.
CLK	19	I	**Clock:** provides the basic timing for the processor and bus controller. It is asymmetric with a 33% duty cycle to provide optimized internal timing.
V_{CC}	40		**V_{CC}:** +5V power supply pin.
GND	1, 20		**Ground**
MN/MX	33	I	**Minimum/Maximum:** indicates what mode the processor is to operate in. The two modes are discussed in the following sections.

The following pin function descriptions are for the 8086/8288 system in maximum mode (i.e., MN/$\overline{MX}$ = V_{SS}). Only the pin functions which are unique to maximum mode are described; all other pin functions are as described above.

Symbol	Pin No.	Type	Name and Function
$\overline{S}_2$, $\overline{S}_1$, $\overline{S}_0$	26-28	O	**Status:** active during T_4, T_1, and T_2 and is returned to the passive state (1,1,1) during T_3 or during T_W when READY is HIGH. This status is used by the 8288 Bus Controller to generate all memory and I/O access control signals. Any change by $\overline{S}_2$, $\overline{S}_1$, or $\overline{S}_0$ during T_4 is used to indicate the beginning of a bus cycle, and the return to the passive state in T_3 or T_W is used to indicate the end of a bus cycle. These signals float to 3-state OFF in "hold acknowledge." These status lines are encoded as shown.
$\overline{RQ}/\overline{GT}_0$, $\overline{RQ}/\overline{GT}_1$	30, 31	I/O	**Request/Grant:** pins are used by other local bus masters to force the processor to release the local bus at the end of the processor's current bus cycle. Each pin is bidirectional with $\overline{RQ}/\overline{GT}_0$ having higher priority than $\overline{RQ}/\overline{GT}_1$. $\overline{RQ}/\overline{GT}$ has an internal pull-up resistor so may be left unconnected. The request/grant sequence is as follows (see Figure 9): 1. A pulse of 1 CLK wide from another local bus master indicates a local bus request ("hold") to the 8086 (pulse 1). 2. During the CPU's next T_4 or T_1 a pulse 1 CLK wide from the 8086 to the requesting master (pulse 2), indicates that the 8086 has allowed the local bus to float and that it will enter the "hold acknowledge" state at the next CLK. The CPU's bus interface unit is disconnected logically from the local bus during "hold acknowledge." 3. A pulse 1 CLK wide from the requesting master indicates to the 8086 (pulse 3) that the "hold" request is about to end and that the 8086 can reclaim the local bus at the next CLK. Each master-master exchange of the local bus is a sequence of 3 pulses. There must be one dead CLK cycle after each bus exchange. Pulses are active LOW.
$\overline{LOCK}$	29	O	**$\overline{LOCK}$:** output indicates that other system bus masters are not to gain control of the system bus while $\overline{LOCK}$ is active LOW. The $\overline{LOCK}$ signal is activated by the "LOCK" prefix instruction and remains active until the completion of the next instruction. This signal is active LOW, and floats to 3-state OFF in "hold acknowledge."
QS_1, QS_0	24, 25	O	**Queue Status:** The queue status is valid during the CLK cycle after which the queue operation is performed. QS_1 and QS_0 provide status to allow external tracking of the internal 8086 instruction queue.

Encoding table for status lines:

$\overline{S}_2$	$\overline{S}_1$	$\overline{S}_0$	Characteristics
0 (LOW)	0	0	Interrupt Acknowledge
0	0	1	Read I/O Port
0	1	0	Write I/O Port
0	1	1	Halt
1 (HIGH)	0	0	Code Access
1	0	1	Read Memory
1	1	0	Write Memory
1	1	1	Passive

(Continued on following page)

TABLE 15.9 (Continued)

The following pin function descriptions are for the 8086 in minimum mode (i.e., MN/MX = V_{CC}). Only the pin functions which are unique to minimum mode are described; all other pin functions are as described above.

Symbol	Pin No.	Type	Name and Function
M/$\overline{IO}$	28	O	**Status line:** logically equivalent to S_2 in the maximum mode. It is used to distinguish a memory access from an I/O access. M/$\overline{IO}$ becomes valid in the T_4 preceding a bus cycle and remains valid until the final T_4 of the cycle (M = HIGH, IO = LOW). M/$\overline{IO}$ floats to 3-state OFF in local bus "hold acknowledge."
$\overline{WR}$	29	O	**Write:** indicates that the processor is performing a write memory or write I/O cycle, depending on the state of the M/$\overline{IO}$ signal. $\overline{WR}$ is active for T_2, T_3 and T_W of any write cycle. It is active LOW, and floats to 3-state OFF in local bus "hold acknowledge."
$\overline{INTA}$	24	O	$\overline{INTA}$ is used as a read strobe for interrupt acknowledge cycles. It is active LOW during T_2, T_3 and T_W of each interrupt acknowledge cycle.
ALE	25	O	**Address Latch Enable:** provided by the processor to latch the address into the 8282/8283 address latch. It is a HIGH pulse active during T_1 of any bus cycle. Note that ALE is never floated.
DT/$\overline{R}$	27	O	**Data Transmit/Receive:** needed in minimum system that desires to use an 8286/8287 data bus transceiver. It is used to control the direction of data flow through the transceiver. Logically DT/$\overline{R}$ is equivalent to $\overline{S_1}$ in the maximum mode, and its timing is the same as for M/$\overline{IO}$. (T = HIGH, R = LOW.) This signal floats to 3-state OFF in local bus "hold acknowledge."
$\overline{DEN}$	26	O	**Data Enable:** provided as an output enable for the 8286/8287 in a minimum system which uses the transceiver. $\overline{DEN}$ is active LOW during each memory and I/O access and for INTA cycles. For a read or $\overline{INTA}$ cycle it is active from the middle of T_2 until the middle of T_4, while for a write cycle it is active from the beginning of T_2 until the middle of T_4. $\overline{DEN}$ floats to 3-state OFF in local bus "hold acknowledge."
HOLD, HLDA	31, 30	I/O	**HOLD:** indicates that another master is requesting a local bus "hold." To be acknowledged, HOLD must be active HIGH. The processor receiving the "hold" request will issue HLDA (HIGH) as an acknowledgement in the middle of T_4 or T_1. Simultaneous with the issuance of HLDA the processor will float the local bus and control lines. After HOLD is detected as being LOW, the processor will LOWer HLDA, and when the processor needs to run another cycle, it will again drive the local bus and control lines. (See Figure 10.) HOLD is not an asynchronous input. External synchronization should be provided if the system cannot otherwise guarantee the setup time.

TABLE 15.10
Absolute maximum ratings*

Ambient Temperature Under Bias . . . . 0°C to 70°C
Storage Temperature . . . . − 65°C to + 150°C
Voltage on Any Pin with
  Respect to Ground . . . . − 1.0 to + 7V
Power Dissipation . . . . 2.5 Watt

*NOTICE: Stresses above those listed under "Absolute Maximum Ratings" may cause permanent damage to the device. This is a stress rating only and functional operation of the device at these or any other conditions above those indicated in the operational sections of this specification is not implied. Exposure to absolute maximum rating conditions for extended periods may affect device reliability.

TABLE 15.11
DC characteristics

8086-1: $T_A = 0°C$ to $70°C$, $V_{CC} = 5V \pm 5\%$

Symbol	Parameter	Min.	Max.	Units	Test Conditions
V_{IL}	Input Low Voltage	-0.5	$+0.8$	V	
V_{IH}	Input High Voltage	2.0	$V_{CC} + 0.5$	V	
V_{OL}	Output Low Voltage		0.45	V	$I_{OL} = 2.0$ mA
V_{OH}	Output High Voltage	2.4		V	$I_{OH} = -400 \mu A$
I_{CC}	Power Supply Current 8086-1		360	mA	$T_A = 25°C$
I_{LI}	Input Leakage Current		± 10	μA	$0V < V_{IN} < V_{CC}$
I_{LO}	Output Leakage Current		± 10	μA	$0.45V \leqslant V_{OUT} \leqslant V_{CC}$
V_{CL}	Clock Input Low Voltage	-0.5	$+0.6$	V	
V_{CH}	Clock Input High Voltage	3.9	$V_{CC} + 1.0$	V	
C_{IN}	Capacitance of Input Buffer (All input except $AD_0 - AD_{15}$, $\overline{RQ}/\overline{GT}$)		15	pF	fc = 1 MHz
C_{IO}	Capacitance of I/O Buffer ($AD_0 - AD_{15}$, $\overline{RQ}/\overline{GT}$)		15	pF	fc = 1 MHz

Courtesy of Intel Corporation

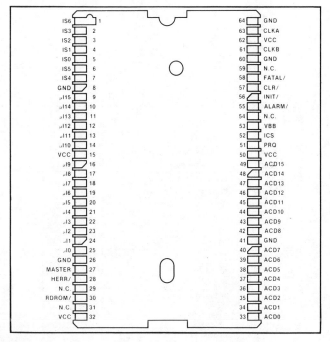

FIGURE 15.21
43201 pin assignment instruction decoder–microinstruction sequencer. (Courtesy of Intel Corporation)

TABLE 15.12
AC characteristics

**8086 MINIMUM COMPLEXITY SYSTEM
TIMING REQUIREMENTS**

8086-1: $T_A = 0\,°C$ to $70\,°C$, $V_{CC} = 5V \pm 5\%$

Symbol	Parameter	8086-1		Units	Test Conditions
		Min.	Max.		
TCLCL	CLK Cycle Period — 8086	100	500	ns	
TCLCH	CLK Low Time	(⅔ TCLCL) – 14		ns	
TCHCL	CLK High Time	(⅓ TCLCL) + 6		ns	
TCH1CH2	CLK Rise Time		10	ns	From 1.0V to 3.5V
TCL2CL1	CLK Fall Time		10	ns	From 3.5V to 1.0V
TDVCL	Data In Setup Time	5		ns	
TCLDX	Data In Hold Time	10		ns	
TR1VCL	RDY Setup Time into 8284A (See Notes 1, 2)	35		ns	
TCLR1X	RDY Hold Time into 8284A (See Notes 1, 2)	0		ns	
TRYHCH	READY Setup Time into 8086	53		ns	
TCHRYX	READY Hold Time into 8086	20		ns	
TRYLCL	READY Inactive to CLK (See Note 3)	– 10		ns	
THVCH	HOLD Setup Time	20		ns	
TINVCH	INTR, NMI, TEST Setup Time (See Note 2)	15		ns	

TIMING RESPONSES

Symbol	Parameter	8086-1		Units	Test Conditions
		Min.	Max.		
TCLAV	Address Valid Delay	10	50	ns	
TCLAX	Address Hold Time	10		ns	
TCLAZ	Address Float Delay	10	40	ns	
TLHLL	ALE Width	TCLCH–10		ns	
TCLLH	ALE Active Delay		40	ns	
TCHLL	ALE Inactive Delay		45	ns	
TLLAX	Address Hold Time to ALE Inactive	TCHCL–10		ris	
TCLDV	Data Valid Delay	10	50	ns	$C_L = 20\text{-}100$ pF for all 8086-1 Outputs (In addition to 8086-1 self-load)
TCHDX	Data Hold Time	10		ns	
TWHDX	Data Hold Time After WR	TCLCH–25		ns	
TCVCTV	Control Active Delay 1	10	50	ns	
TCHCTV	Control Active Delay 2	10	45	ns	
TCVCTX	Control Inactive Delay	10	50	ns	
TAZRL	Address Float to READ Active	0		ns	
TCLRL	RD Active Delay	10	70	ns	
TCLRH	RD Inactive Delay	10	60	ns	
TRHAV	RD Inactive to Next Address Active	TCLCL–35		ns	
TCLHAV	HLDA Valid Delay	10	60	ns	
TRLRH	RD Width	2TCLCL–40		ns	
TWLWH	WR Width	2TCLCL–35		ns	
TAVAL	Address Valid to ALE Low	TCLCH–35		ns	

NOTES: 1. Signal at 8284A shown for reference only.
2. Setup requirement for asynchronous signal only to guarantee recognition at next CLK.
3. Applies only to T2 state. (8 ns into T3)

Courtesy of Intel Corporation

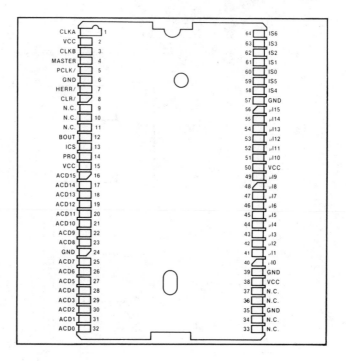

FIGURE 15.22

43202 pin assignment execution unit. (Courtesy of Intel Corporation)

companion devices (shown in Figures 15.21 and 15.22) provide the general data-processing facility of the iAPX 432 micromainframe. The combination of VLSI technology and advanced architecture in the iAPX 432 system results in mainframe functionality with a microcomputer form factor. The new object-based architecture significantly reduces the cost of large software systems and enhances their reliability and security.

Software-transparent multiprocessing allows the user to configure systems matched to the required performance and provides an easy growth path. Hardware support for operating systems and high-level languages eases their implementation.

The GDP provides 2^{40} bytes of virtual address space with capability-based addressing and protection. In addition, a hardware-implemented functional redundancy checking mode is provided for the detection of hardware errors.

The iAPX 43201 and iAPX 43202 block diagrams are shown in Figures 15.13 and 15.24. They are fabricated with Intel's highly reliable +5-V, depletion load, N-channel, silicon gate HMOS technology, and each is packaged in a 64-pin quad in-line package (QUIP).

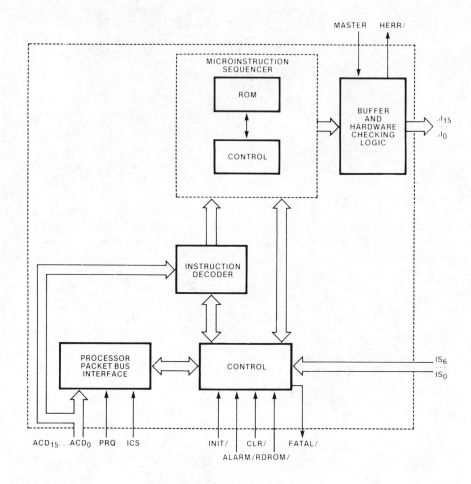

FIGURE 15.23
43201 block diagram. (Courtesy of Intel
Corporation)

15.7
APPLICATION OF THE
MICROPROCESSOR

There are many applications for the μP. In this section we shall discuss a
μP-based true-rms line voltage monitor, a μP-based automated noninvasive
blood-pressure device, and other uses. There are thousands of applications
for μPs.

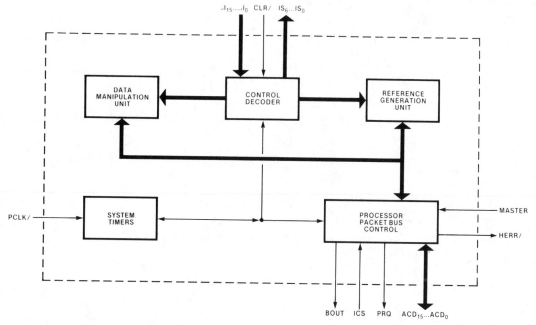

FIGURE 15.24
43202 block diagram. (Courtesy of Intel
Corporation)

15.7.1 A Microprocessor-Based True-RMS
Line Voltage Monitor[i]

The microprocessor-based true-rms line voltage monitor incorporates
analog and digital circuitry and μC hardware and software to obtain high
accuracy, high speed, and a minimal loss of data. The μC forms the heart of
this system. The remaining circuitry serves to input data and to output data
to and from the μC. Analog input circuitry attenuates and squares the input
voltage as shown in Figure 15.25. A V/F converter produces logic pulses at
a rate proportional to the instantaneous value of the squared signal. A 16-bit
counter integrates V^2 by accumulating pulses from the V/F converter.

The μC is comprised of an M68B00 microprocessor (μP), MCM68B10L
128 × 8 random-access memory (RAM), 2716 2k × 8 programmable, read-
only memory (PROM), MC6870A 1-MHz crystal clock oscillator, and ad-

---

[i]The material in Section 15.7.1 is from Jeffrey L. Silberberg, "A Microprocessor-Based
True-RMS Line Voltage Monitor," HHS Publication FDA 80-8179, Bureau of Radiological
Health, FDA, August 1981, and is used courtesy of the Food and Drug Administration of the
U.S. Department of Health and Human Services.

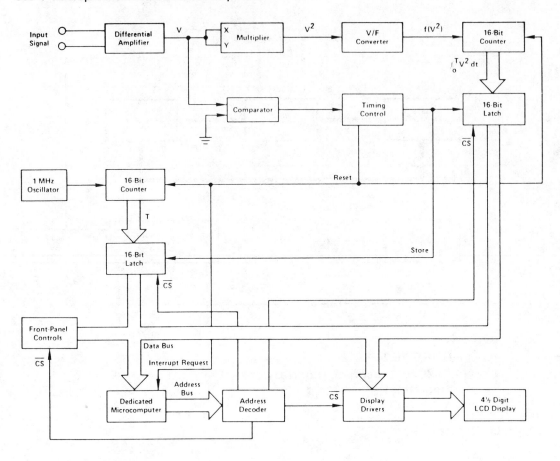

FIGURE 15.25
Block diagram of a microprocessor-based true-rms line voltage monitor. (Courtesy of the Food and Drug Administration)

dress decoder circuitry (Figure 15.26). The address $(A_0 \ldots A_{15})$, data $(D_0 \ldots D_7)$, and control lines are connected as indicated in the figure. The HALT line of the μP is tied high with a pull-up resistor so that the μP may be halted by a clip-on logic analyzer probe during troubleshooting. The interrupt signal from the interrupt latch is inverted by a single transistor stage to provide the *proper* signal polarity and sink current for the $\overline{\text{IRQ}}$ input of the μP. An RC circuit on the $\overline{\text{RESET}}$ input of the μP holds the $\overline{\text{RESET}}$ line low for multiple machine cycles after either power up or actuation of S_2. The RC stage is buffered by CMOS inverters to reduce loading effects on the time constant and to square up the signal.

The address decoder is comprised of 74LS139 and 74LS138 decoders and several gages as shown in Figure 15.26. Low-power Schottky (LS) TTL

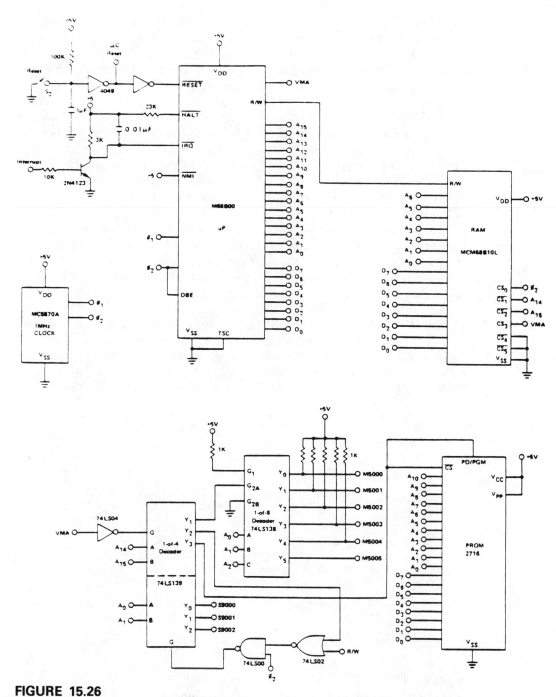

FIGURE 15.26
Microcomputer hardware of a line voltage monitor. (Courtesy of the Food and Drug Administration)

was used rather than CMOS for speed in address decoding. Pull-up resistors (1 k) are used wherever the microprocessor or any LSTTL circuitry drives CMOS, such as the outputs of the 1-of-8 decoder. Half of the 74LS139 (dual 1-of-4 decoder) is used to partition the memory into four segments: RAM, input, output, and PROM. When $A_{15}A_{14}$ is 01 (binary), Y_1 goes low, enabling the 74LS138 to select an 8-bit input byte or to reset the interrupt latch. When $A_{15}A_{14}$ is 11, Y_3 goes low, enabling the PROM. When $A_{15}A_{14}$ is 10, Y_2 goes low. Y_2 is subsequently gated with the ϕ_2 clock and read–write (R/W) to enable the second 1-of-4 decoder. Store pulses are then generated for the 3 bytes of display latches (S9000 ... S9002) as determined by the A_0 and A_1 address lines. The MCM68B10L contains sufficient decoding logic to detect the 00 state of $A_{15}A_{14}$. Thus, it is not necessary to use the Y_0 output of the first 1-of-4 decoder.

The addresses were assigned so there would be no interference with memory resident in the software development facility used in writing and testing the line voltage monitor program. For this reason, several "don't-care" bits are assigned to logic 1 in the input, output, and PROM address designations. Hexadecimal notation is used throughout in referring to the address of memory locations.

The line voltage monitor program occupies less than 2k of ROM, consisting primarily of general-purpose library subroutines. Less than $\frac{1}{2}$k of ROM is occupied by the software specific to the LVM. General-purpose subroutines provide such functions as basic arithmetic, square-root, and format conversion. The LVM-specific routines read data from the latches

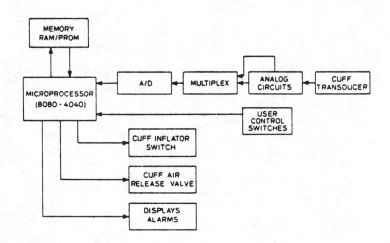

FIGURE 15.27
Architecture of microprocessor-based automated noninvasive blood-pressure device. (Courtesy of *Medical Electronics*, 2994 W. Liberty Ave., Pittsburgh, Pa. 15216)

designated M5000 through M5004, process the data, and store the results in the display circuitry located from S9000 through S9002. The software was written in assembly language.

When the processor is interrupted at the end of each sample period (nT), it responds within 1 ms under normal conditions. In response, the processor first sets the interrupt mask so that it can process one sample completely before reading in another. This is a preventive measure only. Successive samples do not interfere under the voltage and frequency constraints. After setting the interrupt mask, the μP acknowledges the interrupt by addressing location 5005. The setting of the sample–hold switch (S_3) is then read from location 5004. If S_3 is set to hold, the sample is not processed and the μP resumes the display routine described. If S_3 is set to sample, however, the μP proceeds by reading $\int V^2$ from locations 5000 and 5001 as a 16-bit integer and converting it to a floating-point binary number.

15.7.2 A Microprocessor Automated Non-invasive Blood-Pressure Device[j]

Measurement of arterial blood pressure by oscillometry is a new noninvasive technique that provides a new dimension in critical patient care. Using a conventional pressure cuff without microphones or ultrasound transducers, the unit automatically adjusts to a wide variation in operating parameters. Through use of the oscillometric measurement method and adaptive programs in an internal μP (Figure 15.27), dependable measurements of mean arterial pressure can be routinely made on most patients, using any limb and over dressings.

The microprocessor serves as a controller: it samples the cuff baseline pressure and the oscillation magnitude; it controls the cuff inflation and deflation; it performs the required logic decisions and displays the results. This architecture is shown in Figure 15.27.

15.7.3 Microprocessor Applications with Bus Interfaces

Major applications of the ICS Electronic Model 4500 Speech Synthesizer with a microprocessor are shown in Figure 15.28. This device is used in broad tests in the diagnostic area where audio test data are simplified for speed-up fault isolation and repair. As an illustration, the engineer can use the speech synthesizer with an IEEE-488 bus controller (discussed in detail in Section 15.10) and a signature analyzer for diagnosis of a printed circuit board or an extended microprocessor.

---

[j] The material in Section 15.7.2 is from Joseph Looney, Jr., "Blood Pressure by Oscillometry," issue of April 1979 of *Medical Electronics*, 2994 W. Liberty Ave., Pittsburgh, Pa. 15216. Courtesy of the publisher, Mr. Milt Aronson.

Figure 15.29 shows the functional block diagram of the ICS Electronics Model 4871 and its applications. The 4871 uses a microprocessor as an automatic-calibration scheme. Also in Figure 15.29 we see the internal autocalibration mode (dc); external autocalibration mode (dc), and the software autocalibration mode for any parameter.

Multiple use of a microcomputer is not unusual today. It is possible to interface electronic instrumentation wherever the user has to be. Microprocessor systems can also be put in where required. Figure 15.30 shows the ICS 4885A Remote Computer System.

Features of the ICS Model 4886 (an IEEE bus extender) provide software extension of the IEEE 488 bus via twisted pair cables or the use of modems and standard line circuits, discussed briefly in Chapter 10. Through unique error detection software, the data are continuously checked for accuracy, and retransmission is automatically requested should the data not pass the accuracy test. This system provides movement of electronic instrumentation using microprocessor devices.

15.8
EMULATION FOR MICROPROCESSORS [k]

Emulation is an effective means to check how the software modules work with the target microprocessor and hardware. The emulation environment is valuable for debugging hardware and software at interim states. It provides down-load function, a RAM environment, and run controls with easy setup.

Emulation in the Logic Development System is controlled by the host processor across a medium-speed bus. The target microprocessor in the emulation pod and emulation control cards, and emulation memory in the development stations card cage, use a separate, high-frequency bus (Figure 15.31). The high-speed emulation memory uses 8k, 16k, 32k, or 64k of static RAM for 8-bit emulation, with additional options of 96k or 128k RAM for 16-bit emulation. Emulators for 8-bit microprocessors support 8080/8085, 6800/6802, Z80, 8048, and 6809 processors. For 16-bit microprocessors, emulators include 68000, Z8001/Z8002, and 8086/8088 units.

Ideally, emulation would be functionally and electrically transparent to the target system. With functional transparency, the emulator would impose no demands on the target system, such as reserved address space, interrupt inputs, or direct memory access. An emulator with electrical transparency would not alter the electrical specifications of the target system, such as drive capability, capacitance, timing, clock speeds, and thresholds. Practically, one form of transparency is achieved at the expense of the other

---

[k] The material in Section 15.8 is from *1982 Electronic Instrumentations and Systems Catalogue*, Hewlett-Packard, 1982, and is used courtesy of Hewlett-Packard Co., Palo Alto, Calif.

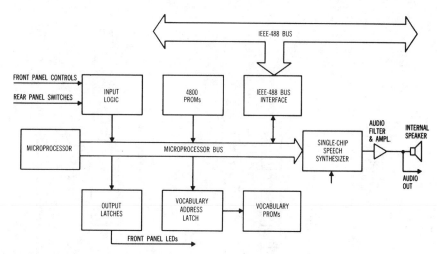

FIGURE 15.28
Hooked onto the IEEE-488 bus, the 4800 Speech
Synthesizer speaks instructions for troubleshoot-
ing and repairing devices undergoing tests.
(Courtesy of the ICS Electronics Corp., San Jose,
Calif.)

form, and trade-offs must be made. In Model 64000, functional transparency
has been given primacy. All emulators have a mode that is functionally
transparent to the target system. However, placing the prototype target
processor in the emulation pod introduces some electrical changes in the
target system; processor drive, logic levels, and loading are altered by the
buffers, and the bus cables introduce some capacitance. Some electrical
transparency was sacrificed to gain better functional transparency.

A necessary condition to achieve functional transparency is the isolation of emulation buses and memory from the operating system buses and
memory. A major benefit gained by functional transparency is real-time
emulation. This means the target system and microprocessor run at operational speeds without inserting wait states.

There are three possible modes of emulation. When all program memory is assigned to the target system, only the microprocessor activity is
emulated. All execution may be performed from the emulation memory of
the development system. A third mode combines these two modes, and
program memory is mapped to both memories, and program execution is
switched between host and emulation memory as specified by the memory
assignment map. Real-time operation is possible in all three modes. Fast-
access memory chips are used to minimize memory board access time.
Model 64000 Logic Development System allows the user to interleave host
and target memories in noncontiguous 1k-byte blocks.

APPLICATIONS

Power Supply Programmer

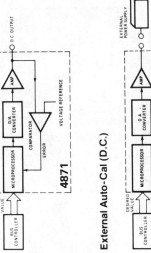

D.C. OUTPUT

D.C. OUTPUT

AUTO CALIBRATION

Instrument Programmer

1 2 3 4 5 6 7

D.C. PROGRAMMED INSTRUMENTS

X-Y Recorder

Y-AXIS

X-AXIS

PEN CONTROL

Process Control

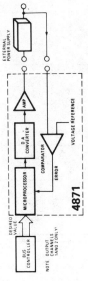

SET TEMP

TEMPERATURE CHAMBER

124

DIGITAL THERMOMETER

BCD DATA

Internal Auto-Cal (D.C.)

BUS CONTROLLER — DESIRED VALUE — MICROPROCESSOR — D/A CONVERTER — AMP — D.C. OUTPUT

COMPARATOR

ERROR

VOLTAGE REFERENCE

4871

External Auto-Cal (D.C.)

EXTERNAL POWER SUPPLY

BUS CONTROLLER — DESIRED VALUE — MICROPROCESSOR — D/A CONVERTER — AMP

NOTE OUTPUT CHANNELS 1 AND 2 ONLY

COMPARATOR

ERROR

VOLTAGE REFERENCE

4871

Software Auto-Cal (Any Parameter)

FREQUENCY GENERATOR (TYP)

FREQUENCY COUNTER (TYP)

BUS CONTROLLER — DESIRED VALUE — MICROPROCESSOR — D/A CONVERTER — AMP

OFFSET AND SCALE FACTOR

COMPARATOR

ERROR

VOLTAGE REFERENCE

ACTUAL VALUE

IEEE 488 BUS

BCD TO IEEE CONVERTER (IF NECESSARY)

4871

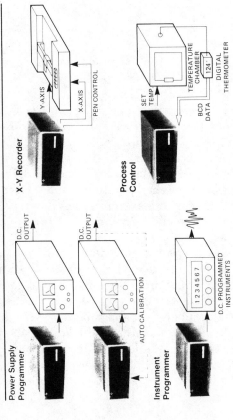

A — 0 TO -10V
B — 10 TO -10V
C — 4 TO 20mA

12 BIT D/A CONVERTER

CHANNEL 1 (TYP)

OPTICAL ISOLATION

MICROPROCESSOR WITH RAM AND EPROM

IEEE 488 BUS

COMPARATOR

OPTICAL ISOLATION

AUTO CAL

CALIBRATION REFERENCE -10V

0V

EXTERNAL AUTO-CAL INPUTS CHANNELS 1, 2

BCD DATA INPUT FROM DPM OR OTHER INSTRUMENT

BCD INTERFACE

RELAY CONTROL

4 ISOLATED SPST RELAYS

FIGURE 15.29

Applications, functional block diagram, and automatic-calibration modes of the ICS Electronic Model 4871. (Courtesy of the ICS Electronics Corp., San Jose, Calif.)

The emulator is an important link between the development phase and the final product. As a rule, most of the program for a microprocessor-based product is stored in ROM. In the absence of any other tools, EPROMs are used during development, but even simple changes can involve a long process. Down-loading to the RAM of the emulator is automatic in the 64000 system, and a great time-saver in developing software that will finally reside on the processor ROM. Address space can be allocated to emulation RAM, emulation ROM, target RAM, target ROM, and an illegal address space. Error messages are displayed anytime the target processor executes an illegal operation, such as a write to ROM or a reference to an illegal address. The emulator allows mapping in 1k-byte blocks for 8-bit emulators, and 1k or 4k blocks for 16-bit emulators. Any set of defined blocks can be mapped, in any order, from target address space into available emulation memory. Even if all of the program is transferred to the target ROM, and one more bug is found, the suspected area can be mapped back to the emulator, revised, and checked without removing the target ROM from the circuit.

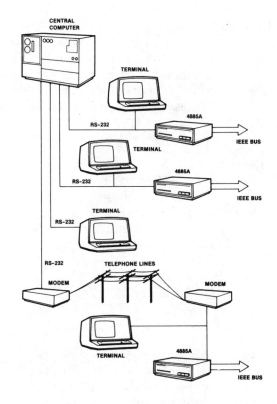

FIGURE 15.30
ICS Model 4885A remote computer system.
(Courtesy of the ICS Electronics Corp., San Jose, Calif.)

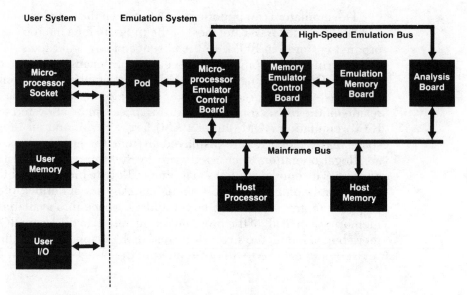

FIGURE 15.31
The emulation function in Model 64000 uses a separate bus for emulation control and memory, logic analysis, and interface to the target system. This enhances the functional transparency of the emulator. (Courtesy of Hewlett-Packard Co., Palo Alto, Calif.)

Emulators and logic analysis functions of the 64000 system can be added to existing mainframe development systems with HP 64005S Emulation Terminal System. Consisting of a development station, 64000 emulation hardware and software, the system can be used as a stand-along emulator and an ASCII terminal for the host mainframe.

15.9
MICROPROCESSOR-BASED TESTING PROBLEMS[1]

The use of microprocessors has produced a new set of production testing problems. These primarily are the result of the bidirectional bus employed by the microprocessor and its high on-board clock rate. The objective of this section is to describe some of those problems and explain how they are

---

[1] The material in Section 15.9 is from Dynamic Digital Board Testing with the HP Model 3060A Option 100, Application Note 308-1, Nov. 1980, and is used courtesy of Hewlett-Packard Co., Palo Alto, Calif.)

solved by the HP 3060A Board Test System. This section will also serve as an introduction to the HP 3060A test hardware, since many system features are described when the test solutions are explained.

Processors often have minimum clock rate specifications. These rates are usually factors of 10 faster than most test systems could go prior to the introduction of microprocessors. The test problem is further complicated by the fact that the microprocessor board generally has an on-board clock to which the test system must be synchronized. Several alternatives may be used to handle this on-board clock problem, including turning off the clock or synchronizing test system stimulus and response detection to the on-board clock.

The option of turning off the clock is undesirable since it creates an unnatural condition on the board under test. However, test engineers often choose this approach for reasons of expedience. The alternate approach of synchronizing the tester to an external clock can be done provided that the tester's pin electronics have memory. The memory is essential since reasonably priced system controllers cannot communicate with the test system hardware fast enough to keep up with on-board clock rates of 1 MHz or more. The memory allows a more leisurely communication rate by storing stimulus patterns, as well as UUT responses. Communication with the UUT may then occur in short, but high-speed bursts. Unfortunately, RAM-backed pin electronics are very expensive. Their use does, however, allow testing at a high clock rate using clock signals coming from the board under test. See Figure 15.32 for how data transfer is synchronous to an on-board check.

The 3060A uses the second approach of synchronizing RAM-backed pin electronics either to external or internal high-speed clock signals. Its stimulus and response hardware is designed specifically for interfacing to the standardized electronic architecture of the microprocessor-based product. Its combination of features results in a test system cost that is half the cost of the traditional high-speed functional test approach.

The test stimulus combines RAM-backed pins for interfacing with the UUT's bidirectional data bus and a binary sequence generator (a counter)

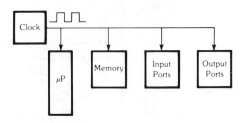

FIGURE 15.32
Data transfer is synchronous to an on-board block. (Courtesy of Hewlett-Packard Co., Palo Alto, Calif.)

for interfacing with UUT address buses. Both can be synchronized to external clock signals. Lower-cost static drivers are also provided for interfacing to UUT inputs, which need not change state synchronously with the on-board clock.

The 3060A pins that record UUT response have almost no memory behind them. Instead they use a data-compression technique called *signature analysis*. Due to compression, this approach actually has the advantage of being able to handle arbitrarily long streams of data. In contrast, only a few thousand bits are normally available using RAM-backed receivers. Furthermore, signature analysis can be synchronized to an external clock signal at rates up to 10 MHz.

Data transfer between devices in the microprocessor-based product takes place on a bidirectional bus. Many devices, including ROMs, RAMs, and the microprocessor, are connected in parallel as shown in Figure 15.33. This creates a new kind of fault-isolation problem, which confounds the older style test systems; given that a bidirectional bus node is failing, which of the components connected to the bus is actually failing? Furthermore, the test system must be able to participate in the complex protocol of the bus structure. In this structure, devices using the data bus must be able to switch their outputs off or on within one clock cycle to avoid any conflict with other devices on the bus. Likewise, test system hardware must also have this capability. If it does not, the response of devices having bidirectional pins could be obscured by a conflict between the test system drivers and the device under test.

The dynamic stimulus of the 3060A system can be placed in the high-impedance state by external signals from the board under test. Because the 3060A uses a high-visibility bed-of-nails test adapter, it can monitor UUT logic signals, which indicate bidirectional bus protocol and usage. The signals can be used to control the stimulus output.

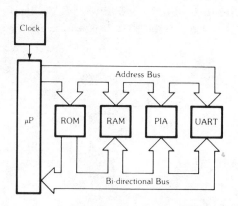

FIGURE 15.33
Typical microprocessor architecture. (Courtesy of Hewlett-Packard Co., Palo Alto, Calif.)

To solve the fault-isolation problem presented by the bidirectional bus, the 3060A relies on the power of the signature analysis digital test technique. Because of the ability of the signature analyzer to selectively observe digital information, it can be used to focus the test on specific devices on the bidirectional bus. Similarly, it can differentiate between data being written into a device and the data output of that device.

Since the product design is now dependent on software, it stands to reason that the test engineer faces the problem of testing devices that are software driven. Indeed, a microprocessor must execute elements of its own instruction set in order to be tested. For example, consider the Motorola 6800 microprocessor. Its control output, READ/WRITE, takes on the logic 0 state only during the execution of instructions that transfer data from the 6800 to memory or peripheral devices. At other times, READ/WRITE is in the logic 1 state. To properly test a digital device, the test must show that all device outputs can take on both a logic 1 and a logic 0 state. Therefore, a proper test of the 6800 must include the execution of at least one store instruction in order to test READ/WRITE. Similarly, its interrupt inputs cannot be tested without allowing the processor to respond to interrupts using software.

Similar situations occur with other brands of microprocessors. The Intel 8085 has five interrupts, five control lines, and two serial I/O lines, which can only be tested if the 8085 under test executes elements of its instruction set. The Zilog Z80 has two interrupt inputs and four control outputs that fall into this category. See Figure 15.34 for the names of these lines.

The solution to this problem is to allow the unknown microprocessor under test to execute some elements of its instruction set. If this problem is considered in the design phase of the microprocessor-based product, it can be solved by adding self-test capability to the product design. Otherwise, the test engineer is obligated to retrofit some sort of external test routine. The 3060A is endowed with a feature that will aid the test engineer in this later situation.

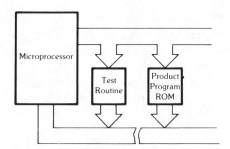

Examples of Processor Pins Testable Only With Software

	8085		6800	Z80
Interrupts	INTR	RST 7.5	$\overline{\text{IRQ}}$	$\overline{\text{INT}}$
	TRAP	RST 6.5	$\overline{\text{NMI}}$	$\overline{\text{NMI}}$
		RST 5.5		
Control Outputs	S_0	$\overline{\text{RD}}$	VMA	$\overline{\text{M1}}$
	S_1	$\overline{\text{WR}}$	R/W	$\overline{\text{RD}}$
	IO/$\overline{\text{M}}$			$\overline{\text{WR}}$
				$\overline{\text{HALT}}$
Other	S1D			
	S0D			

FIGURE 15.34

Testing software-driven devices. (Courtesy of Hewlett-Packard Co., Palo Alto, Calif.)

To hold test instructions, the 3060A dynamic stimulus is configurable as a ROM. The microprocessor under test may fetch its instructions from the test system's RAM-backed dynamic drivers. An additional advantage associated with this capability is that it allows the processor to not only test itself, but also create test stimulus for the rest of the UUT. The response of the UUT may be tested by using signature analysis.

15.10
IEEE-488 BUS INTERFACE SYSTEM AND COMPUTER-AUTOMATED MEASUREMENT AND CONTROL INTERFACE (CAMAC)[m]

In 1978, the Institute of Electrical and Electronic Engineers revised a 1975 standard designed to make it possible to transfer digital data among a number of instruments, using standardized signals, techniques, and a party-line bus. The standard is called IEEE Standard 488-1978, Digital Interface for Programmable Instrumentation. The American National Standards Institute approved it as a National Standard on July 18, 1979. Thus the standard is properly called ANSI/IEEE Standard 488-1978. The International Electrotechnical Commission also has adopted the standard as IEC Std. 625-1.

The word "programmable" in the title of the standard implies that the instruments to be accommodated are programmable by digital signals. Today, however, the standard applies to interface systems for both "programmable" and "nonprogrammable" equipments, including calculators, plotters, memory discs, computers, and the like, and, as we shall see later, another interface standard (CAMAC) has been promulgated especially for computer interface.

The party-line bus used to interconnect equipments is called a general-purpose instrumentation bus, GPIB, or IEEE-488 bus.

Today, the GPIB allows up to 14 instruments to be connected to a computer (the computer is the fifteenth instrument), using only one interface. Furthermore, the GPIB permits connection of several computers to a common set of instruments. A single computer can also be used to control several instrumentation buses with the addition of extra GPIB interfaces.

An example of a computerized GPIB measurement system is shown in Figure 15.35. Since the IEEE-488 specification carefully defines the electrical interface characteristics and the communications protocol used by the

---

[m] The material in Section 15.10 is from *IEEE-488 and CAMAC*, by James J. Truchard, William C. Nowlin, Jr., and Jeffrey L. Kodowsky of the National Instruments Corp., Austin, Texas. This material also appears in *Medical Electronics* in "Microprocessor Course No. 10." It is reprinted with permission of the National Instruments Corp. and *Medical Electronics*, 2994 W. Liberty Ave., Pittsburgh, Pa. 15216.

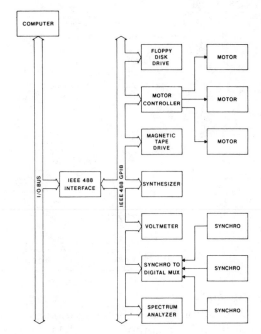

FIGURE 15.35
IEEE-488 measurement system with GPIB.
(Courtesy of National Instruments Corporation,
Austin, Texas)

instruments connected to the GPIB, the addition of new instruments to the measurement system is simplified. If comprehensive computer software support packages are provided with the interface, the user's manpower expenditures for developing software and interfacing new instruments to the measurement system are greatly reduced.

IEEE-488 bus interface defines mechanical, electrical, and functional standards for achieving proper and efficient interfacing (interconnecting) of equipments and includes the following rules:

1. All data must be digital.

2. The number of devices that may be connected to one bus must not exceed 15.

3. The length of interconnecting cable (i.e., the transmission path length) must not exceed 20 m (or 2 × number of instruments, in meters).

4. The rate at which data can be transmitted over the interface must not exceed 1 megabit/s (1 Mb/s = 10^6 bits/s).

The interface is byte serial/bit parallel; that is, each byte of 8 bits is transmitted serially, but all 8 bits of each byte are transmitted simultaneously in parallel.

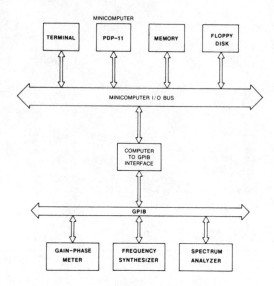

FIGURE 15.36
Implementation of a GPIB-based measurement
system in which a computer (PDP-11, in this case)
is used as the 488 Interface Controller. (Courtesy
of National Instruments Corporation, Austin,
Texas)

Terminology used with the IEEE-488 bus interface systems includes the following:

1. **Handshake.** A protocol (proper sequence of status and control signals) for effecting the transfer of data via the interface using data ready–data received technique.

2. **Device-dependent message.** Messages used by interconnected devices, but not processed by the interface, that is, data transmitted via the interface but not processed or used by the interface in any other manner. For example, to effect a change in the voltage gain of an amplifier, the *message* that does this is a "device-dependent message."

3. **Listen.** Ability to receive a device-dependent message from another device.

4. **Listener.** Device addressed by an interface message so as to "listen" or receive a device-dependent message.

5. **Talk.** Ability to send a device-dependent message via the interface to another device.

6. **Talker.** Device addressed by an interface message so as to "talk."

7. **Control.** Refers to management (control) of the interface system, that is,

commanding specific devices (addressing) to listen, talk, or command an action in a device.

8. **Controller.** Device that can "address" other devices to listen or talk or to command specified actions within devices.

In Figure 15.36 we see a typical computer-based measurement system, including the usual complement of computer hardware—disk, terminal, plotters and printers, and the GPIB with a number of instruments. In this example we show a Digital Equipment Corporation PDP-11 computer, which is being used as the IEEE-488 interface *controller*. This computer interface provides a user with all the GPIB functions applicable to a computerized measurement system.

A more sophisticated system is shown in Figure 15.37. This system has multiple computers, in this case a PDP-11 minicomputer and two microcomputers (LSI-11).

The interface shown in Figure 15.38 provides high-speed operation for demanding applications. A multiple-computer system allows flexibility because instrumentation can be shared by several processors.

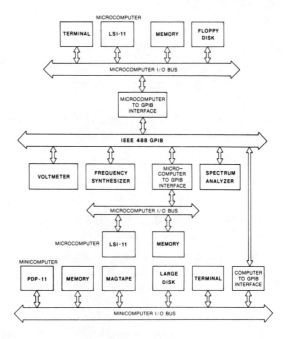

FIGURE 15.37

A multicomputer system using two microcomputers (LSI-11) and one minicomputer (PDP-11). The PDP-11 would be the system controller; the microcomputers could also function as controllers. (Courtesy of the National Instruments Corporation, Austin, Texas)

FIGURE 15.38
PDP-11 Minicomputer-to-GPIB interface; Model
GPIB 11-2 Direct Memory Access (DMA).
(Courtesy of National Instruments Corporation,
Austin, Texas)

15.10.1 Advantages of a Computerized Measurement System

In changing from a manual to an automated measurement system, the user gains many benefits. These include the following:

1. Greater measurement speed.

2. More accurate results, since calibration of the system can be automated.

3. Reduced random errors caused by operator fatigue.

4. Reduced manual handling of data because data are automatically entered into the computer.

5. More interaction with theoretical calculations because measured data can be used directly as input for such calculations.

6. Faster turnaround between time of measurement and final output of processed data in the form of tables, graphs, or other results.

7. Greater ability to process, transfer, or store vast quantities of data.

8. Better diagnostic capabilities because errors can normally be detected and corrected before excessive time has elapsed.

Many of these objectives can be achieved with a calculator-based measurement system; however, there are many situations in which a computer-based measurement system can provide advantages. The computer-based measurement system is ideal when:

1. Both high-speed and low-speed data ports are needed.

2. Sophisticated software is required to process measurement data.

3. Complex theoretical calculations are required, with measured data possibly being used as input to the calculations.

4. Vast quantities of data are being handled and stored for later processing.

5. A computer system is already in use and the capabilities of the system are suitable for using it in a measurement environment.

GPIB Fundamentals.   The GPIB system provides hardware protocol to implement a complete measurement and control system, including a means for communication among a group of interconnected devices. In Figure 15.39 we show the basic functions of the GPIB, using a functional partition of the hardware interfacing a particular device (computer or instrument) to the GPIB. Note that there are two types of messages: (1) GPIB *function messages* used for bus management, called *interface messages,* and (2) device function messages or *device-dependent messages,* which are communicated between the various devices via the GPIB, but which are not used or processed by the bus.

The device functions and messages are determined by the user (for the computer) and by the equipment manufacturers (for the instruments) to provide the necessary measurement and operational capability. The interface functions (provided by the IEEE-48 specification) make it possible to operate the devices as desired.

Figure 15.40 shows a typical system using the GPIB. Three types of devices are required to organize and manage the flow of information on the bus: (1) listener, (2) talker and (3) controller. A *listener* has the capability of being "addressed" by an interface message to receive (device-dependent) messages. A *talker* has the capability of being "addressed" by an interface

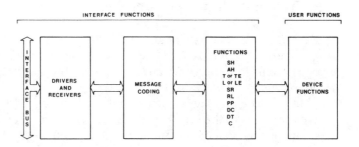

FIGURE 15.39
Interface Functions include SH (source handshake); T (talker), L. (listener), and so on. (Courtesy of National Instrument Corporation, Austin, Texas)

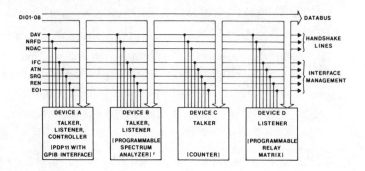

FIGURE 15.40
Components of a GPIB system. (Courtesy of National Instruments Corporation, Austin, Texas)

message to send device-dependent messages. A *controller* can address instruments to talk or listen, and can also command other specific actions within the instrument, via interface messages.

A GPIB-based measurement and control system may have one or more controllers. If more than one controller is connected to the GPIB, one and only one is designated the *system controller*. The system controller may temporarily pass control to any other controller.

The term *interface*, when used with "function" and "message," should not be confused with the computer interfaces mentioned earlier. *Interface function* and *interface message* refer to GPIB specifications for instruments. *Computer interface*, as used here, refers to the circuitry required to allow communication between a computer and other device or bus, in this case, the GPIB.

15.10.2 Interface Functions

Each message consists of an 8-bit bytes and can be of the following types:

The *source handshake* (sh) and *acceptor handshake* (ah) interface functions are used to transmit and receive the bit-parallel, byte-serial messages (*multiline messages*) on the bus.

Multiline messages on the databus (Figure 15.39) are interpreted either as *commands* to the interface or *data bytes* for the device, depending on the status of the bus signal line *attention* (atn) in the interface management group of lines. Attention (atn) is asserted (i.e., 1) for interface commands; it is 0 for a data byte.

The *talker* (t) and *extended talker* (te) functions allow a device to transmit data bytes on the data bus.

The t (talker) function uses a 1-byte address; the te (talker extended) function uses a 2-byte address. In all other respects, t and te are the same.

When an interface has been addressed by a controller as a talker, the device is allowed to use the interface source handshake function to transmit data bytes.

The listener (L) and extended listener (LE) function allows a device to receive data. It uses the acceptor handshake function to receive the data bytes. Only one device at a time may be the talker; however, the talker can communicate simultaneously with one or more listeners.

15.10.3   System Control

As shown in Figure 15.39, each device has its own built-in GPIB interface and could, conceivably, act as a controller. The controller function (C) is the originator of all interface messages (i.e., commands).

However, only one interface at a time (among the many interfaces) may be the active controller. While there may be several interfaces with controller capability (Figure 15.40), there can be only one interface that has the capability to "make itself" the controller; it is referred to as the *system controller*. When in control it can also be called the active controller or controller-in-charge. Any controller can become an active controller, using a transfer of control protocol (described later).

The active controller, or controller-in-charge, is responsible for selecting (addressing or unaddressing) talkers and listeners, as well as performing other bus-management operations, which include the following:

1. Conducting a *parallel poll* (PP) of the instruments on the GPIB (i.e., scanning all status flags) to ascertain equipment status; this is done by using the uniline message *identify* (IDY).

2. Conducting a *serial poll* (i.e., scanning flags) of the instruments to obtain more detailed status information.

3. Passing GPIB control to another controller.

4. Setting device functions in remote or local mode (RL) using the uniline message *remote enable* (REN).

5. Initializing the bus by asserting the uniline message *interface clear* (IFC). The latter two operations are capabilities only of the system controller, and not transferrable to another controller.

15.10.4   GPIB Signal Lines

A total of 24 lines is used to implement the bus (Figure 15.41). Of these lines, 16 are signal lines, 1 is ground, 1 is the cable shield, and 6 are twisted-pair commons for 6 of the signal lines. The 16 signal lines are used to carry all information, interface messages, and device-dependent messages among interconnected devices.

The bus is organized into three sets of signal lines:

DIO1	1	13	DIO5	
DIO2	2	14	DIO6	
DIO3	3	15	DIO7	
DIO4	4	16	DIO8	
EOI	5	17	REN	
DAV	6	18	GND (TW PAIR W/DAV)	
NRFD	7	19	GND (TW PAIR W/NRFD)	
NDAC	8	20	GND (TW PAIR W/NDAC)	
IFC	9	21	GND (TW PAIR W/IFC)	
SRQ	10	22	GND (TW PAIR W/SRQ)	
ATN	11	23	GND (TW PAIR W/ATN)	
SHIELD	12	24	SIGNAL GROUND	

FIGURE 15.41
Pin connections for GPIB cable. The shield must
be grounded only at the system controller.
(Courtesy of National Instrument Corporation,
Austin, Texas)

1. Eight data lines (DIO1 through DIO8, Data InOut 1–8).

2. Three handshake lines (NRFD, DAV, and NDAC).

3. Five interface management lines (IFC, ATN, SRQ, REN, and EOI).

The eight *data lines* carry all multiline messages, including interface messages and device-dependent messages. In many instruments, device messages are based on the 7-bit ASCII character set.

The three handshake lines (NRFD, DAV, and NDAC) are used to effect the transfer of each type of device message data on the DIO (Data InOut) lines from a talker to all addressed listeners, or to effect the transfer of interface messages from the active controller to all interfaces on the GPIB. The three handshake lines thus provide a means to synchronously transfer data between instruments.

The NRFD (not ready for data) line is used to indicate the condition of readiness of all devices to accept data. All instruments drive NRFD "false (0)" as ATN is driven "true (1)." Those instruments capable of receiving interface messages using their acceptor handshake function may drive NRFD true momentarily while interface message bytes are being transmitted. Instruments addressed as listeners may drive NRFD true momentarily while receiving device messages when ATN is false.

The NRFD line is monitored by the controller source handshake function when ATN is true, and by the talker source handshake function when ATN is false. The NRFD line is false (0) when all listeners are ready for data, and true (1) when one or more listeners are not ready for data.

The DAV (data valid) line is used by the source handshake function to indicate the validity of data on the data lines. DAV is driven true (1) by the controller source handshake function when ATN is true, and by the talker source handshake function when ATN is false. The DAV line is monitored by all instruments if ATN is true, and by instruments addressed as listeners when ATN is false.

The NDAC (not data accepted) line is used to indicate acceptance of data by listeners. Listeners indicate acceptance of data by setting NDAC false (0).

When NDAC is true (1), one or more listeners have not accepted the data.

15.10.5 Bus Management

The five *bus-management lines* are IFC, ATN, SRQ, REN, and EOI. IFC and ATN are used by all instruments, while the remaining three may or may not be used by a particular instrument.

The ATN (attention) line is used by the active controller to indicate how data on the data lines are to be interpreted and which devices must respond to data. The ATN line must be monitored at all times by all instruments other than the active controller. When the ATN line is true, the active controller can send interface messages to instruments on the bus. Device-dependent messages can be sent by the active talker to active listeners when the ATN line is false.

The IFC (interface clear) line is used by the system controller to place the GPIB interface functions of all the instruments in a known, quiescent state. The IFC line is driven true (for at least 100 μs) only by the system controller, and must be monitored by all other instruments. IFC may be set true by the system controller at any time.

The REN (remote enable) line is used to enable remote control or remote operation of an instrument. The use of the remote function by an instrument is optional. The REN line is driven true only by the system controller and may be changed at any time. Instruments that use the REN line must monitor it at all times and return to local control whenever it becomes false.

The SRQ (service request) line is used by one or more instruments to asynchronously request service from the active controller. The SRQ line is sensed by the active controller.

The EOI (end or identify) line can be used by a talker to indicate the end of a data string. This is accomplished by setting EOI true at the same time the last byte is enabled onto the data lines. The EOI line also is used by the active controller to conduct a parallel poll. The active controller initiates a parallel poll of all instruments with parallel polling capability by setting ATN and EOI true simultaneously.

15.10.6 GPIB Physical Characteristics

Specially designed bus cables are used for interconnection of instruments. A maximum cable length of either 2 m times the number of instruments on the bus, or 20 m, whichever is less, is allowed. The cables connecting the instruments can be interconnected in a linear, star, or any combination of both linear and star.

Use caution if individual cable lengths exceed 4 m. At high data-transfer rates, exceeding either the 4-m distance between instruments or the 20-m total length could cause errors. Two 24-pin piggy-back connectors, one male and one female, are used on either end of the interlocking cables.

An overall shield is used to reduce susceptibility to noise. This shield is normally grounded only at the system controller. Figure 15.41 shows the pin connections for the GPIB cable.

15.10.7 GPIB11 Series Interfaces

A computer interface to the GPIB would not be completed without a comprehensive user-software package. GPIB11-series interfaces are provided with comprehensive software, including both utility and driver routines.

The *utility routine* provides the high-level interface to a user's programs and contains information about the instruments connected to the GPIB. The *driver routine* conducts the actual GPIB input–output functions.

A user program normally interfaces to the utility routine. This allows the user to call the utility-routine functions to perform all the common measurement system operations without having a detailed understanding of the GPIB protocol.

It is perfectly legal to intermix calls to the utility routine with calls to the driver routine. In the GPIB11-Series, the utility and driver routines are furnished as PDP-11 MACRO source files, which may be assembled as FORTRAN-, BASIC-, or MACRO-callable subroutines. The driver routine is designed to be used in a stand-alone environment or as a device handler under the RT-11 or the RSC-11 operating systems. The GPIB11-Series driver routine includes the following functions for managing GPIB operations:

1. READ. Input data from a selected device.
2. WRITE. Output data to one or more devices selected as listeners.
3. CLEAR. Initialize the bus.
4. REMOTE and LOCAL. Put the instruments on the bus in remote or local mode.
5. PARALLEL POLL. Conduct a parallel poll.
6. SERIAL POLL. Conduct a serial poll.
7. COMMAND. Address devices as talker or listeners, and send other bus-management multiline command messages.
8. PASS CONTROL. Transfer control to another device having controller capability.

9. MONITOR. Monitor command messages on the bus.

10. SET STATUS. Set individual status (for response to a parallel poll from another active controller).

The GPIB11-Series *utility routine* contains a device table that lists information about all devices on the GPIB in order to simplify common operations. The high-level functions provided by the utility routine include WRITE, READ, CLEAR, TRIGGER, REMOTE, LOCAL, POLL, CONFIGURE, and PASS CONTROL. The utility routine calls driver-routine functions to perform the requested operation.

The software provided with the hardware interface reduces the time required for the user to integrate GPIB-based measurement and control equipment into the computer system. With this software the user need only be concerned with the particular application functions, without having to become an expert on the GPIB.

15.10.8 Distributed Computerized GPIB System

In many institutions the computer system has been installed as an integrated facility located a distance from the instrumentation. The IEEE Std. 488-1978 specification requires that there be no more than 4 m between any two devices. This normally would be a severe restriction on the types of computerized measurement systems that could be implemented in a large facility with a central computer and remote or satellite measurement and control station.

15.10.9 Bus Extender

This restriction can be overcome by using a device called a *bus extender* (or repeater) to allow instruments on the GPIB to be located up to hundreds of meters away from the central computer, and also to increase the number of instruments connected to the bus. To be useful for general applications, the extender must be "transparent" (independent) to the user's software and hardware in all modes. For widest application it must be able to operate with one or more controllers on either or both ends of the extension. It must be able to operate at high speeds so that the overall system throughput is not substantially affected by its presence.

One such extender that provides this capability is the National Instruments GPIB-100 (Figure 15.42). This extender provides high-speed capability without changing either hardware or software. The GPIB-100 bus extender converts the IEEE 488-compatible bus signals to and from EIS Std. RS-422* levels for transmission over long distances (up to several

---

*A standard promulgated by the Electronic Industries Association that establishes voltage levels, drive, and speed capabilities versus distance, and so on.

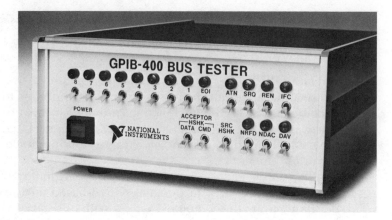

FIGURE 15.42
IEEE-488 bus extender GPIB-100. (Courtesy of National Instruments Corporation, Austin, Texas)

hundred meters). EIA Std. RS-422 provides for excellent high-speed performance, even in an electromagnetically noisy environment. The GPIB-100 extender is also designed so that it can provide relatively inexpensive access to several GPIB instrumentation systems from a single computer controller. Figure 15.43 shows an example of a GPIB-based extended measurement system using high-speed extenders.

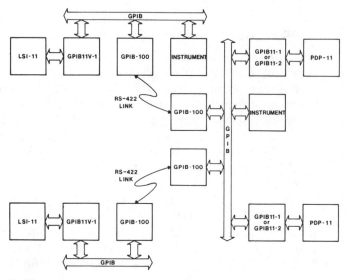

FIGURE 15.43
A distributed PDP-11 system. (Courtesy of National Instruments Corporation, Austin, Texas)

It is important to stress that the GPIB-100 was designed to mesh with the GPIB logic state diagrams in IEEE Std. 488-1978, so any device that works with the GPIB is guaranteed to work with the GPIB-100 extender with no modifications to hardware, and only one modification to software. The one software exception occurs when conducting a parallel poll; since it is physically impossible for a device to respond to a parallel poll request over an extended distance within the time limit allowed by the IEEE standard, two successive polls must be conducted by the active controller to obtain the latest poll results.

15.10.10 CAMAC: Computer Automated Measurement and Control Digital Interface Standards

The IEEE-488 interface uses a general-purpose interface bus that handles bytes (8 bits). The bus thus consists, essentially, of 8 signal lines, which are needed to handle 8 bits in parallel. These are referred to as DIO lines (data input–output), DIO 1–8.

When a computer is to be connected to peripheral devices using a common data bus, it is found that more lines are needed than are available in the IEEE-48 Interface System. Thus several other standards have been developed specifically for computer systems, called *Computer Automatic Measurement and Control (CAMAC) Instructions and Interface Standards:*

- ANSI/IEEE Std. 583-1975, ANSI approved May 4, 1977, "Modular Instrumentation and Digital Interface System (CAMAC)," $10

- ANSI/IEEE Std. 595-1976, ANSI approved Nov. 28, 1977, "Serial Highway Interface System (CAMAC)," $10

- ANSI/IEEE Std. 596-1976, ANSI approved Dec. 12, 1977, "Parallel Highway Interface System (CAMAC)," $6.50

- ANSI/IEEE Std. 683-1975, ANSI approved Nov. 9, 1977, "Block Transfers in CAMAC Systems," $4

Figure 15.44 shows in a general way how sensors, peripherals, controls, and the like, communicate with a computer via CAMAC modules (1, 2, . . .,

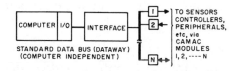

FIGURE 15.44
System with standard data bus (Dataway) and dedicated interface. (Courtesy of National Instruments Corporation, Austin, Texas)

N) through a common dataway and a single interface. The system has a 1-μs Dataway cycle time and a very high addressing capability. The Dataway is parallel, with 24 read lines and 24 write lines, so that words of 24 bits or less can be handled in a single data transfer. Typically, up to 23 stations can be addressed within a "crate," either singly or in any combination, with 16 subaddresses for each station together with 32 function codes. This high addressing capability is further multiplied by 7 in the case of the standard 7-crate parallel highway system, and by 62 in the case of the standard serial highway system.

- Read lines (buses)   24
- Write lines (buses)   24
- Station address   24
- Station demand   24
- Functions   32

Note that CAMAC uses Dataway to describe its bus; IEEE-488 uses Interface Bus or GPIB. Unlike IEEE-488, CAMAC also describes a standard chassis or "crate."

15.11
REVIEW QUESTIONS

1. Discuss the meaning of a microprocessor.
2. Discuss the meaning of VLSI.
3. Discuss the meaning of a coprocessor.
4. Discuss the meaning of a RAM and ROM.
5. Discuss the meaning of a binary digit.
6. Discuss the meaning of an octal number.
7. Discuss the meaning of a hexadecimal number.
8. Discuss the arithmetic and logical functions of a microprocessor.
9. Discuss binary arithmetic, subtraction, multiplication, and division.
10. Discuss the logical operation of a microprocessor.
11. Draw an OR gate circuit and discuss it.
12. Discuss mnemonics.
13. Discuss the Intel 8-, 16-, and 32-bit microprocessors.
14. Define a bit, byte, nibble, and word.
15. Discuss the meaning of PROM.
16. Draw a block diagram of a 32-bit microprocessor.
17. Discuss an application of the microprocessor.

18. Draw a block diagram showing a transducer and the data acquisition and conversion system including a microprocessor.

19. Discuss emulation for microprocessors.

20. Discuss microprocessor-based testing problems.

21. Discuss the IEEE-488 bus interface system.

22. Discuss the meaning of handshake, listener, talk, and controller.

23. Discuss the GPIB signal lines.

24. Discuss the meaning of EPROM and CMOS.

25. Discuss the computer automated measurement and control digital interface standards.

15.12 REFERENCES

1. M. E. Sloan, *Introduction to Minicomputers and Microcomputers*, Addison-Wesley Publishing Co., Reading, Mass., 1980.

2. Rodnay Zaks, *From Chips to Systems: An Introduction to Microprocessors*, Sybex, Berkeley, Calif., 1982.

3. Application Note 222-2 Application Articles on Signature Analysis, Hewlett-Packard, Palo Alto, Calif., Oct. 1980

4. Peter R. Rony, *The 8080A Bugbook*(R): *Microcomputer Interfacing and Programming*, 4th printing, Howard W. Sams and Co., Indianapolis, 1981.

5. Christopher A. Titus, David G. Larsen, and Jonathan A. Titus, *8084A Cookbook*, 2nd printing, Howard W. Sams and Co., Indianapolis, 1981.

6. *Microprocessor Applications*, Special Issue, Proceedings of the IEEE, February 1978.

7. Jim Jarrett, "History of the Microprocessor," *Medical Electronics*, 97–101, Feb. 1982.

8. Eric J. Johnson, "Data Acquisition System on a Chip," *Medical Electronics*, Feb. 1982.

9. "Microprocessor Courses," Nos. 1 through 12, *Medical Electronics*, 2994 West Liberty Ave., Pittsburgh, Pa.

10. *MCS-80/85TM Family User's Manual*, Intel Corp., Santa Clara, Calif., Oct. 1979.

11. D. P. Burton, and A. L. Dexter, *Microprocessor Systems Handbook*, Analog Devices, Inc., Norwood, Mass., 1977.

12. *GPIB-400, Bus Tester Operating and Service Manual*, National Instruments, Austin, Texas, 1981.

13. *Microprocessors and Memories,* Digital Equipment Corp., Maynard, Mass. 1982.

14. "Special Report, IEEE-488 Instruments," *EDN,* 77–91, Oct. 28, 1981.

15. Alfred F. Shackll, "Microprocessors," *IEEE Spectrum,* 32–33, Jan. 1982.

16. *Microcomputer Interfaces Handbook,* Digital Equipment Corp., Maynard, Mass., 1980.

17. *Intel 432 System Summary: Manager's Perspective,* Intel Corp., Santa Clara, Calif., 1981.

18. Andy Santoni, "Instruments," *EDN,* July 28, 1981.

19. *HMOS Technology Expands 8 Bit Design Possibilities,* Intel Corp., Santa Clara Calif., 1980.

20. Jim Jarret, "History of Microprocessors," Intel Corp., Santa Clara, Calif., 1982. Appears in *Medical Electronics,* Issue 73, Feb. 1982.

21. *Microprocessors,* Electronic Book Series, McGraw-Hill Book Co., New York, 1975.

22. *16-Bit Microprocessor User's Manual,* 3rd ed., Prentice-Hall, Inc., Englewood Cliffs, N.J.; (c) 1982 by Motorola Inc.

23. *Mostek 1981 Z80 Microcomputer Data Book,* Mostek Corp., Carrolton, Texas, 1981.

24. *ISBC Applications Handbook,* Intel Corp., Santa Clara, Calif., Sept. 1981.

16

Specialized
Bioelectronic
Instrumentation

16.1
INTRODUCTION

The cell is the basic source of all bioelectric potentials. A bioelectric potential may be defined as the difference in potential (electric charge) between the inside and outside of a cell. A cell (Figure 16.1) consists of an ionic conductor separated from the outside environment by a semipermeable or selectively permeable cell membrane. The internal resting potential within a cell is approximately −90 mV with reference to the outside of the cell. This potential changes to approximately +20 mV for a short period of time during cell activity. Cell activity results from some form of stimulation.

An ionic current is produced by ion movement through a semipermeable membrane as shown in Figure 16.2.

16.2
ELECTRICAL POTENTIALS GENERATED
WITHIN THE HEART (GENERATION OF
THE ELECTROCARDIOGRAM WAVEFORM)

The right atrium contains a bundle of nerves known as the sinoatrial node (SA node). This type of nerve cell is found nowhere else in the body. Its function is to start the heartbeat and set its rhythm or pace. Impulses gener-

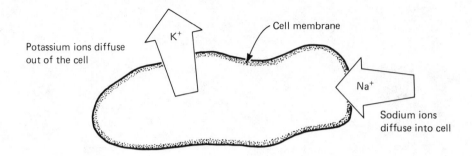

FIGURE 16.1
The bioelectric cell.

ated by the SA node stimulate contraction of the muscles comprising the atria. These impulses also travel along conducting fibers in the atrium to the atrioventricular node or AV node, stimulating depolarization of this node. Stimulation of the atrioventricular node causes impulses to be sent to the myocardium (muscle) comprising the ventricules via the bundle of His and the Purkinje conducting system, resulting in contraction of this muscle. The muscular contractions necessary to maintain the heart's pumping action are initiated by depolarization and repolarization of the SA node, and then the depolarization and repolarization of the AV node.

These depolarizations and repolarizations generate external action potentials that can be recorded at the surface of the body. The record of these action potentials generated from within the heart is known as the electrocardiogram or ECG. (EKG is derived from the German spelling of electrokardiogram.) The ECG and heart action are shown in Figure 16.3.

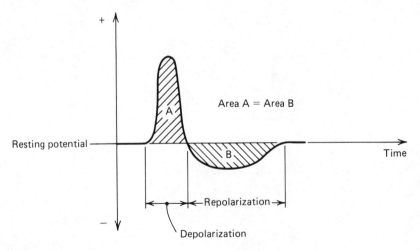

FIGURE 16.2
The action potential.

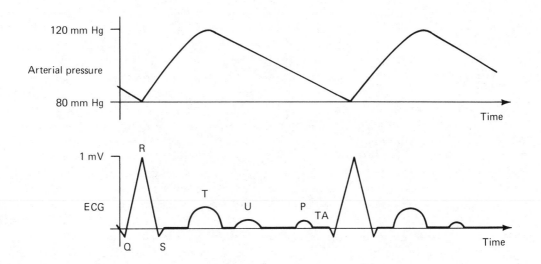

P wave, atrial depolarization
TA wave, atrial repolarization
PR interval, atrioventricular conduction time
QRS segment, ventricular depolarization
T wave, ventricular repolarization
V wave, unknown

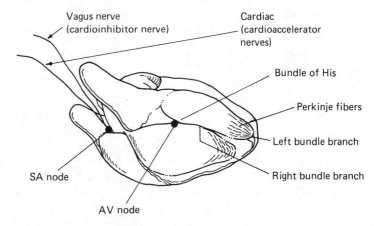

FIGURE 16.3
The ECG and its relationship to heart action.

In another 25 years, every hospital will have a digital imaging department, including cardiology (heart disorders). The computerized body scanner will incorporate all body modalities and examine function as well as structures of the body. A complete body scan will be accomplished with the patient taking a single breath. Basically, a complete body scanner permits a functional and structure analysis of the body, including the heart. The scanning beam rotates around the patient and transmits energy to an array of detectors. The electronic instrument's computer then uses image-construction techniques to build a digital image based on the attenuation of the energy beam as it passes through the patient. It reconstructs entire cross sections made at equally spaced intervals along the patient's length. The energy beam's attenuation coefficients provide information about the density of soft tissue and bone. Today, computer axial tomography (CAT) provides a clear picture of the part of the body but uses radiation. The future uses of bioelectronic instrumentation using nuclear magnetic resolution, positive emission tomography using isotopes, and digital subtraction angiography have not been fully realized but certainly will be in everyday use in another 25 years.

16.3
MEASURING BIOELECTRIC POTENTIALS

Bioelectric potentials are usually less than $1 \, \mathrm{mV} \, (1 \times 10^{-3} \, \mathrm{V})$. This potential is much too small to operate electrical and mechanical devices such as a recorder. The electrical potentials must be amplified (see Figure 16.4). Sixty hertz (power source potentials) interferes with the biopotentials being measured and is eliminated using a 60-Hz filter. The signals are then visualized by viewing on an oscilloscope or recorded by either a tape recorder or strip-chart recorder. A rate meter may also be connected to display the beats-per-minute rate.

16.4
DEFIBRILLATORS

The rapid spread of action potentials over the surface of the atria causes these two chambers of the heart to contract together and pump blood through the two atrioventricular valves into the ventricles. After a critical time delay, the powerful ventricular muscles are synchronously activated to pump blood through the pulmonary and systemic circulatory systems. A condition in which this necessary synchronism is lost is known as *fibrillation*. During fibrillation the normal rhythmic contractions of either the atria or ventricles are replaced by rapid irregular twitching of the muscular wall. The terms are called, respectively, *atrial fibrillation* and *ventricular fibrillation*. Under conditions of atrial fibrillation, the ventricles can still func-

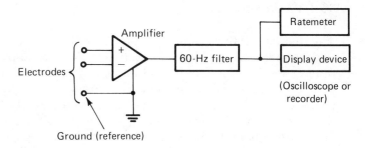

FIGURE 16.4
Bioelectric potential measuring system.

tion normally, but they respond with an irregular rhythm to the nonsynchronized bombardment of electrical stimulation from the fibrillating atria. Since most of the blood flow into the ventricles occurs before atrial contraction, there is still blood for the ventricles to pump. Ventricular fibrillation is far more dangerous, for under this condition the ventricles are unable to pump blood, and if the fibrillation is not corrected, death will usually occur within a few minutes.

Although mechanical methods (heart massage) for defibrillating patients have been tried over the years, the most successful method of defibrillation is the application of an electric shock to the area of the heart. Since the heart muscle fibers respond to electrical excitation, if sufficient current to contract all musculature of the heart simultaneously is applied for a brief period and then released, all heart muscle fibers enter their refractory periods together, after which normal heart action may resume. Best results have been found using direct-current defibrillators. Figure 16.5 shows a defibrillator discharge waveform. In this method a capacitor (a device used to store electrical energy) is charged to a high dc voltage and the capacitor is discharged within a few milliseconds.

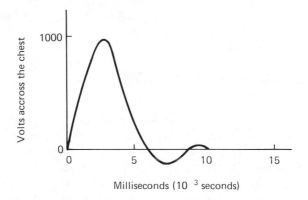

Milliseconds (10^{3} seconds)

FIGURE 16.5
Defibrillator discharge waveform.

16.4.1 Pacemakers

The rhythmic activity of the heart is controlled by a neuronal "pacemaker" as described earlier (SA node). Although the natural pacemaker is capable of self-pacing and independent timing, it is normally controlled by both the parasympathetic and sympathetic nervous systems. When either the natural pacemaker or the controlling innervation to the heart becomes impaired to the point where the heart no longer provides sufficient circulation of the blood throughout the body, an artificial method of pacing the heart may be required. In such cases, periodic stimulation of the appropriate region of the myocardium substitutes for the natural pacing signals and triggers the heart at a rate adequate to maintain proper circulation. Artificial pacemakers come in a variety of forms. They can either be internally implanted for use by patients with permanent heart blocks who may require assisted pacing for the rest of their lives, or they may be worn externally for temporary requirements. Internal pacemakers are surgically implanted beneath the skin, usually in the region of the abdomen. Internal leads (wires) connect electrodes that are inserted directly into the myocardium. Since there are no external connections for applying power, the unit must be completely self-contained with a power source capable of continuously operating the pacemaker for a period of years. External pacemakers may either have electrodes attached directly to the myocardium with the connecting wires penetrating the skin, or the electrodes may be introduced into one of the chambers of the heart through a cardiac catheter called the pacing catheter.

The pacemaker can be worn by an ambulatory patient, or it can be attached to the bed or arm of a patient confined to bed. Most external pacemakers in use today are small, portable, and battery operated. In some cases of acute heart block, when there is insufficient time to insert a catheter or otherwise place electrodes on or in the heart, pacing can be accomplished through the intact chest. In this method, also called external pacing, high intensity pulses of current are applied through large electrodes placed on the chest. This form of pacing is rarely used today because of the discomfort and possible burns caused by the large current levels required.

16.5
BLOOD-PRESSURE TRANSDUCERS

The most common transducers in use today are the liquid-column strain-gage transducer. (Refer to Chapter 2 for a detailed discussion of strain-gage sensors.) The electrical connection to the strain-gage bridge is obtained from the cable at the head of the transducer. Pressure connections are made at the top through Luer fittings in the transparent dome, which offer connection to the catheter system and a means of flushing. The easily removable transparent dome is used so that air bubbles can be seen and eliminated. This form of blood-pressure sensor is available in many types. There are general-purpose models for arterial pressure (0 to 300 mm Hg) and venous pressure (0 to 50 mm Hg), and so on. These transducers must be flushed to remove air bubbles and, also during measurements, to prevent blood from clotting at the end of the catheter. Pressure transducers are normally mounted on the bed frame or on a stand near the patient's bed. It is important to keep the transducer at about the same height as the point at which the measurements are made in order to avoid errors due to hydrostatic pressure. The signal-conditioning and display instruments for these transducers come in a variety of forms. Each basically consists of a method of electrical excitation for the strain gage bridge, a way of zeroing or balancing the bridge, and a display device, such as an oscilloscope, recorder, meter, or digital readout.

Disposable hemodynamic blood pressure transducers developed with a silicon chip are used in hospital critical care units and surgery to prevent patient contamination and reduce cost.

One problem in blood-pressure measurements is the tendency for clot formation in the arterial catheter. Manual periodic flushing of the catheter with saline solution will prevent clotting. This technique of flushing invalidates the pressure measurements. Also, the size and weight of the pressure gage often demand bulky mounting arrangements in an already crowded bedside area. Proper elevation of these gages is essential; otherwise the difference in fluid column head between the gage and patient may cause gross errors.

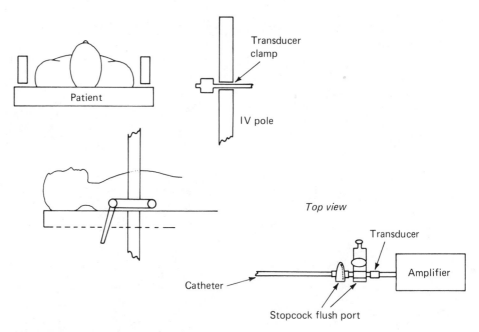

FIGURE 16.6
A pressure transducer mounting scheme.

New, small, lightweight strain-gage pressure gages now offer some solutions to these problems. These gages are simply taped to the patient, or other techniques are used as shown in Figures 16.6, 16.7, and 16.8.

The method of connecting the transducer into the system depends on the size of the transducer as well as the output leads. Each manufacturer may provide its own technique as to how to couple to the transducer via the patient's arm. Each manufacturer can adapt tubing to its pressure transducer. With needle tube of less than $\frac{1}{4}$ in., some manufacturers can fit catheter from patient to transducer. Flushing may be required every 5 min.

Continuous flushing through the rear of the transducer without interrupting the pressure measurement is also feasible. As an example, a micropose in the pressure transducer platform of the Statham/Gould P37 type of Oxnard, California, permits continuous intermittent low-volume flushing without degradation of dynamic signal. This ensures a clot-free system without the need for manual flushing or for anticoagulants.

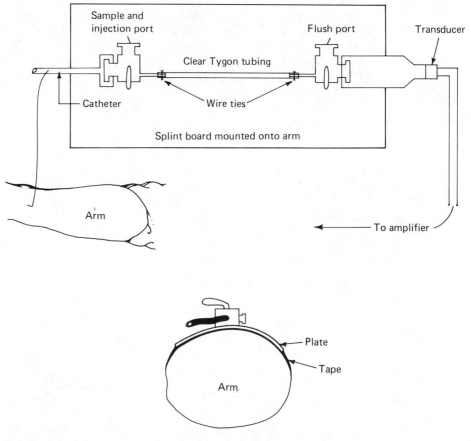

FIGURE 16.7
A second pressure transducer mounting scheme.

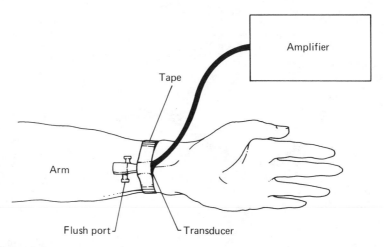

FIGURE 16.8
A third pressure transducer mounting scheme.

Instruments that process pressure data are available for deriving systolic, diastolic, and mean pressures from the arterial pressure waveform. Panel meters or numerical readouts usually provide display in a modular approach. Motion artifacts can be caused by the movement of the patient.

A cuff can be applied to the upper arm and a microphone can be fastened instead of a stethoscope for the detection of the Korotkoff sounds. This microphone features a built-in amplifier with emitter follower so that interference caused by cable movement is eliminated with a considerably higher reliability.

16.6
SWAN–GANZ CATHETER

The Swan–Ganz thermodilution catheter (Figure 16.9) serves as a rapid method of obtaining important, sometimes vital diagnosis information at bedside, without fluoroscopy. A signal catheter insertion provides means for measuring pulmonary artery and pulmonary capillary wedge pressures, right atrial pressure, and sampling from either the right atrium or pulmonary artery, as well as injection of cold solution and detection of temperature change for determination of cardiac output via a computer with a microprocessor in surgery and intensive-care areas.

One large lumen terminates at the tip of the catheter. It is through this lumen that pressure tracings are monitored indicating pulmonary artery pressure or pulmonary capillary wedge pressure. Another lumen terminates 30 cm from the tip, positioning the injection orifice in the right atrium or superior vena cava when the tip is in the pulmonary artery. Central venous pressure may be measured through this lumen as an alternative to injecting

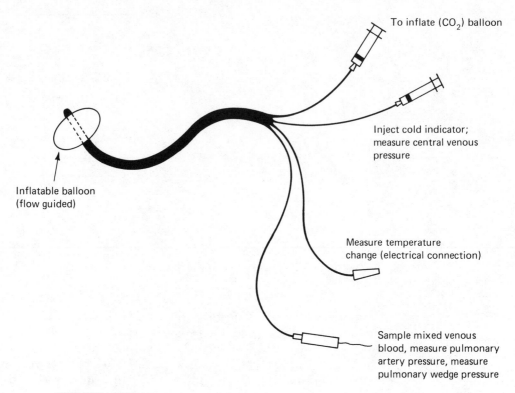

To inflate (CO_2) balloon

Inject cold indicator;
measure central venous
pressure

Inflatable balloon
(flow guided)

Measure temperature
change (electrical connection)

Sample mixed venous
blood, measure pulmonary
artery pressure, measure
pulmonary wedge pressure

FIGURE 16.9
Swan–Ganz catheter.

cold solution. One small lumen serves to inflate or deflate the balloon. The remaining small lumen carries the electrical leads for the thermistor temperature detector, which is 4 cm from the tip on the surface of the catheter.

16.7
ELECTRODE CONTACT

No matter which electrode is used (see Figure 16.10) or for whatever the application, good electrode contact is vital for a good waveform. When electrode contact is poor, 60-Hz artifact, drift, distortion, noise, and poor accuracy will result. Such disturbances may not appear immediately, but will affect the signal as the electrodes remain in place. Not knowing the quality of contact may result in incorrectly attributing the distortion and disturbances to machine malfunction.

Cleansing the body oils and dead tissue (contact layer) are essential for good electrode contact (see Figure 16.11). Several preferred methods of skin preparation are abrasion, alcohol rub, special detergent solutions, and saline wash. The amount of skin preparation will vary widely from one

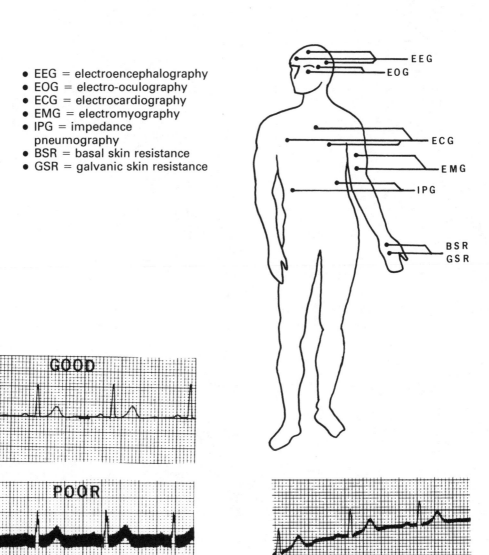

- EEG = electroencephalography
- EOG = electro-oculography
- ECG = electrocardiography
- EMG = electromyography
- IPG = impedance pneumography
- BSR = basal skin resistance
- GSR = galvanic skin resistance

EEG

EOG

ECG

EMG

IPG

BSR
GSR

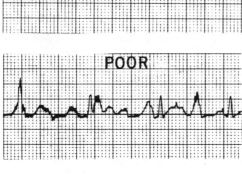

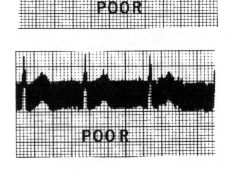

ECG waveforms

FIGURE 16.10
Physiological monitoring electrodes.

TABLE 16.1
Definition of recording terms

Electroencephalography:	Recording of electric currents developed in the brain via scalp electrodes.
Electro-oculography:	Detection of a standing potential between the front and back of the eyeball via electrodes placed on the skin surface near the eye socket.
Electrocardiography:	Recording of electric currents from heart muscles.
Electromyography:	Recording of changes in the electrical potential of muscles.

TABLE 16.2
Properties of important physiological parameters

Physiological Signal	Approximate Amplitude Range	Approximate Frequency Range	Transducer Sensing Device
Electrocardiogram (ECG)	3/4 to 4 mV peak to peak	0.1 to 500 Hz	Electrodes
Phonocardiogram		20 to 2000 Hz	Piezoelectric or magnetic microphone
Blood pressure Venous pressure Arterial pressure	0 to 300 mm Hg 0 to 40 mm Hg 0 to 300 mm Hg	0.5 to 500 Hz	Strain gage
Blood flow	1 to 300 cm/s.	1 to 50 Hz	Electromagnetic or ultrasonic flowmeter
Temperature	90-110°F	0 to 0.1 Hz	Thermistor or thermal expansion device
Respiration rate		0.15 to 6 Hz	Electrode impedance or piezoelectric devices or pneumograph
Tidal volume	10 to 1000 mL per breath	0 to 1 Hz/min	Impedance pneumograph
Stomach pH	1 to 40 g on 0.2 × 0.8 cm strain gage	0 to 1 Hz	Glass electrode or antimony electrodes
Electromyogram (EMG)	0.1 to 4 mV peak to peak	10 to 500 Hz (clinical use)	Electrodes

(Courtesy of M. Clynes and J. H. Milsum, eds., *Biomedical Engineering Systems,* McGraw-Hill Book Co., New York, 1970. Copyright 1970 and used with permission of McGraw-Hill Book Co.)

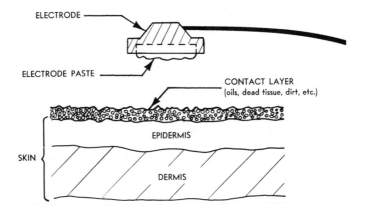

FIGURE 16.11
Electrode, electrode paste, and skin preparation.

patient to another. One patient's skin may be much drier than another's or have more oils on the surface. Excessive preparation will usually cause skin irritation and patient discomfort.

Table 16.1 gives definitions of some of the recording terms shown in Figure 16.10. Table 16.2 gives the properties of physiological parameters, including approximate amplitude range, approximate frequency range, and transducer sensing device.

16.8
ELECTRICAL SAFETY OF MEDICAL EQUIPMENT

For electricity to have any effect on the body, the body must become part of an electric circuit. At least two connections must exist between the body and an external source of voltage to effect a current flow. The strength of the current flow depends on the voltage between the connections and on the electrical resistance of the body. Most body tissue contains a high percentage of water and salt. Therefore, it is a fairly good electrical conductor. The electrical current can affect the tissue in two different ways. First, the electrical energy dissipated in the tissue resistance can cause a temperature increase. If a high enough temperature is reached, tissue damage (burns) will occur. Second, the transmission of impulses through sensory and motor nerves involves electrochemical action potentials. An extraneous electric current of sufficient strength can cause local voltages that can trigger action potentials and stimulate nerves. A high enough intensity of the stimulation can cause tetanus of the muscle, in which all possible fibers are contracted, and the maximal possible muscle force is exerted. An electric current flowing through the body in sufficient magnitude will interfere with the func-

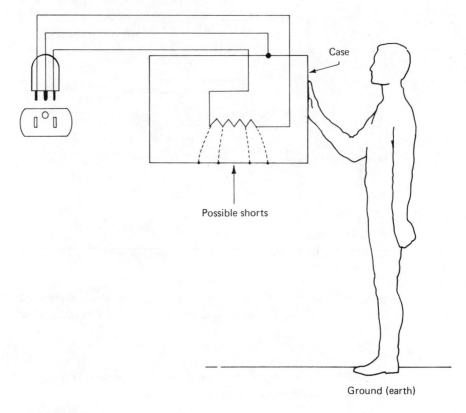

FIGURE 16.12
Illustration of possible shorts caused by bioelectronic instruments.

tioning of organs. The organ most susceptible to electric current is the heart.
A tetanizing stimulation of the heart results in complete myocardio contraction, which stops the pumping action of the heart and interrupts the blood
circulation. If circulation is not resumed within a few minutes, first brain
damage, then death results, due to a lack of oxygen supplied to the brain's
tissue. If the intensity is low, it may excite only a portion of the heart's

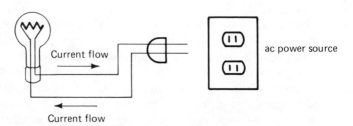

FIGURE 16.13
Complete circuit current flow for a light bulb.

muscle fiber. The partial excitation can change the electrical propagation patterns in the myocardium, desynchronizing the activity of the heart and causing fibrillation. Respiratory paralysis can also occur if the muscles of the thorax are tetanized by an electric current flowing through the chest or through the respiratory control center of the brain.

In all modern installations, each wall receptacle has a third contact called the equipment ground. This contact is connected separately to the steel conduit that protects the other conductors or through a separate ground conductor. The purpose of the equipment ground on the wall receptacle is to reduce the danger of exposure to contact between one of the *hot* conductors and *neutral*. If the equipment case is connected to the equipment ground (Figure 16.12), any accidental contact between the hot connector and case can return the current to ground.

FIGURE 16.14
One of the two computer displays in a typical ACS arrhythmia central is devoted to status display for all monitored patients. Values for all vital signs monitored, which could include heart rate, blood pressure, temperature, and respiration, are indicated for each patient. Arrhythmias, or abnormal heartbeats, are identified by type for each patient, and the rate per minute indicated for each arrhythmia. Keyboard enables operator-to-system communications. (Courtesy of Abbott Medical Electronics Co., Houston, Texas)

Without the case grounded as shown in Figure 16.12, any possible short would cause an electrical connection from the case through the body to ground, completing the electrical current flow. See Figure 16.13 for a complete circuit.

Critical-care areas also require special attention to electrical safety (see Figure 16.14). Because of the multitude of electrical and mechanical instruments, the condition and location of the patient and the deluge of concerns of which the staff must be cognizant, some critical-care units have isolated power systems much like that in the operating room. Other important hazard areas are patient catheters, cords, electrically controlled beds, wire grounding, and electrical isolation of patients.

All critical-care areas are equipped with special grounding connectors for the attachment of auxiliary ground wires. This system allows for the connection of a green grounding wire, which is attached to the chassis of all portable instruments and large stationary metal objects (beds, tables, etc.). The ground plugs should be connected to the special grounding system to allow accidental electrical currents to flow harmlessly out of the room.

Patient-owned equipment such as TVs and radios should never be allowed in critical-care units. The risk of accident is too great to allow an untested device into the patient vicinity. Loose or dry electrodes or defective patient cable should be avoided, since the ECG waveforms are difficult or impossible to interpret.

16.9
CARDIAC-CARE SYSTEMS

Cardiac-care units (CCUs) permit close monitoring of critically ill patients. Examples include those who have just undergone operations, victims of heart attacks, and survivors of bad car accidents. In these cases, the systems provide early signs of impending danger, act as sensitive monitors of the efficacy of therapeutic procedures, relieve nurses of routine observation and data-collection tasks, and possibly reduce costs while increasing the quality of care.

To perform all these tasks, the system requires sensors, processors, alarms, and displays. The sensors measure vital physiological variables like heart performance, blood pressure, respiration, and body temperature. Instruments then either display these directly or process them further to derive other essential variables. If a signal falls above or below safe limits, an alarm pinpoints the trouble and calls for attention.

Monitoring as used in cardiac-care systems categorizes all continuously applied measurement or surveillance systems of those physiological functions of a subject that are required to safeguard life. Such systems are characterized by minimum demands on the equipment operator's attention and alarm provisions for out-of-range physiological variables.

Computers can also affect activities in the CCU. Such automated patient monitoring should relieve nurses of some routine functions such as charting and checking vital signs and should increase the time available to render direct patient care.

In the CCU, reliable disposable electrodes for ECG evaluation of vital signs are required.

In the CCU, the following electronic instruments appear at each bed:

1. Monitor of electrocardiogram

2. Heart rate monitor

3. Pulse rate monitor

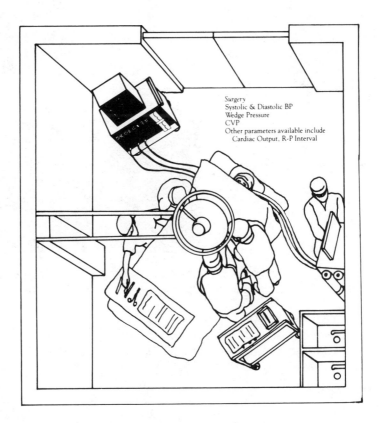

FIGURE 16.15
Surgery: monitor multiple pressures with preamps in same Abbott cabinet. Systolic and diastolic, arterial wedge, and CVP (or intracranial) pressures with digital displays, nonfade wave-forms and readout via chart recorder. (Courtesy of Abbott Medical Electronics, Houston, Texas)

Monitoring multiple pressures in Surgery
- *Up to three channels of pressure from modules housed in same monitoring cabinet, plus EKG and heart rate.*
- *EKG and arterial, wedge and central venous (or intracranial) pressures displayed simultaneously on non-fade cardioscope.*

Abbott monitors are especially versatile in a Surgery where critical clinical procedures are performed which require monitoring multiple pressures. Up to three blood pressures can be monitored, and values displayed via easy-to-read digits, by preamps contained in same Abbott cabinet. No special cables or factory alterations are necessary.

At the same time, EKG and three pressure waveforms can be displayed on Abbott's associated four-channel Non-Fade cardioscope.

Monitoring instruments

	Model MT-40	Non-Fade Cardioscope, four-channel.
	Model EK-4	EKG Preamp with full lead selection.
	Model HR-5	Heart Rate Computer with digital display.
(2)	Model PR-4	Arterial Pressure Computer/Display.
	Model PR-3	Central Venous Pressure Computer/Display.
	Model CR-10	Chart Recorder, single channel.
	Model TE-4	Temperature monitor, °C in digits.
	Model FA-11	Cabinet with CO-3 Elapsed Timer/Control.

Alternate Instruments

Model MS-4	Cardioscope, four-channel, conventional.
Model EK-5	EKG Preamp with Leads I and II.
Model HR-4	Heart Rate Computer/Display.
Model RP-7	R-P Interval monitor with digital display.

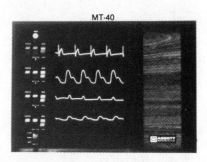

MT-40

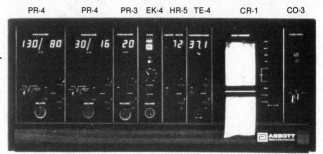

MULTIPLE PRESSURE MONITORING

EIGHT PARAMETERS INCLUDING THREE PRESSURES IN ONE CABINET

FIGURE 16.16
Monitoring multiple pressures in surgery.
(Courtesy of Abbott Medical Electronics, Houston, Texas)

4. Readouts and alarm systems

5. Temperature monitor

6. Central venous pressure monitor

7. Dysrhythmia monitor to handle ectopic beat problems.

It should be noted that many of these monitors are combined in one module.

A patient's progress in the progressive coronary care unit environment can be monitored through recovery to the outpatient status by use of an ambulatory monitor, which includes (1) a telemetry system, and (2) automatic recording of heart rate. All physiological instrument systems are fed to a central control station.

The prime purpose of acute coronary care is to prevent dysrhythmias, particularly life-threatening dysrhythmias, to treat arrhythmias if they occur, and to treat cardiac failure.

Figures 16.15 through 16.18 show bioelectronic instruments used in

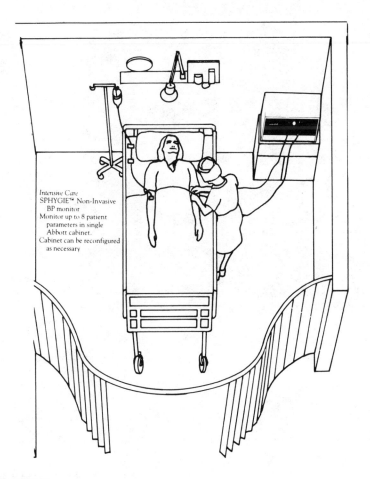

Intensive Care
SPHYGIE™ Non-Invasive
BP monitor
Monitor up to 8 patient
parameters in single
Abbott cabinet.
Cabinet can be reconfigured
as necessary

FIGURE 16.17
Intensive care: With Abbott system up to eight patient parameters can be monitored in single cabinet, with space for chart recorder. Sphygie$^{(TM)}$ Non-Invasive Blood Pressure Monitor is especially useful in intensive-care situations. (Courtesy of Abbott Medical Electronics, Houston, Texas)

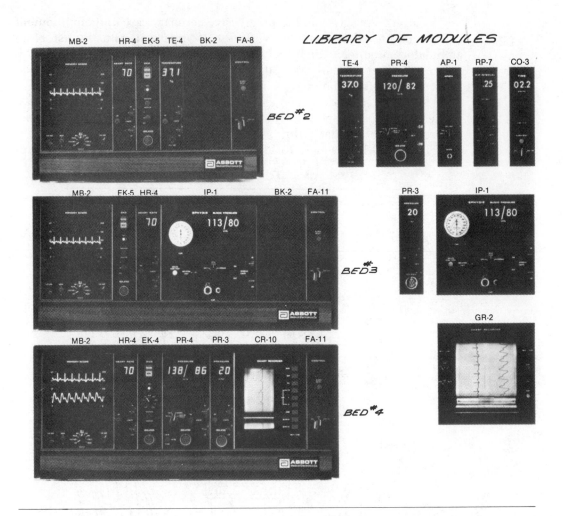

LIBRARY OF MODULES

BED #2

BED #3

BED #4

Help in planning computerized monitoring system

The new Computerized Patient Monitoring System developed by Abbott Medical Electronics features an arrhythmia detection and alarm system; patient trend information; and coordination of medications.

The computer system is easy to operate, and a self-teaching aspect simplifies nurses training.

Unique feature is the use of dual computer terminal displays. This allows a continuous overview of all patients on one screen while calling for trend data or detailed information for a specific patient on the other.

Computer console design ideas available from Abbott include the angled, 16-bed center shown at left.

FIGURE 16.18
Bedside system used in a flexible intensive care unit (ICU). (Courtesy of Abbott Medical Electronics, Houston, Texas)

A versatile system for a flexible ICU
- *EKG and blood pressure waveform displays*
- *Digital displays of heart rate, systolic and diastolic blood pressure, venous pressure, temperature and R-P interval.*
- *Alarm protection for each parameter.*

System 3 is a suggested Intensive Care capability which indicates the considerable flexibility possible through Abbott's modular instrumentation.

In general, each of the four monitors shown in our proposed ICU system is equipped with a Non-Fade dual trace cardioscope, an EKG Preamp and either digital Heart Rate/Alarm or BP/Alarm monitor.

A library of mobile modules for monitoring other parameters can be inventoried at the Central station and issued (or moved from one bed to another) as the patient census and needs change. These mobile modules include temperature, R-P interval, apnea and blood pressure monitors.

NOTE: The sophisticated electronics of the Abbott system permit most modules to be inserted into any open space in a cabinet. All interconnections are predesigned into the cabinet.

The bedside monitors shown in this suggested ICU system can be linked to slave central station display/alarm and recording instruments as shown in System 1 and 2 examples.

Bedside #1 Instruments

Model MB-2	Dual-trace 5-inch Non-Fade Scope.
Model HR-4	Heart Rate/Alarm with digital display.
Model EK-4	EKG Preamp with full 12-lead selection.
Model PR-4	Blood Pressure Preamp/Display.
Model TE-4	Temperature/Alarm with digital display.
Model FA-8	Cabinet & Control with alarm reset.

Bedside #2 Instruments

Model MB-2	Dual-trace 5-inch Non-Fade Scope
Model HR-4	Heart/Alarm with digital display.
Model EK-5	EKG with leads 1 and 2.
Model TE-4	Temperature/Alarm with digital display.
Model FA-8	Cabinet & Control with alarm reset.

Bedside #3 Instruments

Model MB-2	Dual-trace 5-inch Non-Fade Scope.
Model EK-5	EKG Preamp with leads 1 and 2.
Model HR-4	Heart Rate/Alarm with digital display.
Model IP-1	Indirect Pressure Monitor (non-invasive).
Model FA-11	Cabinet & control with alarm reset.

Bedside #4 Instruments

Mqdel MB-2	Dual-trace 5-inch Non-Fade Scope.

Model HR-4	Heart Rate/Alarm with digital display.
Model EK-4	EKG Preamp with full 12-lead selection.
Model PR-4	Blood Pressure Preamp/Display.
Model PR-3	Venous Pressure Preamp/Display.
Model CR-10	Graphic Recorder.
Model FA-11	Cabinet with Model CO-3 Elapsed/Timer.

Library of Modules

Model TE-4	Temperature/Alarm with digital display.
Model PR-4	Blood Pressure Preamp/Display.
Model PR-3	Venous Pressure Preamp/Display.
Model RP-7	R-P Interval Monitor with digital display and alarm.
Model IP-1	Indirect Pressure Monitor
Model AP-1	Apnea Monitor/Alarm.
Model GR-2	Graphic Recorder, 2-channel.
Model CO-3	Elapsed Timer/Control Module.

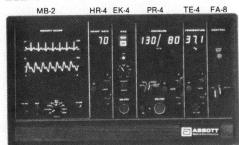

Eight-bed central station for Abbott monitoring equipment is just over 8 feet long and about 46 inches high.

FIGURE 16.18 (Continued)

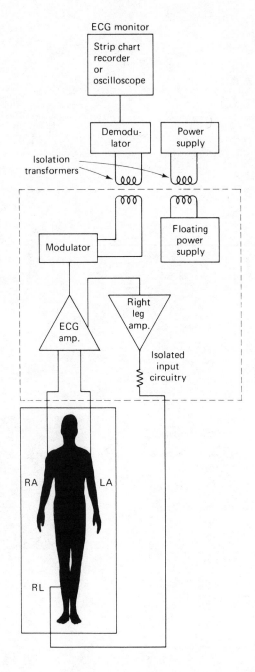

FIGURE 16.19
Isolated input amplifier. (From Harry E. Thomas,
*Handbook of Biomedical Instrumentation and
Measurement,* Reston Publishing Co., Reston, Va.,
1974)

surgery and the intensive-care area. Note all physiological measurements or vital signs must be documented. The ECG presently used has to have an isolated input system for safety protection.

The Abbott monitoring systems shown in Figures 16.15 through 16.18 make use of the following features:

1. Microprocessor technology permitting modules

2. Integrated circuit design

3. Modular design resulting in interchange of bedside modules when they become defective.

4. Digital displays of heart rate, blood pressure, temperature, and R-R interval.

5. Simultaneous bedside and central station displays.

6. Test button for most modules enabling isolation of a service problem and substitution of another module.

7. Patient monitoring instruments that ensure patient safety.

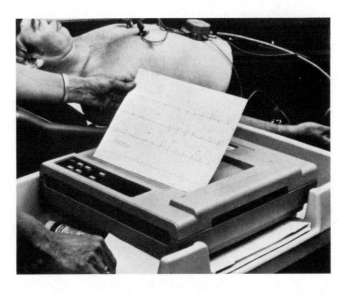

FIGURE 16.20
The HP Model 4700A PageWriter Cardiograph is a state-of-the-art machine that records ECG data in a page-sized format with high trace quality. The page format makes it easier to interpret ECG data and file it for future reference. (Courtesy of the Hewlett-Packard Co., Palo Alto, Calif.)

Furthermore, with the increase of in-dwelling electrodes used in coronary-care units for venous or arterial pressure by direct catheterization, added isolation is attained in much of the newer equipment by transformer coupling of the ECG signals and the power supply for any sensing apparatus. Figure 16.19 shows how the entire input circuitry is isolated. Like instrument amplifiers, amplifiers of the isolated unity gain amplified types (Figure 16.11) have internal feedback. All models of Analog Devices, Norwood, Massachusetts, operate from dc to 2 kHz. They are designed in two parts; the front end isolated amplifier section includes the fixed gain operational amplifier, a demodulator, and a dc regulator circuit, all enclosed in a floating guard-shield. The output section contains a demodulator, filter, and power supply oscillator operating from a 15-V dc supply. Operating power is transformer coupled into the shield input circuits and capacitively or magnetically coupled to the output demodulator circuit. Other isolation techniques include the use of optical isolation. An LED and photodiode or phototransistor is incorporated in a single IC. The LED is modulated and the output of the photodiode or phototransistor is amplified and demodulated.

16.10
A NEW ELECTROCARDIOGRAM USING MICROPROCESSORS AND A MICROCOMPUTER [a]

A low-mass, low-inertia plotting mechanism [6] provides high-quality ECGs in a variety of formats [7]. The HP Model 4700A PageWriter Cardiograph (Figure 16.20) is an electronic instrument that is easy to operate, portable, and reliable. Microprocessors was fitted into the design by allowing some of the hardware with firmware and by making possible a power and flexible instrument.

The recording mechanism is a digitally controlled XY plotter (Figure 16.21). It is required to digitize the ECG by placement of an analog-to-digital converter after the ECG front end. The ECG data also have to be temporarily stored in a memory to enable a single-pen recorder to record multichannel data.

The HP 4700A consists of the four major elements shown in Figure 16.21. The ECG front end is responsible for acquiring the ECG from the patient while also isolating and protecting the patient from electrical shock hazards. The ECG samples are passed to the controller, which stores the data in a 16KX 12-bit dynamic RAM and passes the data to the recorder when it is ready to plot a new point.

---

[a] The material in Section 16.10 is excerpted in part from Peter H. Dorward, Steven J. Koerper, Martin K. Mason, and Steven A. Scampani, "New Plotting Technology Leading to a New Kind of Electrocardiograph," *Hewlett-Packard Journal*, vol. 32, no. 10, 16–23, Oct. 1981, and is used courtesy of the Hewlett-Packard Co., Palo Alto, Calif.

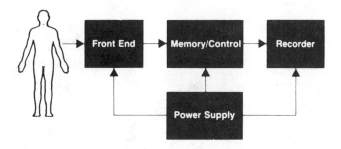

FIGURE 16.21
Basic block diagram of the 4700A PageWriter Cardiograph. (Courtesy of the Hewlett-Packard Co., Palo Alto, Calif.)

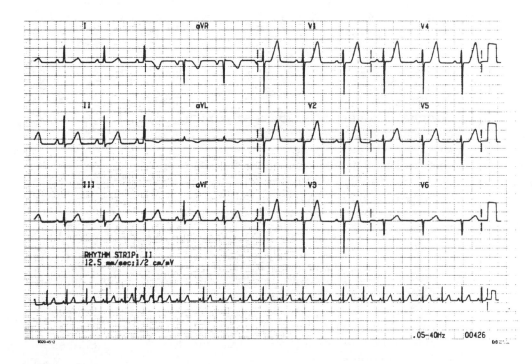

FIGURE 16.22
The 4700A makes it possible to record both morphology (12-lead) and regularity (rhythm) ECG data conveniently on a single piece of paper (shown here at 48% of actual size). (Courtesy of the Hewlett-Packard Co., Palo Alto, Calif.)

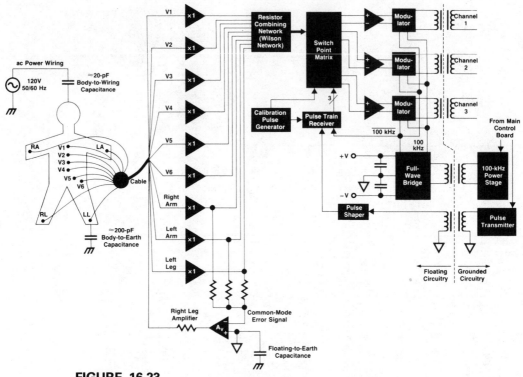

FIGURE 16.23
Diagram of front-end electronics showing the
connections to the patient on the left and the isolation boundary on the right. Signals and data are
communicated across the isolation boundary to
the rest of the 4700A by using high-frequency,
low-interwinding-capacitance transformers.
(Courtesy of the Hewlett-Packard Co., Palo Alto,
Calif.)

The controller is based on a 6800 microprocessor and contains 38K
bytes of firmware to implement the various operating modes of the instrument. The recorder is controlled by a 6801 microprocessor and two custom
servomechanism controllers.

The power supply uses a switching supply to improve efficiency. The
switching frequency is chosen so that modulation of the frequency used in
the ECG front end is avoided. The ac power supply module can be configured by jumpers to operate at any of the line voltages and frequencies used
throughout the world.

The plastic cart form and package are designed to protect the instrument and improve reliability in the rugged use environment of the hospital,
including the cardiac care areas where it can be used.

The HP 4700A uses the traditional operating modes of cardiographs plus several enhancements that draw on the strengths of the new low-mass, low-inertia plotting technology [6]. The recording 12 ECG modes results of the new technology are shown in Fig. 16.22. Note the quality of trace and annotation that can be added to the HP 4700A to identify various sections of the ECG clearly. For theory of ECG waveforms, refer to a good physiology text.

The new recorder technology allows a great deal of flexibility in specifying the output data produced by the cardiograph. However, instrumentation flexibility often complicates the operation of the instrument. The design problem includes the following:

1. Deciding what capabilities of the instrument are useful clinically.

2. Implementing the instrument capabilities in a manner that is easy to operate.

Patient safety is corrected by surrounding isolated circuitry with a plastic enclosure and transferring power and signals via the transformers shown in Figure 16.23.

Power transfer and modulation of the analog channels in Figure 16.23 are operated at a frequency of 100 kHz to allow use of small, low-winding capacitance transformers. The total capacitance between the isolated section and the rest of the instrument which is ground-referenced is less than 200 pF. This ensures no more than 10 μA rms of current will flow for an ac voltage differential of 120 V rms.

A single chip microcomputer controls the recorder. It interfaces to the 4700A controller for data and commands, drives the recorder in an orderly fashion throughout the servomechanism integrated circuits and the optical shaft encoder, and monitors the recorder for error conditions.

Beside the fields of cardiology, microprocessors and microcomputers have become a prime factor in bioelectronic instrumentation design.

16.11
REVIEW QUESTIONS

1. What is a bioelectric potential?

2. What is an action potential?

3. Discuss the electrical potentials generated within the heart.

4. Discuss how to measure bioelectrical potentials.

5. Discuss defibrillators.

6. Discuss pacemakers.

7. Discuss blood-pressure transducers.

8. Discuss the Swan–Ganz catheter.

9. Discuss electrode contacts.

10. Discuss electrical safety of medical equipment.

11. Discuss the use of microprocessors in a cardiograph.

16.12
REFERENCES

1. Harry E. Thomas, *Handbook of Biomedical Instrumentation and Measurement*, Reston Publishing Co., Reston, Va., 1979.

2. Joseph A. Ross and Wilford I. Wummers, *The National Electrical Code Handbook*, NFPA, Boston, Mass., 1981.

3. Electrical Safety Procedures for Clinical Engineers, Ohmic Instrument Co., St. Michaels, Md., 1977.

4. Hospital Safety, Application Note 718, Hewlett-Packard, Waltham, Mass., Feb. 1971.

5. UL544, Standards for Safety (Medical and Dental Equipment), Underwriters Laboratories, Melville, N.Y., (c) 1980.

6. Wayne D. Baron and others, "Development of a High-Performance, Low-Mass, Low-Inertia Plotting Technology," *Hewlett-Packard Journal*, vol. 32, no. 10, 3–4, Oct. 1981.

7. Peter H. Dorward and others, "New Plotting Technology Leads to a New Kind of Electrocardiograph," *Hewlett-Packard Journal*, vol. 32, no. 10, 16–23, Oct. 1981.

16.13
MAGAZINES AND PERIODICALS
RELATED TO MEDICAL SAFETY

1. *Clinical Engineering News*
   Association for the Advancement of Medical Instrumentation, 1901 N. Ft. Myer Drive, Arlington, Va. 22209.

2. *Health Devices*
   Emergency Care Research Institute, 5200 Butler Pike, Plymouth Meeting, Pa. 19462.

3. *Journal of Clinical Engineering*
   Quest Publishing Co., P.O. Box 4141, Diamond Bar, Calif. 91765.

4. *Medical Electronics*
   2994 W. Liberty Ave., Pittsburgh, Pa. 15216.

5. *Medical Instrumentation*, JAAMI and AAMI
   Association for the Advancement of Medical Instrumentation, 1901 N. Ft. Myers Drive, Arlington, Va. 22209.

6. *Biomedical Technology Information Service* and *Biomedical Safety and Standards*
   Quest Publishing Co., P.O. Box 4141, Diamond Bar, Calif. 91765.

7. *IEEE Transactions on Biomedical Engineering*
   Institute of Electrical and Electronic Engineers, Inc., 345 E. 47th Street, New York, N.Y. 10017.

Appendixes

Appendix 1
ABBREVIATIONS AND SYMBOLS*

The use of symbols, prefixes, and abbreviations follows the recommendations of the International Electrotechnical Commission, the American National Standards Institute, Inc., the Institute of Electrical and Electronics Engineers, and other scientific and engineering organizations. Where there is not agreement among these groups, the usage favored by the majority is chosen.

a	atto (10^{-18})	F	farad, Faraday
A	ampere	°F	degrees Fahrenheit
Å	angstrom	f	frequency, femto (10^{-15})
ac	alternating current	fm	frequency modulation
afc	automatic frequency control	FOB	free on board
am	amplitude modulation		
		G	conductance, giga (10^9)
ANSI	American National Standards Institute, Inc.	g	gram, gravitational constant
		g_m	transconductance
APS	American Physical Society		
ASA	Acoustical Society of America	H	henry
ASTM	American Society for Testing and Materials	h	hour, Planck's constant, hecto (10^2)
avc	automatic volume control	hf	high frequency
avg	average	h_f	forward current-transfer ratio
B	susceptance	h_i	short-circuit input impedance
bar	bar ($10^5 N/m^2$)		
BCD	binary-coded decimal	h_o	open-circuit output admittance
c	speed of light, centi (10^{-2})	h_r	reverse voltage-transfer ratio
C	capacitance, coulomb		
°C	degrees Celsius (Centigrade)	Hz	hertz (cycle per second)
cd	candela	HTL	hearing threshold level
CIF	cost, insurance, freight		
CML	current-mode logic	I	current
COD	cash on delivery	IC	integrated circuit
cw	continuous wave	ID	inside diameter
		IEC	International Electrotechnical Commission
d	deci (10^{-1})		
D	dissipation factor	IEEE	Institute of Electrical and Electronics Engineers
da	deka (10)		
dB	decibel	if	intermediate frequency
dBm	decibel referred to one milliwatt	in.	inch
		ISA	Instrument Society of America
dc	direct current		
DCTL	direct-coupled transistor logic	ISO	International Standards Organization
dia	diameter	j	$\sqrt{-1}$
DTL	diode-transistor logic	J	joule
DUT	device under test		
		k	kilo (10^3)
e	electronic charge	°K	degrees Kelvin
E	voltage		
EIA	Electronic Industries Association	l	liter (10^{-3} m^3)
		L	inductance
		lb	pound
emf	electromotive force	LC	inductance-capacitance

620

lm	lumen	s	second, series (as L$_s$)
log	logarithm	shf	super-high frequency
lx	lux	sq	square
m	meter, milli (10^{-3})	sync	synchronous, synchronizing
M	mega (10^6)	T	period, Tesla, tera (10^{12})
max	maximum	t	time
mbar	millibar	TTL	transistor-transistor logic
mil	0.001 inch	TSA	times series analysis
min	minimum, minute		
mo	month	uhf	ultra-high frequency
n	nano (10^{-9})	v	velocity
N	newton	V	volt
		VA	volt ampere
oz	ounce	vhf	very-high frequency
		vlf	very-low frequency
p	page, parallel (as L$_p$), pico (10^{-12})	W	watt
P	poise (10^{-5}N · s/m^2)	Wb	Weber
PF	power factor	wt	weight
ppm	parts per million		
pps	pulses per second	X	reactance
pk-pk	peak-to-peak	Y	admittance
PRF	pulse repetition frequency	yr	year
		Z	impedance
Q	quality factor (storage factor)	α	short-circuit forward current-transfer ratio (common base)
R	resistance		
®	registered trademark	β	short-circuit forward current-transfer ratio (common emitter)
rad	radian		
RC	resistance-capacitance	Γ	reflection coefficient
RCTL	resistor-capacitor-transistor logic	Δ	increment
re	referred to	δ	loss angle
rf	radio frequency	θ	phase angle
RH	relative humidity	λ	wavelength
rms	root-mean-square	μ	micro (10^{-6})
		Ω	ohm
rpm	revolutions per minute	℧	mho
RTL	resistor-transistor logic	ω	angular velocity (2πf)

---

* Courtesy of GenRad, Inc.

Appendix 2
THE INTERNATIONAL (SI) METRIC SYSTEM

- **BASE UNITS:** A set of well-defined units which are regarded as dimensionally independent. That is, each has but one characteristic.

- **DERIVED UNITS:** Units that can be formed by combining base units according to the algebraic relations linking the corresponding quantities. Each derived unit combines at least two base unit characteristics.

- **SUPPLEMENTARY UNITS:** These supplementary units may be regarded as either base units or as derived units.

SI BASE UNITS

Quantity	Unit Name	Symbol
Length	meter	m
Mass	kilogram	kg
Time	second	s
Electric current	ampere	A
Temperature	kelvin	K
Molecular substance	mole	mol
Light intensity	candela	cd

SI DERIVED UNITS

Quantity	Unit Name	Symbol	Expression in Terms of SI Base Units
Acceleration	Meter per second squared		m/s^2
Activity (radioactive)	1 per second		$1/s$
Area	Square meter		m^2
Capacitance	Farad	F	$s^4 \cdot A^2/m^2/kg$
Concentration of amount of substance	Mole per cubic meter		mol/m^3
Conductance	Siemens	S	$s^3 \cdot A^2/m^2/kg$
Current density	Ampere per square meter		A/m^2
Density, mass density	Kilogram per cubic meter		kg/m^3
Dynamic viscosity	Pascal second	Pa·s	$kg/s/m^3$
Electrical potential, potential difference, electromotive force	Volt	V	$m^2 \cdot kg/s^3/A$
Electric charge density	Coulomb per cubic meter	C/m³	$s \cdot A/m^3$
Electric field strength	Volt per meter	V/m	$m \cdot kg/s^3/A$
Electric flux density	Coulomb per square meter	C/m²	$s \cdot A/m^2$
Electric resistance	Ohm	Ω	$m^2 \cdot kg/s^3/A^2$
Energy density	Joule per cubic meter	J/m³	$kg/m/s^2$
Energy, work, quantity of heat	Joule	J	$m^2 \cdot kg/s^2$
Force	Newton	N	$m \cdot kg/s^2$
Frequency	Hertz	Hz	$1/s$
Heat capacity, entropy	Joule per kelvin	J/K	$m^2 \cdot kg/s^2/K$
Heat flux density, irradiance	Watt per square meter	W/m²	kg/s^3
Illuminance	Lux	lx	$cd \cdot sr/m^2$
Inductance	Henry	H	$m^2 \cdot kg/s^2/A^2$
Luminance	Candela per square meter		cd/m^2
Luminous Flux	Lumen	lm	$cd \cdot sr$
Magnetic field strength	Ampere per meter		A/m
Magnetic flux	Weber	Wb	$m^2 \cdot kg/s^2/A$
Magnetic flux density	Tesla	T	$kg/s^2/A$
Molar energy	Joule per mole	J/mol	$m^2 \cdot kg/s^2/mol$
Molar entropy, molar heat capacity	Joule per mole kelvin	J/(mol·K)	$m^2 \cdot kg/s^2/mol/K$
Moment of force	Meter newton	N·m	$m^2 \cdot kg/s^2$
Permeability	Henry per meter	H/m	$m \cdot kg/s^2/A^2$
Permittivity	Farad per meter	F/m	$s^4 \cdot A^2/m^3/kg$
Power, radiant flux	Watt	W	$m^2 \cdot kg/s^3$
Pressure	Pascal	Pa	$kg/m/s^2$
Quantity of electricity, electric charge	Coulomb	C	$A \cdot s$
Specific energy	Joule per kilogram	J/kg	m^2/s^2
Specific heat capacity, specific entropy	Joule per kilogram kelvin	J/(kg·K)	$m^2/s^2/K$
Specific volume	Cubic meter per kilogram		m^3/kg
Speed, velocity	Meter per second		m/s
Surface tension	Newton per meter	N/m	kg/s^2
Thermal conductivity	Watt per meter kelvin	W/(m·K)	$m \cdot kg/s^3/K$
Volume	Cubic meter		m^3
Wave number	1 per meter		$1/m$

SI SUPPLEMENTARY UNITS

Quantity	Unit Name	Symbol
Plane angle	radian	rad
Solid angle	steradian	sr

DERIVED UNITS FORMED BY USING SUPPLEMENTARY UNITS

Quantity	Unit Name	Symbol
Angular velocity	radians per second	rad/s
Angular acceleration	radians per second squared	rad/s²
Radiant intensity	watt per steradian	W/sr
Radiance	watt per square meter steradian	W/m²/sr

Appendix 3
PHYSICS FORMULAS

POWERS OF TEN

Prefixes and symbols to form decimal multiples and/or submultiples.

Power of Ten	E Notation	Decimal Equivalent	Prefix	Phonic	Symbol
10^{12}	E+12	1 000 000 000 000	tera	ter'a	T
10^9	E+09	1 000 000 000	giga	ji'ga	G
10^6	E+06	1 000 000	mega	meg'a	M
10^3	E+03	1 000	kilo	kil'o	k
10^2	E+02	100	hecto	hek'to	h
10	E+01	10	deka	dek'a	da
10^{-1}	E−01	0.1	deci	des'i	d
10^{-2}	E−02	0.01	centi	sen'ti	c
10^{-3}	E−03	0.001	milli	mil'i	m
10^{-6}	E−06	0.000 001	micro	mi'kro	μ
10^{-9}	E−09	0.000 000 001	nano	nan'o	n
10^{-12}	E−12	0.000 000 000 001	pico	pe'ko	p
10^{-15}	E−15	0.000 000 000 000 001	femto	fem'to	f
10^{-18}	E−18	0.000 000 000 000 000 001	atto	at'to	a

MECHANICS

Coefficient of Friction: $\mu = \dfrac{F}{N}$
μ = coefficient of friction
F = force of friction   N = force normal to surface

Velocity: $V_{av} = \dfrac{d}{t}$
V_{av} = average velocity
d = distance traveled t = elapsed time

Acceleration: $a = \dfrac{V_f - V_i}{t}$
a = acceleration V_i = initial velocity
V_f = final velocity t = elapsed time

Newton's 2nd Law of Motion: $F = m \cdot a$
F = force m = mass a = acceleration

Law of Universal Gravitation: $F = G \dfrac{m_1 \cdot m_2}{d^2}$
F = force of attraction $m_1 \cdot m_2$ = product of masses
G = gravitational constant d = distance between their centers

Centripetal Force: $F = \dfrac{m \cdot v^2}{r}$
F = centripetal force
m = mass v = velocity r = radius of path

Pendulum: $T = 2\pi \sqrt{\dfrac{l}{g}}$
T = period l = length g = acceleration of gravity

Work: $W = F \cdot d$
W = work   F = force d = distance

Mechanical Advantage: $IMA = \dfrac{F_E \cdot d}{F_R \cdot d}$ $AMA = \dfrac{F_R}{F_E}$
IMA = ideal mechanical advantage   AMA = actual mechanical advantage
F_E = effort force
F_R = resistance force d = distance

Mechanical Equivalent of Heat: $W = J \cdot Q$
W = work   J = mechanical equivalent of heat   Q = heat

ENERGY

Kinetic Energy: $K.E. = \frac{1}{2} m \cdot v^2$
K.E. = kinetic energy m = mass v = velocity

Potential Energy: $P.E. = m \cdot g \cdot h$
P.E. = potential energy g = acceleration of gravity
m = mass h = vertical distance (height)

Relationship between Mass and Energy: $E = m \cdot c^2$
E = energy m = mass c = velocity of light

LIGHT

Wave Formula: $v = f \cdot \lambda$
v = wave speed f = frequency λ = wave length

Uniformly Illuminated Surface: $E = \dfrac{\Psi}{A}$
E = illumination
Ψ = luminous flux   A = uniformly illuminated area

Images in Mirrors and Lenses: $\dfrac{S_o}{S_i} = \dfrac{D_o}{D_i}$
S_o = object size D_o = object distance
S_i = image size D_i = image distance

Focal Length of Mirrors and Lenses: $\dfrac{1}{f} = \dfrac{1}{D_o} + \dfrac{1}{D_i}$
f = focal length
D_o = object distance D_i = image distance

Index of Refraction: $n = \dfrac{\sin \theta_i}{\sin \theta_r}$
n = index of refraction
θ_i = angle of incidence θ_r = angle of refraction

ELECTRICITY

Electric Current: $I = \dfrac{q}{t}$
I = current q = quantity of charge t = time

Coulomb's Law of Electrostatics: $F = k \dfrac{q_1 \cdot q_2}{d^2}$
F = force between two charges
k = proportionality constant
$q_1 \cdot q_2$ = product of charges
d = distance separating charges

Capacitance of a Capacitor: $C = \dfrac{q}{V}$
C = capacitance of a capacitor
V = potential difference between plates
q = charge on either plate

Ohm's Law of Resistance: $E = I \cdot R$
E = emf of source   I = current in the circuit
R = resistance of the circuit

Joule's Law: $H = I^2 \cdot R \cdot t$
H = heat energy   I = current   R = resistance t = time

Faraday's Law of Electrolysis: $m = z \cdot I \cdot t$
m = mass z = electrochemical equivalent
I = current t = time

Induced emf: Coil in a Magnetic Field: $E = -N \dfrac{\Delta\Phi}{\Delta t}$
E = induced emf   N = number of turns
$\Delta\Phi/\Delta t$ = the change in flux linkage in a given interval of time

Induced emf: Conductor in a Magnetic Field: $E = B \cdot l \cdot v$
E = induced emf   B = flux density of the magnetic field
l = length of conductor
v = velocity of conductor across magnetic field

Instantaneous Voltage: $e = E_{max} \sin \theta$
e = instantaneous voltage
E_{max} = maximum voltage
θ = angle between the plane of the conducting loop and the perpendicular to the magnetic flux (displacement angle)

Instantaneous Current: $i = I_{max} \sin \theta$
i = instantaneous current
I_{max} = maximum current
θ = displacement angle

DECIBEL CONVERSION TABLES*

In communications systems the ratio between any two amounts of electric or acoustic power is usually expressed in units on a logarithmic scale. The decibel (1/10th of the bel) on the briggsian or base-10 scale and the neper on the napierian or base-e scale are in almost universal use for this purpose.

Since voltage and current are related to power by impedance, both the decibel and the neper can be used to express voltage and current ratios, if care is taken to account for the impedances associated with them. In a similar manner the corresponding acoustical quantities can be compared.

From Table I and Table II on the following pages conversions can be made in either direction between the number of decibels and the corresponding power, voltage, and current ratios. Both tables can also be used for nepers by application of a conversion factor.

Decibel — The number of decibels N_{dB} corresponding to the ratio between two amounts of power P_1 and P_2 is

$$N_{dB} = 10 \log_{10} \frac{P_1}{P_2}$$

When two voltages E_1 and E_2 or two currents I_1 and I_2 operate in identical impedances,

$$N_{dB} = 20 \log_{10} \frac{E_1}{E_2} \quad \text{and} \quad N_{dB} = 20 \log_{10} \frac{I_1}{I_2} .$$

If E_1 and E_2 and I_1 and I_2 operate in unequal impedances,

$$N_{dB} = 20 \log_{10} \frac{E_1}{E_2} + 10 \log_{10} \frac{Z_2}{Z_1} + 10 \log_{10} \frac{k_1}{k_2}$$

and $$N_{dB} = 20 \log_{10} \frac{I_1}{I_2} + 10 \log_{10} \frac{Z_1}{Z_2} + 10 \log_{10} \frac{k_1}{k_2} ,$$

where Z_1 and Z_2 are the absolute magnitudes of the corresponding impedances and k_1 and k_2 are the values of power factor for the impedances. E_1, E_2, I_1, and I_2 are also the absolute magnitudes of the corresponding quantities. Note that Table I and Table II can be used to evaluate the impedance and power factor terms, since both are similar to the expression for power ratio.

Neper — The number of nepers N_{nep} corresponding to a power ratio $\frac{P_1}{P_2}$ is

$$N_{nep} = \frac{1}{2} \log_e \frac{P_1}{P_2} .$$

For voltage ratios $\frac{E_1}{E_2}$ or current ratios $\frac{I_1}{I_2}$ working in identical impedances,

$$N_{nep} = \log_e \frac{E_1}{E_2} \quad \text{and} \quad N_{nep} = \log_e \frac{I_1}{I_2} .$$

Relations Between Decibels and Nepers

Multiply decibels by 0.1151 to find nepers
multiply nepers by 8.686 to find decibels

TO FIND VALUES OUTSIDE THE RANGE OF CONVERSION TABLES
Table I: Decibels to Voltage and Power Ratios

Number of decibels positive (+): Subtract +20 decibels successively from the given number of decibels until the remainder falls within range of Table I. To find the voltage ratio, multiply the corresponding value from the right-hand voltage-ratio column by 10 for each time you subtracted 20 dB. To find the power ratio, multiply the corresponding value from the right-hand power-ratio column by 100 for each time you subtracted 20 dB.

Example — Given: 49.2 dB
49.2 dB − 20 dB − 20 dB = 9.2 dB

Voltage ratio: 9.2 dB ⟶ 2.884
2.884 × 10 × 10 = 288.4

Power ratio: 9.2 dB ⟶ 8.318
8.318 × 100 × 100 = 83180

Number of decibels negative (−): Add +20 decibels successively to the given number of decibels until the sum falls within the range of Table I. For the voltage ratio, divide the value from the left-hand voltage-ratio column by 10 for each time you added 20 dB. For the power ratio, divide the value from the left-hand power-ratio column by 100 for each time you added 20 dB.

Example — Given: −49.2 dB
+49.2 dB + 20 dB + 20 dB = −9.2 dB

Voltage ratio: −9.2 dB ⟶ 0.3467
0.3467 × 1/10 × 1/10 = 0.003467

Power ratio: −9.2 dB ⟶ 0.1202
0.1202 × 1/100 × 1/100 = 0.00001202

Table II: Voltage Ratios to Decibels

For ratios smaller than those in table — Multiply the given ratio by 10 successively until the product can be found in the table. From the number of decibels thus found, subtract +20 decibels for each time you multiplied by 10.

Example — Given: Voltage ratio = 0.0131
0.0131 × 10 × 10 = 1.31

From Table II, 1.31 ⟶ 2.345 dB
2.345 dB − 20 dB − 20 dB = −37.655 dB

For ratios greater than those in table — Divide the given ratio by 10 successively until the remainder can be found in the table. To the number of decibels thus found, add +20 dB for each time you divided by 10.

Example — Given: Voltage ratio = 712
712 × 1/10 × 1/10 = 7.12

From Table II, 7.12 ⟶ 17.050 dB
17.050 dB + 20 dB + 20 dB = 57.050 dB

TABLE I

GIVEN: Decibels   **TO FIND: Power and $\begin{Bmatrix} Voltage \\ Current \end{Bmatrix}$ Ratios**

TO ACCOUNT FOR THE SIGN OF THE DECIBEL

For positive (+) values of the decibel — Both voltage and power ratios are greater than unity. Use the two right-hand columns.

For negative (−) values of the decibel — Both voltage and power ratios are less than unity. Use the two left-hand columns.

Example — Given: ±9.1 dB; *Find:*

	Power Ratio	Voltage Ratio
+9.1 dB	8.128	2.851
−9.1 dB	0.1230	0.3508

← −dB+ →

Voltage Ratio	Power Ratio	dB	Voltage Ratio	Power Ratio
1.0000	1.0000	0	1.000	1.000
.9886	.9772	.1	1.012	1.023
.9772	.9550	.2	1.023	1.047
.9661	.9333	.3	1.035	1.072
.9550	.9120	.4	1.047	1.096
.9441	.8913	.5	1.059	1.122
.9333	.8710	.6	1.072	1.148
.9226	.8511	.7	1.084	1.175
.9120	.8318	.8	1.096	1.202
.9016	.8128	.9	1.109	1.230
.8913	.7943	1.0	1.122	1.259
.8810	.7762	1.1	1.135	1.288
.8710	.7586	1.2	1.148	1.318
.8610	.7413	1.3	1.161	1.349
.8511	.7244	1.4	1.175	1.380
.8414	.7079	1.5	1.189	1.413
.8318	.6918	1.6	1.202	1.445
.8222	.6761	1.7	1.216	1.479
.8128	.6607	1.8	1.230	1.514
.8035	.6457	1.9	1.245	1.549
.7943	.6310	2.0	1.259	1.585
.7852	.6166	2.1	1.274	1.622
.7762	.6026	2.2	1.288	1.660
.7674	.5888	2.3	1.303	1.698
.7586	.5754	2.4	1.318	1.738
.7499	.5623	2.5	1.334	1.778
.7413	.5495	2.6	1.349	1.820
.7328	.5370	2.7	1.365	1.862
.7244	.5248	2.8	1.380	1.905
.7161	.5129	2.9	1.396	1.950
.7079	.5012	3.0	1.413	1.995
.6998	.4898	3.1	1.429	2.042
.6918	.4786	3.2	1.445	2.089
.6839	.4677	3.3	1.462	2.138
.6761	.4571	3.4	1.479	2.188
.6683	.4467	3.5	1.496	2.239
.6607	.4365	3.6	1.514	2.291
.6531	.4266	3.7	1.531	2.344
.6457	.4169	3.8	1.549	2.399
.6383	.4074	3.9	1.567	2.455
.6310	.3981	4.0	1.585	2.512
.6237	.3890	4.1	1.603	2.570
.6166	.3802	4.2	1.622	2.630
.6095	.3715	4.3	1.641	2.692
.6026	.3631	4.4	1.660	2.754
.5957	.3548	4.5	1.679	2.818
.5888	.3467	4.6	1.698	2.884
.5821	.3388	4.7	1.718	2.951
.5754	.3311	4.8	1.738	3.020
.5689	.3236	4.9	1.758	3.090

← −dB+ →

Voltage Ratio	Power Ratio	dB	Voltage Ratio	Power Ratio
.5623	3162	5.0	1.778	3.162
.5559	.3090	5.1	1.799	3.236
.5495	.3020	5.2	1.820	3.311
.5433	.2951	5.3	1.841	3.388
.5370	.2884	5.4	1.862	3.467
.5309	.2818	5.5	1.884	3.548
.5248	.2754	5.6	1.905	3.631
.5188	.2692	5.7	1.928	3.715
.5129	.2630	5.8	1.950	3.802
.5070	.2570	5.9	1.972	3.890
.5012	.2512	6.0	1.995	3.981
.4955	.2455	6.1	2.018	4.074
.4898	.2399	6.2	2.042	4.169
.4842	.2344	6.3	2.065	4.266
.4786	.2291	6.4	2.089	4.365
.4732	.2239	6.5	2.113	4.467
.4677	.2188	6.6	2.138	4.571
.4624	.2138	6.7	2.163	4.677
.4571	.2089	6.8	2.188	4.786
.4519	.2042	6.9	2.213	4.898
.4467	.1995	7.0	2.239	5.012
.4416	.1950	7.1	2.265	5.129
.4365	.1905	7.2	2.291	5.248
.4315	.1862	7.3	2.317	5.370
.4266	.1820	7.4	2.344	5.495
.4217	.1778	7.5	2.371	5.623
.4169	.1738	7.6	2.399	5.754
.4121	.1698	7.7	2.427	5.888
.4074	.1660	7.8	2.455	6.026
.4027	.1622	7.9	2.483	6.166
.3981	.1585	8.0	2.512	6.310
.3936	.1549	8.1	2.541	6.457
.3890	.1514	8.2	2.570	6.607
.3846	.1479	8.3	2.600	6.761
.3802	.1445	8.4	2.630	6.918
.3758	.1413	8.5	2.661	7.079
.3715	.1380	8.6	2.692	7.244
.3673	.1349	8.7	2.723	7.413
.3631	.1318	8.8	2.754	7.586
.3589	.1288	8.9	2.786	7.762
.3548	.1259	9.0	2.818	7.943
.3508	.1230	9.1	2.851	8.128
.3467	.1202	9.2	2.884	8.318
.3428	.1175	9.3	2.917	8.511
.3388	.1148	9.4	2.951	8.710
.3350	.1122	9.5	2.985	8.913
.3311	.1096	9.6	3.020	9.120
.3273	.1072	9.7	3.055	9.333
.3236	.1047	9.8	3.090	9.550
.3199	.1023	9.9	3.126	9.772

*Courtesy of GenRad, Inc.

TABLE I (continued)

Voltage Ratio	Power Ratio	dB	Voltage Ratio	Power Ratio
.3162	.1000	10.0	3.162	10.000
.3126	.09772	10.1	3.199	10.23
.3090	.09550	10.2	3.236	10.47
.3055	.09333	10.3	3.273	10.72
.3020	.09120	10.4	3.311	10.96
.2985	.08913	10.5	3.350	11.22
.2951	.08710	10.6	3.388	11.48
.2917	.08511	10.7	3.428	11.75
.2884	.08318	10.8	3.467	12.02
.2851	.08128	10.9	3.508	12.30
.2818	.07943	11.0	3.548	12.59
.2786	.07762	11.1	3.589	12.88
.2754	.07586	11.2	3.631	13.18
.2723	.07413	11.3	3.673	13.49
.2692	.07244	11.4	3.715	13.80
.2661	.07079	11.5	3.758	14.13
.2630	.06918	11.6	3.802	14.45
.2600	.06761	11.7	3.846	14.79
.2570	.06607	11.8	3.890	15.14
.2541	.06457	11.9	3.936	15.49
2512	.06310	12.0	3.981	15.85
.2483	.06166	12.1	4.027	16.22
.2455	.06026	12.2	4.074	16.60
.2427	.05888	12.3	4.121	16.98
.2399	.05754	12.4	4.169	17.38
.2371	.05623	12.5	4.217	17.78
.2344	.05495	12.6	4.266	18.20
.2317	.05370	12.7	4.315	18.62
.2291	.05248	12.8	4.365	19.05
.2265	.05129	12.9	4.416	19.50
.2239	.05012	13.0	4.467	19.95
.2213	.04898	13.1	4.519	20.42
.2188	.04786	13.2	4.571	20.89
.2163	.04677	13.3	4.624	21.38
.2138	.04571	13.4	4.677	21.88
.2113	.04467	13.5	4.732	22.39
.2089	.04365	13.6	4.786	22.91
.2065	.04266	13.7	4.842	23.44
.2042	.04169	13.8	4.898	23.99
.2018	.04074	13.9	4.955	24.55
.1995	.03981	14.0	5.012	25.12
.1972	.03890	14.1	5.070	25.70
.1950	.03802	14.2	5.129	26.30
.1928	.03715	14.3	5.188	26.92
.1905	.03631	14.4	5.248	27.54
.1884	.03548	14.5	5.309	28.18
.1862	.03467	14.6	5.370	28.84
.1841	.03388	14.7	5.433	29.51
.1820	.03311	14.8	5.495	30.20
.1799	.03236	14.9	5.559	30.90
.1778	.03162	15.0	5.623	31.62
.1758	.03090	15.1	5.689	32.36
.1738	.03020	15.2	5.754	33.11
.1718	.02951	15.3	5.821	33.88
.1698	.02884	15.4	5.888	34.67
.1679	.02818	15.5	5.957	35.48
.1660	.02754	15.6	6.026	36.31
.1641	.02692	15.7	6.095	37.15
.1622	.02630	15.8	6.166	38.02
.1603	.02570	15.9	6.237	38.90

Voltage Ratio	Power Ratio	dB	Voltage Ratio	Power Ratio
.1585	.02512	16.0	6.310	39.81
.1567	.02455	16.1	6.383	40.74
.1549	.02399	16.2	6.457	41.69
.1531	.02344	16.3	6.531	42.66
.1514	.02291	16.4	6.607	43.65
.1496	.02239	16.5	6.683	44.67
.1479	.02188	16.6	6.761	45.71
.1462	.02138	16.7	6.839	46.77
.1445	.02089	16.8	6.918	47.86
.1429	.02042	16.9	6.998	48.98
.1413	.01995	17.0	7.079	50.12
.1396	.01950	17.1	7.161	51.29
.1380	.01905	17.2	7.244	52.48
.1365	.01862	17.3	7.328	53.70
.1349	.01820	17.4	7.413	54.95
.1334	.01778	17.5	7.499	56.23
.1318	.01738	17.6	7.586	57.54
.1303	.01698	17.7	7.674	58.88
.1288	.01660	17.8	7.762	60.26
.1274	.01622	17.9	7.852	61.66
.1259	.01585	18.0	7.943	63.10
.1245	.01549	18.1	8.035	64.57
.1230	.01514	18.2	8.128	66.07
.1216	.01479	18.3	8.222	67.61
.1202	.01445	18.4	8.318	69.18
.1189	.01413	18.5	8.414	70.79
.1175	.01380	18.6	8.511	72.44
.1161	.01349	18.7	8.610	74.13
.1148	.01318	18.8	8.710	75.86
.1135	.01288	18.9	8.811	77.62
.1122	.01259	19.0	8.913	79.43
.1109	.01230	19.1	9.016	81.28
.1096	.01202	19.2	9.120	83.18
.1084	.01175	19.3	9.226	85.11
.1072	.01148	19.4	9.333	87.10
.1059	.01122	19.5	9.441	89.13
.1047	.01096	19.6	9.550	91.20
.1035	.01072	19.7	9.661	93.33
.1023	.01047	19.8	9.772	95.50
.1012	.01023	19.9	9.886	97.72
.1000	.01000	20.0	10.000	100.00

Voltage Ratio	Power Ratio	dB	Voltage Ratio	Power Ratio
3.162×10^{-1}	10^{-1}	10	3.162	10
10^{-1}	10^{-2}	20	10	10^2
3.162×10^{-2}	10^{-3}	30	3.162×10	10^3
10^{-2}	10^{-4}	40	10^2	10^4
3.162×10^{-3}	10^{-5}	50	3.162×10^2	10^5
10^{-3}	10^{-6}	60	10^3	10^6
3.162×10^{-4}	10^{-7}	70	3.162×10^3	10^7
10^{-4}	10^{-8}	80	10^4	10^8
3.162×10^{-5}	10^{-9}	90	3.162×10^4	10^9
10^{-5}	10^{-10}	100	10^5	10^{10}

TABLE II

GIVEN: {Voltage / Current} **Ratio**   **TO FIND: Decibels**

POWER RATIOS

To find the number of decibels corresponding to a given power ratio — Assume the given power ratio to be a voltage ratio and find the corresponding number of decibels from the table. The desired result is exactly one-half of the number of decibels thus found.

Example — *Given:* a power ratio of 3.41.

Find: 3.41 in the table:

3.41 → 10.655 dB (voltage)
10.655 dB × ½ = 5.328 dB
(power)

Voltage Ratio	.00	.01	.02	.03	.04	.05	.06	.07	.08	.09
1.0	**.000**	**.086**	**.172**	**.257**	**.341**	**.424**	**.506**	**.588**	**.668**	**.749**
1.1	.828	.906	.984	1.062	1.138	1.214	1.289	1.364	1.438	1.511
1.2	1.584	1.656	1.727	1.798	1.868	1.938	2.007	2.076	2.144	2.212
1.3	2.279	2.345	2.411	2.477	2.542	2.607	2.671	2.734	2.798	2.860
1.4	2.923	2.984	3.046	3.107	3.167	3.227	3.287	3.346	3.405	3.464
1.5	3.522	3.580	3.637	3.694	3.750	3.807	3.862	3.918	3.973	4.028
1.6	4.082	4.137	4.190	4.244	4.297	4.350	4.402	4.454	4.506	4.558
1.7	4.609	4.660	4.711	4.761	4.811	4.861	4.910	4.959	5.008	5.057
1.8	5.105	5.154	5.201	5.249	5.296	5.343	5.390	5.437	5.483	5.529
1.9	5.575	5.621	5.666	5.711	5.756	5.801	5.845	5.889	5.933	5.977
2.0	**6.021**	**6.064**	**6.107**	**6.150**	**6.193**	**6.235**	**6.277**	**6.319**	**6.361**	**6.403**
2.1	6.444	6.486	6.527	6.568	6.608	6.649	6.689	6.729	6.769	6.809
2.2	6.848	6.888	6.927	6.966	7.008	7.044	7.082	7.121	7.159	7.197
2.3	7.235	7.272	7.310	7.347	7.384	7.421	7.458	7.495	7.532	7.568
2.4	7.604	7.640	7.676	7.712	7.748	7.783	7.819	7.854	7.889	7.924
2.5	7.959	7.993	8.028	8.062	8.097	8.131	8.165	8.199	8.232	8.266
2.6	8.299	8.333	8.366	8.399	8.432	8.465	8.498	8.530	8.563	8.595
2.7	8.627	8.659	8.691	8.723	8.755	8.787	8.818	8.850	8.881	8.912
2.8	8.943	8.974	9.005	9.036	9.066	9.097	9.127	9.158	9.188	9.218
2.9	9.248	9.278	9.308	9.337	9.367	9.396	9.426	9.455	9.484	9.513
3.0	**9.542**	**9.571**	**9.600**	**9.629**	**9.657**	**9.686**	**9.714**	**9.743**	**9.771**	**9.799**
3.1	9.827	9.855	9.883	9.911	9.939	9.966	9.994	10.021	10.049	10.076
3.2	10.103	10.130	10.157	10.184	10.211	10.238	10.264	10.291	10.317	10.344
3.3	10.370	10.397	10.423	10.449	10.475	10.501	10.527	10.553	10.578	10.604
3.4	10.630	10.655	10.681	10.706	10.731	10.756	10.782	10.807	10.832	10.857
3.5	10.881	10.906	10.931	10.955	10.980	11.005	11.029	11.053	11.078	11.102
3.6	11.126	11.150	11.174	11.198	11.222	11.246	11.270	11.293	11.317	11.341
3.7	11.364	11.387	11.411	11.434	11.457	11.481	11.504	11.527	11.550	11.573
3.8	11.596	11.618	11.641	11.664	11.687	11.709	11.732	11.754	11.777	11.799
3.9	11.821	11.844	11.866	11.888	11.910	11.932	11.954	11.976	11.998	12.019
4.0	**12.041**	**12.063**	**12.085**	**12.106**	**12.128**	**12.149**	**12.171**	**12.192**	**12.213**	**12.234**
4.1	12.256	12.277	12.298	12.319	12.340	12.361	12.382	12.403	12.424	12.444
4.2	12.465	12.486	12.506	12.527	12.547	12.568	12.588	12.609	12.629	12.649
4.3	12.669	12.690	12.710	12.730	12.750	12.770	12.790	12.810	12.829	12.849
4.4	12.869	12.889	12.908	12.928	12.948	12.967	12.987	13.006	13.026	13.045
4.5	13.064	13.084	13.103	13.122	13.141	13.160	13.179	13.198	13.217	13.236
4.6	13.255	13.274	13.293	13.312	13.330	13.349	13.368	13.386	13.405	13.423
4.7	13.442	13.460	13.479	13.497	13.516	13.534	13.552	13.570	13.589	13.607
4.8	13.625	13.643	13.661	13.679	13.697	13.715	13.733	13.751	13.768	13.786
4.9	13.804	13.822	13.839	13.857	13.875	13.892	13.910	13.927	13.945	13.962
5.0	**13.979**	**13.997**	**14.014**	**14.031**	**14.049**	**14.066**	**14.083**	**14.100**	**14.117**	**14.134**
5.1	14.151	14.168	14.185	14.202	14.219	14.236	14.253	14.270	14.287	14.303
5.2	14.320	14.337	14.353	14.370	14.387	14.403	14.420	14.436	14.453	14.469
5.3	14.486	14.502	14.518	14.535	14.551	14.567	14.583	14.599	14.616	14.632
5.4	14.648	14.664	14.680	14.696	14.712	14.728	14.744	14.760	14.776	14.791
5.5	14.807	14.823	14.839	14.855	14.870	14.886	14.902	14.917	14.933	14.948
5.6	14.964	14.979	14.995	15.010	15.026	15.041	15.056	15.072	15.087	15.102
5.7	15.117	15.133	15.148	15.163	15.178	15.193	15.208	15.224	15.239	15.254
5.8	15.269	15.284	15.298	15.313	15.328	15.343	15.358	15.373	15.388	15.402
5.9	15.417	15.432	15.446	15.461	15.476	15.490	15.505	15.519	15.534	15.549

TABLE II (continued)

Voltage Ratio	.00	.01	.02	.03	.04	.05	.06	.07	.08	.09
6.0	15.563	15.577	15.592	15.606	15.621	15.635	15.649	15.664	15.678	15.692
6.1	15.707	15.721	15.735	15.749	15.763	15.778	15.792	15.806	15.820	15.834
6.2	15.848	15.862	15.876	15.890	15.904	15.918	15.931	15.945	15.959	15.973
6.3	15.987	16.001	16.014	16.028	16.042	16.055	16.069	16.083	16.096	16.110
6.4	16.124	16.137	16.151	16.164	16.178	16.191	16.205	16.218	16.232	16.245
6.5	16.258	16.272	16.285	16.298	16.312	16.325	16.338	16.351	16.365	16.378
6.6	16.391	16.404	16.417	16.430	16.443	16.456	16.469	16.483	16.496	16.509
6.7	16.521	16.534	16.547	16.560	16.573	16.586	16.599	16.612	16.625	16.637
6.8	16.650	16.663	16.676	16.688	16.701	16.714	16.726	16.739	16.752	16.764
6.9	16.777	16.790	16.802	16.815	16.827	16.840	16.852	16.865	16.877	16.890
7.0	16.902	16.914	16.927	16.939	16.951	16.964	16.976	16.988	17.001	17.013
7.1	17.025	17.037	17.050	17.062	17.074	17.086	17.098	17.110	17.122	17.135
7.2	17.147	17.159	17.171	17.183	17.195	17.207	17.219	17.231	17.243	17.255
7.3	17.266	17.278	17.290	17.302	17.314	17.326	17.338	17.349	17.361	17.373
7.4	17.385	17.396	17.408	17.420	17.431	17.443	17.455	17.466	17.478	17.490
7.5	17.501	17.513	17.524	17.536	17.547	17.559	17.570	17.582	17.593	17.605
7.6	17.616	17.628	17.639	17.650	17.662	17.673	17.685	17.696	17.707	17.719
7.7	17.730	17.741	17.752	17.764	17.775	17.786	17.797	17.808	17.820	17.831
7.8	17.842	17.853	17.864	17.875	17.886	17.897	17.908	17.919	17.931	17.942
7.9	17.953	17.964	17.975	17.985	17.996	18.007	18.018	18.029	18.040	18.051
8.0	18.062	18.073	18.083	18.094	18.105	18.116	18.127	18.137	18.148	18.159
8.1	18.170	18.180	18.191	18.202	18.212	18.223	18.234	18.244	18.255	18.266
8.2	18.276	18.287	18.297	18.308	18.319	18.329	18.340	18.350	18.361	18.371
8.3	18.382	18.392	18.402	18.413	18.423	18.434	18.444	18.455	18.465	18.475
8.4	18.486	18.496	18.506	18.517	18.527	18.537	18.547	18.558	18.568	18.578
8.5	18.588	18.599	18.609	18.619	18.629	18.639	18.649	18.660	18.670	18.680
8.6	18.690	18.700	18.710	18.720	18.730	18.740	18.750	18.760	18.770	18.780
8.7	18.790	18.800	18.810	18.820	18.830	18.840	18.850	18.860	18.870	18.880
8.8	18.890	18.900	18.909	18.919	18.929	18.939	18.949	18.958	18.968	18.978
8.9	18.988	18.998	19.007	19.017	19.027	19.036	19.046	19.056	19.066	19.075
9.0	19.085	19.094	19.104	19.114	19.123	19.133	19.143	19.152	19.162	19.171
9.1	19.181	19.190	19.200	19.209	19.219	19.228	19.238	19.247	19.257	19.266
9.2	19.276	19.285	19.295	19.304	19.313	19.323	19.332	19.342	19.351	19.360
9.3	19.370	19.379	19.388	19.398	19.407	19.416	19.426	19.435	19.444	19.453
9.4	19.463	19.472	19.481	19.490	19.499	19.509	19.518	19.527	19.536	19.545
9.5	19.554	19.564	19.573	19.582	19.591	19.600	19.609	19.618	19.627	19.636
9.6	19.645	19.654	19.664	19.673	19.682	19.691	19.700	19.709	19.718	19.726
9.7	19.735	19.744	19.753	19.762	19.771	19.780	19.789	19.798	19.807	19.816
9.8	19.825	19.833	19.842	19.851	19.860	19.869	19.878	19.886	19.895	19.904
9.9	19.913	19.921	19.930	19.939	19.948	19.956	19.965	19.974	19.983	19.991

Voltage Ratio	0	1	2	3	4	5	6	7	8	9
10	20.000	20.828	21.584	22.279	22.923	23.522	24.082	24.609	25.105	25.575
20	26.021	26.444	26.848	27.235	27.604	27.959	28.299	28.627	28.943	29.248
30	29.542	29.827	30.103	30.370	30.630	30.881	31.126	31.364	31.596	31.821
40	32.041	32.256	32.465	32.669	32.869	33.064	33.255	33.442	33.625	33.804
50	33.979	34.151	34.320	34.486	34.648	34.807	34.964	35.117	35.269	35.417
60	35.563	35.707	35.848	35.987	36.124	36.258	36.391	36.521	36.650	36.777
70	36.902	37.025	37.147	37.266	37.385	37.501	37.616	37.730	37.842	37.953
80	38.062	38.170	38.276	38.382	38.486	38.588	38.690	38.790	38.890	38.988
90	39.085	39.181	39.276	39.370	39.463	39.554	39.645	39.735	39.825	39.913
100	40.000	—	—	—	—	—	—	—	—	—

Appendix 5
REACTANCE CHARTS*

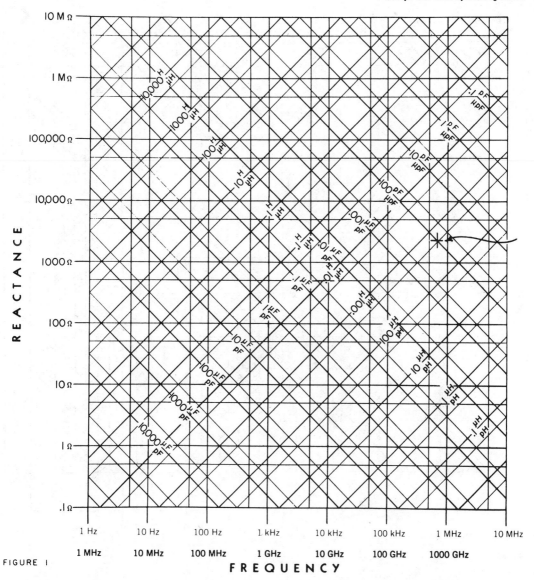

FIGURE I

FREQUENCY

Figure 1 is the complete chart, used for rough calculations. Figure 2, which is a single decade of Figure 1 enlarged approximately 7 times, is used where two or three significant figures are to be determined.

TO FIND REACTANCE

Enter the charts vertically from the bottom (frequency) and along the lines slanting upward to the left (capacitance) or to the right (inductance). Corresponding scales (green or black) must be used throughout. Project horizontally to the left from the intersection and read reactance.

TO FIND RESONANT FREQUENCY

Enter the slanting lines for the given inductance and capacitance. Project downward and read resonant frequency from the bottom scale. **Corresponding scales (green or black) must be used throughout.**

---

*Courtesy of GenRad, Inc.

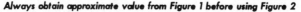

Always obtain approximate value from Figure 1 before using Figure 2

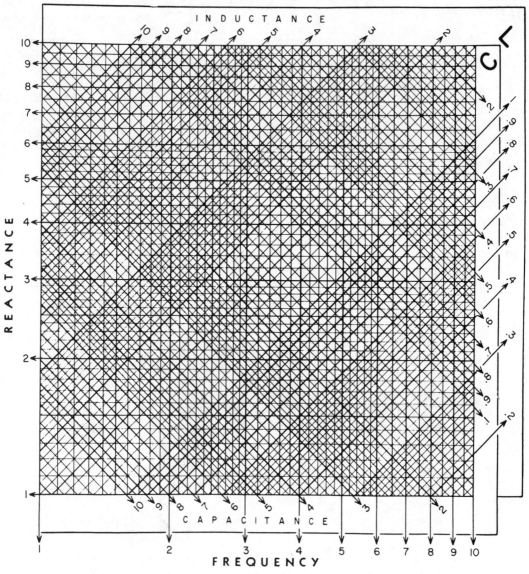

FIGURE 2

Example: The point indicated in Figure 1 corresponds to a frequency of about 700 kHz and an inductance of 500 μH, or a capacitance of 100 pF, giving in either case a reactance of about 2000 ohms. The resonant frequency of a circuit containing these values of inductance and capacitance is, of course, 700 kHz, approximately.

USE OF FIGURE 2

Figure 2 gives additional precision but does not place the decimal point, which must be located from a prelim-inary entry on Figure 1. Since the chart necessarily requires two logarithmic decades for inductance and ca-pacitance for every single decade of frequency and reac-tance, unless the correct decade for L and C is chosen, the calculated values of reactance and frequency will be in error by a factor of 3.16. In Figure 2, the capacitance scale is green; inductance scale is black.

Example: (Continued) The reactance corresponding to 500 μH or 100 pF is 2230 ohms at 712 kHz, their reso-nant frequency.

Appendix 6
GLOSSARY OF DATA CONVERSION TERMS*

This glossary defines the most often used terms in the field of data conversion technology.

ABSOLUTE ACCURACY: The worst-case input to output error of a data converter referred to the NBS standard volt.

ACCURACY: The conformance of a measured value with its true value; the maximum error of a device such as a data converter from the true value. See *relative accuracy* and *absolute accuracy.*

ACQUISITION TIME: For a sample-hold, the time required, after the sample command is given, for the hold capacitor to charge to a full scale voltage change and then remain within a specified error band around final value.

ACTIVE FILTER: An electronic filter which uses passive circuit elements with active devices such as gyrators or operational amplifiers. In general, resistors and capacitors are used but no inductors.

ACTUATOR: A device which converts a voltage or current input into a mechanical output.

ADC: Abbreviation for analog-to-digital converter. See *A/D converter.*

A/D CONVERTER: Analog-to-digital converter. A circuit which converts an analog (continuous) voltage or current into an output digital code.

ALIAS FREQUENCY: In reconstructed analog data, a false lower frequency component which is the result of insufficient sampling rate, i.e., less than that required by the sampling theorem.

ALIASING: See *Alias Frequency.*

ANALOG MULTIPLEXER: An array of switches with a common output connection for selecting one of a number of analog inputs. The output signal follows the selected input within a small error.

ANTI-ALIAS FILTER: See *Pre-Sampling Filter.*

APERTURE DELAY TIME: In a sample-hold, the time elapsed from the hold command to the actual opening of the sampling switch.

APERTURE JITTER: See *Aperture Uncertainty Time.*

APERTURE TIME: The time window, or time uncertainty, in making a measurement. For an A/D converter it is the conversion time; for a sample-hold it is the signal averaging time during the sample-to-hold transition.

APERTURE UNCERTAINTY TIME: In a sample-hold, the time variation, or time jitter, in the opening of the sampling switch; also the variation in aperture delay time from sample to sample.

AUTO-ZERO: A stabilization circuit which servos an amplifier or A/D converter input offset to zero during a portion of its operating cycle.

BANDGAP REFERENCE: A voltage reference circuit which is based on the principle of the predictable base-to-emitter voltage of a transistor to generate a constant voltage equal to the extrapolated bandgap voltage of silicon ($\approx$1.22V).

BANDWIDTH: The frequency at which the gain of an amplifier or other circuit is reduced by 3 dB from its DC value; also the range of frequencies within which the attenuation is less than 3 dB from the center frequency value.

BCD: See *Binary Coded Decimal.*

BINARY CODE: See *Natural Binary Code.*

BINARY CODED DECIMAL (BCD): A binary code used to represent decimal numbers in which each digit from 0 to 9 is represented by four bits weighted 8-4-2-1. Only 10 of the 16 possible states are used.

BIPOLAR MODE: For a data converter, when the analog signal range includes both positive and negative values.

BIPOLAR OFFSET: The analog displacement of one half of full scale range in a data converter operated in the bipolar mode. The offset is generally derived from the converter reference circuit.

BREAK-BEFORE-MAKE SWITCHING: A characteristic of analog multiplexers in which there is a small time delay between disconnection from the previous channel and connection to the next channel. This assures that no two inputs are ever momentarily shorted together.

BUFFER AMPLIFIER: An amplifier employed to isolate the loading effect of one circuit from another.

BURIED ZENER REFERENCE: See *Subsurface Zener Reference.*

BUSY OUTPUT: See *Status Output.*

BUTTERFLY CHARACTERISTIC: An error versus temperature graph in which all errors are contained within two straight lines which intersect at room temperature, or approximately 25°C.

CHARGE BALANCING A/D CONVERTER: An analog-to-digital conversion technique which employs an operational integrator circuit within a pulse generating feedback loop. Current pulses from the feedback loop are precisely balanced against the analog input by the integrator, and the resulting pulses are counted for a fixed period of time to produce an output digital word. This technique is also called *quantized-feedback.*

CHARGE DUMPING: See *Charge Transfer.*

CHARGE INJECTION: See *Charge Transfer.*

*Appendix 6 appears on pages 231 to 236 in *Data Acquisition and Conversion Handbook,* edited by Eugene Zuck,© 1979 by Datel-Intersil Inc., 4th printing, Sept. 1981. Reprinted with permission of Datel-Intersil, Inc.

CHARGE TRANSFER: In a sample-hold, the phenomenon of moving a small charge from the sampling switch to the hold capacitor during switch turn-off. This is caused by the switch control voltage change coupling through switch capacitance to the hold capacitor. Also called *charge dumping* or *charge injection.*

CHOPPER-STABILIZED AMPLIFIER: An operational amplifier which employs a special DC modulator-demodulator circuit to reduce input offset voltage drift to an extremely low value.

CLOCK: A circuit in an A/D converter that generates timing pulses which synchronize the operation of the converter.

CLOCK RATE: The frequency of the timing pulses of the clock circuit in an A/D converter.

COMMON-MODE REJECTION RATIO: For an amplifier, the ratio of differential voltage gain to common-mode voltage gain, generally expressed in dB.

$$CMRR = 20 \log_{10} \frac{A_D}{A_{CM}}$$

where A_D is differential voltage gain and A_{CM} is common mode voltage gain.

COMPANDING CONVERTER: An A/D or D/A converter which employs a logarithmic transfer function to expand or compress the analog signal range. These converters have large effective dynamic ranges and are commonly used in digitized voice communication systems.

COMPLEMENTARY BINARY CODE: A binary code which is the logical complement of straight binary. All 1's become 0's and vice versa.

CONVERSION TIME: The time required for an A/D converter to complete a single conversion to specified resolution and linearity for a full scale analog input change.

CONVERSION RATE: The number of repetitive A/D or D/A conversions per second for a full scale change to specified resolution and linearity.

COUNTER TYPE A/D CONVERTER: A feedback method of A/D conversion whereby a digital counter drives a D/A converter which generates an output ramp which is compared with the analog input. When the two are equal, a comparator stops the counter and output data is ready. Also called a *servo type* A/D converter.

CREEP VOLTAGE: A voltage change with time across an open capacitor caused by dielectric absorption. This causes sample-hold output error.

CROSSTALK: In an analog multiplexer, the ratio of output voltage to input voltage with all channels connected in parallel and off. It is generally expressed as an input to output attenuation ratio in dB.

DAC: Abbreviation for digital-to-analog converter. See *D/A Converter.*

D/A CONVERTER: Digital-to-analog converter. A circuit which converts a digital code word into an output analog (continuous) voltage or current.

DATA ACQUISITION SYSTEM: A system consisting of analog multiplexers, sample-holds, A/D converters, and other circuits which process one or more analog signals and convert them into digital form for use by a computer.

DATA AMPLIFIER: See *Instrumentation Amplifier.*

DATA CONVERTER: An A/D or D/A Converter.

DATA DISTRIBUTION SYSTEM: A system which uses D/A converters and other circuits to convert the digital outputs of a computer into analog form for control of a process or system.

DATA RECOVERY FILTER: A filter used to reconstruct an analog signal from a train of analog samples.

DATA WORD: A digital code-word that represents data to be processed.

DECAY RATE: See *Hold-Mode Droop.*

DECODER: A communications term for D/A converter.

DEGLITCHED DAC: A D/A converter which incorporates a deglitching circuit to virtually eliminate output spikes (or glitches). These DAC's are commonly used in CRT display systems.

DEGLITCHER: A special sample-hold circuit used to eliminate the output spikes (or glitches) from a D/A converter.

DIELECTRIC ABSORPTION: A voltage memory characteristic of capacitors caused by the dielectric material not polarizing instantaneously. The result is that not all the energy stored in a charged capacitor can be quickly recovered upon discharge, and the open capacitor voltage will creep. See also *Creep Voltage.*

DIFFERENTIAL LINEARITY ERROR: The maximum deviation of any quantum (LSB change) in the transfer function of a data converter from its ideal size of $FSR/2^n$.

DIFFERENTIAL LINEARITY TEMPCO: The change in differential linearity error with temperature for a data converter, expressed in ppm/°C of FSR (Full Scale Range).

DIGITIZER: A device which converts analog into digital data; an A/D converter.

DOUBLE-LEVEL MULTIPLEXING: A method of channel expansion in analog multiplexers whereby the outputs of a group of multiplexers connect to the inputs of another multiplexer.

DROOP: See *Hold-Mode Droop.*

DUAL SLOPE A/D CONVERTER: An indirect method of A/D conversion whereby an analog voltage is converted into a time period by an integrator and reference and then measured by a clock and counter. The method is relatively slow but capable of high accuracy.

DYNAMIC ACCURACY: The total error of a data converter or conversion system when operated at its maximum specified conversion rate or throughput rate.

DYNAMIC RANGE: The ratio of full scale range (FSR) of a data converter to the smallest difference it can resolve. In terms of converter resolution:

Dynamic Range (DR) = 2^n

It is generally expressed in dB:

DR = $20 \log_{10} 2^n = 6.02 n$

where n is the resolution in bits.

EFFECTIVE APERTURE DELAY: In a sample-hold, the time difference between the hold command and the time at which the input signal equalled the held voltage.

ELECTROMETER AMPLIFIER: An amplifier characterized by ultra-low input bias current and input noise which is used to measure currents in the picoampere region and lower.

ENCODER: A communications term for an A/D converter.

E.O.C.: End of Conversion. See *Status Output*.

ERROR BUDGET: A systematic listing of errors in a circuit or system to determine worst case total or statistical error.

EXTRAPOLATIVE HOLD: See *First-Order Hold*.

FEEDBACK TYPE A/D CONVERTER: A class of analog-to-digital converters in which a D/A converter is enclosed in the feedback loop of a digital control circuit which changes the D/A output until it equals the analog input.

FIRST-ORDER HOLD: A type of sample-hold, used as a recovery filter, which uses the present and previous analog samples to predict the slope to the next sample. Also called an *extrapolative hold*.

FLASH TYPE A/D CONVERTER: See *Parallel A/D Converter*.

FLYING-CAPACITOR MULTIPLEXER: A multiplexer switch which employs a double-pole, double-throw switch connected to a capacitor. By first connecting the capacitor to the signal source and then to a differential amplifier, a signal with a high common-mode voltage can be multiplexed to a ground-referenced circuit.

FRACTIONAL-ORDER HOLD: A type of sample-hold, used as a recovery filter, which uses a fixed fraction of the difference between the present and previous analog samples to predict the slope to the next sample.

FREQUENCY FOLDING: In the recovery of sampled data, the overlap of adjacent spectra caused by insufficient sampling rate. The overlapping results in distortion in the recovered signal which cannot be eliminated by filtering the recovered signal.

FREQUENCY-TO-VOLTAGE (F/V) CONVERTER: A device which converts an input pulse rate into an output analog voltage.

FSR: Full Scale Range.

FULL POWER FREQUENCY: The maximum frequency at which an amplifier, or other device, can deliver rated peak-to-peak output voltage into rated load at a specified distortion level.

FULL SCALE RANGE (FSR): the difference between maximum and minimum analog values for an A/D converter input or D/A converter output.

F/V CONVERTER: See *Frequency-To-Voltage Converter*.

GAIN-BANDWIDTH PRODUCT: The product of gain and small signal bandwidth for an operational amplifier or other circuit. This product is constant for a single-pole response.

GAIN ERROR: The difference in slope between the actual and ideal transfer functions for a data converter or other circuit. It is expressed as a percent of analog magnitude.

GAIN TEMPCO: The change in gain (or scale factor) with temperature for a data converter or other circuit, generally expressed in ppm/°C.

HIGH-LEVEL MULTIPLEXING: An analog multiplexing circuit in which the analog signal is first amplified to a higher level (1 to 10 volts) and then multiplexed. This is the preferred method of multiplexing to prevent noise contamination of the analog signal.

HOLD CAPACITOR: A high quality capacitor used in a sample-hold circuit to store the analog voltage. The capacitor must have low leakage and low dielectric absorption. Types commonly used include polystyrene, teflon, polycarbonate, polypropylene, and MOS.

HOLD-MODE: The operating mode of a sample-hold circuit in which the sampling switch is open.

HOLD-MODE DROOP: In a sample-hold, the output voltage change per unit of time with the sampling switch open. It is commonly expressed in V/sec. or $\mu V/\mu \sec$.

HOLD-MODE FEEDTHROUGH: In a sample-hold, the percentage of input sinusoidal or step signal measured at the output with the sampling switch open.

HOLD-MODE SETTLING TIME: In a sample-hold, the time from the hold-command transition until the output has settled within a specified error band.

HYSTERESIS ERROR: The small variation in analog transition points of an A/D converter whereby the transition level depends on the direction from which it is approached. In most A/D converters this hysteresis is very small and is caused by the analog comparator.

IDEAL FILTER: A low pass filter with flat pass-band response, infinite attenuation at the cutoff frequency, and zero response past cutoff; it also has linear phase response in the passband. Ideal filters are mathematical filters frequently used in textbook examples but not physically realizable.

INDIRECT TYPE A/D CONVERTER: A class of analog-to-digital converters which converts the unknown input voltage into a time period and then measures this period.

INFINITE-HOLD: A sample-hold circuit which converts an analog voltage into digital form which is then held indefinitely, without decay, in a register.

INPUT DYNAMIC RANGE: In an amplifier, the maximum permissible peak-to-peak voltage across the input terminals which does not cause the output to slew rate limit or distort. Mathematically it is found as

$$\text{IDR (Input Dynamic Range)} \frac{\text{SR}}{\pi \text{GB}}$$

where SR is the slew rate and GB is gain bandwidth.

INSTRUMENTATION AMPLIFIER: An amplifier circuit with high impedance differential inputs and high common-mode rejection. Gain is set by one or two resistors which do not connect to the input terminals.

INTEGRAL LINEARITY ERROR: The maximum deviation of a data converter transfer function from the ideal straight line with offset and gain errors zeroed. It is generally expressed in LSB's or in percent of FSR.

INTEGRATING A/D CONVERTER: One of several types of A/D conversion techniques whereby the analog input is integrated with time. This includes dual slope, triple slope, and charge balancing type A/D converters.

INTERPOLATIVE HOLD: See *Polygonal Hold.*

ISOLATION AMPLIFIER: An amplifier which is electrically isolated between input and output in order to be able to amplify a differential signal superimposed on a high common-mode voltage.

LEAST SIGNIFICANT BIT (LSB): The rightmost bit in a data converter code. The analog size of the LSB can be found from the converter resolution:

$$\text{LSB Size} = \frac{\text{FSR}}{2^n}$$

where FSR is full scale range and n is the resolution in bits.

LINEARITY ERROR: See *Integral Linearity Error* and *Differential Linearity Error.*

LONG TERM STABILITY: The variation in data converter accuracy due to time change alone. It is commonly specified in percent per 1000 hours or per year.

LOW-LEVEL MULTIPLEXING: An analog multiplexing system in which a low amplitude signal is first multiplexed and then amplified.

LSB: Least Significant Bit.

LSB SIZE: See *Quantum.*

MAJOR CARRY: See *Major Transition.*

MAJOR TRANSITION: In a data converter, the change from a code of 1000...000 to 0111...1111 or vice-versa. This transition is the most difficult one to make from a linearity standpoint since the MSB weight must ideally be precisely one LSB larger than the sum of all other bit weights.

MISSING CODE: In an A/D converter, the characteristic whereby not all output codes are present in the transfer function of the converter. This is caused by a nonmonotonic D/A converter inside the A/D.

MONOTONICITY: For a D/A converter, the characteristic of the transfer function whereby an increasing input code produces a continuously increasing analog output. *Nonmonotonicity* may occur if the converter differential linearity error exceeds ±1 LSB.

MOST SIGNIFICANT BIT (MSB): The leftmost bit in a data converter code. It has the largest weight, equal to one half of full scale range.

MSB: Most Significant Bit.

MULTIPLYING D/A CONVERTER: A type of digital-to-analog converter in which the reference voltage can be varied over a wide range to produce an analog output which is the product of the input code and input reference voltage. Multiplication can be accomplished in one, two, or four algebraic quadrants.

MUX: Abbreviation for multiplexer. See *Analog Multiplexer.*

NATURAL BINARY CODE: A positive weighted code in which a number is represented by

$$N = a_0 2^0 + a_1 2^1 + a_2 2^2 + a_3 2^3 + \ldots + a_n 2^n$$

where each coefficient "a" has a value of zero or one. Data converters use this code in its fractional form where:

$$N = a_1 2^{-1} + a_2 2^{-2} + a_3 2^{-3} + \ldots a_n 2^{-n}$$

and N has a fractional value between zero and one.

NEGATIVE TRUE LOGIC: A logic system in which the more negative of two voltage levels is defined as a logical 1 (true) and the more positive level is defined as a logical 0 (false).

NOISE REJECTION: The amount of suppression of normal mode analog input noise of an A/D converter or other circuit, generally expressed in dB. Good noise rejection is a characteristic of integrating type A/D converters.

NONMONOTONIC: A D/A converter transfer characteristic in which the output does not continuously increase with increasing input. At one or more points there may be a dip in the output function.

NORMAL-MODE REJECTION: The attenuation of a specific frequency or band of frequencies appearing directly across two electrical terminals. In A/D converters, normal-mode rejection is determined by an input filter or by integration of the input signal.

NOTCH FILTER: An electronic filter which attenuates or rejects a specific frequency or narrow band of frequencies with a sharp cutoff on either side of the band.

NYQUIST THEOREM: See *Sampling Theorem*.

OFFSET BINARY CODE: Natural binary code in which the code word 0000 . . . . 0000 is displaced by one-half analog full scale. The code represents analog values between $-FS$ and $+FS$ (full scale). The code word 1000 . . . . 0000 then corresponds to analog zero.

OFFSET DRIFT: The change with temperature of analog zero for a data converter operating in the bipolar mode. It is generally expressed in ppm/°C of FSR.

OFFSET ERROR: The error at analog zero for a data converter operating in the bipolar mode.

ONE'S COMPLEMENT CODE: A bipolar binary code in which positive and negative codes of the same magnitude sum to all one's.

PARALLEL TYPE A/D CONVERTER: An ultra-fast method of A/D conversion which uses an array of $2^n - 1$ comparators to directly implement a quantizer, where n is the resolution in bits. The quantizer is followed by a decoder circuit which converts the comparator outputs into binary code.

PARALLEL TYPE D/A CONVERTER: The most commonly used type of D/A converter in which upon application of an input code, all bits change simultaneously to produce a new output.

PASSIVE FILTER: A filter circuit using only resistors, capacitors, and inductors.

POLYGONAL HOLD: A type of sample-hold, used as a signal recovery filter, which produces a voltage output which is a straight line joining the previous sample value to the present sample. This results in an accurate signal reconstruction but with a one sample-period output delay.

POSITIVE TRUE LOGIC: A logic system in which the more positive of two voltage levels is defined as a logical 1 (true) and the more negative level is defined as a logical 0 (false).

POWER SUPPLY SENSITIVITY: The output change in a data converter caused by a change in power supply voltage. Power supply sensitivity is generally specified in %/V or in %/% supply change.

PRECISION: The degree of repeatability, or reproducibility of a series of successive measurements. Precision is affected by the noise, hysteresis, time, and temperature stability of a data converter or other device.

PRE-SAMPLING FILTER: A low pass filter used to limit the bandwidth of a signal before sampling in order to assure that the conditions of the Sampling Theorem are met. Therefore frequency folding is eliminated or greatly diminished in the recovered signal spectrum.

PROGRAMMABLE GAIN AMPLIFIER: An amplifier with a digitally controlled gain for use in data acquisition systems.

PROGRAMMER-SEQUENCER: A digital logic circuit which controls the sequence of operations in a data acquisition system.

PROPAGATION TYPE A/D CONVERTER: A type of A/D conversion method which employs one comparator per bit to achieve ultra-fast A/D conversion. The conversion propagates down the series of cascaded comparators.

QUAD CURRENT SWITCH: A group of four current sources weighted 8-4-2-1 which are switched on and off by TTL inputs. They are used to implement A/D and D/A converter designs up to 16 bits resolution by using multiple quads with current dividers between each quad.

QUANTIZATION NOISE: See *Quantization Error*.

QUANTIZATION UNCERTAINTY: See *Quantization Error*.

QUANTIZED FEEDBACK A/D CONVERTER: See *Charge Balancing A/D Converter*.

QUANTIZER: A circuit which transforms a continuous analog signal into a set of discrete output states. Its transfer function is the familiar stair-case function.

QUANTIZING ERROR: The inherent uncertainty in digitizing an analog value due to the finite resolution of the conversion process. The quantized value is uncertain by up to $\pm Q/2$ where Q is the quantum size. This error can be reduced only by increasing the resolution of the converter. Also called *quantization uncertainty* or *quantization noise*.

QUANTUM: The analog difference between two adjacent codes for an A/D or D/A converter. Also called *LSB size*.

R-2R LADDER NETWORK: An array of matched resistors with series values of R and shunt values of 2R in a standard ladder circuit configuration.

RATIOMETRIC A/D CONVERTER: An analog-to-digital converter which uses a variable reference to measure the ratio of the input voltage to the reference.

RECONSTRUCTION FILTER: See *Data Recovery Filter*.

RECOVERY FILTER: See *Data Recovery Filter*.

REFERENCE CIRCUIT: A circuit which produces a stable output voltage over time and temperature for use in A/D and D/A converters. The circuit generally uses an operational amplifier with a precision Zener or bandgap type reference element.

RELATIVE ACCURACY: The worst case input to output error of a data conversion, as a percent of full scale, referred to the converter reference. The error consists of offset, gain, and linearity components.

RESOLUTION: The smallest change that can be distinguished by an A/D converter or produced by a D/A converter. Resolution may be stated in percent of full scale, but is commonly expressed as the number of bits n where the converter has 2^n possible states.

SAMPLE-HOLD: A circuit which accurately acquires and stores an analog voltage on a capacitor for a specified period of time.

SAMPLE-HOLD FIGURE OR MERIT: The ratio of capacitor charging current in the sample-mode to the leakage current off the capacitor in the hold-mode.

SAMPLE-MODE: The operating mode of a sample-hold circuit in which the sampling switch is closed.

SAMPLER: An electronic switch which is turned on and off at a fast rate to produce a train of analog sample pulses.

SAMPLE-TO-HOLD OFFSET ERROR: For a sample-hold, the change in output voltage from the sample-mode to the hold-mode, with constant input voltage. This error is caused by the sampling switch transferring charge onto the hold capacitor as it opens.

SAMPLE-TO-HOLD STEP: See *Sample-to-Hold Offset Error.*

SAMPLE-TO-HOLD TRANSIENT: A small spike at the output of a sample-hold when it goes into the hold mode. It is caused by feedthrough from the sampling switch control voltage.

SAMPLING THEOREM: A theorem due to Nyquist which says if a continuous bandwidth-limited signal contains no frequency components higher than f_c, then the original signal can be recovered without distortion if it is sampled at a rate of at least $2f_c$ samples per second.

SAR: Successive approximation register. A digital control circuit used to control the operation of a successive approximation A/D converter.

SCALE FACTOR ERROR: See *Gain Error.*

SERIAL TYPE D/A CONVERTER: A type of digital-to-analog converter in which the digital input data is received in sequential form before an analog output is produced.

SERVO-TYPE A/D CONVERTER: See *Counter-Type A/D Converter.*

SETTLING TIME: The time elapsed from the application of a full scale step input to a circuit to the time when the output has entered and remained within a specified error band around its final value.

This term is an important specification for operational amplifiers, analog multiplexers, and D/A converters.

SHORT CYCLING: The termination of an A/D conversion process at a resolution less than the full resolution of the converter. This results in a shorter conversion time for reduced resolution in A/D converters with a short cycling capability.

SIGNAL RECONSTRUCTION FILTER: A low pass filter used to accurately reconstruct an analog signal from a train of analog samples.

SIGN-MAGNITUDE BCD: A binary coded decimal code in which a sign bit is added to distinguish positive from negative in bipolar operation.

SIGN-MAGNITUDE BINARY CODE: The natural binary code to which a sign bit is added to distinguish positive from negative in bipolar operation.

SIMULTANEOUS SAMPLE-HOLD: A system in which a series of sample-hold circuits are used to sample a number of analog channels, all at the same instant. This requires one sample-hold per analog channel.

SIMULTANEOUS TYPE A/D CONVERTER: See *Parallel Type A/D Converter.*

SINGLE-LEVEL MULTIPLEXING: A method of channel expansion in analog multiplexers whereby several multiplexers are operated in parallel by connecting their outputs together. Each multiplexer is controlled by a digital *enable* input.

SINGLE-SLOPE A/D CONVERTER: A simple A/D converter technique in which a ramp voltage generated from a voltage reference and integrator is compared with the analog input voltage by a comparator. The time required for the ramp to equal the input is measured by a clock and counter to produce the digital output word.

SKIPPED CODE: See *Missing Code.*

SLEW RATE: The maximum rate of change of the output of an operational amplifier or other circuit. Slew rate is limited by internal charging currents and capacitances and is generally expressed in volts per microsecond.

SPAN: For an A/D or D/A converter, the full scale range or difference between maximum and minimum analog values.

START-CONVERT: The input pulse to an A/D converter which initiates conversion.

STATIC ACCURACY: The total error of a data converter or conversion system under DC input conditions.

STATUS OUTPUT: The logic output of an A/D converter which indicates whether the device is in the process of making a conversion or the conversion has been completed and output data is ready. Also called *busy output* or *end of conversion* output.

STRAIGHT BINARY CODE: See *Natural Binary Code.*

SUBSURFACE ZENER REFERENCE: A compensated voltage reference diode in which avalanche breakdown occurs below the surface of the silicon in the bulk region rather than at the surface. This results in lower noise and higher stability. The reversed biased diode is temperature compensated by a series connected, forward biased signal diode.

SUCCESSIVE APPROXIMATION A/D CONVERTER: An A/D conversion method that compares in sequence a series of binary weighted values with the analog input to produce an output digital word in just n steps, where n is the resolution in bits. The process is efficient and is analogous to weighing an unknown quantity on a balance scale using a set of binary standard weights.

TEMPERATURE COEFFICIENT: The change in analog magnitude with temperature, expressed in ppm/°C.

THREE-STATE OUTPUT: A type of A/D converter output used to connect to a data bus. The three output states are logic 1, logic 0, and off. An *enable* control turns the output on or off.

THROUGHPUT RATE: The maximum repetitive rate at which a data converison system can operate to give specified output accuracy. It is determined by adding the various times required for multiplexer settling, sample-hold acquisition, A/D conversion, etc. and then taking the inverse of total time.

TRACK-AND-HOLD: A sample-hold circuit which can continuously follow the input signal in the sample-mode and then go into hold-mode upon command.

TRACKING A/D CONVERTER: A counter-type analog-to-digital converter which can continuously follow the analog input at some specified maximum rate and continuously update its digital output as the input signal changes. The circuit uses a D/A converter driven by an up-down counter.

TRANSDUCER: A device which converts a physical parameter such as temperature or pressure into an electrical voltage or current.

TRANSFER FUNCTION: The input to output characteristic of a device such as a data converter expressed either mathematically or graphically.

TRIPLE-SLOPE A/D CONVERTER: A variation on the dual slope type A/D converter in which the time period measured by the clock and counter is divided into a coarse (fast slope) measurement and a fine (slow slope) measurement.

TWO'S COMPLEMENT CODE: A bipolar binary code in which positive and negative codes of the same magnitude sum to all zero's plus a carry.

TWO-STAGE PARALLEL A/D CONVERTER: An ultra-fast A/D converter in which two parallel type A/D's are operated in cascade to give higher resolution. In the usual case a 4-bit parallel converter first makes a conversion; the resulting output code drives an ultra-fast 4-bit D/A, the output of which is subtracted from the analog input to form a residual. This residual then goes to a second 4 bit parallel A/D. The result is an 8 bit word converted in two steps.

UNIPOLAR MODE: In a data converter, when the analog range includes values of one polarity only.

V/F CONVERTER: See *Voltage-to-Frequency Converter.*

VIDEO A/D CONVERTER: An ultra-fast A/D converter capable of conversion rates of 5 MHz and higher. Resolution is usually 8 bits but can vary depending on the application. Conversion rates of 20 MHz and higher are common.

VOLTAGE DECAY: See *Hold-Mode Droop.*

VOLTAGE REFERENCE: See *Reference Circuit.*

VOLTAGE-TO-FREQUENCY (V/F) CONVERTER: A device which converts an analog voltage into a train of digital pulses with frequency proportional to the input voltage.

WEIGHTED CURRENT SOURCE D/A CONVERTER: A digital-to-analog converter design based on a series of binary weighted transistor current sources which can be turned on or off by digital inputs.

ZERO DRIFT: The change with temperature of analog zero for a data converter operating in the unipolar mode. It is generally expressed in $\mu V/°C$.

ZERO ERROR: The error at analog zero for a data converter operating in the unipolar mode.

ZERO-ORDER HOLD: A name for a sample-hold circuit used as a data recovery filter. It is used to accurately reconstruct an analog signal from a train of analog samples.

AND gate   An electronic logic circuit with single or multiple inputs. This logic circuit will operate only when the signals at all inputs are present.

Audio frequencies   Frequencies that can be heard by the human ear (approximately between 15 and 20,000 Hz).

Baseband   In the process of modulation, the baseband is the frequency band occupied by the aggregate of the transmitted signals when first used to modulate a carrier.

Baud   In general, a unit of signaling speed. In data processing, it is a group of tracks on a magnetic drum or on a side of a magnetic disk. In data communication the band is the frequency spectrum between two defined limits.

Binary   The number system consisting of only the digits 0 and 1.

Bit   Binary digit equal to 0 or 1.

Bus   One or more conductors used for transmitting signals or power. These buses include a microprocessor interface bus, address bus, data bus, and control bus.

Byte   An 8-bit binary number.

Channel   A channel is part of a communication system with a frequency bandwidth sufficient for a one-way system. A channel in computers is a circuit, link, or path for the flow of information.

Character   A member of a set of elements for which agreement on meaning has been reached and that is used for the organization, control, or representation of data. Characters may be letters, digits, punctuation marks, or other symbols.

Comparator   The purpose of comparator circuits, using operational amplifiers, is to clamp large output dc voltage levels with the use of smaller control dc levels, while comparing the input with a reference voltage. These comparators, of five basic types, are used in logic or automatic control circuits where specific clamping voltage dc levels are required. (The word *clamping* in electrical terms means to hold a voltage level at some specific value. This level can be positive, negative, or zero, such as $+4$ V, -3 V, or 0).

Crosstalk   Unwanted energy transferred from one circuit, called the disturbing circuit, to another circuit, called the disturbed circuit.

Data   Any representation of information such as characters or analog quantities to which meaning is or might be assigned.

Data communications   The transmission and reception of data via an electrical transmission system.

Data processing   The systematic performance of operations upon data, for example handling, emerging, sorting, and computing.

Flip-flop   A circuit or device containing active elements capable of assuming either one of two stable states at a given time.

Interface   A shared boundary. As an example, the interface can be the boundary that links two devices.

Laser   A maser that amplifies radiation of frequencies within or near the range of visible light. It is used in data communication and industrial processes.

Maser   Abbreviation for microwave amplification by stimulated emission of radiation.

Microwaves   Any electromagnetic wave in the radio-frequency spectrum above 890 MHz. Usually, the wavelength is from 1 mm to 50 cm.

Modem   A device that modulates and then demodulates signals transmitted over communication facilities.

OR function   A function (or output) for a gate circuit that is true when one or more of the logical quantities are true and false when all the variables are false.

Telemetry   The complete measuring, transmitting, and receiving apparatus for indicating, recording, or integrating at a distance, by electrical translating means, of a quantity.

Word   A group of digits handled by a digital computer as a single unit. A word is usually 8 bits in a microprocessor.

Index